DEVELOPMENTS IN
EARTH & ENVIRONMENTAL SCIENCES 1

GEOSCIENCES, ENVIRONMENT AND MAN

Developments in Earth & Environmental Sciences, 1

GEOSCIENCES, ENVIRONMENT AND MAN

HERVÉ CHAMLEY
University of Lille, Villeneuve d'Ascq, France

2003

ELSEVIER

Amsterdam – Boston – Heidelberg – London – New York – Oxford – Paris
San Diego – San Francisco – Singapore – Sydney – Tokyo

ELSEVIER SCIENCE B.V.
Sara Burgerhartstraat 25
P.O. Box 211, 1000 AE Amsterdam, The Netherlands

First edition 2003 (English translation of original)

Library of Congress Cataloging in Publication Data
A catalog record from the Library of Congress has been applied for.

British Library Cataloguing in Publication Data
A catalogue record from the British Library has been applied for.

This title was originally published in French under the title:
Environnement Géologiques Activités Humaines © Vuibert, Paris 2002
ISBN: 0-444-51422-8 (hardcover)
0-444-51425-2 (paperback)
ISSN: 1571-9197 (series)

♾ The paper used in this publication meets the requirements of ANSI/NISO Z39.48-1992 (Permanence of Paper).
Printed in The Netherlands.

Abridged Contents

Contents

Foreword

The concept of environmental geology was defined in 1962 by J. E. Hackett (Betz, 1975). The first textbook devoted to this emergent discipline was published in 1970 by P. T. Flawn, who also wrote the founding paper *Environmental Geology*, a scientific journal launched in 1975 in New York City (Springer). In the two following decades, several books dealing with this subject were written and printed in the USA. Some of them integrated environmental geology with the other environmental sciences (e.g. G. T. Miller Jr, *Living in the Environment*; 12th edition, 2002), and new scientific journals were progressively launched with a similar aim: *The Science of the Total Environment* (Elsevier), *Water, Air & Soil Pollution* (Kluwer), *Progress in Environmental Science* (Pergamon), *Environmental Geosciences* (AAPG-Blackwell), etc. A significant number of fascinating books and journals dealing with solid Earth and environmental sciences therefore came onto library shelves during the last quarter of the 20th century. During the same period, I came progressively to share my interest in deep-sea clay sedimentation with that of coastal sedimentary processes. This shift forced me to broaden my scientific field of interest and become more involved in environmental questions. This also allowed me to meet many colleagues who specialized in scientific domains other than mine and to discover the richness of multidisciplinary approaches and programmes. I also realized the difficulty faced by scientists from different disciplines to actually communicate and work with each other, and even to share a common vocabulary. It became evident to me that the headlong rush into always more-specialized scientific expertise, to which I had participated so much and which I had taken pleasure in over several decades, should be counterbalanced by transversal approaches favouring both interdisciplinarity and scientific opening up. I felt involved by participating in such re-balancing actions.

These are the reasons responsible for me writing an essay on the interactions linking geological sciences, human activities and pressure, and the environment. Such a transverse and crossed approach cannot be commonly accessible without being somewhat superficial, and I beg the reader's indulgence if I have avoided providing enough in-depth information. Many additional data may be obtained from the rather extensive reference list provided, as well as in the books and websites quoted at the end of each chapter.

I have tried to reach three objectives, which determined the division into three parts of this volume: I, to consider the main natural geological processes interfering with and therefore threatening human activities: earthquakes, volcanic eruptions, land movements, floods, wind and coastal risks. Main prevention and mitigation measures against these natural hazards are presented; II, to examine the exploitation of Earth's natural resources such as materials, ores and minerals, fossil fuels, water, radioactivity, and the resulting consequences on the Earth's balance and future; III, to assess the influential level reached by human activities on planet surface envelopes through agriculture, urbanization, industrialization, and communications; the local to global effects of human influence triggered by recent demographic growth on underground terrains, soil, water and air characteristics are taken into account. Both deteriorating and beneficial aspects of Earth–Man interactions have been emphasized, as well as mitigation or restoration measures and perspectives.

This volume comprises 11 chapters, which cover the whole range of Earth–Man interactions, from most natural and internal phenomena (i.e. earthquakes and volcanic eruptions) to most anthropogenic and external processes (i.e. rock, soil, water and atmosphere disruption and contamination). The Introduction presents the current situation where Earth is facing the phenomenal expansion of our species, and the Epilogue summarizes the main facts and perspectives presently available. Each chapter has a similar organization: presentation of one or two case studies, description of basic processes, current state of the Earth–Human dependency and impacts, measures of forecasting and prevention or of mitigation and rehabilitation, perspectives, and general considerations. I hope this organization will be agreeable to the readership, which is expected to first consist of future environmental geologists and other environmentalists, as well as of Ph.D. students and young researchers interested in Earth–environment interactions. I am confident that the transverse and hopefully independent approach followed will also be of interest for some people having a scientific concern for the environmental changes affecting our planet due to 20th century exponential demographic growth.

In addition to various encouragements received for embarking on the perilous exercise of writing a trans-disciplinary essay on geoenvironments and human activities, I have benefited from a critical and valued reading of each chapter by a scientist specializing in the considered domain. I am deeply grateful to each of these colleagues: Marc Tardy (University of Savoie), earthquakes; Thierry Juteau (University of Bretagne Occidentale), volcanic eruptions; Robert Meyer (Rouen University), land movements; Claude Larsonneur (Caen University), water, wind and coastal hazards; Francis Meilliez (Lille 1 University) on underground materials and mineral ores; Jacky Mania (EUDIL, Lille), water resources; Alain Trouiller (ANDRA, Châtenay-Malabray), radioactivity; Emmanuel Choisnel (Météo-France, Paris), soil resources and hazards; Paul Lecomte (BRGM, Orléans), urbanization, industrialization and communications impact; Michel Meybeck (Paris 6 University), land and sea contamination; Michel Crochet (University of Toulon and Var), regional to global phenomena affecting primarily Earth's fluid envelopes. I also would like to thank the librarians at Paris Earth and Universe Sciences CADIST, at London's Imperial College, and at the University College of London. All illustrations have been redrawn by Martine Bocquet, art technician at CNRS-Lille.

Introduction

The Earth–Man Encounter

Environmental geology deals with interactions linking the solid and fluid Earth envelopes and the various human activities. Born about three decades ago and actively documented for less than a dozen years, this scientific discipline is rapidly growing. It belongs to the handful of sciences issued from observations proving the increasing disruption of terrestrial environments due to human pressure. The extraordinary growth and hold of our species during the two last centuries, a time interval insignificant relative to the geological history and very short relative to that of human history, is clearly underlined when considering the evolution, behaviour and economy of our societies. Relevant indicators include the population explosion, the rapid modifications of life styles, our growing exposure to natural hazards, the exploitation of natural resources and resulting changes at the Earth surface. Let us briefly consider these different indicators.

1. Population Explosion

Man first appeared in eastern Africa around three million years ago, then settled and progressively spread in Europe and Asia, and probably occupied Australia and North America 60,000 and 20,000 years ago, respectively. The human population is estimated to have comprised between 40,000 and 500,000 until the last glacial maximum (around 20,000 years before present, BP). The invention and development of agriculture, synchronous with the glacial-postglacial transition, gave way to the first rapid population increase. Our species probably attained one hundred million people (0.1 billion) around 12,000–10,000 years BP (Fig. 1). Then the augmentation slowed down and became more regular until the 18th century: 0.25 billion around the beginning of the Christian era, 0.5 billion in 1650. A second, much more important *population explosion* arose *during the 19th century*, due to the conjunction of the industrial revolution, technical, scientific and medical progress, and successive steps in new territories' conquest and colonization. This second demographic development rapidly acquired an exponential regime: 1 billion people in 1830, 1.6 billion in

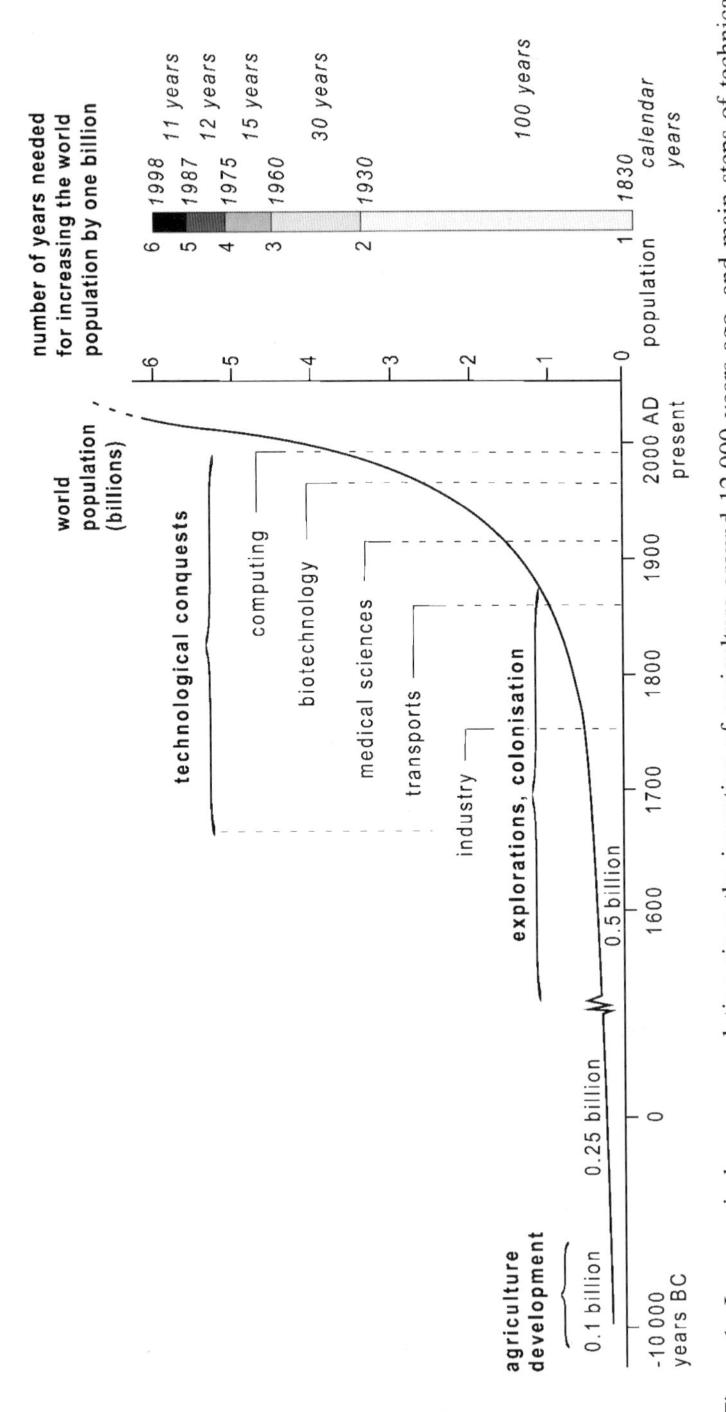

Figure 1: Increase in human population since the invention of agriculture around 12,000 years ago, and main steps of technical development (modified after Park, 1997).

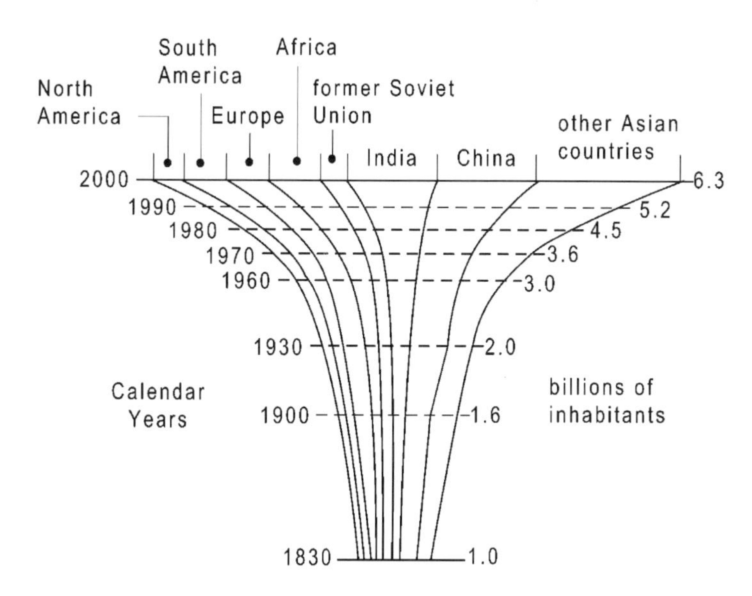

Figure 2: Demographic flows since 1830 in the main regions of the world (modified after Valdiya, 1987).

1900, 2.5 billion in 1950, 5.2 billion in 1990, and more than 6.0 billion in 2000. The continuation of the current rate of mankind's increase should lead to a world population exceeding 8 billion people around 2025.

The prodigious, somewhat anomalous character of the recent population explosion is more easily comprehended by considering some comparative figures. The human species took 120,000 years to reach an amount of one billion people, and only 10 years to gain another billion (from 5 to 6 billion) at the end of the 20th century: the acceleration factor is about 12,000! The present growth rate is marked by 250,000 births each day, i.e. about 91,000,000 each year; this determines every 6 months an increase in people that approaches the total number of soldiers killed on the world's battlefields during the last 500 years!

The phenomenal increase in the world population is amplified by the strong augmentation of the average life. The *life expectancy* fluctuated between 30 and 40 years until 1850 and successively reached 50 years around 1900, 60 years in 1950, and 75 years at the turn of the 21st century. This progression greatly differs according the countries considered. For instance life expectancy approaches 80 years in Japan, whereas that of Benin does not exceed 47 years. Such differences result from disparities in food resources, industrialization level, technological development, pollution and disease control. Remember that three-quarters of the world's energy is consumed by a quarter of the world's population (i.e. North America, western Europe, Japan, and the former Soviet

Union), that a quarter of the world's people lacks healthy water, and that a fifth suffers from serious malnutrition (especially in Africa and South Asia).

There is a *strong demographic contrast between developed and developing countries*. Despite a much lower mortality rate, the former tend towards a stabilization or decline of the population, whereas the latter experience a large, locally exponential increase. This is especially obvious when comparing the North American or European situation since the late 17th century to that of Asian and partly of African countries (Fig. 2). The continuation of such an evolution should lead to a noticeable increase in the proportion of Asian and African populations during the next decades, provided that there is no major change in the immigration influx. Simulations show that by the year 2025 Africans should increase from 12.7 to 18% of the world population, and Europeans decrease from 12.7 to 8.6% (World Resources 1996–1997; see Montgomery, 2000).

2. Human Occupancy Steps

The occupancy and exploitation of terrestrial environments has been punctuated by successive phases, the rate and intensity of which have dramatically increased during the last centuries.

Hunting and fishing constitute the oldest and least-disruptive human activities. Basically destined for feeding, they were progressively also used for clothing, housing and ornamentation, at the same time as societies were developing. These traditional activities have limited implications on deforestation and soil reworking.

Livestock farming and reproduction started more than 10,000 years ago and led to the partial destruction of original vegetation for creating pastures and pens. The need to remove the surface stones from the soil, to irrigate and to fertilize arose progressively. Traditional farming determines moderate damage to the natural environment and even participates through manuring in renewing the nutritive properties of the soils. By contrast the intensive livestock farming developed during the last century has accelerated the physical deterioration, chemical modification and contamination of the soils: compacting and destructuring through animal displacements, acidification and reduction, nitration, etc. Extensive world soils have lost their original permeability, oxygen content and suitability for planting, and have fallen into disuse in the form of erodible wasteland (Chapter 8).

Agriculture, which represents the third step of soil occupancy, was progressively extended on the solid Earth surface, the original vegetation being replaced by plants of domestic or economic interest. Some classical developments such as slope restyling and terrace building have allowed us to prevent erosion and sliding hazards. On the other hand, some cultivated plants are hardly adapted to natural ecosystems, as for instance the shallow-rooting conifers in dry and windswept regions. More important, the extension of monoculture practices has determined seasonal denudation and an accelerated erosion of surface soils (e.g. North American prairies during the 1930s). The wish to increase the soil productivity induced diverse environmental changes that were intensified during the last decades: irrigation, ploughing, addition of artificial fertilizers, and use of pesticides. In addition the settlement of an increasing number of traditionally nomadic populations led to overexploitation and irredeemably impoverishment of soils in numerous subarid regions (Chapter 8).

The extraction and use of mineral resources have been practised since the early organizational stages of human societies, as recalled by the name of primitive tool- and weapon-manufacturing steps: Stone Age, Bronze Age, Iron Age . . . But in the same way as for other human activities, the extraction rate of Earth's materials and minerals, as well as the resulting nuisance, have dramatically increased as and when industrialization, need of fossil fuels, water consumption and soil occupancy were expanding (Chapters 5–8). For instance, the world consumption of minerals extracted from geological formations was multiplied tenfold between 1900 and 2000. In addition the rock and ore exploitation determines the extraction of various amounts of unused materials as well as the storage of mineral and chemical residues, forming huge mining and industrial waste. Synthetic substitute materials are used increasingly in developed countries. But this is not the case in most countries under development, where the underground exploitations and associated nuisance tend to be more and more extended.

The massive urbanization and communications development have attained giant proportions during the few last decades, inducing various modifications in soil and rock properties and stability: waterproofing of surface formations, development of artificial drainage, aquifer over-exploitation, subsidence, soil and slope destabilization by inappropriate building or restyling, and underground pollution (Chapters 9 and 10). These changes are associated with important modifications in life style, marked by a growing need for goods and conveniences. Such a trend progressively extends in developing countries, exacerbating the consumption of natural resources and energy.

Urban concentrations constitute one of the most recent and worrying phenomena. In 1850 about 50 cities of more than 100,000 inhabitants were counted, among which only 3 were exceeding one million. In 1973 more than 1,000 cities exceeded populations of 100,000 people, and more than 200 exceeded one million. In 2000 half of the world population was concentrated in urban complexes, the total surface of which represents only 0.4% of continental surfaces! Simulations show that ten very large cities hosting more than five

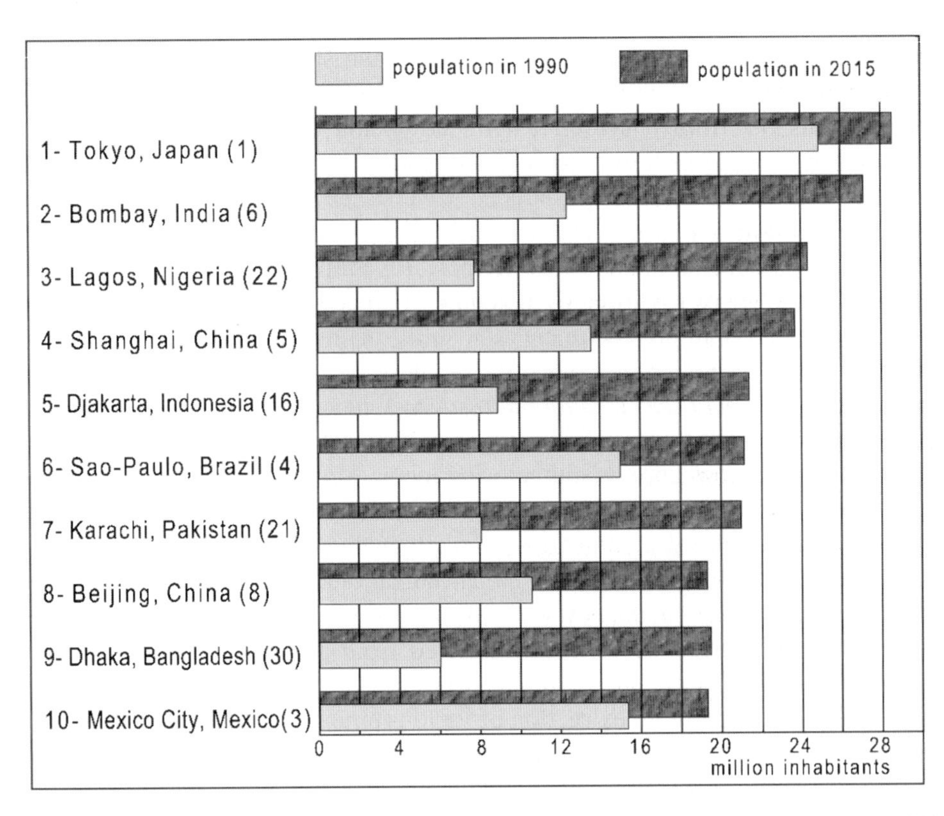

Figure 3: Population expected for the ten world's largest cities in 2015 relative to 1990 (United Nations document). 1990 rank in parentheses.

million inhabitants in the 1990s will constitute in 2015 conurbations of 18.5 million to more than 28 million people (Fig. 3). Most present and future conurbations are situated in developing countries such as India (Bombay, Bangalore, Dehli, Hyderabad), Pakistan (Karachi, Lahore), Bangladesh (Dacca), and Central Africa (Kinshasa, Lagos). This highly unbalanced human hold constitutes another major challenge for the next decades, including: urban building and development, soil and underground pollution, water and atmosphere quality, and traffic density (Chapter 9).

3. Main Natural Hazards

Usually we hardly accept the evidence of unexpected and harsh naturally occurring phenomena, which are potentially responsible for numerous deaths and goods losses. Our increasing hold on the environment leads us unconsciously to the need for understanding and controlling all natural processes, or at least to identify some responsible people to blame for not having predicted or prevented the hazards. In past times the responsibility was borne by the gods, whose clemency was implored to push away the nature-induced catastrophes. Today the responsibility is usually attributed to political decision-makers, whose lack of foresight and poor management ability is castigated, and who are required to urgently indemnify the victims for the damage and losses. The subject of natural, including geological, hazards is all the more sensitive (see Betz, 1975; Allègre, 1990, 1993; Martin, 1997; Bell, 1999) because we often unintentionally weaken our memory of previous catastrophic events and easily consider the current events as unique or at least extraordinary. Yet the instantaneous and violent nature of some geological events is eminently natural, their unpredictable character being related to the *complexity* of physical and chemical interactions driving the dynamics of Earth's external envelopes.

A natural hazard represents a risk if it potentially threatens human populations and constructions. For instance, the coastal erosion by storm swells, which contributes to sedimentary feeding of foreshore environments, becomes a risk when threatening the buildings and other constructions established on seashore cliffs. Natural risks cover a wide spectrum of phenomena, from high-intensity and low-frequency events often responsible for numerous casualties (earthquakes . . .) to less intense and more frequent events determining usually few victims but important economic damage (coastal and fluvial erosion . . .). The main natural hazards threatening the Earth's surface issue from three natural domains: the atmosphere, which forms tropical storms, hurricanes, cyclones and major droughts; the planet's external envelopes, which develop flooding, slope movements, coast and soil erosion processes; and the deeper Earth from which arise volcanic eruptions, earthquakes and tsunamis (Chapters 1–4). In addition there exist some extraterrestrial hazards such as meteorite impacts. Different natural hazards may occur successively or in a superimposed manner, giving way to complementary or additional effects: flooding events often follow atmospheric storms; earthquakes favour soil liquefaction, slope movements and tsunamis; volcanic eruptions are frequently associated with earthquakes, etc.

Natural disasters determined the *death* of about three million people during the last quarter of the twentieth century, and associated *damages* reached more than 1,000 billion U.S. dollars (Bennett & Doyle, 1997). The year 1997, marked by 530 important natural catastrophes determining 22,300 victims or missing persons, represents an average annual situation for the last decades (Dagorne & Dars, 1999). The tributes differ considerably according to the type of natural event. Most casualties result from storms and cyclones as well as from earthquakes, whereas damage to goods primarily results from flooding, secondarily from storms and earthquakes.

Most countries exposed to major natural hazards are situated in subtropical to equatorial domains, where the movements of lithospheric plates determine strong seismic and volcanic activity and where major violent and unstable atmospheric events commonly occur (Fig. 4). Most of these countries are presently under development and are still scarcely able to predict, prevent or resist natural blights, nor to overcome the resulting food and health problems. As a consequence the number of victims of natural disasters is roughly in inverse proportion to the level of economic development of the countries affected (Table 1). By comparison with tropical-equatorial countries and even with the industrialized countries of North America and Japan, *Western Europe is little exposed* to natural hazards, which are mostly of moderate intensity and frequency (Lumsden, 1994). In France the number of victims and the financial cost of damages following natural catastrophes is usually 100–1,000 times lower than in many developing countries. Let us recall that the 1992 flood of the Ouvèze River in south-eastern France, which was considered as a major disaster, was responsible for 43 deaths and for goods losses approaching 300 million euros. By comparison the cyclone of April 1991 in the northern Indian Ocean determined in Bangladesh the death of 140,000 people and a 10% drop in the gross national product. The French regions most exposed to natural hazards comprise essentially the overseas departments and territories, which are located in tropical zones susceptible to cyclones, earthquakes and volcanic eruptions (West Indies, Polynesia, Réunion).

Investigations into the temporal variability of natural disasters show an *increase in frequency and resulting damage since the 1960s*. The augmentation of casualties is close to 6% per year, and that of economic losses close to 5% per decade (Fig. 5). Such a trend seems to result from four main causes (Bennett & Doyle, 1997), the respective importance of which is difficult to assess because of the lack of long-term statistics: (1) the doubling of the world population, determining an accentuated exposure to risks; (2) the human concentration in regions and especially cities severely exposed to hazards, poorly equipped, and the vulnerability of which is exacerbated (e.g. earthquakes in Mexico city, flooding in Bangladesh); (3) the nature or man-induced deterioration of surface soils, favouring massive erosion, flooding and landslides (e.g. the tropical rainforest has been deteriorated by about 40% since World War II); (4) the improvement of information techniques allowing the availability of more precise and complete data, and therefore improved and better spread counting. As for other domains, *the increasing vulnerability of people and property concerns especially the developing countries*, where the population growth is higher, the deforestation more widely distributed and the urbanization barely controlled. For instance, the deforestation in India between 1970 and 1980 led to a doubling of the land surface liable to flooding, which multiplied by 14 the damages relative to 1950 (Ledoux, 1995). By contrast the developed countries that face natural hazards much more efficiently experience both a strong

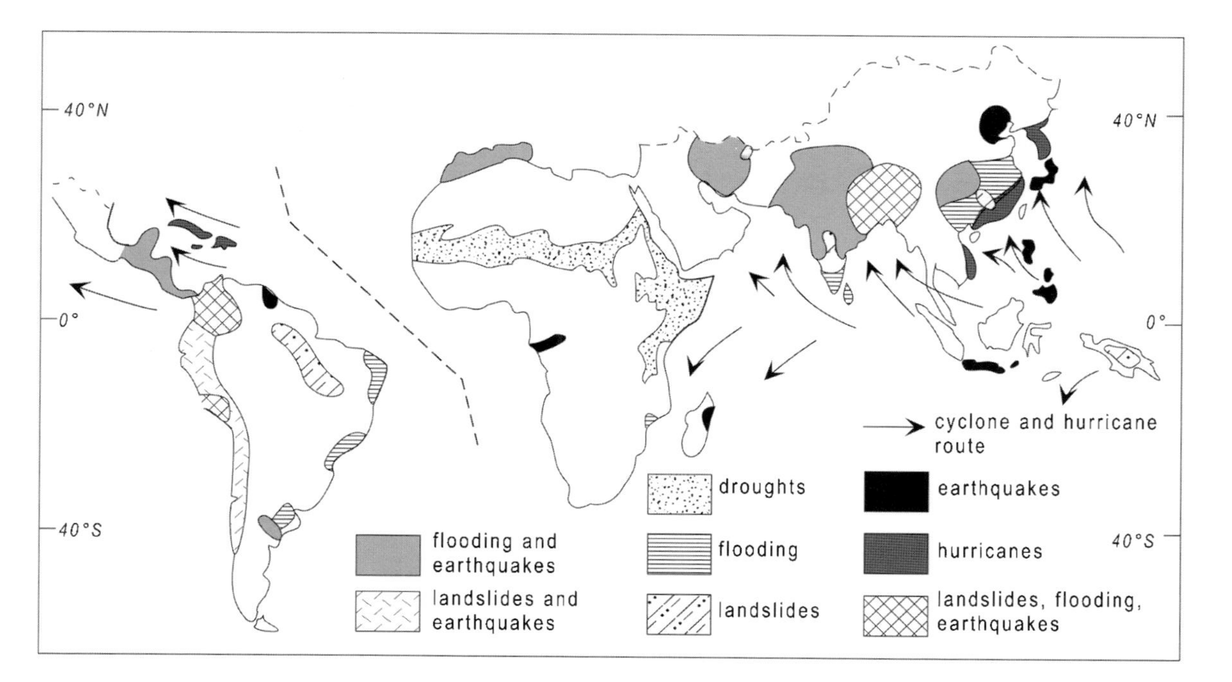

Figure 4: Distribution of developing countries of subtropical to equatorial regions where natural disasters determined more than one hundred victims in half a century (1935–1985; after Cooke & Doornkamp, 1990).

Table 1: Distribution of natural disasters, 1960–1998, according to the economic-development level of different countries (after Lumsden, 1995).

Gross national product	Number of disasters	Victims	
		Number	Proportion
Low	329	1,090,900	76
Medium	451	335,000	23.3
High	79	10,000	0.7
Total	**859**	**1,435,900**	**100**

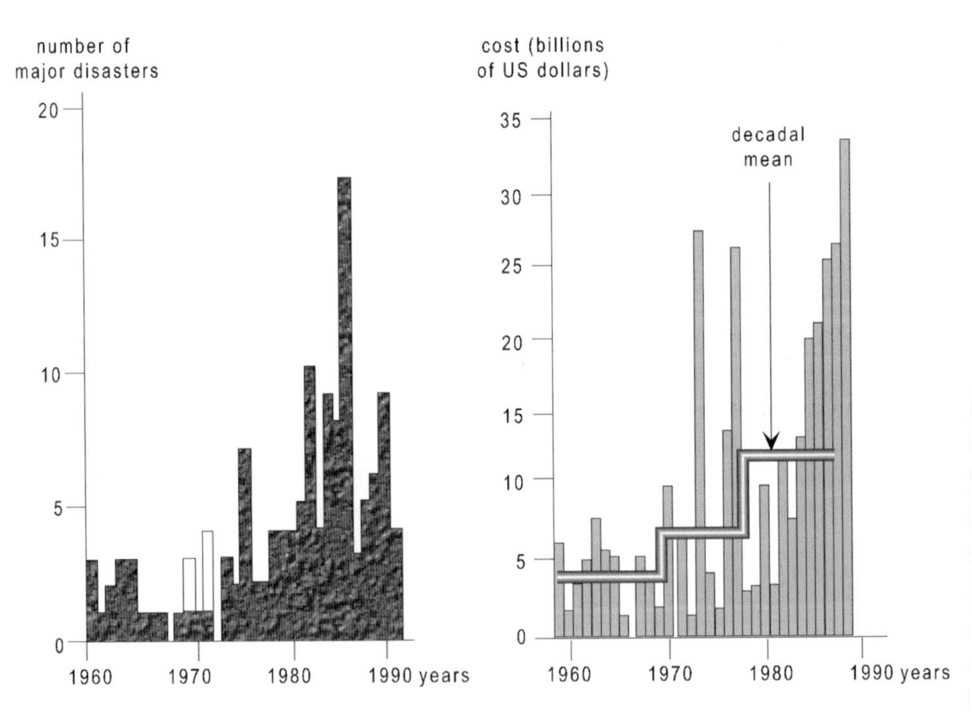

Figure 5: Evolutional trend in the number of natural disasters number after the mid-20th century, and of the inferred economic impact (after Rosenfeld, see Bennett & Doyle, 1997). The '10-year mean' is indicative.

diminution of human deaths and a spectacular increase in financial compensations.

4. Main Man-induced Impacts and Risks

While natural risks periodically threaten mankind, in parallel, *human activities are endangering, in a growing way, the surface environments* of our planet. By developing livestock farming, soil exploitation, freshwater control, building constructions and surface-to-underground developments, *humans modify the functioning of geo- and biosystems.* Agricultural developments are particularly responsible for impairing and weakening the surface formations over very large areas, especially through permanent searches for more intensive exploitation and higher productivity: large-scale irrigation determines noticeable changes in hydrographical networks, landscapes (canals, dams and reservoirs) and ambient humidity (soils, air); the deeper and more mechanized the ploughing, the more destructured the soils; an excessive or unadapted use of fertilizers and pesticides leads to pollution of underground and downstream waters, favouring environmental and sanitary disorders (e.g. eutrophication); and extreme pressure on soils and vegetation cover can end up in desertification (Chapter 8).

The increasing use of fertilizers may even have counter-effects on plant fertility. For instance, and in spite of a continuously increasing supply of fertilizers since 1950, the world cereal production started to level off during the 1980s (Owen et al., 1998). The curves of cereal and fertilizer amounts grew progressively closer and finally crossed each other (Fig. 6), the productivity becoming stable in the 1990s. An indirect consequence was to weaken the fertilizer's efficacy. Information gathered at the end of 2001 shows that if the fertilizer used in the Mississippi basin had been diminished by 12% between 1960 and 1998 the risk of eutrophication in the northern Gulf of Mexico would have been noticeably lowered, the cereal production remaining roughly the same. Such examples underline the need both to better understand the complex interactions linking vegetables and fertilizer assimilation and to improve the training in agriculture jobs.

The exploitation of terrestrial resources concerns primarily solid and liquid mineral materials (Fig. 7) that were formed during the geological history and therefore tend statistically to exhaustion (Chapter 5). The permanent or renewable character of terrestrial resources on a human time-scale concerns only water that changes in state and quality but not in abundance (Chapter 6), some geomaterials rapidly shaped by external geodynamical forces (stone, pebble, sand, silt, clay) and the lesser-evolved types of plant-derived fuels (peat). Various materials such as coal, oil, natural gas and some metalliferous ores (e.g. deep-sea Fe–Mn nodules) still form within geological formations, but

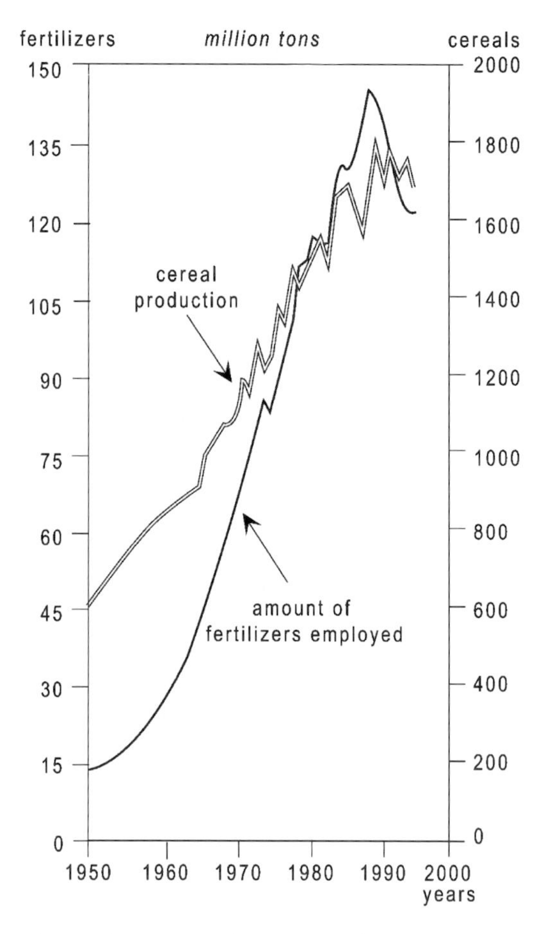

Figure 6: Amounts of cereals harvested and of fertilizers used in the world from 1950 to 1995 (after Owen et al., 1998).

their very slow growth, low abundance, and/or difficult accessibility, extraction and concentration prevent us from envisaging any significant extra availability in the centuries to come. Because of the demographic explosion, most useful rock-derived materials are currently the subject of a very high demand, which concerns the industrialized countries and also more and more the developing countries. Each year people move, exploit and carry more nature-derived materials than does the world's water system constituted by all glaciers, streams and rivers. The mean annual use per capita of earth-derived materials was the following in the USA at the end of the 20th century: 3,700 kg of stone, 2,700 kg of gravel and sand, 800 kg of iron and steel, 300 kg of cement, 150 kg of clay, phosphates and salt, 3,100 kg of crude oil, 2,700 kg of coal, and 1,900 kg of natural gas.

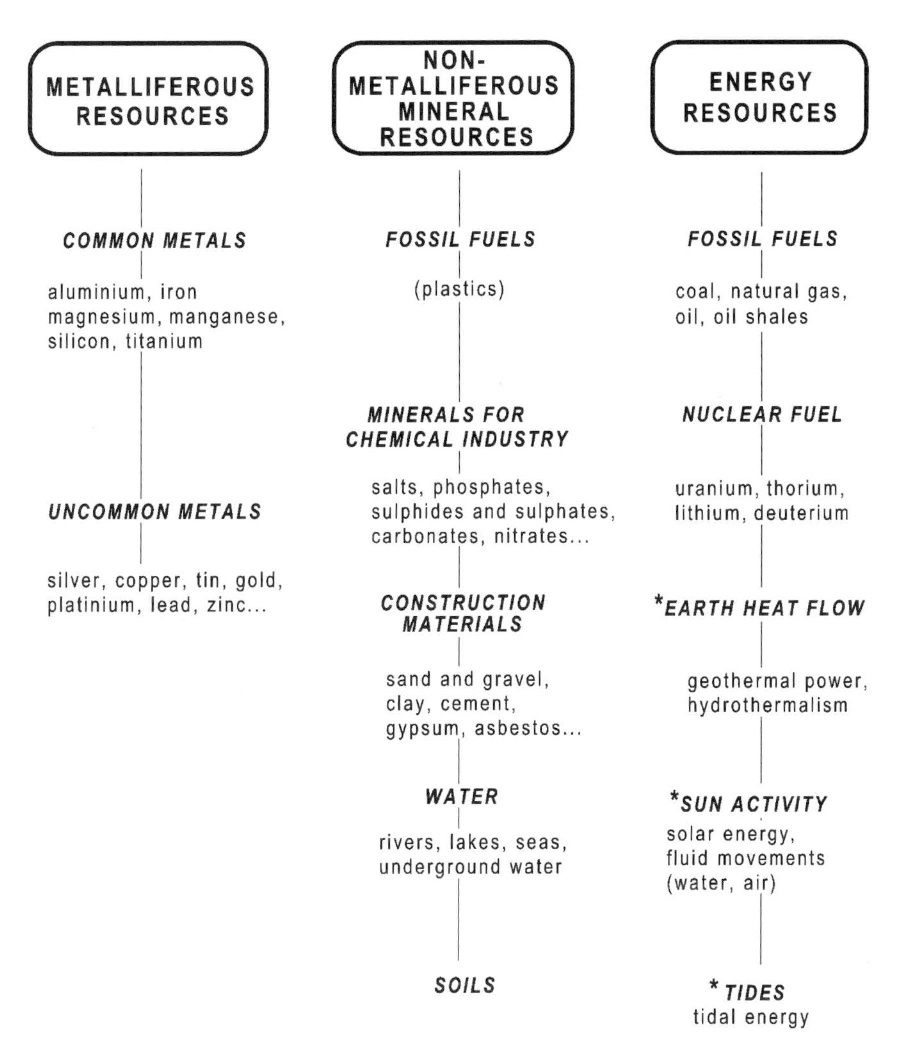

Figure 7: Main world mineral and energy resources (after Craig et al., 1988). The three items preceded by an asterisk correspond to non-mineral energy resources.

The accelerated consumption of our natural resources is accompanied by the *production of* huge quantities of sometimes toxic *waste*, and considerable *disposal of particulate or dissolved matter* responsible for air, water, soil and underground pollution (Chapter 10). This results in noticeable modifications of the Earth's surface: reduction of the vegetation cover, formation of artificial relief (quarries, slag heaps) and slope instability, expansion of erosion and subsidence, diversion, level change and potability loss of surface and underground water, etc. The extraction of non-combustible materials solely induces each year landscape modifications covering one million hectares (Panizza, 1996).

An enlightening example quoted by Keller (1999) deals with the Copper Basin, a mining area from Tennessee (east USA) where extraction and processing activities led to deterioration of a surface exceeding 130 km^2. Exploited since 1843 by gold-diggers followed by copper ore-extracting companies, this region has been riddled with numerous 200-m-wide and 30-m-deep excavations, and the placing of big ovens where copper, zinc, iron and sulphides were separated. The exploitation of nearby forests for firewood, the massive evacuation of dark and noxious smokes, the sulphur dioxide emanations causing acid dust and rain, progressively induced the vegetal cover destruction, soil erosion, underground outcropping and desertification. Restoration works started after mines closed during the first decades of the 20th century, but the complete environmental rehabilitation of the Copper Basin will take several hundred years.

Numerous similar cases dating back to the 19th century are recorded in most industrialized countries. For a few decades, the developing countries have been exposed to comparable landscape deteriorations. The risk control and environmental management have considerably progressed in developed countries, whereas the technological backwardness impedes comparable improvements in many regions recently industrialized (Central and South America, Southeast Asia) or having maintained for a long time high pollution levels (Eastern Europe, the former Soviet Union).

On a global scale, the environmental disruptions expressed by atmosphere physicochemistry and glacial and sedimentary records appear to also depend on human activities (Chapter 11). Noticeable changes of global environmental parameters have been evidenced, which are related to the recent augmentation of human pressure, energy needs, and industrial and urban development. These modifications are particularly emphasized by the temperature value and rhythm records obtained for the recent geological times (Fig. 8). The temperature deduced from oxygen isotope composition of marine carbonate shells displays progressive and sub-periodical cooling and warming during the last million years (Fig. 8A), the last climatic cycle for 150,000 years (Fig. 8B), and the post-glacial warming having started 15,000 years ago (Fig. 8C). These successive periods were not appreciably influenced by human activities. By contrast a strong and rapid acceleration of terrestrial warming appeared before the end of the 19th century (Fig. 8D) and increased during the whole of the 20th century (Fig. 8E). The temperature augmentation begun about 130 years ago reaches 0.5°C, and parallels the increase in atmospheric CO_2 content that passed from 280 to 355 ppm (parts per million) between 1850 and 1990. The CO_2 augmentation in less than 150 years is higher than the changes recorded during

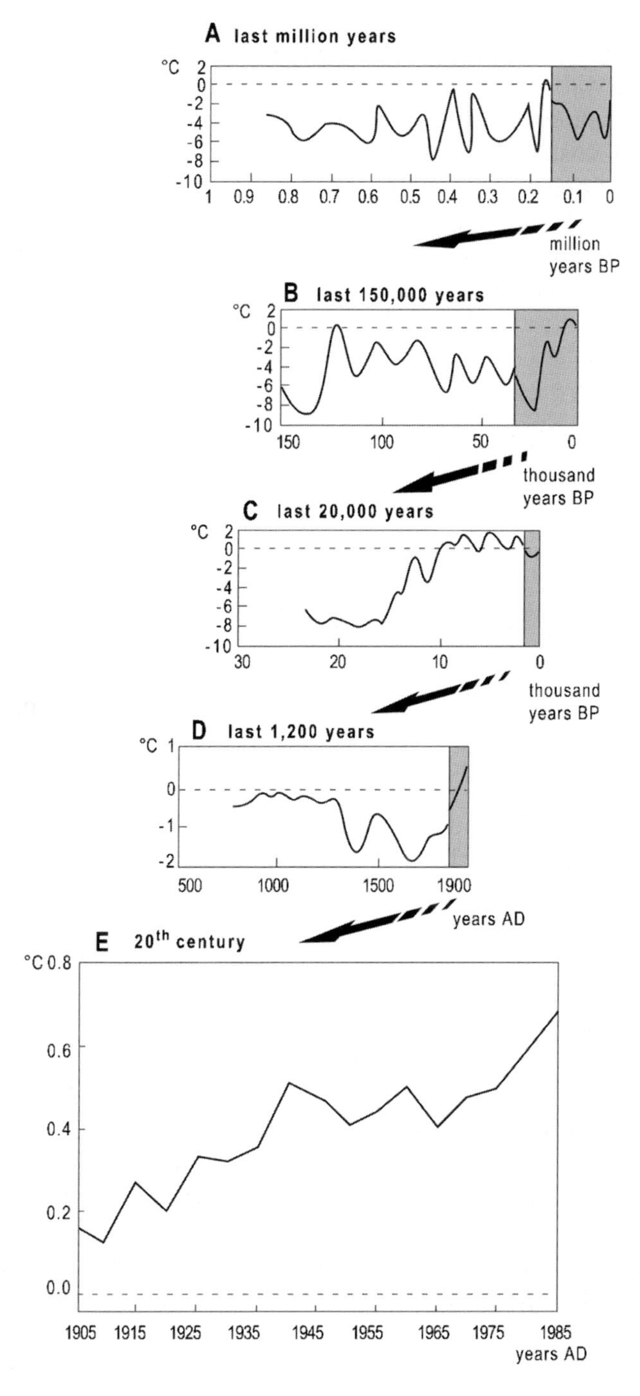

Figure 8A–E: Relative change of world temperature relative to 1900, at different time scales during the last million years (diverse sources, after Keller, 1999).

the last 15,000 years, during which a major natural warming developed. The CO_2 content has varied only between 200 and 300 ppm during the last 160,000 years, and did not experience any noticeable change during the millennium preceding the industrial era. Even the Little Glacial Age (1500–1850) is not reflected by significantly increasing values of atmospheric CO_2, in contrast to what has occurred since human population and needs have exponentially increased. Man therefore most probably plays a significant role in the global evolution of Earth's climate.

5. Perspectives

The unprecedented, pluri-decadal increase of demographic pressure, which determines an extraordinary augmentation of consumption of natural resources and some physicochemical deterioration of solid and fluid terrestrial envelopes, will unquestionably continue during at least the first century of the third millennium. As most energy, mineral and food resources exist in finite amounts, the question arises of the future of our species. Pessimistic people who establish projections based on actual data predict a collapse of the human population in the second half of the 21st century; the progressive dwindling of food, raw materials and industrial activity is supposed to determine a subsequent decrease in the pollution rate (Fig. 9). The availability of food would progressively constitute the main limiting factor; some historians recall the cannibalism once developed on Easter Island after food and wood exhaustion. On the other hand, the optimists, who feel as realistic as their opponents, expect a technological and managerial revolution stimulated by famine and energy depletion, and leading to a wide renewal of our living conditions.

During the past century, mankind has indisputably become the very dominant terrestrial species, responsible for major and long-term landscape, physicochemical and other environmental changes. Such fundamental modifications seem not to have been assimilated and taken into account by many inhabitants and even decision-makers. Obviously humans tend to extract too much from a world marked by finite resources, and to modify its surface in a partly non-reversible way. As early as 1974, M. K. Hubbert emphasized in a striking way the very short duration of the fossil fuels using period relative to the length of human history (Fig. 10). As early as 1976, W. von Engelhardt, J. Goguel and four other scientists forming the founding group of the Geosciences and Man Committee of the International Union of Geological Sciences (IUGS) predicted and explained an inevitable, geologically imminent end of the exponential development of our species; the current period was shown as being transitory and inevitably destined to pass into a stable, slowly evolving period similar to that having characterized the previous centuries and millennia.

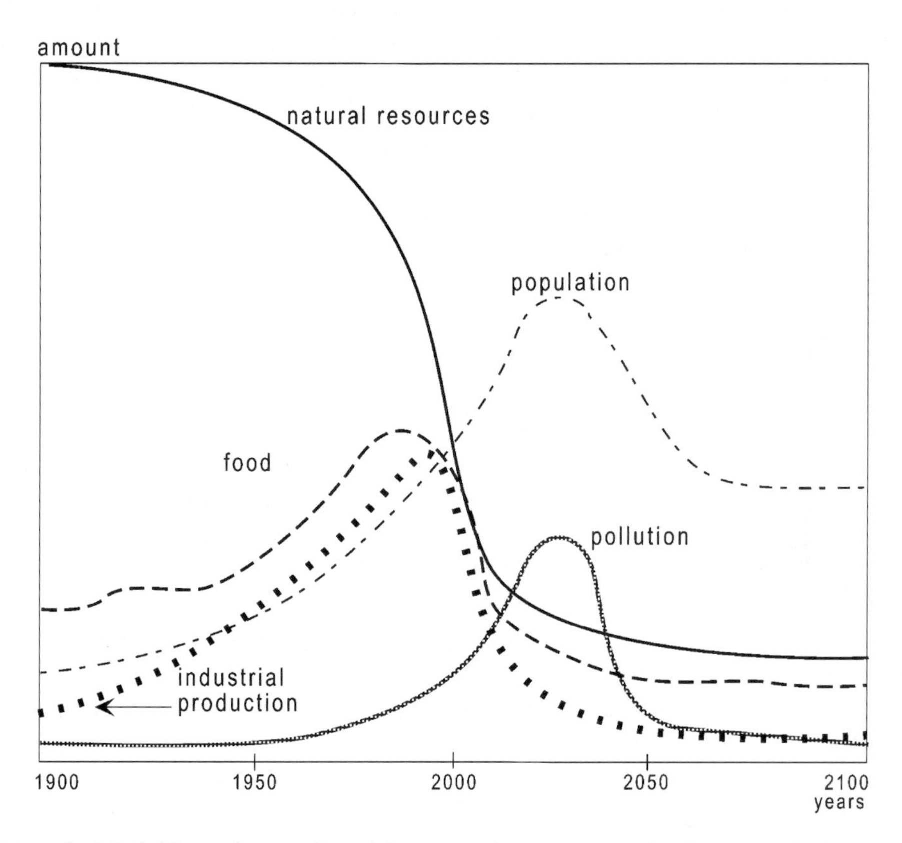

Figure 9: Modelling of several social–economic parameters in the current context of human expansion and consumption (after Owen et al., 1998). The ordinate scale is deliberately omitted.

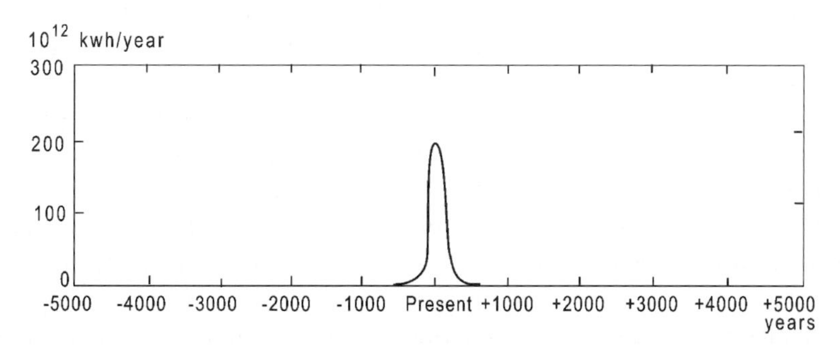

Figure 10: Duration of the period of fossil fuel use relative to the human history (after Hubbert, 1974).

The purpose of this introduction is to introduce the environmental geology concept into the economic and demographic context that marks the dawn of the third millennium. The unprecedented hold of our species on terrestrial environments cannot continue to grow as it does currently. Realizing such evidence as well as its consequences is each individual's responsibility and not only that of specific professional categories (e.g. politicians, ecologists, economists, tipsters, scientists). As far as *environmental geologists* are concerned, their major task is to *better understand the processes, balances and breaking points of the planet's functioning relative to human activities, and then to inform the greatest number of people* by developing public communication. Such a purpose implies a detailed and multidisciplinary approach, necessitating pooling the researcher's complementary expertises. Beside the use of more and more sophisticated analysis and modelling techniques, it is crucial to maintain and develop the practice of field observations and experimentations as well as mapping syntheses, which are critical for ensuring concrete and coherent interpretations. In his 1970 founding book of environmental geology, P. T. Flawn stressed the need for geologists to develop a horizontal approach and to combine quality observations and environmental management questions, if necessary to the detriment of systematic, sometimes disputably useful quantitative measurements: "*Geologists [who] during the last two decades have been in a frenzy to quantify the classical descriptive natural science of geology . . . must not neglect their responsibilities in the areas of resources and environment problems*". Finally, the horizontal approach imposed by the complexity of environmental problems necessitates one to combine highly specialized expertises and wide scientific views: this is one of the most difficult and exciting challenges faced by present-day researchers.

Various *methodologies and techniques* are being developed to progress the understanding of the interactions between geoscience and man's activity. For instance, the monitoring and anticipation of natural hazards is ensured in an increasing way by the use of continuously working, instrumented field sites: geophysical and volcanic observatories, flooding- and landslide-prediction systems, automatic recording devices of ozone and carbon dioxide atmospheric content, etc. In situ measurements are developed by the mean of sophisticated devices transported on cars, boats or airplanes (e.g. underground hydrocarbon diffusion – Davis et al., 1997; radionuclide sedimentary dispersion – Noakes et al., 1999); extensive and repeated observations are performed, e.g. from satellites. Indirect measurements have become more and more reliable: magnetometry for locating underground waste (e.g. Gibson et al., 1996), polarized electric conductivity for assessing the permeability of sedimentary rocks (Well & Börner, 1996), 3D resistivity for identifying the nature of buried waste (Chamber et al., 1999), tomographic detection of subterraneous cavities for controlling the acid water drainage (Madrussani et al., 1998, 1999). The increased use of transfer functions allows one to improve the evaluation of hazard factors: geo-indicators helping to predict the brutal or progressive character of natural risks (Berger, 1997), lichen species used for recording the volcanic activity variations (Notcutt & Davies, 1999), magnetic susceptibility values indicative of the abundance of toxic heavy metals (Chan et al., 1998), etc. Modelling techniques are increasingly used in most projected investigations, e.g.: slope instabilities (Casale & Margotini, 1999), earthquake effects on the stability of radioactive waste (Davies & Archambeau, 1997), potential contamination of superimposed aquifers (Rine et al., 1998). Special critical care should be applied to the reliability of certain analytical techniques, the precision of the results (e.g. cumulated approximations when measuring the sediment-trapped mercury, Gottgens et al., 1999), the clear distinction between natural and anthropogenic origin or control of potentially toxic and noxious contaminants (e.g. Zhang et al., 1999), and the validation and credibility of geostatistical and modelling approaches (e.g. Toy et al., 1993).

Website

http://www/webdirectory.com: Comprehensive site managed by the Environment World Organization, providing information on most environmental domains and disciplines.

Part I

Man Facing Earth's Hazards

Chapter 1

Earthquakes

1.1. 1999, Major Earthquakes in Turkey

During the night of 2 to 3 August, 1999, a very violent earthquake of 45 sec duration hit the Izmit-Gölcük area, situated in northwestern Turkey, south of Marmara Sea and east of Istanbul (Fig. 11). Official statistics reported 17,000 people dead and 44,000 injured, 85,000 destroyed or uninhabitable buildings,

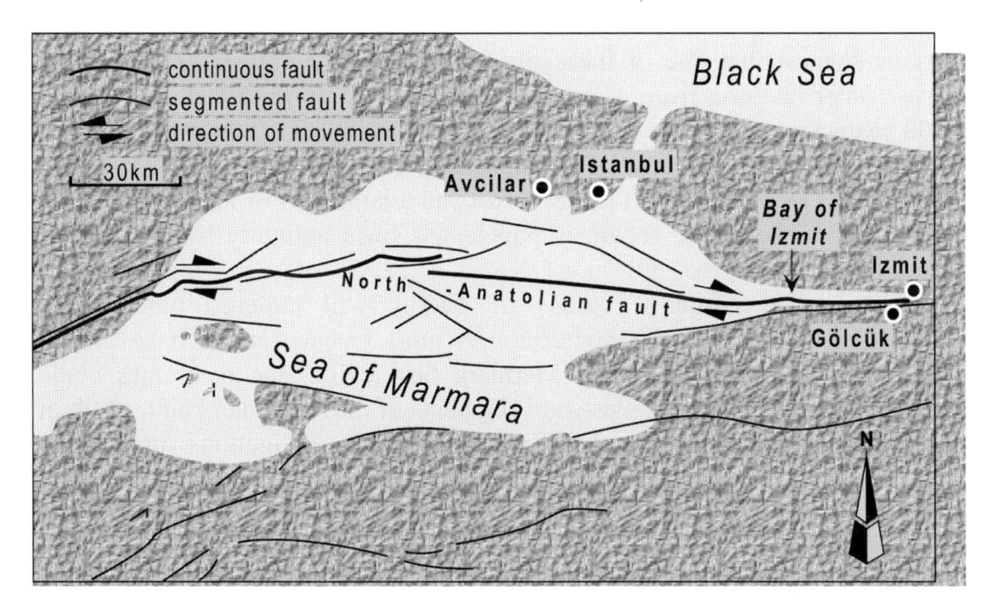

Figure 11: Location of the North Anatolian Fault in the northern region of Turkey affected by major earthquakes in 1999. The continuous bold line represents the general route of the fault beneath the Marmara Sea; dotted lines correspond to the major fault segments identified (after National Geographic, 2000).

and several tens of thousands of stricken families. The damages incurred were all the more serious because many recent buildings had been constructed with materials of poor quality and strength, and did not fit earthquake-resistant standards. On November 12 of the same year, a second, almost as violent earthquake occurred 130 km eastwards, in the less densely populated Cevizlik area, where 850 dead and 5,000 injured people were counted. These two major seismic tremors were initiated underground at about 15 km deep, along a large, tectonically unstable zone called the North Anatolian Fault. This fault system extends east-west over 1,600 km, from north Turkey alongside the Black Sea and across the Marmara Sea, to north Greece across the Aegean Sea. It constitutes the northern boundary of a small, mobile lithospheric plate called the Anatolian Plate.

The Anatolian Plate is wedged between the two large African and Eurasian plates, the converging movement of which determines both compression and a westward expulsion. The friction forces resulting from this oblique collision were responsible for the Iszmit and Cevizlik earthquakes. The 1999 earthquakes belong to a series comprising 13 major tremors since 1939 in the region bordering the southern Black Sea (Fig. 12). Archives and dating reveal that the collision started about 5 million years ago (Early Pliocene) and induced during the last thousand years more than 600 earthquakes, among which 40 were very violent. Recent sedimentological studies on western and southern Black Sea margins suggest that one of these major earthquakes could have favoured a marine water invasion from Aegean–Marmara Seas into the Black Sea around 7,500 years ago, and might be indirectly related to biblical Deluge descriptions (investigations by W. Ryan, W. Pitman, R. Ballard, among others). In 1509 and 1766 of our era, violent earthquakes destroyed a large part of Istanbul city, and the history of the whole region is punctuated since antiquity by comparable tremors.

Obviously the northern regions of Turkey will undergo other major earthquakes during the next centuries. As most recent tremors have mainly affected the eastern part of the Marmara Sea region, the next ones could preferentially concern more westward segments of the Anatolian Fault, south of the Istanbul conurbation (Fig. 11). Some statistical investigations suggest that the earthquakes affecting northern Turkey tend to occur more frequently during a few decades within 200- to 300-year cycles. The intensity, location and potential danger of future earthquakes depend on the continuous or discontinuous nature of the North Anatolian Fault, on other diverse geological and geodynamical characteristics, and on the way and density by which the ground surface is occupied by human constructions. Let us consider from a general point of view the mechanisms responsible for major seismic events and how humans are progressing in predicting and mitigating the earthquake hazards.

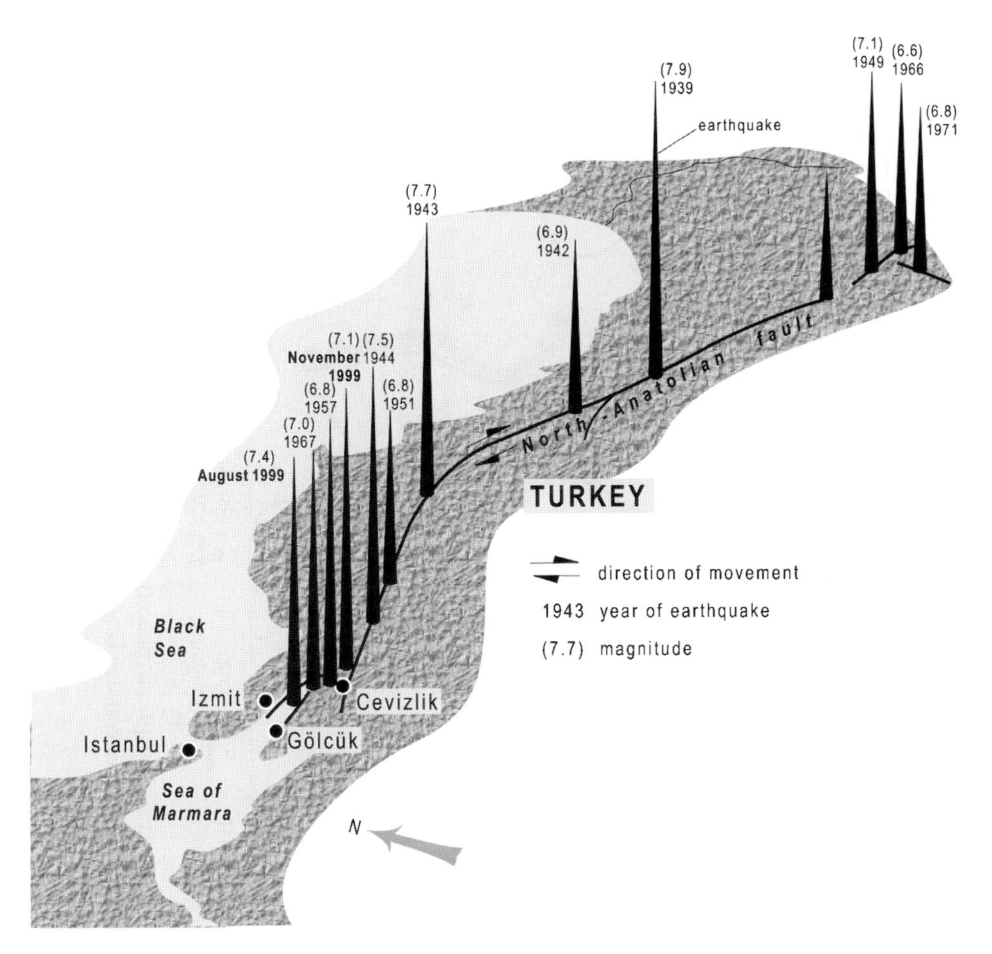

Figure 12: Distribution of major recent earthquakes along the North Anatolian Fault (after National Geographic, 2000). The peak height is proportional to the seismic magnitude.

1.2. The Seismic Hazard

Location Earthquakes occur systematically in *crust areas subject to important tectonic activity and crossed by fault systems.* These areas are mainly located near the border of moving lithospheric plates, especially in subduction, collision and sliding zones where intense compressions occur: Mediterranean Basin and its borders, Himalayan suture zone, Southeast Asian and West Pacific, west of the American continents (Fig. 13). Ninety-five percent of the world's earthquakes form in peri-Pacific (80%) and Mediterranean (15%) tectonic belts. The regions where major seismic tremors commonly occur are often subject to

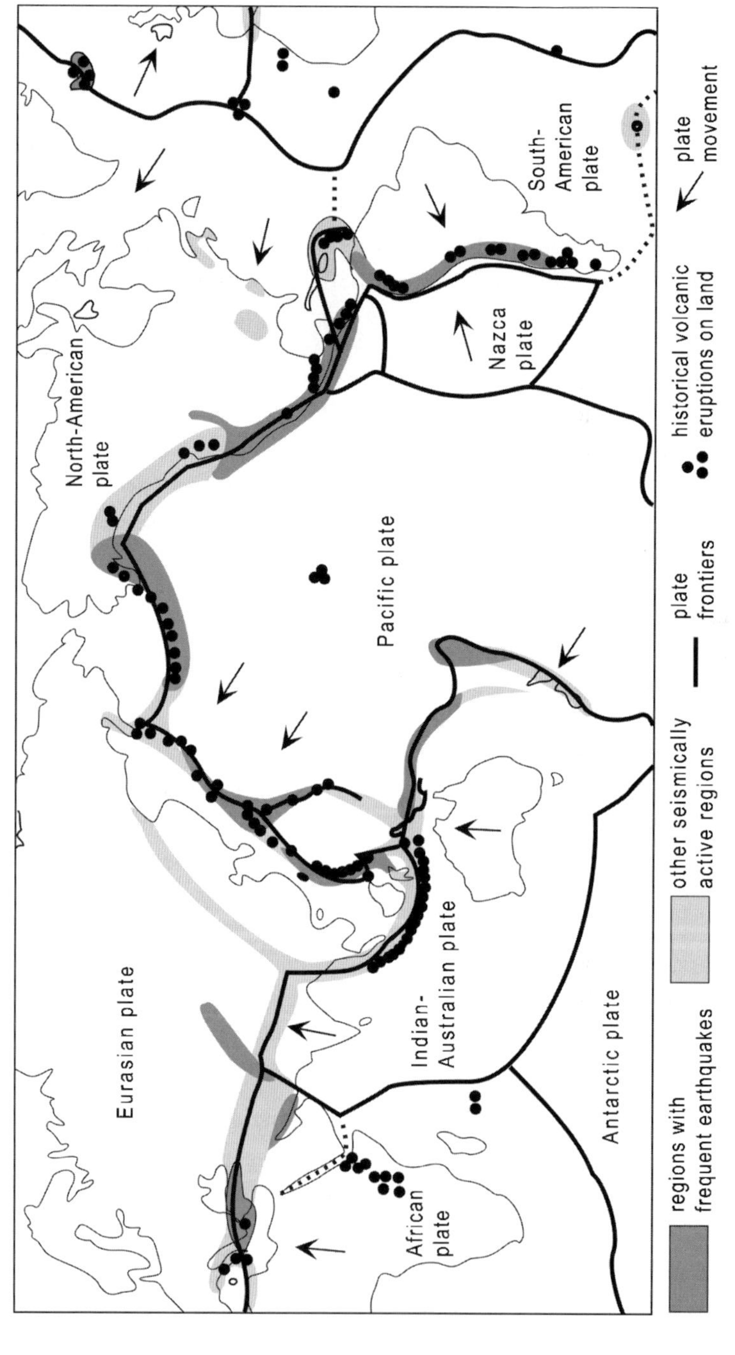

Figure 13: Location of main world seismic regions, of main lithospheric plate frontiers, and of subaerial volcanoes responsible for major historical eruptions (after Bryant, 1991).

volcanic eruptions, which also express the ground dynamics along plate frontiers. Some violent earthquakes may also affect the active fault zones of intraplate domains, as, for example, in northeast China, west-central USA and northwest Europe.

Damage Earthquakes constitute the natural hazard responsible for the *greatest number of deaths and material loss*. The two earthquakes that struck north Turkey in 1999 are not representative of the average damage caused by major seismic shakes. For instance the 1976 earthquake of Tangshan in north China caused the death of about 750,000 people and that of Tokyo–Yokohama in 1923, 143,000 victims. The total number of deaths during the last 2,000 years is estimated to be 8 million. Considering the considerable error calculation of such estimation because of insufficiently reliable archives, the actual number of casualties could be double. The effect of seismic tremors is often amplified by fires caused by short-circuits or spilled fuel, and by broken pipes causing flooding or water shortage for the fire fighters. This is particularly the case in densely urbanized areas. Ninety percent of the damage to buildings after the 1923 Tokyo–Yokohama earthquake was caused by fire following the tremors.

Geographic distribution The high proportion of tectonic zones marked by *converging plate boundaries in South-east Asia and West Pacific regions* (Fig. 13) causes high numbers of deaths and important damage. This wide domain includes some of the most populated and helpless world countries, which aggravates the quake consequences (Table 2). In *Europe* important seismic hazards are threatening southern countries, because of the African and Eurasian plates collision responsible for the Alpine chain's uplift. Italy, former Yugoslavia and Greece are especially concerned with earthquakes, the number and gravity of which tend to increase in the direction of Turkey and the Middle East, where Anatolian and Arabian plates add to collision and convergence phenomena.

In *France* the seismic activity is usually low. The highest values are registered in the Cenozoic mountains (Alps, Pyrenees), along the Variscan Palaeozoic basement (south Britain, Vendée, the Massif Central and southern Vosges), and in the pre-rift valleys of the Rhine River and Limagne. The southeastern French regions situated at the Africa–Europe collision front are particularly vulnerable and have already experienced serious quakes (e.g. Lambesc, June, 1909). The overseas departments and territories display a higher average seismic activity than metropolitan regions, namely in the French Indies, where subduction, volcanism and seismicity combine in an active manner (e.g. the 1843 Pointe-à-Pitre earthquake in Guadeloupe: 15,000 deaths).

Quantification The *intensity* of an earthquake corresponds to the nature of damage recorded at the ground surface. It comprises 12 degrees (Roman

Table 2: Examples of major historical earthquakes (modified from Lumsden, 1992).

Outside Europe		Europe	
Year	**Location**	**Year**	**Location**
30,000–50,000 deaths			
342	Antioch, Turkey	856	Corinth, Greece
1641	Tabriz, Iran	1456	Naples, Italy
1935	Quetta, Pakistan	1915	Avezzano, Italy
1939	Chillau, Chile		
50,000–100,000 deaths			
1269	Seyhan, Turkey	1755	Lisbon, Portugal
1667	Semakha, Russia	1908	Messina-Reggio, Italy
1727	Tabriz, Iran		
1927	Tsinghaï, China		
More than 100,000 deaths			
1290	Chihli, China	1202	Areas surrounding the Aegean Sea
1556	Shanxi-Henan, China		
1730	Beijing, China		
1737	Calcutta, India		
1923	Tokyo-Yokohama, Japan		
1976	Tangshan, China		

numerals), first defined by Mercalli in 1956, further clarified in 1964, and called as the *MSK scale* (Table 3). Intensity values do not result from instrumental measurements but from qualitative *observations* performed during and after the earthquakes, and concerning especially the damage to buildings. For a given earthquake the intensity values vary with the location: they are maximum where the tremor has the highest effects at ground surface (i.e. the epicentre), and diminish with increasing distance by forming roughly concentric zones. The information collected from the different sectors of a region hit by an earthquake allows mapping of the distribution of seismic intensities (Fig. 14).

The *magnitude* represents a value *calculated* from seismometer recordings. It expresses the amount of energy released at the seismic focus (i.e. the place in the deep rock basement where the tremor is initiated, which is called the hypocentre) and is specific for each earthquake. Magnitude values allow us to compare different earthquakes, provided that adequate calibration has been accomplished: type of seismographs, distance to focus, location and depth of focus, nature of geological formations, tremor duration, etc. Called the *Richter scale* because it was defined by Charles Richter in 1935, the magnitude scale is

Table 3: MSK scale of earthquake intensities.

Scale	Description
I.	Non-perceptible tremor, recording by seismographs only
II.	Tremor felt only on upper floors of buildings
III.	Slightly felt shake, glass vibrations
IV.	Widely felt shake; window, door and dish vibrations
V.	Sleeper's awakening, displacement of items
VI.	Dish-breaking, superficial fissuring of walls, tile-fall, soil cracks, heavy furniture movement
VII.	Building fissuring, chimney-breaking and -fall, difficulty standing
VIII.	Furniture, tower and monument collapse, partial fall of masonry, change in well and lake water aspect
IX.	Extensive damage to constructions, rail torsion, landslides, breaking of buried pipes
X.	Widespread building destruction, large soil cracks, breaking of roads, railways and dams
XI.	Disruption of all man-made developments, general soil deformation
XII.	Total topography and hydrography disruption

logarithmic and theoretically without upper limit. The augmentation by one degree on the Richter scale corresponds to a maximum seismic wave of amplitude 10 times higher than that of the preceding degree, and to energy 30 times larger. For instance an earthquake of magnitude 7 transmits energy 900 times higher than an earthquake of magnitude 5 (i.e. 30×30). The earthquake of August 1999 in north Turkey had a magnitude of 7.4, a value exceeded by only 3 of the 13 major tremors having affected the North Anatolian Fault since 1939 (Fig. 12). The highest magnitude value was recorded in 1960 in Chile and attained 9.5 on the Richter scale. In France about 20 earthquakes annually only reach a magnitude of 3.5, a value attained each year several thousand times in the whole Mediterranean range. Contrary to intensity values, the magnitude does not provide any information on the damage affecting material goods, not on people themselves. Depending on the development and population level in a given region, a high magnitude earthquake can cause little or disastrous damage. This is the key problem of both seismically active and densely urbanized regions (see Chapter 1.6 and 1.7).

The *duration* of an earthquake usually increases with its magnitude, which often aggravates the potential damage. A building able to resist violent tremors during several 10-sec periods may collapse if the ground movements exceed 100 sec. The 1995 earthquake of Kobe (Japan) was marked by highly intense tremors

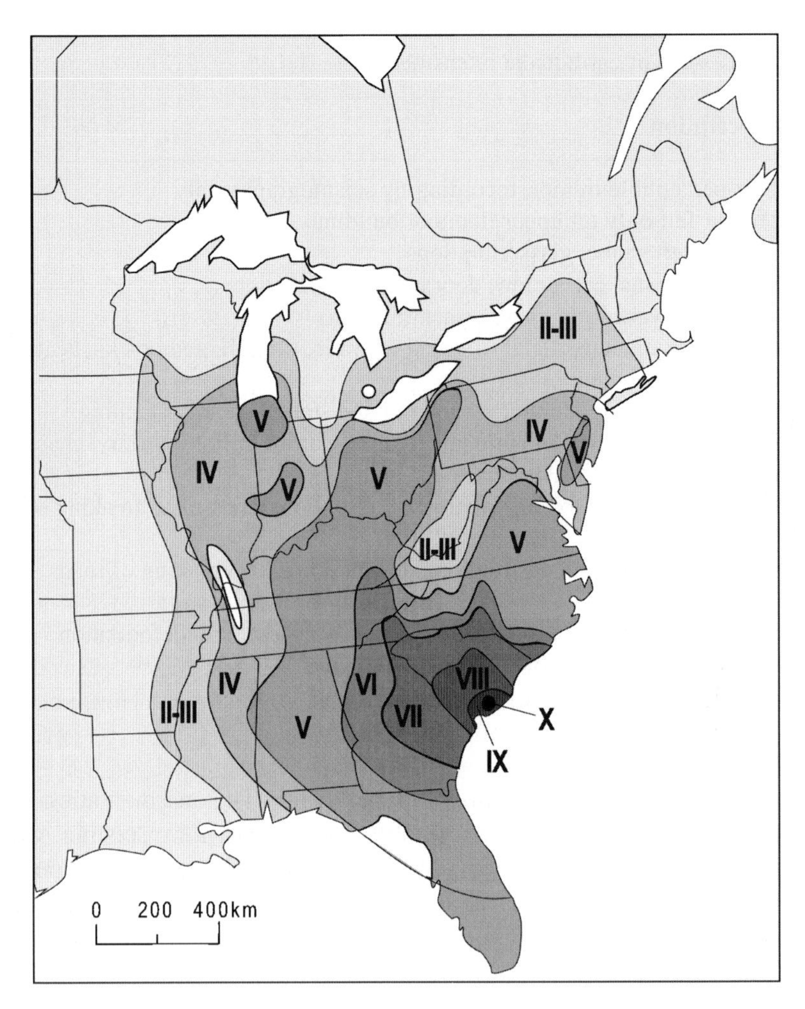

Figure 14: Seismic intensity map established after the 1886 Charleston earthquake (August 31, South Carolina, USA). The earthquake which was perceived in most eastern states of the USA and as far as Cuba and Bermuda, extended over 10 degrees of the 12 of the intensity scale (Murck et al., 1996).

during only 11 sec; but the movements continued for about 2 min, determining widespread ground destabilization and extensive collapsing of constructions.

The *earthquake frequency* statistically varies in inverse proportion to intensity and magnitude. Worldwide compilations show that about 800,000 tremors of magnitude lower than 3.5 (intensity I) occur annually, 4,800 of magnitude 4.3–4.8 (intensity IV), 500 of magnitude 5.5–6.1 (intensity VI, VII), and less than 1 of magnitude exceeding 8 (intensity XII). The earthquake of August,

1999, in Turkey constituted one of the four annual tremors of magnitude ca. 7.4 affecting our planet on average.

1.3. Mechanism

Tectonic context Earthquakes occur in *active fault zones* separating rock compartments submitted to strong compression or tension driven by tectonic forces. Mostly induced by lithospheric plate movements, these forces determine increasing frictions along the faults, issuing in instantaneous, often violent rupture. The rupture, which expresses the earthquake proper, is all the more violent and energy-emitting as the progressive rock deformation has been more important and longer, and has been acting on greater distance and thickness. Such a linkage between driving mechanisms explains the correlation observed between earthquake intensity, magnitude and frequency.

Functioning The explanation of earthquake release is generally based on the principle of *elastic rebound.* Two rigid rock compartments in contact with each other and submitted to very slow, opposite movements are deformed in a progressive, elastic way, until the deformation energy exceeds the friction energy. At this moment a sudden, instantaneous rupture occurs: it constitutes the "elastic rebound" of the deformed system, the physical expression of which forms the earthquake. As soon as the rupture has taken place and the deformation energy has been emitted within the ground, the two compartments recover their previous shape and rigidity, but are shifted relative to the initial respective position (Fig. 15). If the tectonic friction continues the deformation movement progressively increases again, until new rupture, elastic rebound and earthquake successively take place. Most earthquakes (95%) occur in the Earth's crust at depths shallower than 60 km, where the underground rocks are more rigid and cannot be deformed on a long-term basis without periodical rupture. At greater depth the rocky formations get both hotter, owing to the terrestrial heat flow, and more compressed, because of the increased lithostatic burden; plastic deformation processes then become easier, elastic properties diminish, and rebound phenomena are less frequent. Earthquakes do not form at focus depths greater than 700 km.

Seismic waves Frequently preceded by minor tremors ("precursory" shocks) and usually followed by more or less intense and numerous tremors ("replica" shocks), the earthquakes induce the emission of seismic waves, which are responsible for the damages affecting the Earth surface. The seismic waves are elastic and propagate either inside the geological formation or at their surface. The former group constitutes the *volume waves*, the speed of which increases

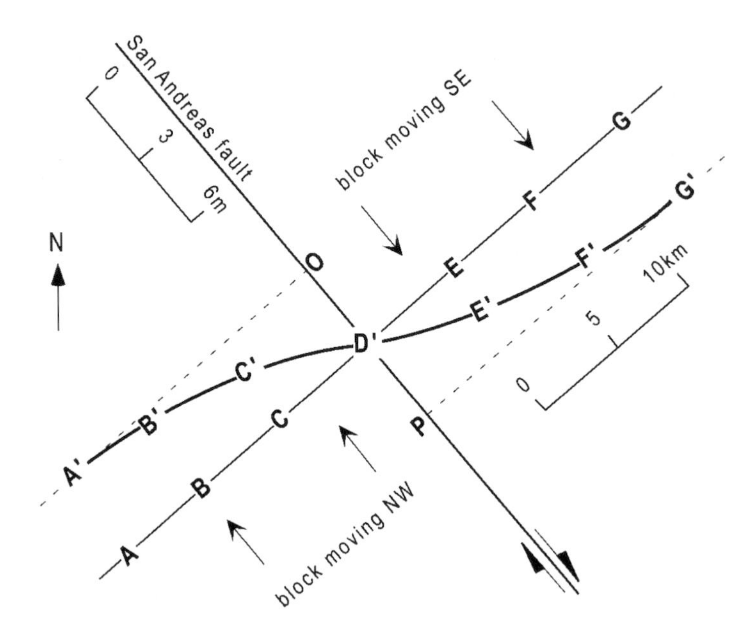

Figure 15: Application of the elastic rebound principle to the major 1906 earthquake that affected the San Andreas fault in California (after Murck et al., 1996). The seven points A–G initially aligned have undergone a deformation along the A′ to G′ line. After the rupture they have been re-aligned on A′O and PG′ segments.

with the rock density. Volume waves comprise primary (**P**), longitudinal waves called compression waves, and secondary (**S**), transverse waves called shear waves. The P waves move faster than S waves, which are absorbed by liquids and propagate only in Earth's solid envelopes. Measuring the respective speed of P waves (V_P) and S waves (V_S) and calculating the $V_P : V_S$ ratio allows us to determine the distance separating the earthquake epicentre and focus, and therefore to locate the focus.

The group of seismic waves propagating at the rock surface forms the *surface waves*, which move like ripples on a water plane, more slowly and with less ample wavelength than volume waves. Surface waves include Love (**L**) waves that progress transversely like S waves, and Rayleigh (**R**) waves that display a rather slow, ellipsoidal displacement with a retrograde resultant. The shearing induced during earthquakes by S- and L-wave propagation is responsible for most damage affecting human constructions.

Geological factors The physical properties of geological formations exposed to earthquakes play a crucial role in the propagation of seismic vibrations. The *small indurated rocks* tend to amplify the effect of earthquake-induced shock waves, especially if they are soaked in water. The rigid, compact, homogeneous

Table 4: Value of the seismic intensity for different types of rocks relative to non-weathered granite (after Mevdedev; see Bell, 1999).

Dominant lithology	Relative seismic intensity
granite	0
limestone, sandstone	1.0–1.5
gravel	1.0–1 6
dry sand	1.2–1.8
dry clay	1.1–2.1
wet sand	1.7–2.8
wet clay	3.3–3.9

rocks such as non-fractured granites are much less sensitive to earthquakes (Table 4). The influence of the geological and lithological underground characteristics was clearly illustrated by the effects of the 1985 Mexico City earthquake, which induced the damage or destruction of 7,400 buildings, some of them being located several hundred kilometres away from the epicentre. A large part of the town was built on ancient lake deposits, which are loosely consolidated and soaked in water. The buildings located in these areas were extensively damaged, whereas most constructions established on hard, water-free rocky substrates remained unaffected. Calculation showed that the amplification of the shock waves within unconsolidated formations had reached values 8 to 50 times higher than in hard rocks.

Secondary effects Some indirect phenomena triggered by seismic tremors may induce people and goods damages exceeding those directly determined by the earthquakes themselves. Such effects are to be added to those arising from human activity: building collapse in badly constructed urban areas, short-circuit induced fire, flooding due to pipe- or dam-breaking, etc. The *soil liquefaction* (Chapter 3.2) results in a loss of cohesion and strength within surface formations soaked in water or marked by weak inter-particle bonds. Earthquake-triggered liquefaction may have several negative consequences: (1) softening of building-supporting ground, favouring house-collapsing; (2) oscillatory movements of horizontal surface determining repeated opening and closure of cracks, the size of which may reach 1 m wide and 10 m deep; (3) lateral expansion and distortion of liquefied surface soils, which may flow above firm substrate on a less than 3° slope; (4) mass sliding of liquefied formations on more than 3° slopes, favouring mudslides, landslides and other gravity movements with potentially devastating effects. For instance the combination of liquefied, partly frozen soils and of landslides was responsible for most building destructions in Anchorage after the Good Friday earthquake that struck southern Alaska in 1964.

Other effects are indirectly driven by earthquakes:

- *Ground uplift or downlift* can exceed meter to decametre heights and affect huge areas. The 1964 Alaska earthquake induced considerable topographic changes; the line of Pacific Ocean coast was displaced over a distance approaching 1,000 km, some sectors becoming submarine and others subaerial. The distortion of coastal plains and adjacent submarine bottom affected a surface of 260,000 km^2.
- Earthquake-triggered *tsunamis* constitute very large and rapid marine waves, the curve of which dramatically increases when approaching coastal areas (Chapter 4.3.2). The tsunami determined by the 1946 Unimak Island earthquake in Alaska ran right across the North Pacific at 800 km/h and reached Hawaii after only 4.5 h, forming a giant wave at 18 m above the normal high tide level.
- *Floods and avalanches* may result from earthquake-induced changes in hydrographic networks and destabilization of mountainous sectors. This was the case after major tremors affected the northern part of the Andes cordillera at the end of the 1990s.

1.4. Artificial Earthquakes

Dam building To the geologists' surprise, the usually seismically inactive area of the lower Colorado river basin (USA) displayed more than 600 low-to-medium magnitude earthquakes during the decade following the 1935 Hoover Dam construction and subsequent filling with Mead Lake water. The most important tremor reached a value of 5 on the Richter scale. Such earthquakes of unintentional human origin result from *water overload* on the local ground, as well as from an *increase in the interstitial pressure* within the bedrock hydraulically connected with the lake (cracks, fractures). Many other examples are reported of seismic instability induced or intensified by filling large artificial reservoirs with water: Greece, India, USA, Zambia, Zimbabwe, etc. (Costa & Baker, 1981). In India the settlement of the Koyna dam and reservoir in 1967 caused an earthquake of magnitude 5 felt as far as Bombay, 230 km from the epicentre. This earthquake was responsible for 177 victims, 2,200 stricken people and considerable damage.

Liquid burying During the 1960s a lot of minor tremors affected the usually stable region of Denver (Colorado, USA). The investigations performed to

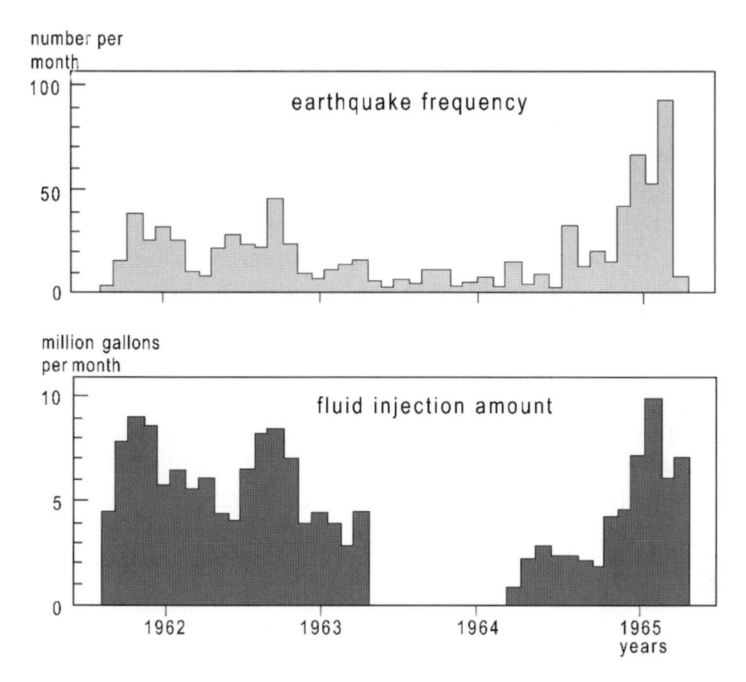

Figure 16: Correlation between liquid-waste burying and earthquake frequency in the U.S. army Rocky Mountain arsenal site, Denver region (after Evans; see Montgomery, 2000).

understand this unexpected seismic activity revealed the existence of close temporal relationships between the number of earthquakes recorded and the amount of toxic liquid waste buried in the deep rocky formation (3,671 m) of a military storage site located in the eastern Rocky Mountains (Fig. 16). Minor earthquakes have also been registered in some deeply buried oilfields. These different seismic events have been attributed to *increased pore pressure* determined by the artificial injection of liquids in the bedrock, favouring "elastic rebounds" by diminishing the friction forces. Some scientists have envisaged using this phenomenon to prevent some potentially dangerous earthquakes: the liquid injection in seismically active geological formations would allow us to initiate minor, harmless earthquakes by attenuating the internal rock frictions, and therefore eliminating the energy stored in excess. Such mitigation attempts could nevertheless also determine unexpected, dangerous seismic effects. It is, for instance, difficult to exclude the possibility that the fluid injection might accentuate the pressure release in already strongly compressed formations and determine an unanticipated, violent earthquake.

Nuclear explosions Underground nuclear shots induce artificial seismic tremors that may potentially trigger *earthquake replicas*. Geological monitoring in the late 1960s revealed the formation in the Nevada pilot region of dozens of replica-tremors, which lasted for 3 weeks after the underground nuclear explosions. The magnitude was usually lower than 5, the focus being located at a depth of 7 km and the epicentres up to 13 km from the shooting point (see Murck et al., 1996). Seismic signals induced by nuclear explosions display shorter wavelengths than those resulting from important natural earthquakes, and are similar to those induced by smaller tremors (magnitude < 4.5). This property was used during the Cold War to conceal the nuclear tests, by various tricks amplifying the signal transmitted within the bedrock, e.g. quick successions of shots, and shots in loudly echoing cavities.

Remember that earthquake analysis and monitoring have allowed, together with other geophysical investigations, to better *know and understand the Earth's internal structure and dynamics*: average types, physical properties and chemical composition of solid and liquid rock envelopes ; discontinuities, 3D heterogeneity, temperature and pressure gradients ; material movements and exchanges, etc. Such indirectly obtained knowledge is of crucial importance for progress in assessing and predicting the seismic hazards threatened by man and his environment.

1.5. Earthquake Forecasting

1.5.1. Indicators

Seismicity Assessing the seismic hazards necessitates first a good knowledge of *local-to-regional seismic activity*: location, magnitude, duration and frequency of recent earthquakes; distance between epicentres and focus; speed and propagation conditions of the different wave systems, etc. Such information is mainly based on seismograph measurements, which are more useful as the records become more numerous, continuous, real-time processed, and reliably compared. This implies extensive seismic measurement networks and very rapid computer processing of huge amounts of data. The various national seismic monitoring networks (e.g. the French RéNaSS located at the Strasbourg Institute of Earth Physics) have become progressively integrated in a standardized international network of seismic stations and centres (World-Wide Standard Seismological Network, WWSSN). The aim of this global network is to ensure a worldwide watch and to provide updated information and projected trends.

The results of seismic activity monitoring are ambivalent in earthquake forecasting. Counting the number and intensity of precursory tremors has been successful in predicting the major earthquake (magnitude 7.3) responsible for the destruction of half of the city of Haicheng (China) in February 1975: preliminary tremors with increased intensity were recorded during the days

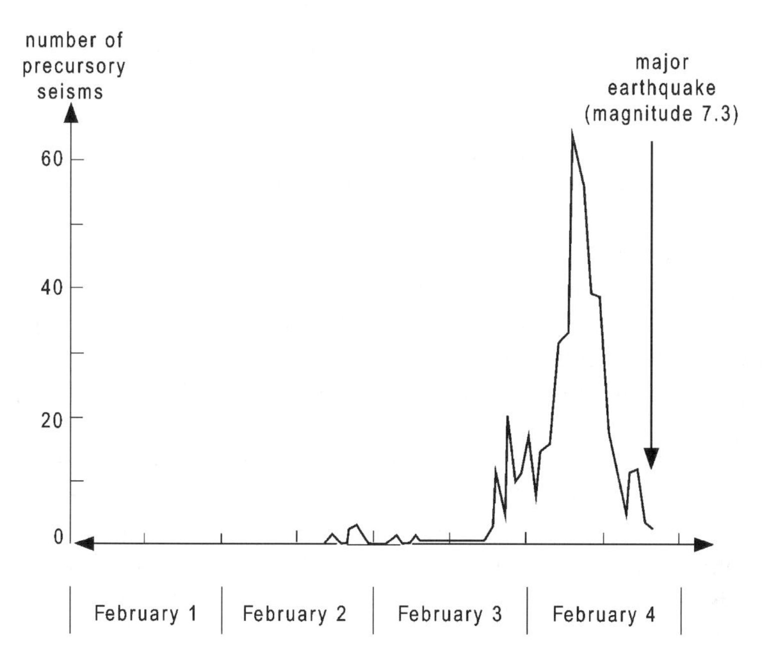

Figure 17: Frequency of precursory vibrations linked to the 1975 Haicheng earthquake, China (see Murck et al., 1996).

preceding the disaster (Fig. 17), which allowed evacuation of more than one million people and limited the death toll to a few hundred. In most other cases, the precursory tremors were too low or too late to permit reliably-based forecasting. Another potential criterion concerns the variations of the P- and S-waves speed ratio ($V_P : V_S$). This ratio tends to diminish before an earthquake occurs, because of the rock fissuring by pressure forces, increasing the volume of cracks and allowing interstitial water to seep. Some consistent results have been obtained by processing the seismogram data obtained for past earthquakes. But the duration of the $V_P : V_S$ ratio variations seems to depend dominantly on the magnitude; the time interval is of a few days for minor tremors but may attain several years before a major earthquake occurs (Pipkin, 1994). This criterion is therefore not enough for discriminating real-time processing.

Palaeoseismicity Information gathered on past earthquakes is helpful for understanding the linkage existing between tectonic and seismic activities, assessing the seismic recurrence controlled by potential cyclicity, and identifying in a given seismic area the tectonic segments that have been inactive for a long time. Such structural segments characterized by "*seismic gaps*" may have accumulated much internal energy that could be released shortly through violent earthquakes. Using such an approach by compiling the data available on Pacific

Ocean peripheral regions has allowed the location of high seismic risks in southeast Indonesia, east Kamtchatka, south California, and along the coast of Peru.

Geology, geomorphology Assessing seismic hazards implies detailed, thorough *field observation*: location, number, displacement direction and speed, and continuous or discontinuous character of *active faults*; location of recent and ancient epicentres relative to these faults; rigid or soft nature of the bedrock subject to vibrations, with additional information on homogeneity or fracturing, freshness or deterioration, interstitial water absence or presence; and possibility of landslides or flooding. The field observations and measurements are sometimes completed by *satellite data*, which allows us to precisely compare successive images for a given area (interferograms). But these data are hardly useful for predicting an earthquake, such images being unable to identify precursory movements at a local or even regional scale. On the other hand the pressure acting at depth and inducing fissuring, water seepage and expansion underground may determine *surface distortions*. Such surface movements are quantified with tiltmeters, extension meters, and other precision instruments similar to those used for studying landslides (Chapter 3.1). A nice illustration of the efficiency of such forecasting tools concerns the June 1963 Niigata earthquake, in west Japan (magnitude 7.5). Continuous centimetric upheaval or subsidence movements were recorded during the sixty years preceding the earthquake, systematic uplift movements having affected the area close to the future epicentre. After a slowing down of these vertical movements at the end of the 1950s, a general ground subsidence occurred when the earthquake happened (Fig. 18).

Physical properties The tendency of both the slowdown of P-waves relative to S seismic waves and the ground dilatation mainly results from the increase in pore water amount and pressure, due to micro-fracturing development and fissuring of underground rocks. The influx of interstitial water also tends to induce a *progressive diminution of the electric resistivity*, the monitoring of which participates in suspecting the imminence of an earthquake. As soon as the seismic shock has allowed the internal pressure to be released, cracks and micro-fractures close up and resistivity values increase rapidly again. Other investigations show that when an earthquake becomes imminent the rock dilatation by water influx affects the *magnetic susceptibility*, because of a temporary increase in the magnetic field.

An original approach tested by Greek geophysicists *V*arotsos, *A*lexopoulos and *N*omicos (i.e. VAN method) is based on the occurrence of an electric current emission, that precedes earthquakes of a magnitude greater than 3.5. Called *electro-seismic signals*, these electric impulses are measured with electrodes

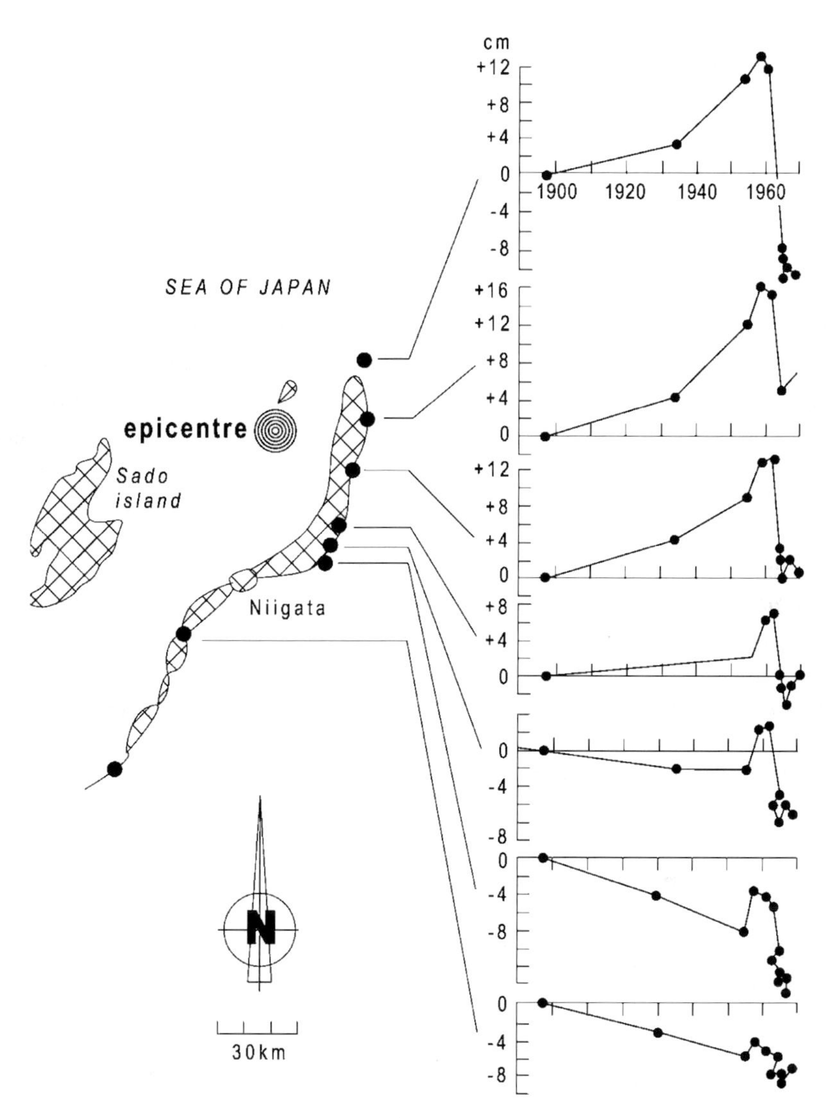

Figure 18: Surface morphological distortions having preceded the 1964 Niigata earthquake, Japan (see Bell, 1999). Black dots correspond to measurement stations.

implanted close to the ground surface (ca. 1 m depth). The electrodes and signal receivers constitute a network fitting the local geological configuration and calibrated according to local seismic activity and intensity. The VAN method has been tested with some success in Greece since the 1980s, but it is still strongly controversial. Some unsuccessful results have been attributed to the presence of unexpected faults preventing normal propagation of earth currents, or to an excessive depth of the seismic focus location within the Earth's lithosphere.

Chemical changes When an earthquake is approaching, some inert gas such as radon produced by natural radioactivity (Chapter 7), as well as argon, helium, neon or xenon, tends to be progressively expelled from geological formations submitted to increasing internal pressure. Potentially evacuated within aquiferous rocks these gassy products can easily be identified by chemical analysis and their quantitative variations measured within well or spring water. The monitoring of the nature and abundance of gas in underground water can therefore participate in forecasting a seismic event. The 1966 Tashkent earthquake in Russia (magnitude 5.3) was preceded 1960 by an increase in $5–15 \times 10^{-10}$ Ci/l of radon dissolved in water of a nearby borehole; initial values were recovered immediately after the earthquake had occurred (see Bell, 1999). Other reports indicate some changes in the chemical composition, transfer speed and flow rate of underground water. The pressure undergone by the rocks located close to active faults may induce a rise in the well level, an injection in water of clay or organic suspended matter, or an augmentation of water temperature due to geothermal activity. The *monitoring of chemical and associated physical characteristics* of underground water constitutes therefore a useful tool to help in predicting seismic tremors.

Animal behaviour In the morning of July 18, 1969, various animals of the Tianjin zoo (China) displayed abnormal behaviour, such as cries by usually silent pandas, refusal by swans to go to swim or by snakes to nest, and no appetite for feeding. Scientists, who had been alerted by zoo managers, immediately performed a precise calculation. At noon a violent earthquake occurred (magnitude 7.4). During the last decades similar cases were reported in Asia, Europe, North and South America, about various animal groups: fish, batrachians and reptiles, domestic or wild birds and mammals. Experiments have been undertaken by Japanese scientists on the animal behaviour in relation to seismic vibrations, but the results do not bear any significant forecasting value. The data obtained are generally scarcely reproducible and systematically concern very short time intervals only.

Statistical approach As earthquake events are characterized by a succession of slow elastic distortions and instantaneous rock ruptures, statistics may help to forecast major tremors by using *probabilistic models*. By integrating all geophysical, geological and historical information for a given area, it is possible to assess the likelihood for an active fault to reach, after a given time interval, a state of tension approaching the breaking point, and to be able to determine an earthquake of a given magnitude. Such predictions have been established for the San Andreas fault system in California, along which the earthquake probability during the 1988–2018 period was assessed and quantified for various tectonic segments extending between the Mexican border and Oregon State (Fig. 19).

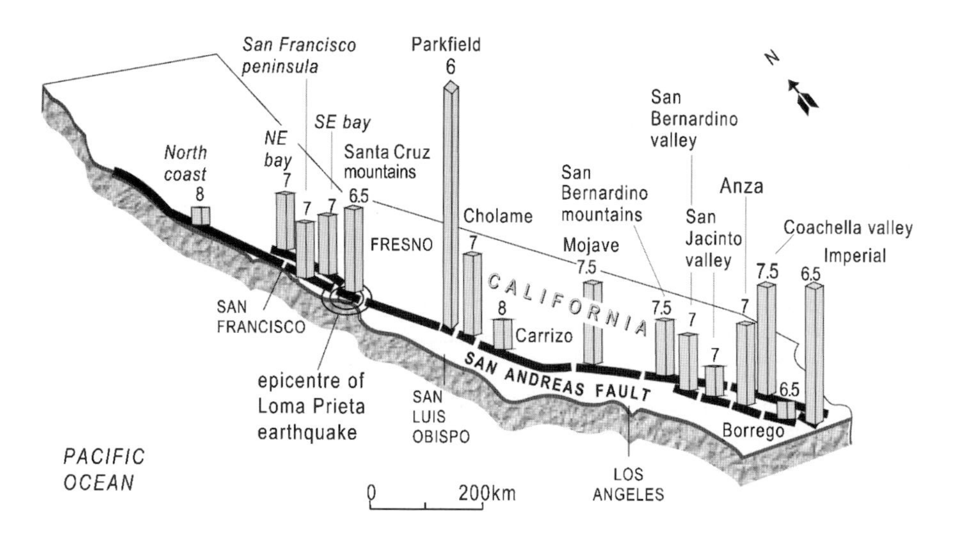

Figure 19: Earthquake probability for the 1988–2018 period along the different tectonic segments of the San Andreas fault system, California (after Lundgren, 1999). The degree of probability is expressed by the bar height. The figures at the bar top indicate the expected magnitude.

The usefulness of the method was proved after the 1989 Loma Prieta earthquake had occurred in the Santa Cruz Mountains, a segment estimated to be seismically the most sensitive of the San Francisco region

1.5.2. Combination of Seismic Indicators, Hazard Maps, Assessments

As seen previously (Chapter 1.5.1) the search for predicting seismic risks gives way to a large diversity of markers, each of them being of a limited efficiency. The gravity of damages induced by major earthquakes, together with the difficulty of finding reliable forecasting indicators, led to the establishment and development of *multi-tool sites* in some of the most sensitive regions. Devices are chosen to complement each other according to local conditions: seismographs, gravimeters and magnetometers; topometers (GPS) and laser-controlled equipment; tiltmeters, extension meters, analysers of mechanical properties (shearing, friction, permeability); piezometers and resistivity meters; radon and water chemistry analysers; etc. (Smith, 1992). The comparison of continuously recorded data allows us to improve the prediction level. This is mainly the case when combining data on real-time processed P- and S-waves's speed ratio, underground distortion, radon abundance in underground water, electric resistivity, and abundance of precursory tremors (Fig. 20).

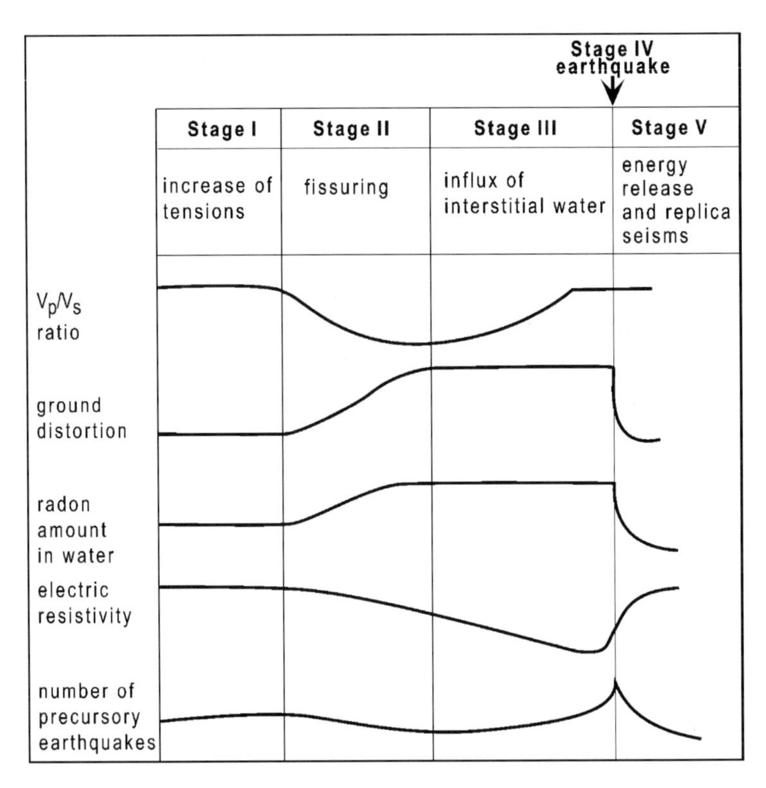

Figure 20: Expected changes of different parameters in the prospect of a major earthquake (after Pipkin & Trent, 2001).

The gathering of all available information, permanently completed by data provided by the mean of local to international seismic networks, allows us to establish *seismic hazards maps*. The map scale differs according to the nature, distribution and importance of the earthquakes expected. On most detailed maps the potential site effects are taken into account by means of seismic micro-zoning. Such maps exist for various countries, for instance the USA, which is exposed to a very large array of risks (Fig. 21). The seismic frequency, not expressed cartographically, is much higher in California (southwest of the country) than in New England (northeast), where earthquakes of similar intensity and magnitude may nevertheless occur.

To summarize, *the forecasting of earthquakes has been considerably improved on a long-term scale*. It is possible to predict with a reasonable likelihood where major earthquakes in world regions will occur during the next decades, what will be their approximate magnitude, which damage intensity may be expected, and in which time interval new tremors could occur again.

Such predictability is somewhat satisfactory from a scientific point of view and determines significant progress in the understanding of earthquake

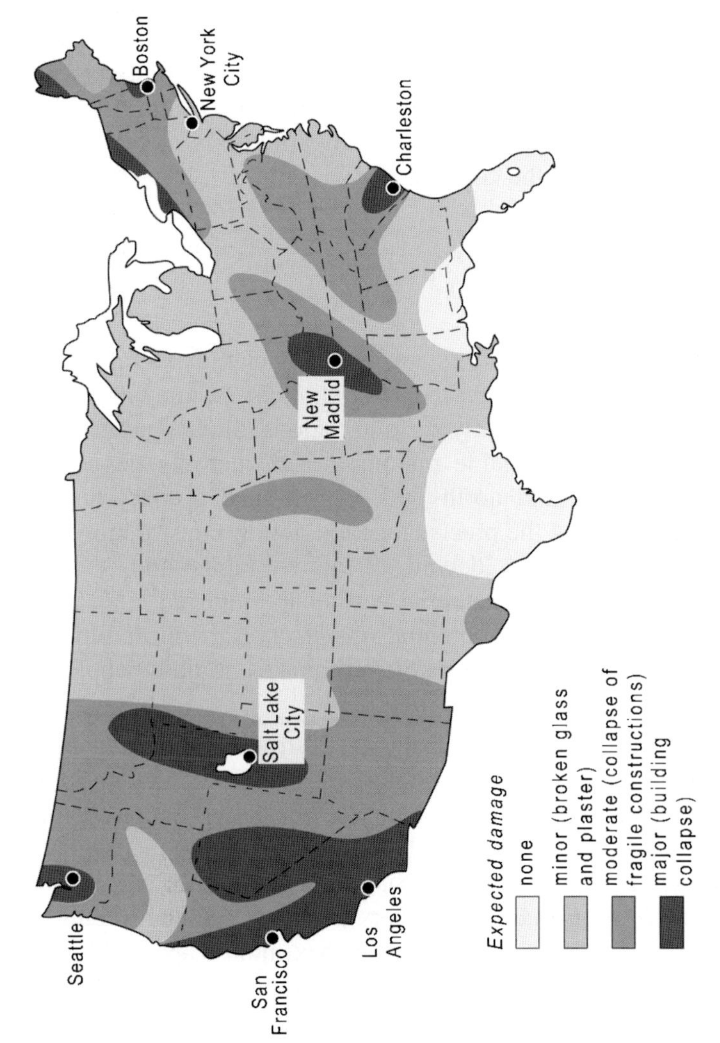

Figure 21: Map of seismic risks in the USA (after Murck et al., 1996).

mechanisms. It is nevertheless of little use for the inhabitants of threatened regions, who are basically concerned with the exact moment of the next seismically induced disasters. Unfortunately *short-term earthquake forecasting remains impossible or at least highly aleatory.* This is mainly due to the high complexity and remote origin of natural processes governing the internal dynamics of Earth's envelopes. Remember that, after the successful prediction of the 1969 Tianjin earthquake in China, the geophysicists and other experts were unable to forecast the terribly destructive earthquake of Tangshan in 1976. In the same way the last major earthquake of the 20th century in Japan (1995) was not expected in the Kobe area but in the particularly mobile tectonic zones located in the southeast of the country.

1.6. Earthquake Prevention

Building location As earthquakes are both ineluctable and rarely predictable, it is only possible to try to *prevent* their *harmful effects*. The basic precaution to adopt is to avoid building in highly seismic regions. Such a purpose is of course impossible to achieve in unstable regions already densely populated such as Japan, south California, and north Turkey. But such a concern is of crucial importance when planning the precise location of new construction, for instance by choosing ground sectors devoid of active faults and soft geological formations, and remote from potential earthquake-induced landsliding slopes. An accurate characterization of potential *site effects* is therefore fundamental for helping to mitigate the consequences of earthquakes on both people and goods.

Building nature and types The shape, size, structure, bonding and nature of the materials employed for construction are of crucial importance for helping in earthquake prevention. Buildings, which are basically conceived for resisting the Earth's gravity forces, are usually more sensitive to earthquake-induced horizontal vibrations (S and L waves) than to vertical ones.

The intrinsic *vibration frequency* differs greatly according to building nature and design. The risk of destruction becomes maximum when building vibrations are in phase with seismic vibrations. High buildings are particularly sensitive to low-frequency shock waves, and small buildings to high-frequency, lesser-magnitude waves. For this reason the building rules in seismically active zones are primarily based on the building size and relative position: preferred design of low and massive constructions; choice of independent, non-adjoining high buildings to prevent any interference in their respective vibration frequency; disconnection of the buildings of different size and shape that are potentially prone to react independently when facing seismic vibrations.

A second important parameter concerns the *nature of materials* employed for construction. An increasing resistance against seismic shocks characterizes the following materials: (1) cob or adobe, a sun-dried mixture of straw and clay, common in some countries where building stone is rare. Houses with adobe walls are frequent in the Middle East and South America. In Peru they constitute two-thirds of countryside and one third of urban buildings; (2) wood, which is more resistant when buildings are high and their structure is flexible; (3) bricks, stone and cavity blocks, more resistant when assembled with quality mortar and corner support pieces; (4) concrete, if possible prestressed and reinforced, with 3D metal support pieces; and (5) steel.

The *type of foundations and architecture* constitutes a third significant parameter. Building on loose or heterogeneous ground necessitates long and deeply cramped piles, in order to prevent the risk of liquefaction, sliding or subsidence induced by seismic activity. Earthquake-resistant constructions are built without large picture windows, mezzanines or verandas. They include a hard and flexible internal frame, and are sometimes placed above soft rubber or plastic blocks. In a general way the competence for buildings to resist earthquakes increases if the following characteristics are gathered together: lightness, stability, geometric continuity, three-dimensional chaining, material homogeneity.

The para-seismic engineering techniques have considerably progressed during the last decades. The seismically active regions correctly designed, built and equipped, including urban areas densely populated and with sky-scrapers, are potentially protected in an efficient way against most damage. Only the very high intensity earthquakes are responsible for extensive damage in such adequately designed regions (Table 5). For economic reasons the recent progress acquired in civil engineering techniques is usually put into practice much better in industrialized than in developing countries. Some ancient buildings made of flexibly adjusted wood or with vaulted stone roofs prove nevertheless to be more earthquake-resistant than many recent buildings (e.g. the Turkish earthquakes of 1999; Chapter 1.1).

1.7. Perspectives

The number of human deaths induced by earthquakes and of associated damage to property has dramatically increased during the last two millennia. Documented in a particularly reliable manner for the last 400–500 years, this fact is largely explained by the world population increase (Fig. 22) exacerbating the seismic risks. It does not result from an intrinsic augmentation of the devastating earthquake frequency. *The relation observed is nevertheless not strictly proportional to the demographic explosion*, suggesting that additional

Table 5: Percentage of damage expected for constructions made of different materials, built or not according to para-seismic standards, and exposed to earthquakes of various MSK scale intensity values (after Bennett & Doyle, 1997).

	MSK				
	VI	**VII**	**VIII**	**IX**	**X**
Common constructions					
cob, adobe	8	22	50	100	100
masonry	3.5	14	40	80	100
reinforced concrete	2.5	11	33	70	100
steel	1.8	6	18	40	60
Para-seismic constructions					
reinforced masonry	0.3	1.5	5	13	25
super-reinforced concrete	0.9	4	13	33	58
reinforced steel	0.4	2	7	20	40

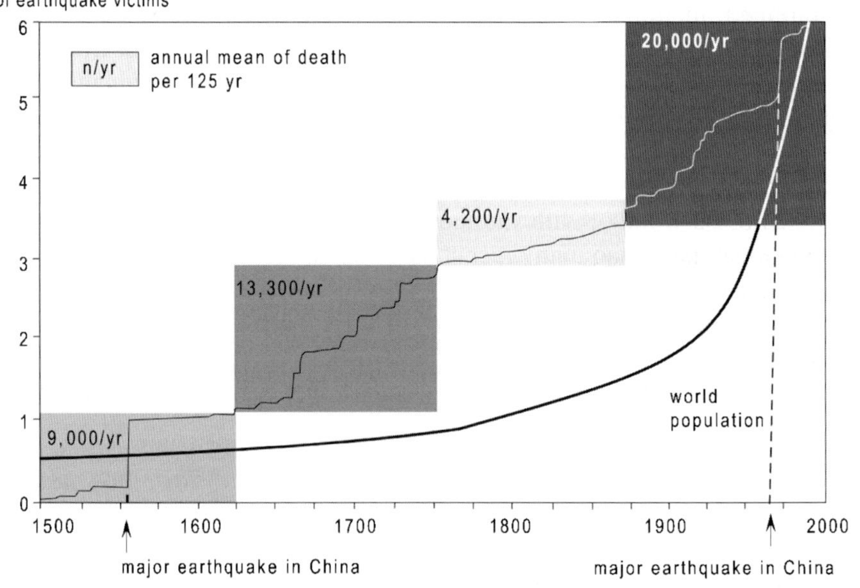

Figure 22: Earthquake-induced human deaths accumulated during the last five centuries, and correlative evolution of the world population (after Bilham, 1996).

causes participate in increasing the seismic risk. The 125 year averaged number of annual casualties effectively reaches maximum values for the most recent period (20,000 deaths/year between 1875 and 2000), where the world population has risen from 1.25 to more than 6 billion inhabitants. But the 1623–1750 period was already struck by 13,000 annual victims, whereas the following period (1750–1875) displayed "only" 4,200 annual victims among a much more abundant population (ca. 800 million, Fig. 22). Natural and artificial additional causes are probably to be taken into account, such as earthquake cyclicity and countryside vs urban human occupancy.

As far as the most recent period is concerned (1975–2000), the exponential increase in the world population is associated with a *decrease in the proportion of human victims* relative to previous periods, although the absolute number of human deaths has increased. This basically reflects the progress accomplished in preventing the earthquake effects. In *addition earthquake-induced disasters have augmented in a lesser percentage in some regions than in others, in spite of similar seismic risks* (Bilham, 1994). This is due to the *excessive growth of urban conurbations in seismically unstable regions*. It is striking that at the beginning of the 21st century a dominant proportion of the 28 megalopolises hosting more than 8 million people, and of the 325 urban areas of 1–7 million inhabitants, are situated in tectonically active and therefore potentially seismic regions, e.g. in the vicinity of subduction or collision zones, transform faults or intraplate fracturation areas (Fig. 23). Many large to very large towns extend and are still expanding close to major tectonic and seismic belts. These huge urban concentrations, where more than one-fifth of the world population lives in often poor-quality and little-resistant buildings, leads one to anticipate major human disasters in the next 30 years. Apart from the 1923 Tokyo–Yokohama and 1976 Tangshan catastrophes, the large megalopolises have been relatively safe from seismic hazards until now, but statistics-based simulations suggest the occurrence of very large disasters in the coming period. The risk is all the more exacerbated as the earthquakes, which are systematically a source of panic and disorganization, affect regions that are densely urbanized, highly populated, poorly prepared for natural hazards, and under-equipped in rescue resources. In spite of modern and numerous equipment, the city of Kobe (Japan) was paralysed during about 10 h until the first aid workers were able to intervene efficiently after the 1995 major earthquake.

A Few Landmarks

- Earthquakes occur in active fault areas which are principally located at lithospheric plate frontiers (Pacific belt, Mediterranean range, etc.), secondarily in intraplate regions (China, western USA).

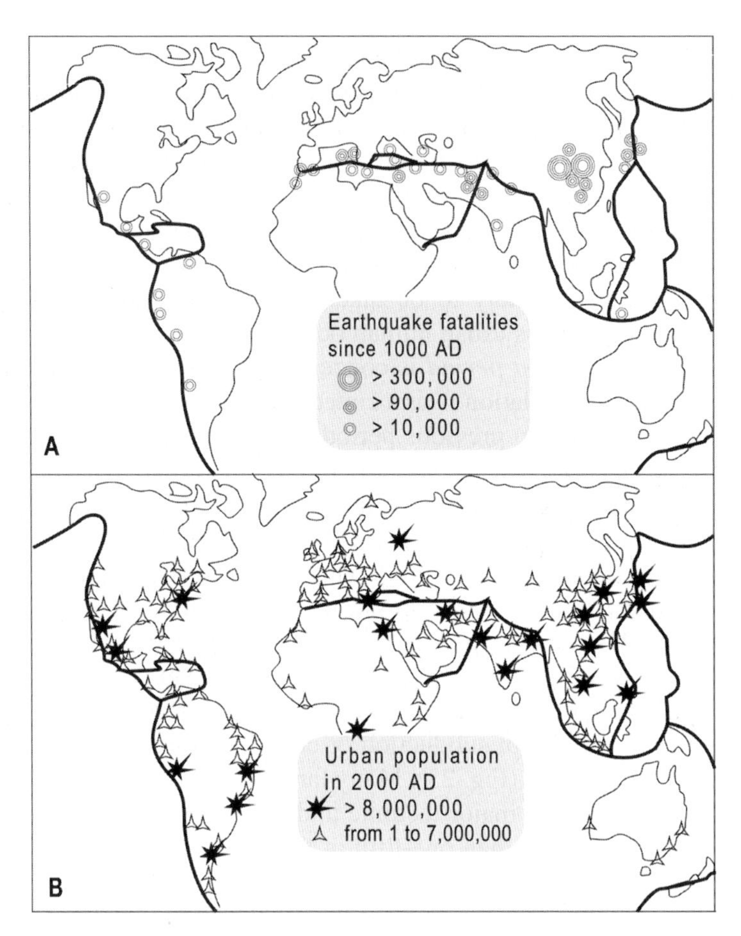

Figure 23: Comparison between (A) the location of earthquakes responsible for a great number of deaths since 1,000 AD, and (B) the location of world megalopolises in 2000 (after Bilham, 1996).

- Earthquakes result from the sudden release of underground tensions accumulated between two tectonic blocks, which are progressively distorted until a rupture state is attained. The energy released by the rupture is estimated either by the intensity of damage incurred at a given point of the ground surface (MSK scale), or by the magnitude of the shock at the focus (Richter scale).
- Major earthquakes are frequently preceded by precursory vibrations and followed by replica tremors. They can be associated with secondary effects, the sum of which is generally more harmful to people and property than the shocks themselves: landslides, soil liquefaction, subsidence, fire, floods, building and other developments collapsing.

- Responsible for the greatest number of natural hazard-induced deaths (8 million in 2,000 years, 20,000/year during the last 125 years), earthquakes are extensively studied and their formation mechanisms replaced in a broad geodynamic context. The investigations developed help in understanding the functioning and predicting the occurrence of earthquakes: seismicity and palaeoseismicity, geomorphology and tectonics, rock and interstitial fluid physics and chemistry, animal behaviour, probabilistic calculations, development of multi-tool field sites and seismic monitoring networks, and drawing of hazard maps at different scales.
- Earthquake forecasting has reached a reasonable level of reliability as far as long-term events are concerned, but remains almost impossible for short-term events because of the complexity of underground mechanisms and their interference with surface processes.
- The prevention of earthquake effects has considerably progressed during the last decades: location of construction according to geological and geomorphological conditions, choice of building materials, and para-seismic engineering. Only high-magnitude earthquakes still have disastrous and hardly bearable effects on both people and property. Unintentionally responsible for moderate tremors due to ground overloading (dam reservoirs) and/or liquid injection, man has endeavoured to prevent major tremors by triggering artificially attenuate precursory earthquakes; such a purpose remains aleatory in the absence of complementary research and modelling.
- If the number of earthquake-induced casualties increases parallel to the demographic expansion, its percentage tends to decrease due to efficient prevention measures. But the results differ depending on the world regions concerned. Seismic hazards remain high in developing countries. In addition the risks increase in densely urbanized areas evolving into megalopolises, a large number of which are located in highly seismic tectonic zones.

Further Reading

Alexander D., 1993. Natural disasters. UCL Press, London, 632 p.

Bell F. G., 1999. Geological hazards : their assessment, avoidance and mitigation. Spon, London, 648 p.

Bolt B. A., 1999. Earthquakes. Freeman & Co, 4th ed., 366 p.

Costa J. E. & Baker V. R., 1981. Surficial geology: building with the Earth. Wiley, 498 p.

Dagorne A. & Dars R., 1999. Les risques naturels. La cyndinique. PUF, coll. que sais-je, Paris, 128 p.

Ledoux B., 1995. Les catastrophes naturelles en France. Payot, Paris, 455 p.

Lefèvre C. & Schneider J.-L., 2003. Risques naturels majeurs. Gordon & Breach.

Madariaga R. & Perrier G., 1991. Les tremblements de terre. CNRS éd., Paris, 210 p.

Some Websites

http://www.ipgp.jussieu.fr/: information on Globe Physics Institutes, seismic watch networks, data base from worldwide field sites

http://www.gsrg.nmh.ac.uk/: extensive information site documented by the global seismology and geomagnetism group of the British Geological Survey

http://quake.wr.usgs.gov/: this U.S. Geological Survey site provides worldwide information on geological parameters, seismic risks, recent earthquakes location and characteristics

http://www.ngdc.noaa.gov/: past and recent information on most natural hazards, provided by the National Geophysical Data Center of the NOAA (National Oceanic and Atmospheric Administration, USA)

http://geohazards.cr.usgs.gov/: information on various natural hazards, especially those affecting North America

http://www.eri.u-tokyo.ac.jp/: detailed data and graphs on past and recent earthquakes and volcanic eruptions having affected Japan and various other world regions

http://www.crustal.ucsb.edu/ics/understanding: educational site on earthquakes: mechanisms, historical aspects, prediction and prevention, etc.

Chapter 2

Volcanic Eruptions

2.1. 1980, Return of Mount St. Helens Activity

Mount St. Helens in the northwest USA is 1 of 12 major volcanoes lining the Cascade Mountains chain parallel to the northeast Pacific active margin (see Fig. 25). Located to the southwest of Washington state the volcano had been inactive for 123 years before returning to activity in March 1980.

After precursory vibrations and minor underground explosions due to warm water–rock interactions, several activity phases have followed one another: (1) moderate earthquakes from March 20 to 27; (2) steam eruption in two successive cycles (March 27 to April 25, May 7–12 to May 18); (3) a major explosion on May 18, followed by eruptions until June 14; (4) progressive formation until December of a central lava dome associated with occasional explosions and flows; (5) slowing down of the activity, local emission of fumarolic gas. The Mount St. Helens activity stretched over 9 months and was chronically accompanied by seismic tremors.

The major eruption of May 18 started very suddenly and violently at 08.32 h, surprising the residents, scientists and managers. In about 1 minute a series of events occurred (Fig. 24, A to D) that followed a phase of increasing seismic tremors, steam jets, and bulge of the crater's northern side:

(1) Earthquake of magnitude 5 causing the breaking of the northern part of the volcano and a huge landslide. The collapse affected 2.3 km^3 of material that propagated as a debris avalanche over a distance of 1,320 m. The volcanic matter inside the volcano was instantly decompressed, which permitted the volcanic eruption to start.
(2) Violent side explosion determining the northward emission of solid and gassy material over a distance of 3,000 m. The strength of the explosion induced the breaking of a second part of the volcanic rim, leading the major eruption to develop in a vertical direction.
(3) Production of an eruptive column of increasing intensity determining the expulsion of a very dense cloud made of blocks, stone, ash, dust and gas.

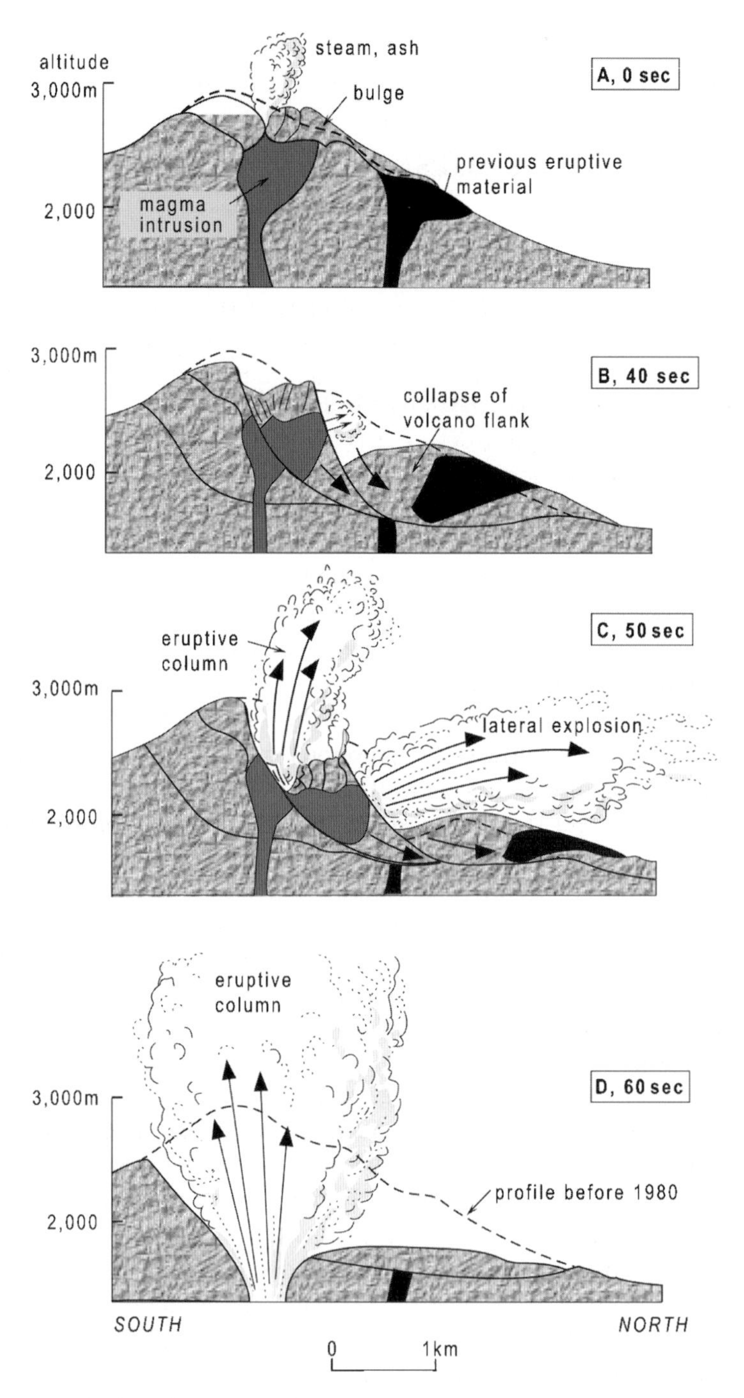

Figure 24A–D: Series of events during the major eruption of Mount St. Helens volcano on May 18, 1980 (after Murck et al., 1996).

> The smaller pyroclastic debris reached an altitude of 19,000 m. The major eruption lasted more than 9 h, causing volcanic materials to be deposited over a surface approaching 600 km^2.

The human loss amounted to 65 people, which is a fairly low number thanks to both the sparse distribution of the local population and the previous official recommendation to evacuate the dangerous area. By contrast the environment was considerably damaged. The wild and domestic fauna around the Mount was instantly killed. The forest was blown out or ruined as far as 30 km from the volcano (more than 50,000 ha were destroyed). Important morphological changes occurred: ground collapse, formation of new relief by slid materials, displacement of river course. Some rivers became silted up (e.g. 8 m of mud accumulated in the Columbia River at 90 km from the volcano) and many constructions were damaged, e.g. roads, bridges, buildings. The loss of property has been estimated at 3 billion U.S. dollars.

2.2. The Volcanic Hazard

2.2.1. Importance and Damage

The damage caused by volcanic eruptions differs considerably according to emission types and characteristics. The proportion of human victims can be very low (only one person has been killed since 1900 due to eruptions on Hawaii Big Island, where volcanic activity is almost permanent), or higher than for all other natural disasters (the 1902 eruption of the Montagne Pelée at Saint-Pierre de Martinique only spared 2 people among 29,000). The number of deaths due to major volcanic eruptions and their secondary effects (Table 6) is, on average, much lower than that induced by high-magnitude earthquakes (see Table 2). There have been thousands and, rarely, tens of thousands of victims due to volcanic catastrophes, relative to several hundred thousands after major earthquakes. Such a difference results from several causes: (1) a noticeable proportion of volcanic eruptions corresponds to lava flows, which usually move slowly and follow a foreseeable route; (2) the precise location of volcanic eruptions at the ground surface is generally well defined (i.e. summit or side of volcanic mountains), in contrast to earthquakes (Chapter 1.3, 1.5); (3) the number of big cities located close to active or subactive volcanoes (Naples, Quito, Sapporo) is much lower than that of urban conurbations threatened by important seismic tremors (e.g. Los Angeles, Manila, Mexico City, San Francisco, Santiago, Tokyo; Chapter 1.7).

The total number of volcanoes likely to erupt is about 1,400 among 10,000 volcanic structures listed on exposed landmasses; *720 volcanoes are presently*

Table 6: Example of major disasters due to volcanic eruptions since the end of the 18th century (diverse sources).

Year	Location	Human fatalities	
		Cause = eruption	Other causes
1783	Laki, Iceland		10,500 (famine)
1792	Unzen, Japan		15,190 (tsunami)
1815	Tambora, Indonesia	12,000	80,000 (famine)
1883	Krakatoa, Indonesia		36,400 (tsunami)
1902	Soufrière, Saint-Vincent	1,560	
1902	Mont-Pelée, Martinique	29,000	
1982	El Chichon, Mexico	1,700	
1985	Nevado del Ruiz, Colombia		25,000 (mud flow)
1986	Lake Nyos, Cameroon		1,750 (toxic gas)

more or less active. Each year 50–70 subaerial eruptions occur worldwide. The number of human deaths due to volcanic disasters is estimated at about 60,000 for the whole 20th century, i.e. an average of about 600/year (compared to the 20,000 deaths/year due to earthquakes; Chapter 1.7). By contrast the damage to goods and property may be very important by affecting livestock farming, agriculture, urban and rural developments (see Chapter 2.3). Some volcanoes may erupt after a lengthy dormancy. For instance the Popocatépetl, located 60 km west of Mexico City, started to erupt on the 18th December 2000, after a virtual inactivity of 8 centuries. The event was responsible for a pyroclastic discharge up to 2,000 m high and within a 4- to 20-km radius of the volcano, which led to the evacuation of 10,000 people.

2.2.2. *Location*

The rising of volcanic magma principally occurs directly along the faulted zones of the Earth's crust. *Most volcanoes are located along the frontiers separating the lithospheric plates* (Fig. 25), and their general distribution is superimposed to that of major seismic zones (see Fig. 13). The volcanic activity is widespread straight along the oceanic ridges, where extension and sea-floor spreading processes develop, and where the oceanic crust is renewed to a distance of 60,000 km. Most submarine eruptions occur at a fairly great water depth (1,500–2,000 m below the sea level) and remain generally unnoticed at the

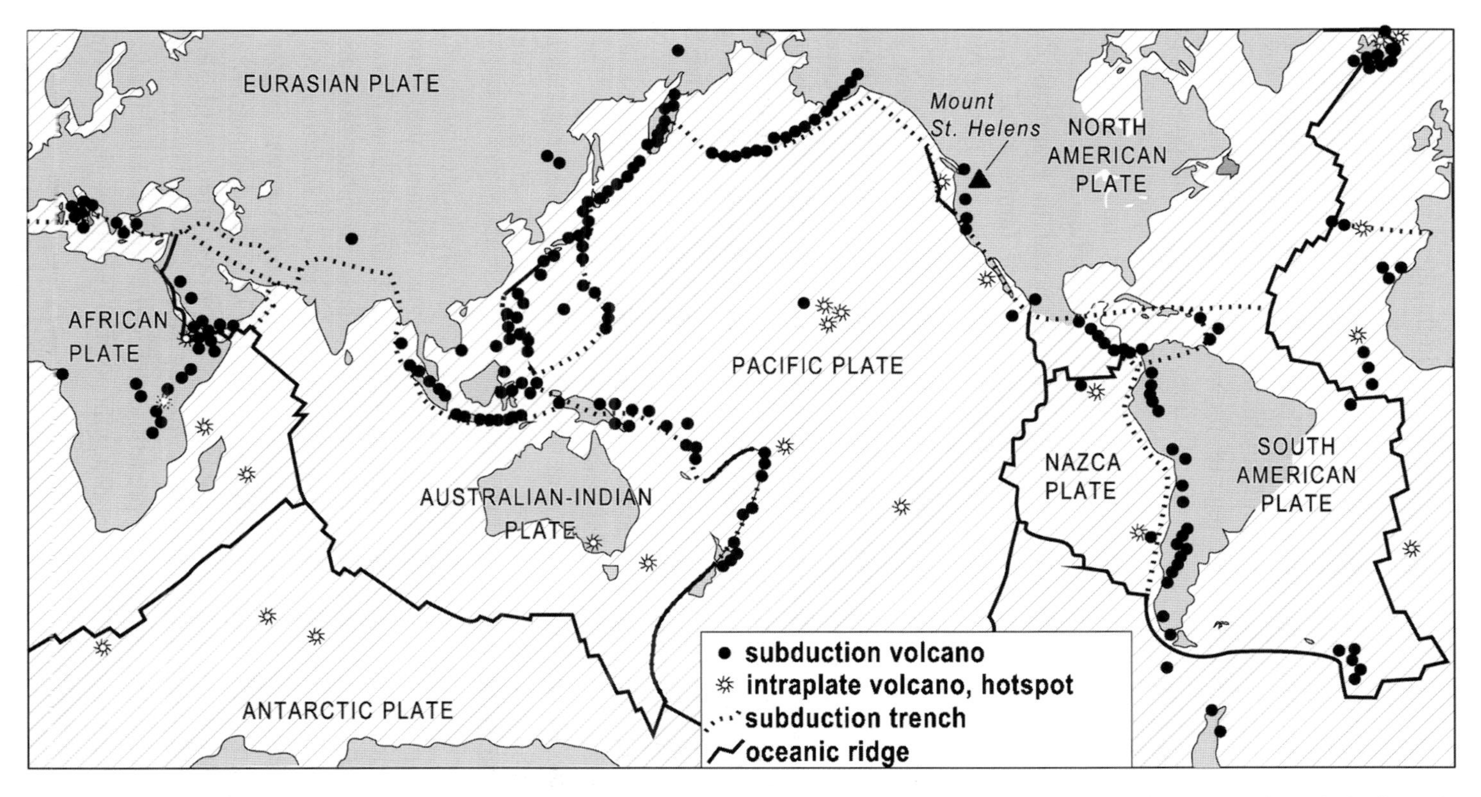

Figure 25: Map showing the main exposed volcanoes presently or recently active. The frontier of lithospheric plates is indicated: ridges, subduction zones.

surface. The largest volcanic range in the world therefore presents low risk for Man's activities. In addition, most submarine eruptions constitute fluid effusions and not explosive projections, which still diminishes the potential risk. Nevertheless, some oceanic ridges locally emerge above the sea surface; if such a situation coincides with hot spots (e.g. Iceland, the Azores; see below), it may constitute serious hazards for island inhabitants.

In active compressive margin regions, the volcanic activity is enhanced by the tectonic fracturing characteristic of colliding, subducting, transpressive plates, as well as by rock melting at the asthenosphere–lithosphere transition. The Pacific "fire belt" is punctuated by an especially large number of volcanoes, among which in the northeastern group is Mount St. Helens (Chapter 2.1). The southern coasts of Indonesia are also widely exposed to eruptions due to subduction of the Indian Ocean floor (e.g. explosion of Tambora and Krakatoa volcanoes; Chapter 2.2.5); a similar situation characterizes the Lesser Antilles (Guadeloupe, Martinique, etc.; Table 6).

In addition to plate frontier zones, some volcanic systems develop above *hotspots*, which constitute local areas where asthenophere matter rises up to the Earth surface. Sometimes connected with very deep zones of Earth envelopes (up to hundreds km), hotspots usually reach the surface within the lithospheric plates and not at their border, and are therefore referred to as *intraplate volcanic processes*. The hotspot activity is characterized by very long durations (up to several tens of million years) and exceptionally large volcanic systems. Some big islands and aligned archipelagos have been built through such a process, Hawaii, French Polynesia, Réunion, for example. An intraplate volcanic activity also occurs in some terrestrial environments, where it is more or less contaminated by material issued from the continental crust melting: Cantal and Mont Dore in Auvergne (France), volcanic provinces of Hoggar (Algerian Saharian) and Yellowstone (northwest USA).

In Europe the main presently active volcanic areas are located westwards on oceanic hotspots rising within the Mid-Atlantic Ridge (Iceland, the Azores) or closer to Africa in intraplate environments (Madeira, Canary Islands), as well as to the east along the Mediterranean colliding/subducting plate frontiers (the Eolian Islands, Naples area, Sicily, Aegean Sea; see Fig. 13). *In France* volcanic systems developed in Palaeozoic (Vosges, Morvan, Massif Central, Esterel) and Cenozoic times (Massif Central–Mont Dore, Cantal, etc.; Languedoc; Provence). Numerous eruptions of limited extension occurred until the late Quaternary (4,000 years ago) in Bas-Vivarais and Les Puys chain, and then slowed down. Volcanic risks are presently almost nil in mainland France, but potentially high in some overseas departments and territories (Lesser Antilles, Réunion, Polynesia).

2.2.3. Eruptive Mechanisms and Magma Composition

The volcanic products emitted from magmatic chambers contain 50–75% silica (SiO_2). *The relative abundance of silica* is associated with specific physicochemical characters of magmas, which partly determine the nature of volcanic risk:

- When silica constitutes about 50% of chemical components the melting temperature is very high (ca. 1,400°C), magma is fluid, volcanic activity is of an *effusive* type and gives way to basalt lavas. *Basalts* constitute most *submarine volcanic rocks* produced at the oceanic ridge axis and above the hotspots where asthenosphere magma rises. Huge amounts of basalt were also produced within some terrestrial domains in some geological periods, e.g.: traprocks of Deccan (India), Siberia, Karoo (South Africa), Columbia River (USA), Paraña (Brazil). Basalts flow down the topographic depressions. The flowing speed (from 1 m/day to 3 m/sec) and distance (up to several tens of kilometres) depend on the viscosity and cooling time of melted rocks. Successive basalt flows constitute broad and slightly sloped cone-shaped shields, illustrated by the Hawaiian archipelago volcanoes (Hawaiian type: Mauna Loa, Kilauea). These slowly moving lavas present a *limited risk* to human life, but may cause important damage such as fire, destruction of buildings and other constructions.
- When the silica content is about 70% of the rock, the more acidic character determines a melting at about 800°C, the production of a viscous magma and a very slow flow rate. Eruptions tend to be of an *explosive or pyroclastic* type (ignimbrites). The material produced belongs to the *rhyolite* group, which characterizes *continental volcanic systems* and sometimes lava flows inserted between more alkaline rocks (e.g. Iceland). The *risk* inferred is *theoretically important*, but is attenuated because of very few present-day occurrences of such type of eruption. Rhyolites occur generally in fairly small areas, except in specific geographic locations: ignimbrites of the Ten Thousand Smokes valley (Katmai, Alaska), of Taupo in New Zealand, of Toba in Sumatra, etc.
- Rocks with an intermediate silica content (about 60%) correspond to other types of volcanic eruptions, which are essentially of an *explosive* type, lead to pyroclastic products of the *andesite* group, and mostly characterize *active margins* close to *subduction zones* (e.g. the circum-Pacific "andesitic line"). The *risk* resulting from volcanic explosions is *highly variable*; it depends mainly on the heterogeneity and solidification degree of emitted magma, as well as on the gas and/or water content. The main magma types are:
  - Magma comprising *heterometric materials* close to solidification and *poor in gas*; emission of volcanic bombs, stone, lapilli and ash in the immediate vicinity of the crater (less than a few kilometres), determining major *local risk* for people and goods. The volcano shape is conical with a steep slope;

it corresponds to the Strombolian type (named after the Stromboli volcano, Eolian Islands).

- *Homogeneous, fairly viscous and gas-rich magma*: emission over rather large distance (several tens of kilometres) of finely vesicular and porous materials (pumice, lapilli, ash) determining an *important regional risk*, especially for human goods: covering and contamination of farming areas, deterioration of water quality, airplane hitches, sanitary problems, etc. Volcanoes are cone-shaped with a gentle slope, are surrounded by fine-grained pyroclastic deposits, and correspond to the Vulcanian type (from Vulcano, Eolian Islands).
- *Highly viscous magma* tending to fill the volcanic chimneys and submitted to *strong internal pressure* by very hot gas: sudden explosion and emission of a very hot pyroclastic cloud at 300–800°C ("nuées ardentes"), which moves very rapidly (several tens to hundreds of metres per second), and induces *major risk on a regional scale*. The volcano summit is topped by a dome (Pelean type, from Montagne Pelée, Martinique Island). The *extreme stage* encountered for such eruptions corresponds to the *crater caldeira collapse* due to a blowout under the combined pressure of rising magma, gas and sometimes boiling water. It is marked by violent successive explosive and effusive eruptions, as well as by major morphological changes (Plinian type). The *risk* is *major* for both people and property. Such volcanic events are driven by complex underground interactions involving different geological formations and magmatic phases. They have characterized the major eruptions of Mount St. Helens in 1980 and of Vesuvius (Italy) in 79 AD.
- *Interaction of magma with large amounts of water*: this constitutes the *phreato-magmatic activity*, inducing strong explosions and instantaneous vaporization of water in contact with incandescent rock. The consequences are potentially catastrophic (e.g. the Krakatoa explosion, 1893).

2.2.4. Secondary Effects

Mud flows and debris avalanches High-gravity forces along volcano slopes induce the reworking of pyroclastic material deposited around the crater, its transfer and resedimentation in downstream valleys. Also called *lahars*, these often heterogeneous and heterometric flows are triggered by various phenomena: collapse of some of the volcano's flanks during the eruption (e.g. phase B of Mount St. Helens major explosion in 1980; Chapter 2.1); excess load of soft pyroclastic material accumulated on the sides of the volcano; flooding by snow melting through magma rising, steam expulsion or crater-lake emptying. Dominantly composed of volcaniclastic material, the moving mass tends to grow in volume by eroding the soft underlying formations. Lahars are able to propagate over several tens of kilometres, to reach a speed of several tens of kilometres per hour, and to finally accumulate by forming metres-thick deposits. Such volcaniclastic gravity flows constitute a *major risk* for both people and goods, especially in downstream valleys marked by morphological narrowing. In 1985 a minor pyroclastic eruption caused the partial melting of snow covering the Nevado del Ruiz volcano in Columbia (South America). Water impregnation of the volcanic deposits led to the formation of a huge mudflow that moved down in the adjacent valley, growing dramatically in volume. The giant flow finished its course in Armero city area located 40 km away from the volcano, where it caused the death of about 25,000 people and considerable damage.

Earthquakes Frequently associated with volcanic eruptions under tectonic pressure, earthquakes may also directly result from eruptive activity. The corresponding tremors are usually of moderate magnitude, but can last for weeks or even months (e.g. the 1980 Mount St. Helens eruption; Chapter 2.1), and be responsible for noticeable damage (subsidence, fissuring; e.g. the villages built on the flanks of Etna, Sicily). Rhythmic seismic vibrations of low frequency and without danger occur sometimes synchronously with magma activity in the depth of the volcano. Called *volcanic tremors*, they result from magma convection, boiling and friction phenomena.

Tsunamis and water problems Tsunamis are usually triggered by submarine earthquakes but may also result from violent explosions affecting *coastal or submarine volcanoes*. Such subaqueous eruptions may trigger giant waves potentially responsible for major damage (e.g. 1883 Krakatoa eruption, Indonesia; Chapters 2.2.5 and 4.3.2). *On land* volcanic eruptions may indirectly induce *flooding* hazards: emptying of crater lakes after the rupture by explosion of the volcano rim; blocking up of river valleys by lava flows; overflowing of rivers where fine pyroclastites have accumulated. The damage to human activities and developments is potentially important, especially in densely urbanized areas. The possibility of a major eruption during the next few decades in the volcanic area of Long Valley, California, has given way to in-depth research and modelling. Several hydrographic basins converge in this area, where 80% of Los Angeles' water is collected. An eruption in the Long Valley caldera would induce major geomorphological change, impeding for months and perhaps years the amount and quality of surface and underground *water resources* (Hopson, 1991).

Emission of noxious gas Volcanic eruptions are often preceded, accompanied or followed by gas escape. *Steam* which is responsible for fumarolic emanations and geysers, constitutes the most widespread volcanogenic gas; it is of limited danger in the absence of explosive eruption. Other gassy products emitted during volcanic events become noxious when their abundance modifies the air composition, inducing possible breathing disorders and sanitary problems. This is the case when hydrochloric and hydrofluoric acids, sulphur dioxide, hydrogen sulphide, and carbon monoxide and dioxide are emitted. The most serious risk consists in the *expulsion of CO and CO_2* progressively accumulated in *crater-lake* sediment and water. The gas, having slowly diffused from underlying magmatic chambers, can be instantaneously expelled in large quantity into the air, due to wind-driven water movements, slight ground vibrations or minor eruptions. As carbon oxides are denser than air, they are hardly diluted and tend to stay in morphologically depressed areas, where they have lethal effects. A massive CO_2 degassing of the Nyos crater lake (West Cameroon) occurred in the form of giant bubbles coalescing during 2 h on August 21, 1986. About 100 million m^3 of the noxious gas spread throughout adjacent valleys, killing 1,750 people and 3,000 head of cattle.

2.2.5. A Few Historical Examples

Various disasters triggered by volcanic eruptions have punctuated historical times, fed chronicles and legends, and sustained myths and religions. Most human losses and damages were determined by secondary or indirect effects rather than by proper pyroclastite fallout. The two major volcanic catastrophes in the 19th century occurred in *Indonesia*. The first one resulted from the 1815 *Tambora* volcano eruption in Sumbawa Island, responsible for 92,000 deaths, of which 12,000 were killed under pyroclastic falling debris and 80,000 by subsequent famine. The second disaster occurred when the *Krakatoa* volcanic island exploded in 1883, causing 36,400 victims, most of whom were killed by a giant tsunami triggered by the caldera collapse. The island was literally blown up by the explosion, the noise of which was perceived for several thousand kilometres in the Indian Ocean. A huge wave formed, which broke on Java's coast with a height of 40 m. About 220 km^3 of volcanic debris were thrown into the atmosphere, the finer particles attaining an altitude of 50 km. Large amounts of volcaniclastic dust were incorporated in latitudinal stratospheric air masses. In less than 2 weeks a circum-terrestrial dust cloud formed and filtered the solar radiation, inducing a 0.5°C drop of the mean world temperature in 1884 (Chapter 2.3.1). Dust fallout continued worldwide for 5 years after the eruption.

Vesuvius (southeast of Naples bay, Italy), which was considered extinct at the beginning of the Christian era, experienced a violent return to activity on August 24, 79 AD. Preceded by seismic tremors and coarse pyroclastite discharge, a huge column of incandescent ash formed at the same time as the crater exploded. The tephras beat down on Pompei and Herculanum, two Roman cities located southeast and south of the volcano. The major eruptive phase ended in the morning of August 25, after additional emission of stone, viscous lava and hot gas had destroyed all life around the volcano. The ash deposit exceeded a thickness of 2 m as far as 20 km to the southeast, and still reached 20 cm 100 km south of Naples. About 20,000 people were killed by both noxious gas and ash fallout. The instantaneous ash coverage and burying of Pompei and Herculanum cities allowed an outstanding preservation of buildings, human developments and items typical of the Roman society 2,000 years ago.

The volcanic eruptions of *Santorini* (Thera) Island in southern Aegean Sea have deeply marked the Greek civilizations in the eastern Mediterranean range. A major pyroclastic eruption occurred in 1450 BC, and was associated with the caldera collapsing. A huge tsunami formed, which was responsible for the destruction of the Minoan fleet anchored 120 km to the south along the north Cretan coast. The development of King Minos's civilization was then definitively compromised. The sandy-silty ash transported by winds to the southeast was spread as far as North Africa offshore and constitutes a precise chronological marker interbedded in common marine sediments. The effects of big replica waves might have been registered as far as the eastern-most Atlantic Ocean, and climatic disturbance might have affected various regions of the planet (see LaMoreaux, 1995). Some scientists infer that the Atlantis legend is related to the Santorini caldera collapse, responsible for the marine submersion of Greek cities build on the volcano flanks. If the regional impact of this major natural disaster is widely admitted from political, military, social, economic and cultural points of view, some major disagreements remain as far as more global changes are concerned (Pyle, 1997).

2.3. Indirect Effects

2.3.1. Climate, Palaeoclimate, Global Environment

The likelihood for volcanic eruptions to have an influence on the Earth's climate results from *the ability of pyroclastic dust released into the atmosphere to obstruct and filter solar radiation over a long period of time*. This depends on very large amounts of fine volcaniclastic material being expelled at *high altitude in the stratosphere*, which implies an eruptive regime of an explosive type, a magma very rich in gas, an upward vertical discharge of fine pyroclastic material, and the formation of large amounts of aerosols (i.e. mixtures of particles and water micro-drops). Aerosols interact especially with visible rays by inducing an increasing reflection towards the interplanetary space. This induces a cooling of the troposphere and Earth's surface, resulting in a "volcanic winter".

The 1991 explosion of the *Pinatubo* volcano (Philippines) led to a vertical discharge of more than 8 km^3 of a dense mixture of fine pyroclastites and sulphur gas, allowing the production of aerosols as high as 31,000 m in the stratosphere. This volcanic eruption could have been responsible for the 0.5°C drop in air temperature at the Earth's surface registered in 1992 and 1993. In 1815 the 3 days following the *Tambora* explosion (Indonesia) were regionally characterized by almost total darkness, half-light being noticeable as far as 500 km from the volcano; the year 1816 was labelled as a "sun-deprived year" and marked by a temperature drop exceeding 1°C. The explosion of *Toba* volcano (Sumatra) 73,500 years ago was probably the biggest in the Quaternary period and led to the discharge of about one billion tons of fine ash and sulphur gas up to an altitude of 27–37 km. This determined a strong decrease in solar radiation, responsible for a worldwide cooling, estimated to have attained 3–5°C and lasting several years (see Pickering & Owen, 1997). In older geological times, diverse, long-lasting climate changes have been at least partly attributed to an intensification of volcanic activity, especially during the Cenozoic (see Kennett, 1982).

Sulphur gases (SO_2, H_2S) emitted during some volcanic eruptions enter the atmosphere, where they combine with steam to form sulphuric acid drops (H_2SO_4). Such products derived from volcanic activity may determine the precipitation of *acid rain*, similar to that induced by industrial discharge of sulphur dioxide (SO_2) in the air. Other *converging effects of volcanic and anthropogenic emissions* potentially affect the regional to global environment: fluoride in soils and water, chemical pollution of underground water by toxic elements derived from ash weathering, deterioration of the ozone layer, etc. The respective natural and anthropogenic contributions to such disruptions are crucial to know and explain for better management of terrestrial environments (Chapter 11.1).

To summarize, the indirect risks induced by volcanic activity are added in a significant way to direct and secondary risks for modifying the Earth's environment, particularly on a global scale (Fig. 26). The mechanisms controlling these various hazards, their relationships with the geochemical and geodynamical context, their geographical extension, the duration of their impact are some of the factors needing to be better understood for improving forecasting and prevention (Chapters 2.4, 2.5).

2.3.2. Beneficial Aspects

Responsible for various natural hazards, volcanic activity also has some positive consequences for human activities. Apart from cultural and visual aspects of volcanic landscapes and events (craters, eruptions, lava flows and fields, geysers,

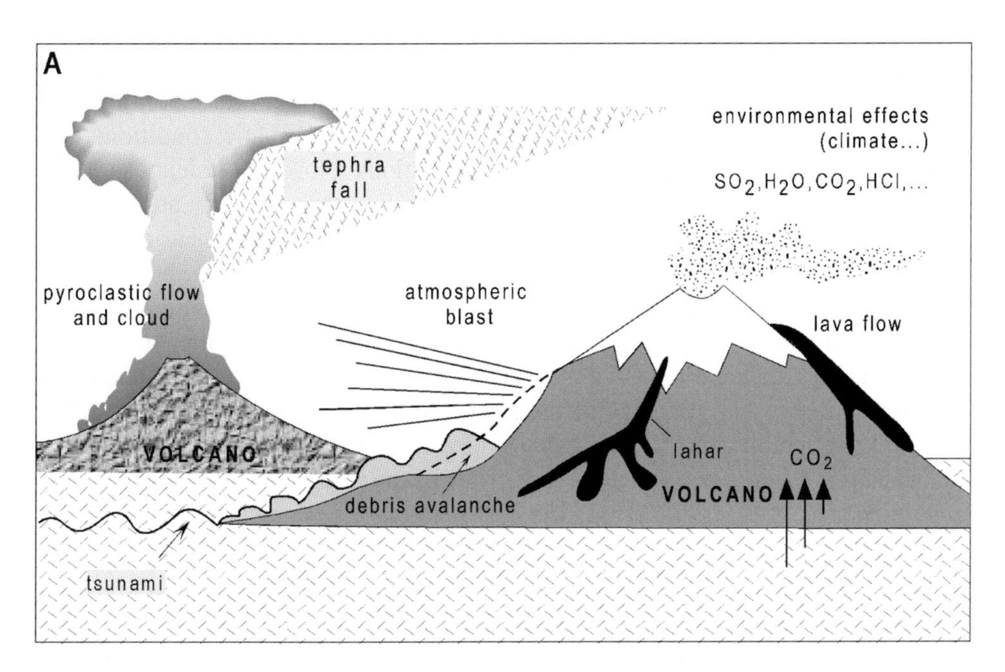

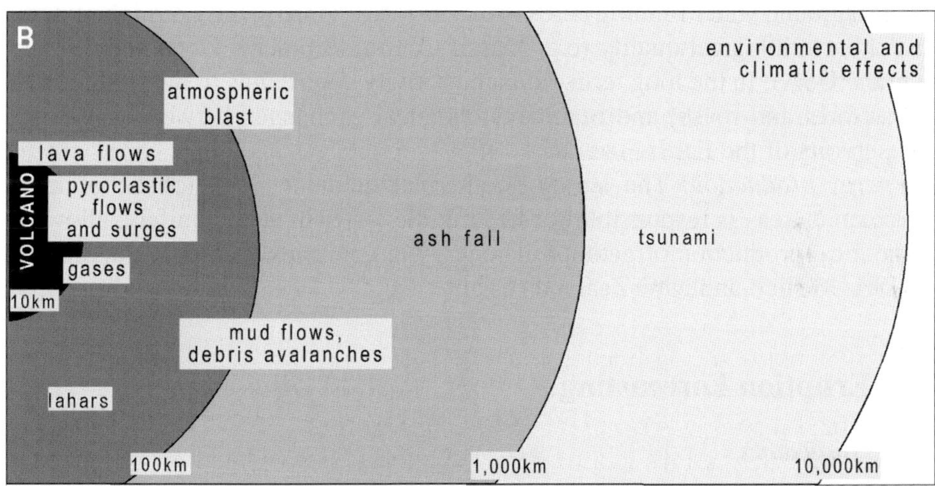

Figure 26: Sketch showing (A) the major events associated with volcanic eruptions, [B] the main hazards and potential impacts at increasing distance from volcanoes (after McGuirre, 1998).

fumarolic gas, etc.), diverse concrete advantages arise from the existence of volcanoes.

– *Soils revitalization.* The usually small-sized, vesicular or porous, and poorly crystalline pyroclastic debris undergoes rapid chemical weathering in surface conditions, and releases various elements that serve as fertilizers for agricultural soils (e.g. K, P, Fe). Pyroclastic rocks often produce fertile soils, which tend to stay permanently wet even under dry climatic conditions and have always ensured the return of populations close to volcanoes after eruptions have ceased. The infertility of Australian soils is partly attributed to the extreme rarity of volcanic events during the geological history of this continent. The supply of pyroclastite-derived sulphur, selenium, etc. participates in restoring the chemical balance of New Zealand soils and favouring the anion retention (Cronin et al., 1998). In Argentina the impact of pyroclastites spread over a distance of 1,000 km east of the Andes chain in the Patagonian Meseta statistically proved after only 2 years to be much more beneficial than detrimental to agricultural soil quality (Inbar et al., 1995).

- *Material exploitation.* Volcanic rocks and by-products are used for different purposes in civil engineering, the chemical industry, etc.: building stone; vesicular pebbles and filter pozzolana for water drainage; pumice and ash for light aggregates; and ammonia, boric acid, etc. An intense, effusive activity determines the formation of new land in some regions; the 1960 Kilauea eruption has added about 1 km^2 to the Big Island of Hawaii archipelago. The underground water running across volcanic areas is frequently of high mineral and bacteriological quality (e.g. Massif Central in France, Oregon and Idaho in the USA). In the long term, volcanic activity contributes to the synthesis of new ores and rocks, and maintains chemical exchanges between the major reservoirs of the Earth.
- *Energy production.* The strong geothermal gradient characteristic of most volcanic areas is responsible for heating the water in deep aquifers, allowing the local production of geothermal energy in countries such as Iceland, Italy, USA, Mexico and New Zealand (Chapter 6).

2.4. Eruption Forecasting

2.4.1. Indicators

Seismicity The rising of magma and related morphological bulges cause an intensification of fissuring and fracturing within volcanic structures. This predisposes earthquake formation during the stages preceding eruptions. *The frequency and intensity of seismic vibrations tend to increase* when the eruption is approaching, which can be measured with seismographs established around volcanoes, the corresponding data being, if possible, real-time processed. Nevertheless the advance of the seismic activity is often irregular (Fig. 27). Some eruptions are not preceded by seismic tremors, and some earthquakes affecting volcanoes are not followed by any eruption. In addition the seismic

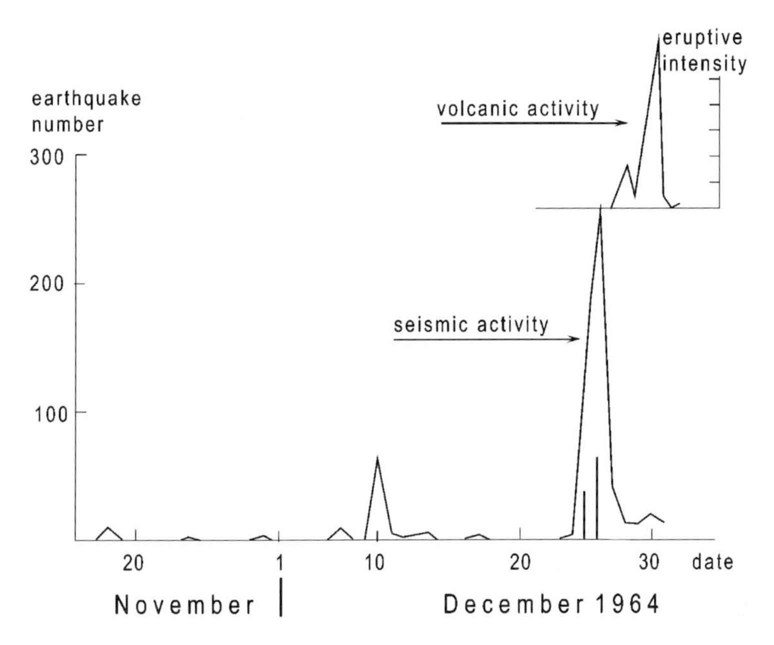

Figure 27: Number and temporal distribution of earthquakes having preceded the 1964 Mihara-yama volcano eruption, Japan (after Shimozuro; see Bell, 1999).

activity may continue during a highly variable period after the eruption, preventing inference of any new prediction. The frequent vagueness of such information gives to seismic monitoring a *moderate confidence*. Another geophysical approach consists of following the progression of the fluid magma within the volcanic structure by measuring the *speed ratio of P and S waves*, which are artificially triggered on volcano flanks: as S waves, in contrast to P waves, do not propagate in liquids (Chapter 1.3), the variations of the $V_P : V_S$ ratio allow one to visualize in three dimensions the successive positions of the magma approaching the volcano surface.

Morphological changes Volcanic eruptions occur when millions of tons of hot magma and associated gas have more or less rapidly risen from magmatic chambers towards the ground surface, inducing morphological changes in volcano flanks and craters. *Bulge and uplift movements* are commonly observed, sometimes compensated for by subsidence in cold areas peripheral to volcanic cones. These morphological distortions can be precisely measured by getting data from a variety of sources such as theodolites, tiltmeters, GPS positioning systems, successive satellite images (interferograms), gravimeters (density changes), and tidal gauges (change of crater-lake level). *Such indicators are among the most reliable* to predict the approach of an eruption. It is particularly the case when magmatic chambers are located relatively close to the ground surface. The Kilauea (Big Island, Hawaii) tends to rise by about 1 m during the month preceding each eruption. The northern side of Mount St. Helens was subject to a bulge of about 2 m/day during spring 1980, the total ground rise having attained 200 m when the explosion happened on May 18 (Chapter 2.1).

Physical properties Measuring the *augmentation of temperature* in the water of crater lakes, wells, re-emergences and fumarolic exhalations may contribute to the rise of magma and the eruption risk. The progressive warming of the volcanic ground itself can be obtained by infrared radiation values, measured either directly in situ or by way of satellite or aeroplane imaging. The *magnetic properties* of some minerals are also likely to undergo some changes when magma is rising, namely when temperature exceeds the Curie point where demagnetization phenomena occur (e.g. 575°C for magnetite Fe_2O_3). The mineral demagnetization is facilitated by the high temperature of some magmas (800–1,100°C; Chapter 2.2.3). Other changes may affect the *electric field* and are expressed by resistivity values.

Animal behaviour Wild fauna has deserted some volcanic areas before major volcanic events happened. This was the case in the 1902 Montagne Pelée eruption, the 'nuées ardentes' which killed 29,000 people on Martinique Island. Non-seasonal migrations of some bird species could result from similar causes.

Statistics, modelling Statistical investigations on past volcanic eruptions allow of developing a *probabilistic approach* that is useful for the mid-term prediction of further volcanic events. The information processed includes the location and frequency of previous volcanic discharges, the nature and chemical type of lavas or pyroclastites successively expelled, the record of morphological changes having occurred, and the depth of associated earthquakes focus. Probability calculation performed in 1975 led the U.S. Geological Survey to predict a major eruption of Mount St. Helens volcano before the end of the 20th century; the eruption occurred in 1980 (Chapter 2.1). Similar calculation has led to the anticipation of the surface extension and the covering by more than 1 m of ash deposit in parts of Tenerife (the Canary Islands) subject to discharge from the Teide volcano (Fig. 28). Field investigations associated with computer simulations have allowed prediction of which areas would be threatened in the case of a major eruption of the Pinatubo volcano (Philippines), and then to establish and implement a population evacuation plan (see Chapter 2.4.2).

2.4.2. Combination of Volcanic Indicators, Hazard Maps, Assessments

As for earthquake forecasting (Chapter 1.5.2), the complementary use of various indicators in *multi-tool field sites*, as well as *satellite imaging* (interferometry),

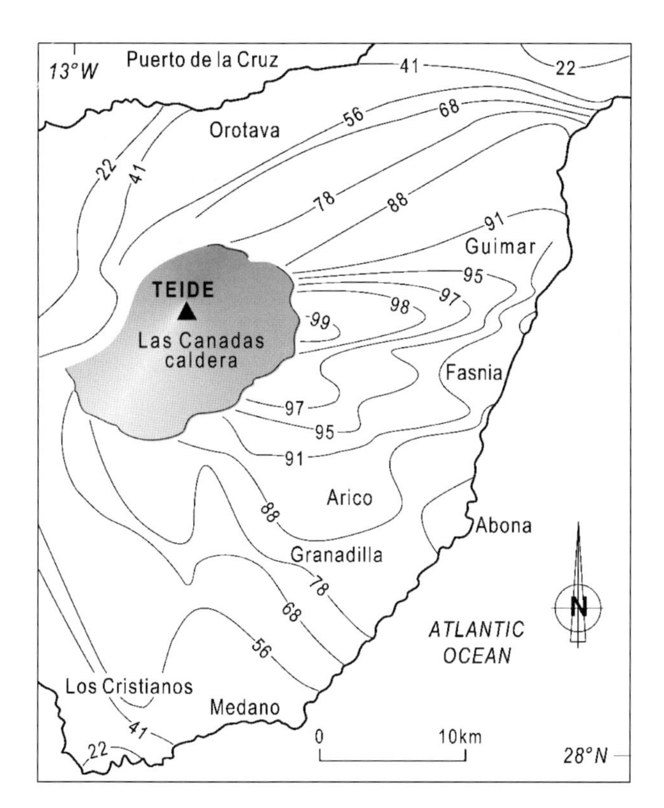

Figure 28: Extension and probability of ash covering in the southeastern part of Tenerife (Canary Islands) by a layer thicker than 1 m, in the case of a major explosion of the Teide volcano (after Booth; see Bell, 1999).

contribute to improve the accuracy of risks determined by volcanic hazards. After the major eruption of Mount St. Helens in 1980, a continuous monitoring was implemented which allowed the prediction with progressively increased precision of the moderate eruption that occurred in March and April 1982 (Fig. 29).

Volcanic hazard maps are often more reliable and easier to implement than seismic hazard maps. This results from the near certainty of the eruption location, from good probability concerning the nature and type of volcanic material to be discharged, and from the knowledge of the approximate route to be followed by gravity-controlled materials (lavas, mud flows, debris avalanches). The good degree of prediction that may be attained is illustrated by comparing the hazard map established for Pinatubo volcano (Philippines) before the June 1991 catastrophe and that showing the actual distribution of pyroclastites and lava flows after the eruption had occurred (Fig. 30). Such forecasting results imply that hazard maps are effectively used and implemented sufficiently in advance. If the maps produced before the Pinatubo eruption have led to the evacuation of more than 58,000 people and allowed us to save hundreds of human lives, those established for the Nevado del Ruiz and Armero areas in Columbia did not lead to practical application before the 1985 disaster, where more than 25,000 people died.

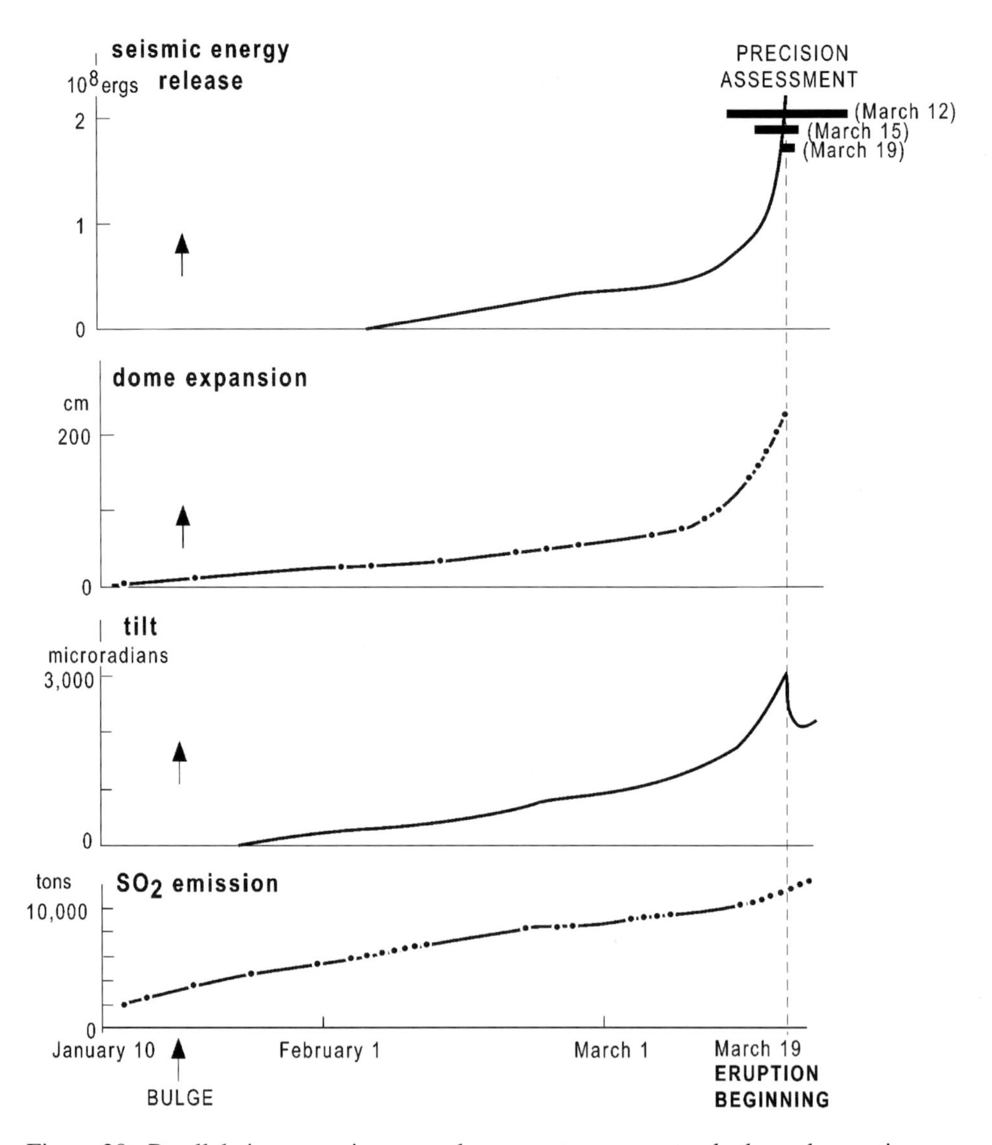

Figure 29: Parallel increase in several parameters measured through continuous monitoring before the March 1982 Mount St. Helens eruption (after Murck et al., 1996).

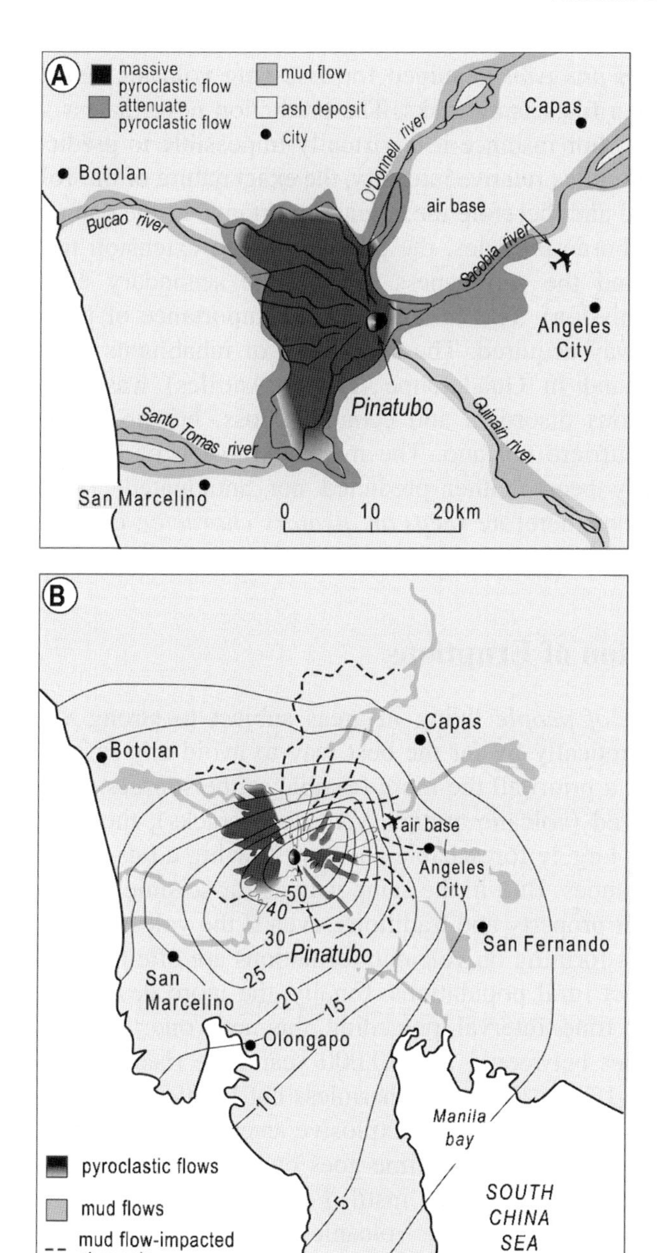

Figure 30: Region of the Pinatubo volcano, Philippines. Maps showing (A) the volcanic hazards and expected volcaniclastic deposits before the June 1991 eruption, (B) the actual distribution after the eruption of pyroclastite and mud flow sediments (after Murck et al., 1996).

The degree of precision obtained for assessing volcanic risks *is often fairly good*, better than for seismic risks. The prediction nevertheless suffers *various rough estimates*. For instance it is virtually impossible to predict precisely the date of an eruption, its relative intensity, the exact nature of the volcanic material to be discharged and therefore the actual importance of the danger, the transport direction of airborne particles, the ground surface extension to be covered by pyroclastites, and the seriousness of possible secondary effects (tsunamis, noxious gas emission). The certainty of the importance of the volcanic event itself is in no way acquired. The evacuation of inhabitants from a part of the Basse-Terre Island in Guadeloupe (Lesser Antilles) was decided in 1976, leading to various disorders and economic loss, but no noticeable eruption affected the Soufrière volcano. The major explosion of Mount St. Helens in 1980 had really been neither predicted nor anticipated. *The forecasting of volcanic eruptions* therefore *keeps an aleatory character*, especially in the case of explosive discharges.

2.5. Prevention of Eruptions

The evacuation of people living in areas subject to strong volcanic hazards represents theoretically by far the best way to avoid any serious risks. People displacement is a priori all the more feasible as the sources of potential danger are well localized (volcano craters, vents and cracks), the route followed by gravity flows is largely foreseeable and very few big cities are situated close to dangerous volcanoes. But, in the same way as for seismic hazards, people tend to return to their property and traditions, even if the volcanic risk remains high. In addition soils forming on volcanic formations are often very fertile (Chapter 2.3.2) and attract rural populations. Finally, the more devastating an eruption, the longer the time interval preceding the next one: the average eruption periodicity varies between 1 and 10,000 years, the risk to loss of human life spreading from 1% for the almost harmless but frequent events, to 31% for the most dangerous but very rare, explosive events (Brown et al., 1992). As memories become dulled when time goes on, the actual perception of a latent danger is difficult to keep or to instil. For all these reasons, most preventive measures taken for fighting the volcanic risk consist in modifying the local topography and anticipating potential hazards. The type of measures deployed depends on the nature of eruptive discharge.

Lava flows Effusive materials discharged from volcanoes flow by following the natural slopes and reach adjacent dales and valleys. Their flow route is therefore largely predictable. Viscous lavas called "aa" (a Hawaiian word) progress slowly, often inside tunnels, the crust of which cools rapidly by contact

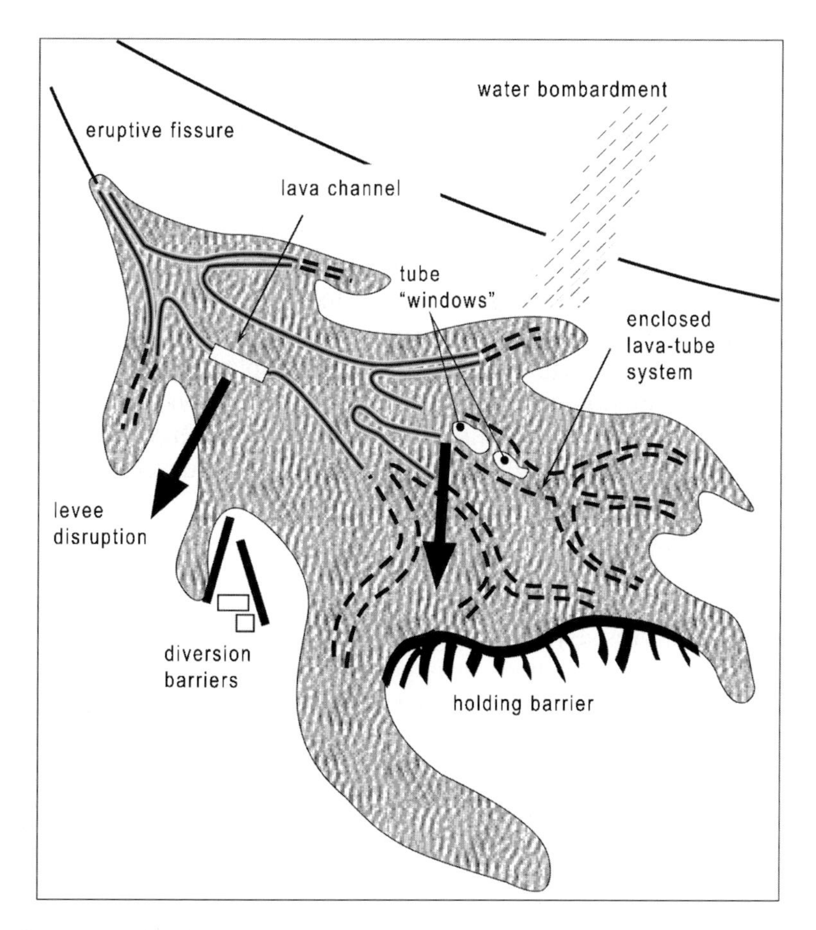

Figure 31: Example of measures taken to dam or divert the lava flows on Etna's slopes (Sicily) during the 1983 and 1991–1993 eruptions (after McGuirre, 1998).

with the air, and their route can be easily anticipated and followed. Fluid lavas called "pahoehoe" move more rapidly but do not exceed a speed of a few tens of kilometres per hour; their route can therefore often be controlled and if necessary artificially modified. Various works and actions can be developed: cooling by watering the lava front; setting up of obstacles by bombarding the hardened lava tunnels or building levees, dykes, and walls; digging of diversion channels; etc. Such developments have been applied with success to various volcanoes, namely Etna in Sicily (Fig. 31).

Coarse, airborne pyroclastites (tephras) The fallout of blocks, volcanic bombs and pebbles generally occurs very close to main or side craters, and commonsense is usually sufficient for preventing their damage. For instance, the constructions built on Stromboli Island (offshore North Sicily) are established away from fallout dangerous areas, which are situated in a few small-sized zones. Accidents may nevertheless happen, mainly because of the spread of tourist pressure.

Fine airborne pyroclastites As for other tephra, the ash and dust fallout does not depend on ground topography. Preventive measures are all the more difficult, as transportation distances may be very large and their direction largely unpredictable (vertical to oblique discharge, varying wind regime, etc.). Apart from agricultural and hydrological problems (Chapter 2.2.3) an immediate danger may arise from the weight of abundant ash falling on houses; a possible solution consists in reinforcing the roofs and frameworks. Risks are particularly difficult to anticipate and to fight when the ash discharge develops within very hot gas (nuées ardentes).

Mud flows, debris avalanches The propagation of pyroclastic flows soaked in water is largely controlled by the local topography, which facilitates the choice of prevention measures. By compiling historic data, establishing hazard maps and applying modelling methods, it is possible to propose *realistic preventive solutions*: demarcation of clearly defined hazard areas (e.g. Fig. 32), construction of artificial relief for diverting the flows, and installation of alarm systems. The settlement of alert services and evacuation plans proves also to be efficient in the case of tsunamis, the usual propagation time of which (i.e. a few hours) allows evacuation of the threatened populations (Chapter 4.2).

Noxious gas The CO_2 diffusion from volcanic structures can be controlled in an efficient way when it is progressive and of moderate extent: measurement at regular time intervals of the gas content in house basements, and appropriate ventilation; degassing by water turbulence or diffusions of the gas seeping within small crater lakes (e.g. Lake Monoun, Cameroon). Preventive actions are more aleatory when the lakes accumulating volcanogenic CO_2 are deep, of many kilometres in size, and marked by stratified water: any physical distortion of water or sediment may induce an immediate ejection of the lethal gas (e.g. Lake Nyos; Chapter 2.2.4).

2.6. Perspectives

Volcanic eruptions constitute a large variety of phenomena causing big differences in the degree of risk for both people and property, in the possibility of prediction, and in prevention measures. The risks, forecasting and prevention

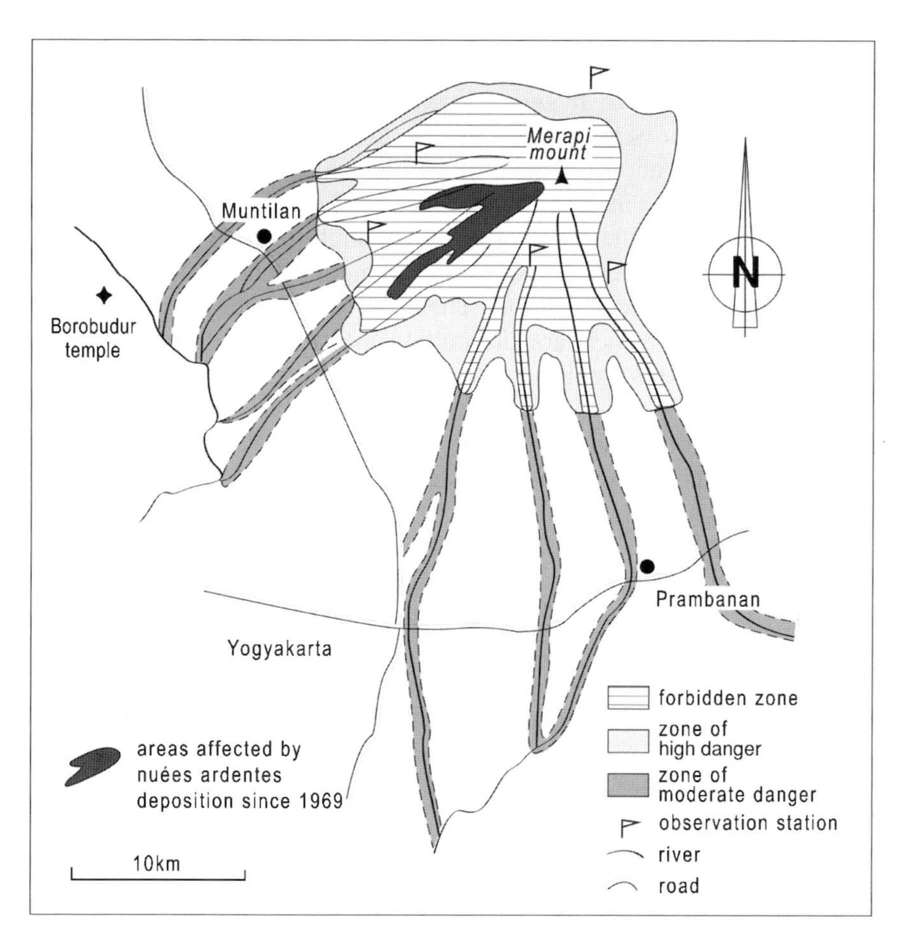

Figure 32: Prevention measures taken for occupying the land subject to pyroclastic emissions of Merapi volcano, Indonesia (after Suryo & Clarke; see Bell, 1999).

are now generally well established and controlled for effusive lava discharges, and even better if the flows are viscous and devoid of accompanying explosive emissions. On the other hand, the risks are the highest and forecasting and prevention possibilities the lowest when the volcanic regime is explosive, induces partial collapse possibly responsible for earthquakes and/or tsunamis, determines the discharge in the atmosphere of a mixture of particles and hot gas, and gives way to massive gravity flows. All intermediate situations may occur, causing various and complex hazards with considerably diverse frequencies and intensities. To summarize, if volcanic hazards are rather better artificially-controlled than seismic hazards, *the better understood and managed situations are also the less dangerous; most worrying situations remain often aleatory and less adequately controlled.*

Improving eruption forecasting and prevention implies *progress in the understanding of volcanic phenomena dynamics*. Such an aim necessitates increasing the number of volcanic sites that are subject to precise observations and measurements made at regular intervals; only 20% of presently active volcanoes give way to a more or less frequent follow-up. It is also necessary to increase the number of sites equipped with multi-tool monitoring systems, especially in volcanic areas where hazards are highest; about 15 volcanoes only are presently under such control. These developments require important financial investments, which are difficult to afford for developing countries, especially when the risks concerned are long term, as is the case for major volcanic hazards.

The research effort needs to be associated with *better communications* from the scientific community, namely towards the populations living close to potentially dangerous volcanoes. It is important to deliver lectures, documents, brochures, movies, etc. that emphasize not only the beauty of the eruptions but also their danger and what to do in case of major events. An exemplary illustration of such actions is provided by the videotape established by Maurice and Katia Krafft, before they died, of the 1991 eruption of Unzen volcano in Japan. The distribution of this tape during the last 20 years, under the auspices of the International Association of Volcanology, has largely contributed to convincing populations to evacuate some particularly dangerous zones (250,00 people have been displaced) and to save about 10,000 lives.

A Few Landmarks

- Volcanic activity gives way to the formation of a large array of rocks and morphologies. It corresponds either to effusive discharge of more or less viscous lavas, or to the emission of variously sized and solidified pyroclastic material. Dominantly controlled by the chemical composition of magmas as well as by temperature and gas content, volcanic events are predominantly of an effusive basaltic type in oceanic domains, of an explosive rhyolitic type in intra-continental areas, and of an andesitic type with flows and explosions on continent borders.
- Volcanic eruptions, together with earthquakes, result from inner-Earth geodynamical processes, which essentially work along lithospheric plate frontiers and depend on active fault systems. Unlike earthquakes the volcanic events are enhanced in accretion and extension zones such as oceanic ridges, occur in well-localized areas by inducing specific morphologies (i.e. various volcano shapes), and induce very variable risks depending on the type of emission and not only on its intensity.

- Direct volcanic risks are minimal for viscous, slowly propagating lava flows, and maximum for pyroclastic explosions leading to an airborne mixture of ash and stone within dense and hot gas. Secondary risks, which often induce worse damage, mainly result from mudflows and debris avalanches, from tsunamis, and from noxious gas expulsion (especially, CO_2). Some indirect effects of volcanic emissions act long term and/or on the global environment, and may partly interact or combine with human activity: air emission of sulphur gas inducing the formation of acid rain, or of dust and aerosols favouring climate changes.
- The forecasting and prevention of volcanic eruptions have reached a satisfactory degree of reliability as far as effusive processes and some secondary effects are concerned. This has been allowed by various developments: implementation of multi-tool monitoring field stations (measurement of seismicity, morphological and physicochemical changes, etc.), satellite imaging and computer modelling, dams and diversion systems, use of precise hazard maps and alarm systems. By contrast the predictability remains aleatory and the prevention very crude for major pyroclastic discharges accompanied by high-temperature fluids and morphological upheaval.
- The major natural catastrophes determined by volcanic activity are responsible for a particularly high proportion of human victims, the number of which tends to increase due to the recent demographic explosion. By contrast the absolute number of casualties is about 30 times lower than for earthquakes. This results from several convergent reasons: a moderate number of highly active volcanoes (50–70 eruptions/year); an eruption rhythm roughly in inverse proportion to the danger; well-localized situations of most dangerous areas; improved forecasting and prevention systems; and limited urbanization in major risk areas. Incidentally the volcanic activity is beneficial for some human activities: revitalization of agricultural soils, ores and construction materials, and geothermal energy.
- The diversity of volcanic events and risks necessitates active development of scientific research, especially by the mean of multi-tool survey and monitoring (volcanic observatories). Information on the actual importance of the volcanic danger should be better developed towards the populations, in spite of the sporadic character of these natural hazards.

Further Reading

Bardintzeff J.-M., 1998. Volcanologie. Doin, Paris, 284 p.

Bell F. G., 1999. Geological hazards: their assessment, avoidance and mitigation. Spon, London, 648 p.

Durieux J., 2000. Volcans. CDrom, Syrinx, Paris.

Fisher R. V., Heiken G. & Hulen J. B., 1997. Volcanoes, crucibles of change. Princeton University Press, 317 p.
Juteau T. & Maury R., 1997. Géologie de la croûte océanique. Masson, Paris, 367 p.
Lefèvre C. & Schneider J.-L., 2003. Risques naturels majeurs. Gordon & Breach.
McGuire W. J., 1998. Volcanic hazards and their mitigation. In Maund J. G. & Eddleston M., Geological Society of Engineering Geologists, spec. public. 15: 79–95.
Tilling R. I. ed., 1989. Volcanic hazards, short course in geology. AGU, Washington, 1, 123 p.

Some Websites

http://www.ipgp.jussieu.fr/: information on Globe Physics Institutes, volcanic watch networks, data base from worldwide field sites
http://193.204.162.114/: site of the Roberto Scandone geophysical group of the University of Roma Tre, devoted to Italian and Mediterranean volcanoes: history, eruptive activity, hazards, forecasting and prevention
http://www.swv2c.org/: worldwide volcanic activity and watch systems, presented by the Southwest Volcano Research Center, Apache Junction, Arizona
http://www.ngdc.noaa.gov/: past and recent information on all major types of natural hazards, presented by the National Geophysical Data Center, NOAA (National Oceanic and Atmospheric Administration, USA)
http://geohazards.cr.usgs.gov/: information on various natural hazards, particularly in North America
http://www.eri.u-tokyo.ac.jp/: abundant data, graphs and figures on earthquakes and volcanic eruptions in Japan and several other world regions, especially for recent times

Chapter 3

Land Movements

3.1. Downslope Gravity Displacements

3.1.1. The Vaiont Landslide, October 9, 1963

Located in the Alpine Piedmont north of Venice (Italy), Mount Toc is known to have been subject to various types of slope movements since prehistoric times. Its northern side is made up of a series of *clay and limestone beds* sloping northwards in the direction of the adjacent river Vaiont (Fig. 33). A reservoir lake, progressively filled from 1960, was established at the foot of the mountain in a glacial valley dug out by the river. On October 9, 1963, a huge landslide developed on the unstable slope extending from the mountain to the river. A mass of about 270 million cubic metres of clay and limestone broke out along an ancient fault plane and started to move northwards through gravity effects.

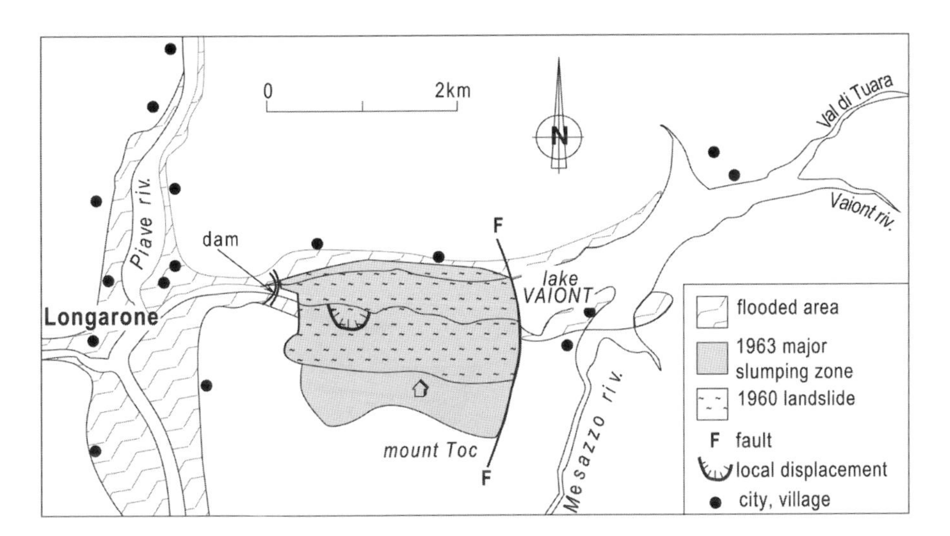

Figure 33: Location of Vaiont landslide and resulting flooding (after Waltham, 1994; Murck et al., 1996).

The sliding mass reached a speed of 20–30 m/sec and finished its 400-m course in the lake, which induced the formation of a giant wave of 100 m high. The wave passed over the double-vaulted Vaiont dam and flooded the downstream valleys, causing the deaths of 2,043 people in Longarone city and surrounding areas.

Due to a major landslide followed by major flooding (Chapter 4.1), the Vaiont disaster resulted from the *combination of natural and anthropogenic causes*:

- The Mount Toc area comprises several south–north-oriented faults, especially on the eastern border of the sliding zone, which facilitates gravity movements along the tectonic planes. The geological formations comprise numerous argillaceous beds that are prone to expand when soaked in water, and therefore to flow and carry the intercalated limestone rocks downstream. The finely stratified character of clay–limestone alternating beds as well as their 10° to 45° northward angle favour gravity movements down the mountain in the direction of the Vaiont river and lake. The rainfall level was particularly important between 1960 and 1963, leading the rocks to be water-impregnated and favouring the layer-on-layer slides.
- The construction and filling along Mount Toc's unstable side of the Vaiont hydroelectric lake exacerbated the natural propensity for geological layers to slide down. Before the artificial lake was filled, the groundwater level prolonged the Vaiont river level and therefore located fairly low in the basal, karstified but massive limestone series of the mountain base (Fig. 34). As the lake was filled, the piezometric level (i.e. the surface of the underground water; Chapter 6.2) was progressively rising, until the superimposed, unstable clay-rich levels were gradually soaked in water. The catastrophe occurred

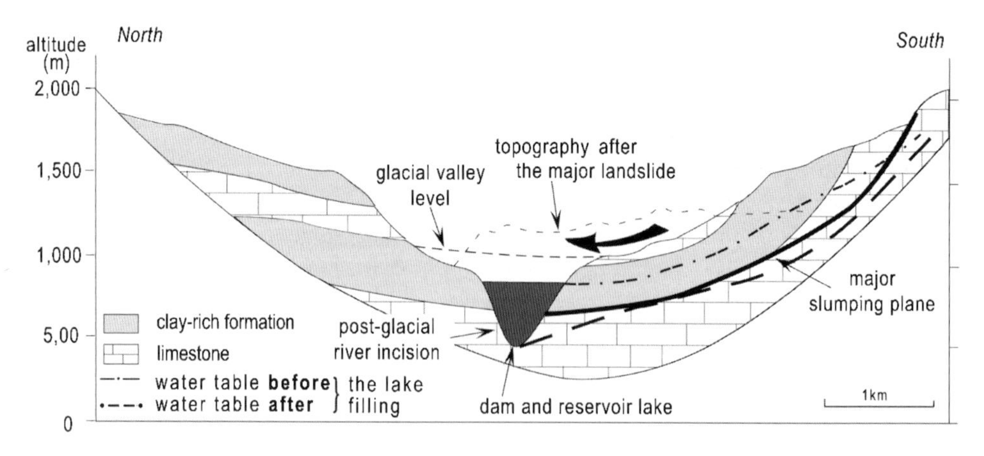

Figure 34: Geological section of the 1963 Vaiont landslide (after Waltham, 1994; Park, 1997).

when the lake level reached an altitude of 701 m, causing the underground water to impregnate most argillaceous formations. Simulations performed after the disaster showed that the major slide would have happened even if climate conditions were dryer, the landslide starting inevitably when the lake would have risen to 722 m; this was the level planned by the engineers for the normal functioning of the dam lake for producing hydroelectricity.

The site of Vaiont was obviously inappropriate for hosting an artificial lake, the water of which tended necessarily to migrate within adjacent geological formations and to reactivate well-known landslides. In fact some *premonitory events* were noticed before the disaster, the effects of which could certainly have been reduced in a significant way. A local landslide happened in 1960 (see Fig. 33) when the lake began to be filled, and very slow gravity movements (i.e. creep) developed chronically on the northern side of Mount Toc. Creep movements were growing during the days preceding the major slide and, on the day before, some preliminary mass movements affected a part of the northern side. These precursory events led to the decision to start to empty the artificial lake, which unfortunately caused an overpressure of underground water and increased ground destabilization. The 1960–1963 monitoring of land movements, rainfall intensity and lake-filling stages showed some positive correlation between the water level and slope instability (Fig. 35). The risk was therefore clearly identified. Abundant rainfall before the disaster was responsible for triggering rapid movements, invalidating the predictions of the site managers, who were expecting slow movements until the massive stabilization.

3.1.2. The Risk

Slope movements are likely to occur in all *steep regions* made up of soft, faulted, fractured geological formations that are subject to noticeable humidity and temperature variations. The very large landslides, mudflows and rock collapses are fairly rare, but potentially responsible for numerous human deaths. *Hazards tend to increase in relation to the last century's demographic explosion*, especially in developing countries where land management practices are often imperfectly standardized and controlled. Major disasters occurred during the 20th century in Italy (1916, 1963: 12,000 victims), China (1920: 200,000 victims), the former Soviet Union (1949: about 16,000 victims), Peru (1962, 1970: 75,000 victims) and Columbia (1985: 25,000 victims). Some of these catastrophic events were triggered by earthquakes, volcanic eruptions or flooding. *Damage to property is generally very serious* and depends only partly on the size of displaced material: deterioration and destruction of buildings, transport links and other developments.

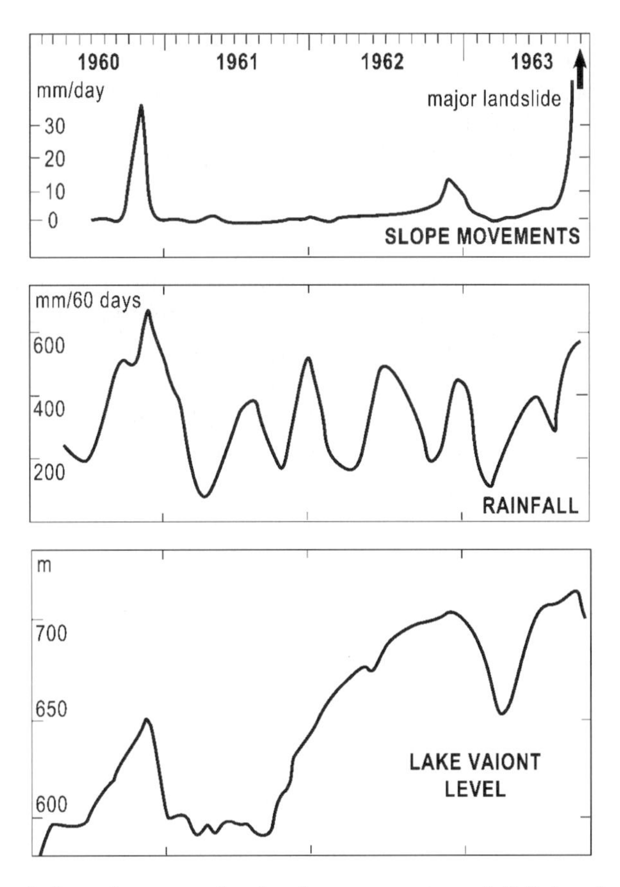

Figure 35: Correlations between the land movements, rainfall level and lake level stages before the major Vaiont sliding (after Waltham, 1994).

The slope-movement hazards tend to increase in mountainous regions, especially those *where clay-rich rocks and soils outcrop*. In the USA major risks characterize the Appalachian mountain range to the east, the Rocky Mountains and coastal chains to the west, as well as some northwestern regions (e.g. Dakota, Montana) characterized by abundant argillaceous badlands (Fig. 36); about 50 major casualties are registered each year. In Europe countless local landslides occur in mountainous regions characterized by swelling clay formations, for instance in southern Italy (*franes*); these movements usually claim few human victims but their repetition induces important material damage. In France most land movements occur in the Alps (e.g. St-Etienne-de-Tinée, Séchilienne) and to a lesser extent in other mountainous areas, as well as along cuestas and coastal cliffs marked by abundant swelling clay (Lorraine, Normandy, Loire Valley, Aquitaine; Ledoux, 1995).

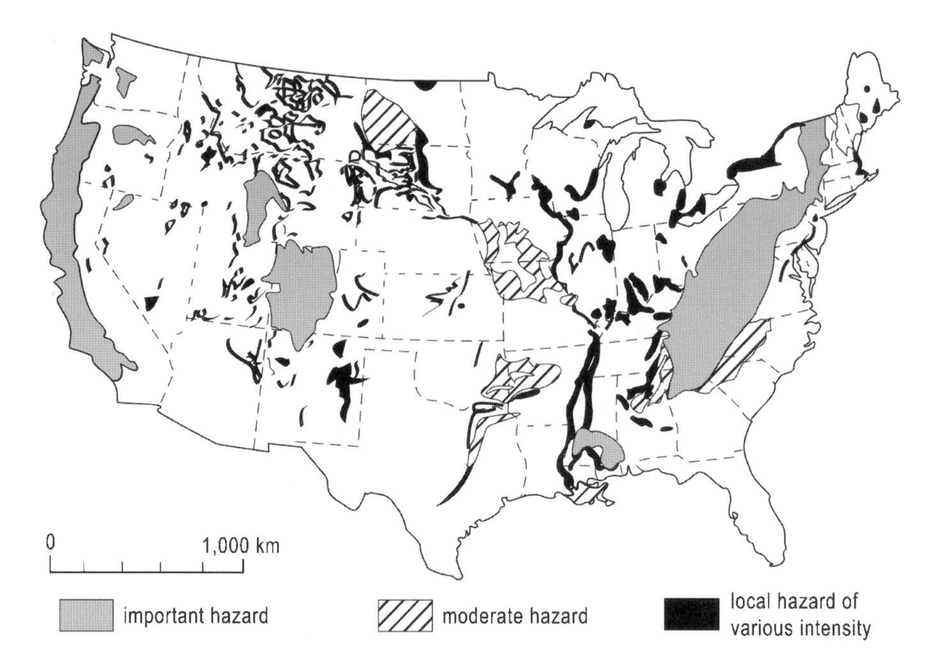

Figure 36: Location and importance of slope movements in the conterminous U.S. (after Hays; see Bryant, 1991).

3.1.3. Types and Effects

Gravity displacements may affect ground materials of very diverse nature and size, include a liquid or even gaseous phase in addition to the solid one, display a highly variable speed (from >30 m/sec to <30 m/500 year), and occur on very low to steep slopes (from 2–3° to upright). As a result the slope movements are characterized by *very different types and complex mechanisms*, the classification of which is difficult and depends on the scientific approach followed (e.g. geomorphological, geological, mechanical classification; see Flageollet, 1989; Cooke & Doornkamp, 1990; Singh, 1996). In the following paragraphs we summarize the functioning of slope movements by distinguishing the displacements of either cohesive rocks or of mixtures of solid material and fluid. The mode and speed of displacement, as well as the abundance of eventual interparticular fluids, are taken into account for each material category (Fig. 37).

Displacement of Coherent Rocks

Fall Rock falls correspond to the extremely rapid displacement of blocks, stone, pebbles, and smaller debris or earth, which detach themselves from a

MECHANISM		MATERIAL		VELOCITY (m/sec)
		COHERENT ROCK	PARTICLE MIXTURE	
FALL		rockfall	debris fall, earth fall	extremely rapid # 10
SLIDE		slumping	debris slumping earth slumping	slow to fairly rapid (10^{-6} to 10^{-3})
		rock slide	debris slide earth slide	rapid (10^{-3})
FLOW		rock avalanche	debris avalanche, mudflow, earthflow, debris flow, grain flow	very rapid (1-10)
		creep, solifluction	creep, solifluction	extremely slow

Figure 37: Simplified classification of slope movements according to displacement mechanism, material nature and speed (after Pipkin, 1994).

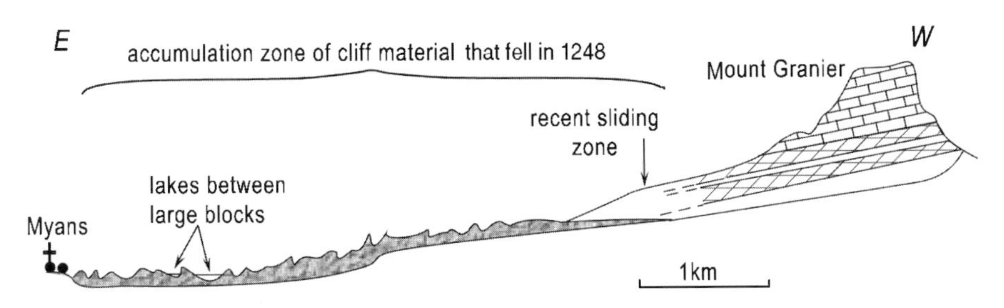

Figure 38: Present-day section of the landscape east of Mount Granier (Isère, SE France) showing the effects of the 1248 limestone cliff collapse (after Goguel, 1980).

vertical to subvertical rock face: cuestas made of fractured or overhanging rocks, coastal cliffs eroded by waves, etc. A *rock collapse* refers to the fall of a large rock section, the pieces of which potentially spread over several hectometres to kilometres, without any grain-size sorting. If the collapsed material is of a large volume and involves large-size blocks, the danger is serious in populated areas. This was the case in 1248 on the northeastern face of the Chartreuse Massif (Isère, SE France), where a large amount of Early Cretaceous limestone and marls forming the eastern cliff of Mount Granier suddenly collapsed. Several thousand people died, and the landscape down the mount was deeply modified for a distance exceeding 5 km (Fig. 38). The Mount Granier collapse apparently resulted from a progressive clearing of soft rocks forming the lower part of the cliff, rather than from an earthquake. A *rockfall* designates the fall of isolated blocks or stone that progressively accumulate around the cliff base and constitute a cone marked by a downward-increasing grain-size and a slope angle tending to become stable. Hazards due to rockfalls are usually low.

Slide Land sliding (or gliding) through *rotation movement*, or *slumping*, occurs on mountains or hillsides marked by a fairly steep slope (15° to 45°) and mainly constituted of clayey or marly material. The ground instability is enhanced by water seepage in most pervious rock layers, as well as by the presence of swelling clay (i.e. the smectite mineral group). It gives way to the formation of a *slump*. In *terrestrial environments* slumps are usually slow (about 1 m/month) and occur over fairly short distances (from about 1 m to 100 m). Successive, more or less independent slumps frequently form in the same area (Fig. 39) and can be reactivated during rainy weather periods. Slumps develop particularly when slopes made of clay or marl are rejuvenated through natural or artificial erosion (e.g. lateral displacement of river meander, action of coastal waves, slope clearing for road, canal or railway track construction). The resulting damage to people is usually very low, contrary to that affecting property (collapse of constructions and other developments).

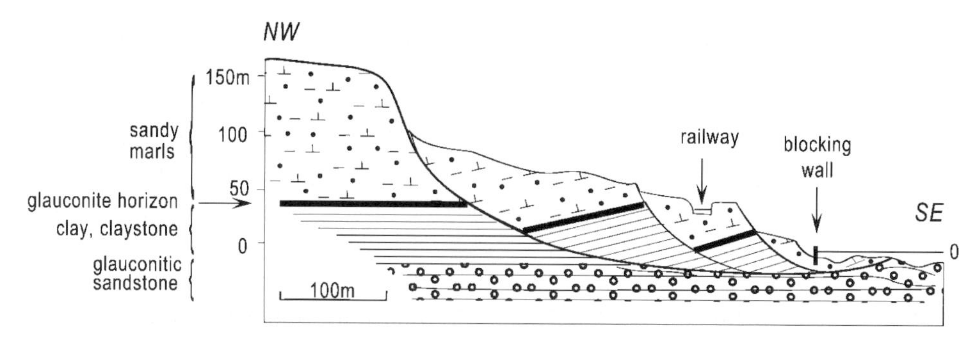

Figure 39: Slumps resulting from rotational movements in Cretaceous marls and clay of Folkestone coast, British Channel, South-East England (after Flageollet, 1988).

Slumps develop frequently in *sub-aquatic environments* characterized by active clay and silt deposition: lacustrine and marine deltas, distal parts of micro-tidal shelves, continental slopes. The submarine displacement of water-impregnated sediments may occur on low slopes (a few degrees; e.g. off the Mississippi and Rhone rivers), attain fairly high speeds ($>$ 1 km/hr) and extend over several kilometres.

Gravity rock displacements through *translation movements* along a plane, called *slides* (or glides), occur without important dissociation of initial material components. The sliding plane corresponds to the surface of a stratigraphic bed or of a tectonic fault, along which gravity effects develop. The sliding material is made of massive rock blocks, of heterometric fractured and broken rock debris (i.e. debris flow) or of fine particles (i.e. earthflow). The movement is accelerated if sliding takes place on a "soap-like" layer made of gypsum (hydrated $CaSO_4$), sodium chloride salt (NaCl) or soft clay, or if it concerns pulverulent earth material. Hazards induced by such movements occur in mountainous regions, are generally of moderate danger, and concern human goods and property rather than people themselves.

Flows of Particle and Fluid Mixtures

Slurry flows, avalanches *Quick gravity displacements of water-saturated material potentially involve all particle types and sizes:* rock and debris avalanches; debris, mud-and earthflows; etc. Snow avalanches proceed from similar mechanisms. The slurry flows usually present both a high fluidity and a high density; this differs from most flood-induced torrential flows, the density of which is closer to that of water. Large blocks may be transported by slurry flows, since sorting processes occur rarely inside the moving mass. The diversity and complexity of the flowing mixtures, which sometimes include gaseous fluid, determine a *wide spectrum of situations and hazards*. Coarse rock avalanches are infrequent and represent a transition stage with rock collapses. Debris to earthflows occur commonly on steep slopes from many regions and can be labelled as *mass flows*. Most devastating mass displacements consist of *debris avalanches*, made up of dense, waterborne heterometric material which moves at a speed approaching potentially 100 km/hr. Debris avalanches tend to erode the soft underlying formations and therefore to grow in volume during displacement, and may move over tens of kilometres. They may be triggered on both natural or artificial slopes such as: mountain and hillsides, embankments and slag heaps. Recent spectacular examples of disastrous debris avalanches were associated volcanic eruptions of Mount St. Helens in 1980, and especially of Nevada del Ruiz in 1985 (25,000 victims; see Chapter 2). The high-speed displacement over a great distance of debris avalanches is still imperfectly understood, and of course difficult to document due to the scarcity, rapidity and danger of such events. The possible role is often mentioned of a thin air layer compressed beneath the moving mass, diminishing the internal frictions.

Mudflows occur more frequently than debris avalanches, especially in mountainous regions where spring warming allows snow to melt. Mudflows move less rapidly than avalanches (ca. 1 m/sec) and represent a local risk only. In sub-aquatic environments, the particle flows mostly constitute *debris flows*, which develop particularly at a right angle to continental slopes in the canyon axis and may propagate over large distances (tens to hundreds of kilometres). The Agulhas submarine flows off the South-East African coast involve a surface of 80,000 km^2, a length of 750 km, a width of 106 km, and a sediment volume estimated at 20,000 km^3. Debris flows potentially represent a noticeable risk for submarine ore extractions, exploitations and developments: breaking of cables and pipelines, associated tidal waves causing bottom disturbance, sediment erosion favouring the discharge of gas hydrates, etc.

Debris flows may also result from the upstream dissociation of a sliding mass (i.e. slump) and pass themselves downstream into turbidity flows giving way to *turbidite* deposition. Turbidites induce the formation of sedimentary lobes, which may extend in the continental rise and abyssal plains adjacent to margin slopes. Submarine debris-flows and turbidites are sometimes triggered by earthquakes, which may induce disruptions affecting both the natural environment and man-made developments. A major earthquake occurred in 1929 off Nova Scotia and Grand Banks in the northwest Atlantic Ocean, determining the formation of huge turbidity currents that expanded at 93 km/hr over a width of 150 km, and broke telephone cables located as far as 470 km from the focus (Fig. 40). Turbidites and other reworking processes on continental margins mostly result from excessive sedimentary accumulation on fairly steep slopes. The sedimentary overload may result from natural or artificial causes. In 1979

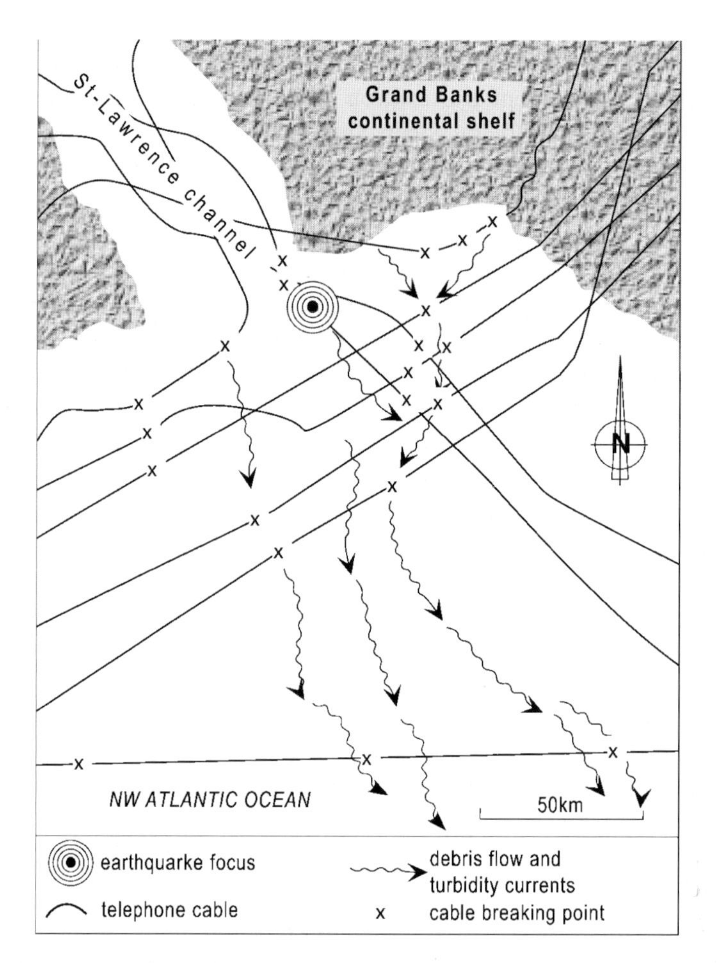

Figure 40: Route and effects of debris flows and turbidites triggered off Grand Banks by the 1929 earthquake in the northwest Atlantic Ocean (after Heezen & Drake in Murck et al., 1996).

the sudden collapse of sedimentary materials, artificially accumulated along the French Riviera coast to enlarge Nice airport, induced a series of submarine and coastal disorders: sedimentary flow extending over 200 km as far as the foot of Corsica's continental margin, breaking of telephone cables linking Genoa to the Balearic Islands and to Sardinia, and water counter-effects causing a mini-tsunami on the French littoral. The moving mass displayed several successive shapes: slumping responsible for strong slope erosion, debris flow causing the formation of bottom furrows, huge turbid bubbles with strong erosive power inducing the deposition of gravel dunes in the bathyal plain, sandy-clayey proximal turbidites responsible for cables breaking, and fine distal turbidite sedimentation (Mulder et al., 1998).

Earthflows carry fine and rather homogeneous material (diameter <2 mm), the characters and effects of which are similar to those of mudflows. They may determine micro-avalanches with low potential danger. They include the *grain flows*, forming either on the steep side of continental dunes subject to overloading or sapping (coastal erosion, human developments), or on submarine sand dunes shaped by tidal currents.

Creep, solifluction Very slow ground-surface flows (ca. a few decimetres per year) occur in most mountainous areas marked by sparse forest cover, clay- and silt-rich soils and subsoils, and noticeable amounts of interstitial water. These very slight gravity movements are referred to as creep or solifluction, and are expressed in the landscape morphology on a long-term scale only: irregular topography, tilted pylons and fences, deformed tree trunks, fallen gravestones, blistered and cracked roadbeds, disaligned control-poles, etc. The downward movement slows down if soils lack liquid water and become cohesive, and accelerates when water adsorption allows clay to swell and to become more fluid. Creep processes therefore increase during *spring snow-melting phases*, when surface layers soaked in water flow slowly above deeper, still-frozen soil. Solifluction also characterizes commonly the cold climate regions where soils are frozen at depth all year round (i.e. permafrost; Chapter 3.3), the surface layer melting partly in summer and becoming unstable (gelifluction). In addition, the seasonal alternation of freezing and thawing stages induces vertical displacements, the results of which determine lateral downslope movements. The various creep processes, acting in high-latitude and high-altitude regions, and which also include the movement of rock boulders inserted in ice (rock glaciers), result in a *very slow but considerable displacement of surface materials*. The corresponding deformations at the ground surface are responsible for important *damage to goods*: deterioration of buildings and transport links, breaking of buried pipes, etc.

3.1.4. The Causes

Natural factors responsible for rock and soil movements along the slopes are of course dominated by *gravity*, the role of which is enhanced by the slope *gradient*. The *water abundance and state* constitute a second factor, which intervenes in different ways: fluidification of unstable geological formations, ground-soaking through rise of the water table, freezing-thawing and wetting-drying alternations modifying the soil volume and consistency, and increase in interstitial pressure. Numerous disorders due to slope movements are triggered by heavy rain. Other intervening factors are largely related to the *nature of movements caused by gravity* (fall, slide or flow):

- *Rock falls* develop under various surface processes: opening of rock fractures and cracks by erosion or gravity, sapping of cliff foots and creation of overhangs through river meander migration or coastal wave erosion; rock receding due to plant root growth and animal burrowing; rock volume increase through salt crystallization oxidation of pyrite into gypsum; earthquake-induced vibrations, etc.
- *Mass sliding* movements affect preferentially the finely laminated or fissile sedimentary and metamorphic series, especially when rock bedding is oriented parallel to the slope. Geological formations comprising impervious, plastic and lubricating beds (gypsum, clay, 'soap-like' layers) or swelling clay minerals (smectite group) are also prone to favour rock slides.
- Slow or rapid *flows* developing both in subaerial and subaquatic environments may be triggered by deep seismic and eruptive vibrations, as well as by neotectonic activity. Surface factors include the lack of plant cover (woods, meadows) and the excess of animal activity (grazing, soil erosion by large herd displacements, burrowing). In addition, sedimentary flows along lake and ocean slopes are facilitated by organic fermentation (gas escape) and especially by the excess river particle discharge and its deposition close to canyon heads. For instance the Mississippi sedimentary supply during the last century has led to the deposition of more than 30-m-thick terrigenous sediments on the continental shelf lying off the river mouth.

Human factors favouring gravity movements first tend to *exacerbate the natural role of steep slopes and abundant water.* Artificial reshaping of hill or mountain slopes for road, canal or railway constructions induces chronic instabilities. This is, for instance, illustrated by the Dover to Folkestone coastal railway (see Fig. 39). The water table rises due either to the filling of lake reservoirs (e.g. Vaiont dam, Italy, Chapter 3.1.1), to the stopping of water pumping after mine closure or to the deterioration of and leaks in buried pipes, and leads to impregnation of permeable rock and soil formations and induction of slope movements. The emptying of lakes and reservoirs for irrigation or maintenance determines the water table to be temporarily "suspended" within adjacent ground formations; the interstitial pressure is therefore no longer compensated by free water, which increases the landslide hazard.

Other human actions may aggravate the slope vulnerability to gravity movements. For instance, it is the case of the excess load determined by large, heavy buildings constructed on the top of potentially unstable cliffs or hills, of shore or shelf embankments established very close to continental slopes (e.g. Nice airport extension works in 1979), of excavations dug at the feet of slopes (e.g. surface quarries), or of mine waste accumulated without adequate slope control (e.g. clay-rich slag heaps). More important, the extension of deforestation and overgrazing in uneven regions leads to the destruction of the soil root frame, allowing the development of widespread slope movements. In the Tyrol (Austrian Alps) two-thirds of land slides and flows are attributed to human actions: inappropriate building location, uncontrolled torrent course in densely inhabited areas, poorly designed road layout, poor forest and soil management practices (Campy & Macaire, 1989).

The continuously growing pressure of man on the planet's surface environments results in the fact that *gravity displacements are more and more caused by a mixture of natural and anthropogenic factors*. This was clearly illustrated by the 1963 disaster triggered by the Vaiont landslide (Chapter 3.1.1). In New Zealand, where slope movements affect numerous clay-rich formations and constitute a national problem, the *frequency* of gravity-induced displacements is directly connected to the extent of deforestation (Fig. 41), associated to that of urbanized areas. The *intensity* of ground movements is nevertheless less dependent on the vegetation cover, some regions marked by a similar degree of deforestation showing different landslide rates (e.g. Wellington in North Island relative to Greymouth in South Island). The differential factor is then natural and consists in the rainfall level (Glade, 1998). In southern Italy, local ground

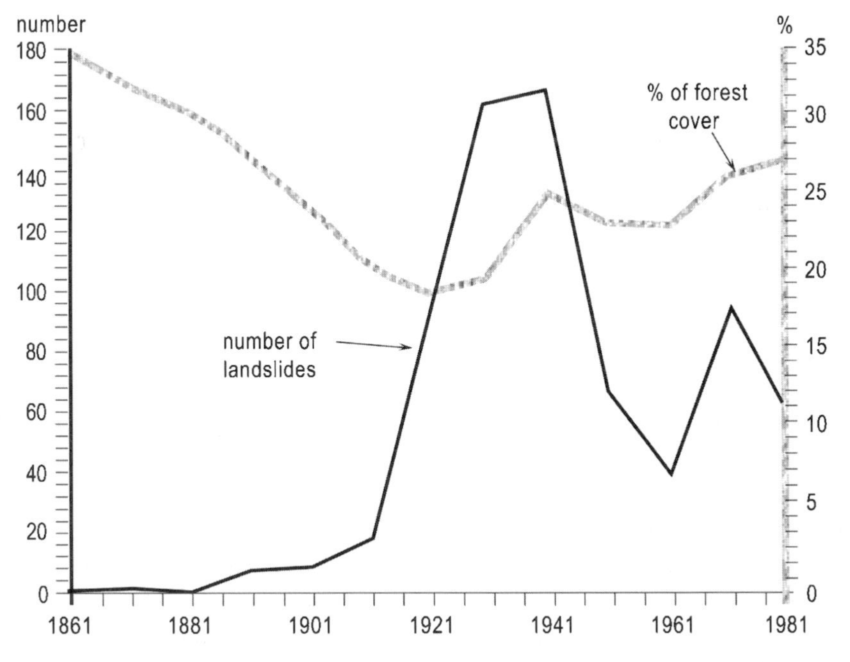

Figure 41: Relationships between the New Zealand forest cover and the recorded number of landslide-triggering rainstorm, from 1861 to 1991 (after Glade, 1998).

instabilities, the chronic character of which are maintained by the neotectonic activity (e.g. Valle dell'Oro), have recently extended because of intensive deforestation practices linked to new land policy and extensive urbanization; large areas are now subject to widespread mud flows, with high potential danger for both people and properties (Alexander, 1992).

3.1.5. Forecasting

Owing to the very large *dispersion* of surface sites potentially subject to slope movements, as well as to the broad *diversity* of possible mechanisms and potentially triggering factors, it is quite impossible to carry out widespread monitoring and forecasting actions. In fact it is only necessary to watch closely the hazards where there is serious danger for people and property, which in itself constitutes a titanic task. This is why some devastating slope movements such as huge mudflows or rock falls may still threaten most world regions (e.g. northern South American countries, 1999–2000; Algiers, November 2001), including high-technology countries. The forecasting of gravity movements comprises various actions:

- *field mapping* of unstable zones, allowing location of those representing an actual danger to human activities (Fig. 42), and to establish a scale of risks: high risk for buildings and development, mid-term potential risk, low risk, or improbable risk (Pickering & Owen, 1997);
- multi-tool, long-term monitoring of *topographical changes* (use of inclinometers, tiltmeters, landmark alignments), associated with photogrammetric reports, statistical calculations, three-dimensional modelling, and issuing in the establishment of fairly easily-applicable expert-systems (e.g. Al-Hamoud & Tahtamoni, 2000);
- laboratory measurement of the *physical properties* of soil and their variations in the course of time: water content, density, Atterberg limits of plasticity and liquidity, compressibility, swelling pressure, etc.;
- assessment of the water influence on stability by measuring the variations of the *water table* level. At a regional level, the temporal forecasting is particularly based on *rainfall* intensity and duration. For instance the number of landslides increases in the wettest Japanese regions when the daily rainfall exceeds 100–150 mm, and in dryer regions of California when rainfall exceeds 250 mm in 2 to 3 months (Flageollet, 1989). Such data sets allow scientists to use simple criteria and to establish *alert plans*.
- *acoustic methods* are sometimes employed for monitoring very slow slope movements (creep, solifluction) that may induce low intensity sound vibrations when the strength limit is attained. The more compact the material, the louder the acoustic emission.

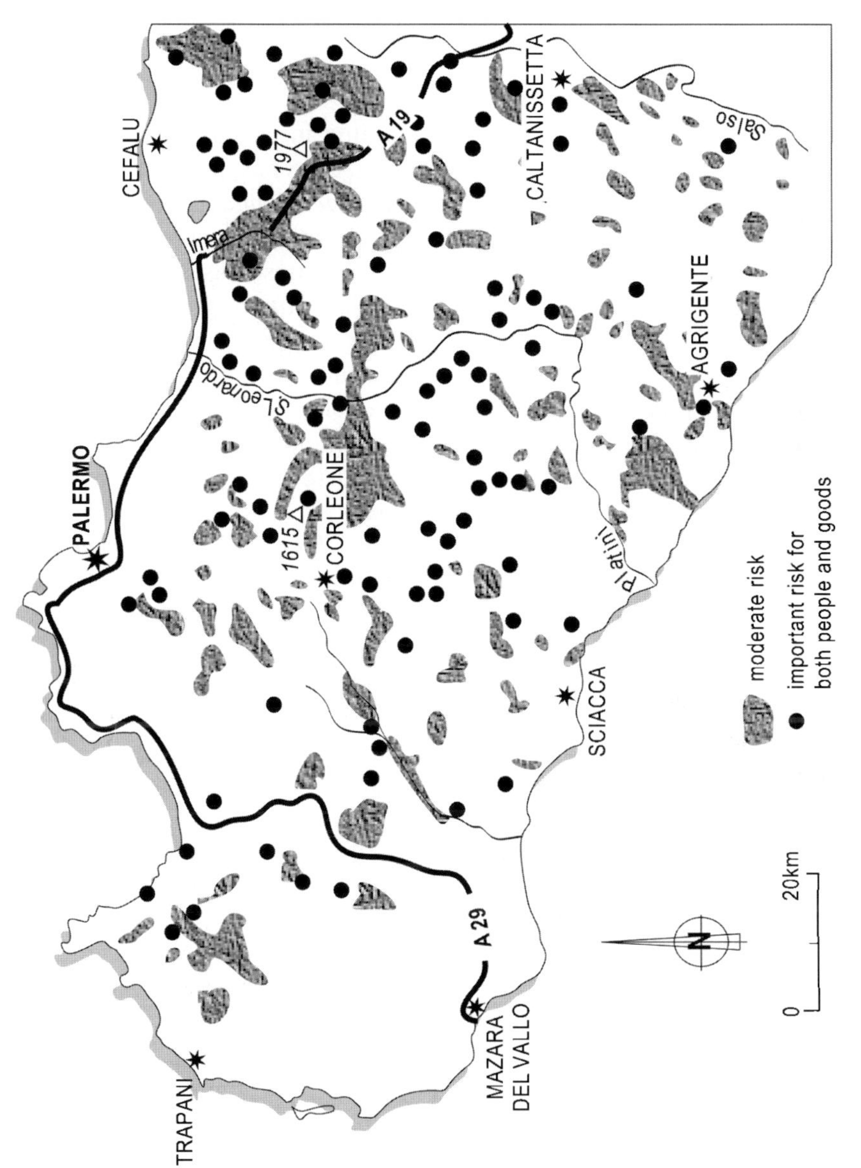

Figure 42: Location of main landslides in Western Sicily (after Agnesi in Flageollet, 1989).

On average *the forecasting of slope gravity movements is efficient as far as their exact location is concerned, but remains very imprecise concerning their starting time.*

3.1.6. Prevention

Apart from the recommendation to avoid building in unstable areas, which of course is difficult to implement in already urbanized zones, preventive measures consist of *stabilizing*, *draining* and practising *geotechnical methods* in the dangerous zones, after in-depth *geological investigations* have been performed. The wide array of existing measures allows in most cases the significant reduction of gravity hazards. *Forestation* or *reafforestation*, sometimes complemented or replaced by turfing, represents the most natural way to stabilize the terrains, as well as to facilitate water seepage and to limit erosion processes (Chapter 8). A few precautions are necessary: choice of tree species adapted to local morphological and climatic conditions, sufficiently deep and wide root system preventing the soil surface sliding, adequate tree spacing to avoid any overloading (e.g. 25 kg/cm^2 for boreal forest trees).

Civil engineering work and geotechnical developments provide additional remediation: slope reshaping (alternation of flat terraces and of low stone walls), building of retaining walls on roadsides, active drainage, etc. Slope profiling destined to reduce the gradient is performed on small hills only; on larger outcrops the excessive volume of material to be displaced, the success of the finished work and its cost, are often out of proportion.

Prevention and protection sometimes include several *combined measures*. The chronic landslide affecting the southeastern British coast and the railway track close to Folkestone has been stopped by associating the drainage of clay-rich local marls and the building of a wall blocking the lower slope (see Fig. 39).

3.2. Vertical Movements

3.2.1. Dissolution – Collapse

The Winter Park Collapse, Florida

On May 8, 1981, a small tree collapsed without apparent reason on a commercial terrain located close to Orlando (USA) at Winter Park, an area built on Miocene carbonate formations that form a large part of subterranean Florida. In a few hours a sort of crater formed, which widened and deepened rapidly. The following day the steeply sloped depression, referred to as a *sinkhole*, had reached a diameter of 100 m and a depth of 30 m, and was progressively filled by water seeping up through the ground. The collapse caused the partial destruction of six commercial buildings, one house, one public swimming pool, two streets and several cars. The Winter Park accident, which fortunately did not cause any casualties, followed a long period of drought, responsible for an important water table drop. Investigations performed after the collapse showed that underground calcareous formations were extensively dissolved, forming a network of large cavities covered by a few metres of permeable and little consolidated sediments (Fig. 43). The lowering of the water table caused high pressure to develop at the cavity roof, leading to the collapse of sand, clay and sandstone sealing the underground voids.

The Risk

The sudden vertical movements affecting surface formations topping partly dissolved calcareous rocks are not frequently of the size and importance of the Winter Park collapse. Their *number and concentration at the ground surface* are nevertheless *locally high*. Some Oligocene and Miocene exposed areas in Florida comprise more than 1,000 sinkholes per 5 km^2, which all result from underground dissolution. Surface depressions controlled by subterranean dissolving processes affect numerous *limestone formations* of any geological age, and also *dolomitic* and diverse *saline*, highly soluble *rocks* (rock salt, potash, gypsum). Some collapse events in the dolomite region of Transvaal, South Africa, have caused many human losses and considerable damage. Dissolution depressions the size of the Winter Park sinkhole are described in the saline formations of Texas and Mediterranean regions. The danger is maximum when a soft sedimentary cover that suddenly collapses obliterates sinkholes.

Let us recall that extraterrestrial hazards may also determine important collapse at the Earth surface. This is particularly so in the case of *meteorite impacts*, sometimes responsible for the instantaneous formation of holes, but also for rock fusion, recrystallization, and sublimation. The risk for 1-m-diameter meteorites to impact our planet is of about 1 per year, and of 1 per million years for 100-m-large meteorites. The very high penetration speed into the atmosphere of most extraterrestrial objects (4 to 40 km/sec) causes the evaporation of small meteorites, which amount to ten million to one billion tons/year. Larger meteorites potentially represent a very big danger. For instance a 20-m-diameter object entering the Earth's atmosphere at a speed of 15 km/sec would produce an energy equivalent to 4 million tons of dynamite and form a several-kilometre-wide and several-hectometre-deep crater (i.e. about the size of the Meteor Crater, caused by a meteor that hit Arizona, North America ca. 25,000 years ago).

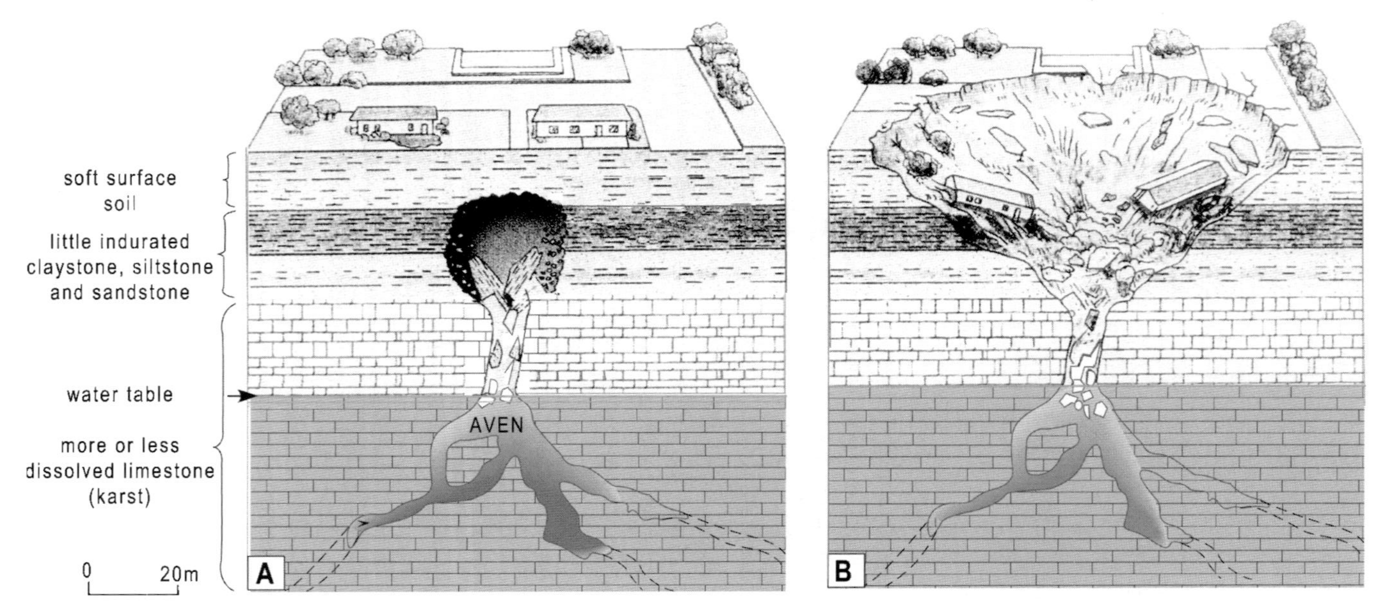

Figure 43: Ground collapse due to the formation of dissolution cavities at Winter Park, Florida. Situation sketch (A) before and (B) after the collapse (after Murck et al., 1996).

Types and Effects

Rock dissolution causing surface collapse results from *chemical weathering* processes induced by water seepage within the ground. Weathering is amplified by microbiological activity, which tends to increase the CO_2 content of permeable soils. Carbonate minerals (e.g. calcite, aragonite, dolomite) are not very soluble in pure water, but actively dissolved by carbonic acid (H_2CO_3) that is commonly present in rainwater. The more porous (e.g. chalk), fractured and faulted, and devoid of impurities (e.g. clay, quartz) the calcareous rock, the more easily it is dissolved. Also, when covered by an argillaceous-siliceous sedimentary material, seeping water is susceptible to acidification. Some carbonate facies are more rapidly destroyed by chemical dissolution than by surface erosion processes.

The subterranean dissolution of calcareous or saline rocks induces the formation of complex underground cavity networks, which are sometimes of a huge size (e.g. more than 500 km of galleries and chambers in Mammoth Cave limestone, Kentucky, USA). The networks constitute the *karst* (named from a region of former Yugoslavia). Karst developments are responsible for surface subsidence, and collapse leads to the formation of closed depressions called *dolines*, which sometimes connect (*polje*) and may be located above sub-vertical cavities forming *avens*. Dolines comprise four different types, according to the nature and thickness of the sedimentary cover, as well as to the importance of cavities in the underlying karst network (Fig. 44):

(1) dissolution dolines where morphological depressions result from progressive chemical dissolution at the ground surface of local calcareous rocks;

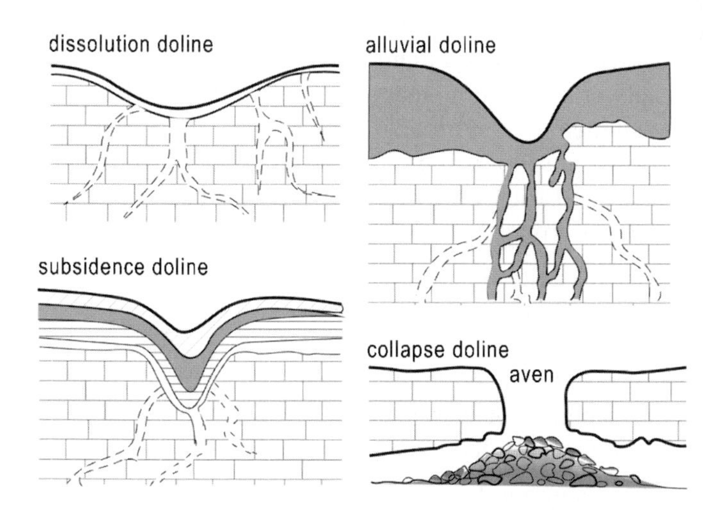

Figure 44: Different types of dolines (after Cooke & Doornkamp, 1990).

(2) subsidence dolines resulting from dissolution at depth in the substrate, inducing the non-calcareous cover to bend;
(3) alluvial dolines where water-transported surface materials invade and plug the underlying karst network, as in the case of the Winter Park accident; and
(4) collapse dolines or avens marked by the sudden breaking of limestone and deposits topping a widely open karst network.

The two latter types present the maximum potential risk. Other karst developments include a large diversity of more or less diffuse forms, inducing very limited danger except as far as their ability to trap and possibly release polluted liquids is concerned (Chapter 10.1).

Causes

The natural factors responsible for the dissolution–collapse coupling consist dominantly in the *carbonic acid content* and the *water action duration.* A secondary factor is represented by the water flow rate, dissolution being maintained and accelerated with increasing water speed. The destruction of karst vaults is proceeded either by a sudden breaking or by the slow propagation of minor collapses along geological faults (formation of *fontis).* The rupture is facilitated if the rock submitted to dissolution is heterogeneous, fractured and heavily overloaded by topping formations (river, aeolian or colluvial deposits; soils, forest, etc.).

Strong alternations in the rainfall regime may participate in triggering the collapse:

(1) abundant rain induces overpressures to develop in aquiferous rocks, the swelling of clay accumulated in underground cavities, and the ejection from the karst network of sedimentary plugs. All these changes are able to weaken the strength of limestone vaults; and
(2) drought periods cause lowering of the water table, the diminution of aquifer pressure relative to rock pressure, and the shrinkage of clay accumulated within the karst (formation of desiccation cracks).

Such changes leave vaults out of plumb, and therefore increase the risk of collapses (e.g. Winter Park accident, see above).

Human factors contribute in an important way to increase dissolution processes. The recent demographic expansion largely explains the strong augmentation of surface collapse events recorded over a few decades:

– intense *aquifer pumping* for irrigation and industrial or domestic needs exacerbates the natural effects of drought periods: water table lowering, rock

pressure increase in cavities, acceleration of flow rate (Fig. 45; Chapter 6). The dissolution processes are sometimes amplified by untimely alternations of pumping cessations and resumptions, of by artificial aquifer refills. In the karstic zones of Alabama (SE USA) about 4,000 dolines that have extended during the past century are attributed to excessive or inappropriate exploitation of underground water; some of these sinkholes exceed a depth of 30 m and a diameter of 3.2 km;
- active *deforestation* allows soil erosion to develop, disrupting the natural water seepage process and exacerbating intense alternation of rainfall and drought;
- *heavy constructions* constitute overloading that amplifies the risk of underground karsts collapsing;
- *mining and quarrying* extract huge amounts of ground material, leading to various disorders in rock stability and subterranean water routes (Chapter 5). Cavities formed artificially for salt extraction may experience further dissolution, responsible for further underground vault fragility.

Forecasting

In the same way as for slope movements, the great number and wide dispersion of ground areas subject to dissolution-collapse effects make the risk assessment impossible to achieve on a general scale. Only a few unstable areas that really threaten human activities and developments may give way to serious forecasting actions. The tools employed first comprise detailed *mapping* of critical sectors: rock nature and components, lithological variations, faulting and fracturing states, importance of dissolution stages of carbonate or saline rocks susceptible to collapse; and water table level and fluctuations, and aquifer flow measurements. *Remote sensing* data associated with high-definition positioning systems (GPS) provide very useful information: vegetation density reflecting the doline position and water table proximity, location of tectonic lineaments where faults and fractures concentrate, progressive change of the surface topography. *Subsurface geophysics* by seismic reflection and electric prospecting is compared with *modelling* experiments for identifying the 3D extension of underground cavities susceptible to collapse; many uncertainties nevertheless remain for applying in an efficient way the model data to natural, often heterogeneous environments. To summarize, *the predictability of rock collapse is rather good as far as the hazard location is concerned, but mediocre about the possible moment of occurrence*. Forecasting is more reliable for young karsts marked by few evolutionary stages than for old ones, the modelling of which is much more aleatory. Predictability is particularly difficult in the absence of concrete expression of topographic lows at the ground surface.

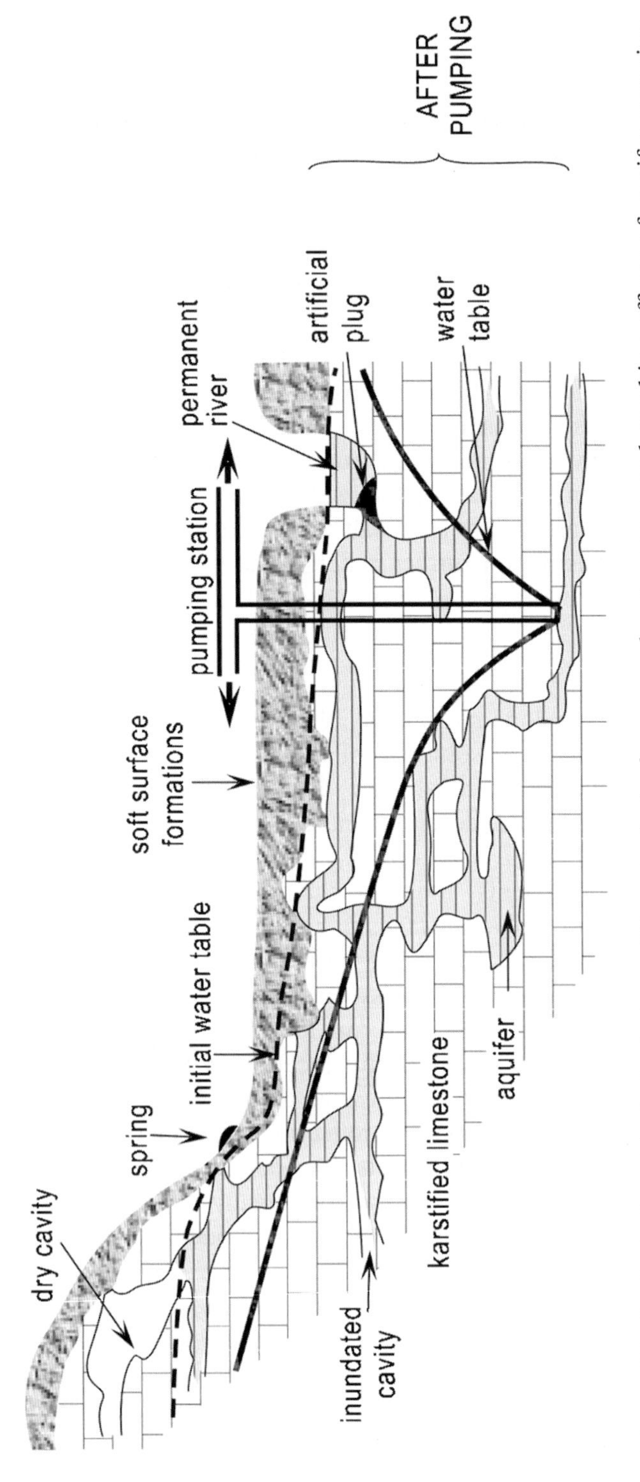

Figure 45: Geological and hydrological conditions typical of a young karst system, and resulting effects of aquifer pumping developments (after LaMoreaux & Newton, 1986).

Prevention

The vagueness issuing frequently from the forecasting of collapse events caused by dissolution is often compensated by the *efficiency* of prevention measures. The complementary use of several actions allows considerable diminishing of the hazards:

- control of underground drainage and projected management of pumping periods;
- land reaforestation and returfing for regulating the water infiltration mechanisms;
- transfer of surface developments out of the most sensitive zones;
- building design fitting the constraints imposed by the tri-dimensional ground configuration (e.g. Sowers, 1986);
- removal of excessive overloads;
- underground strengthening in order to stabilize the buried structures (pipelines, cables);
- concrete or cement injection into underground cavities subject to excessive dissolution and deterioration.

3.2.2. Subsidence

The Long Beach Subsidence

The Long Beach underground along the Californian Pacific coast contains an offshore oilfield, which has been exploited since the 1930s in the Wilmington area. During the 1940s the ground surface started to experience slight subsidence, which progressively propagated around the extraction area (Fig. 46, A). Various disorders were recorded: soil fissuring, inland progression of the coastline, deterioration of buildings and transport links. The subsidence rate reached values up to 70 cm/yr during the 1950s. The total subsidence along the seashore was locally close to 10 m, which led to the raising of the Long Beach harbour walls by 9 m in order to limit the seawater intrusion. The total area affected was 50 km^2.

Investigations conducted at the end of the 1940 decade showed that the subsidence rate roughly paralleled the amount of oil extracted from the ground. These findings led to injection of large quantities of water inside the ground through the existing wells, in order to compensate the voids determined by oil

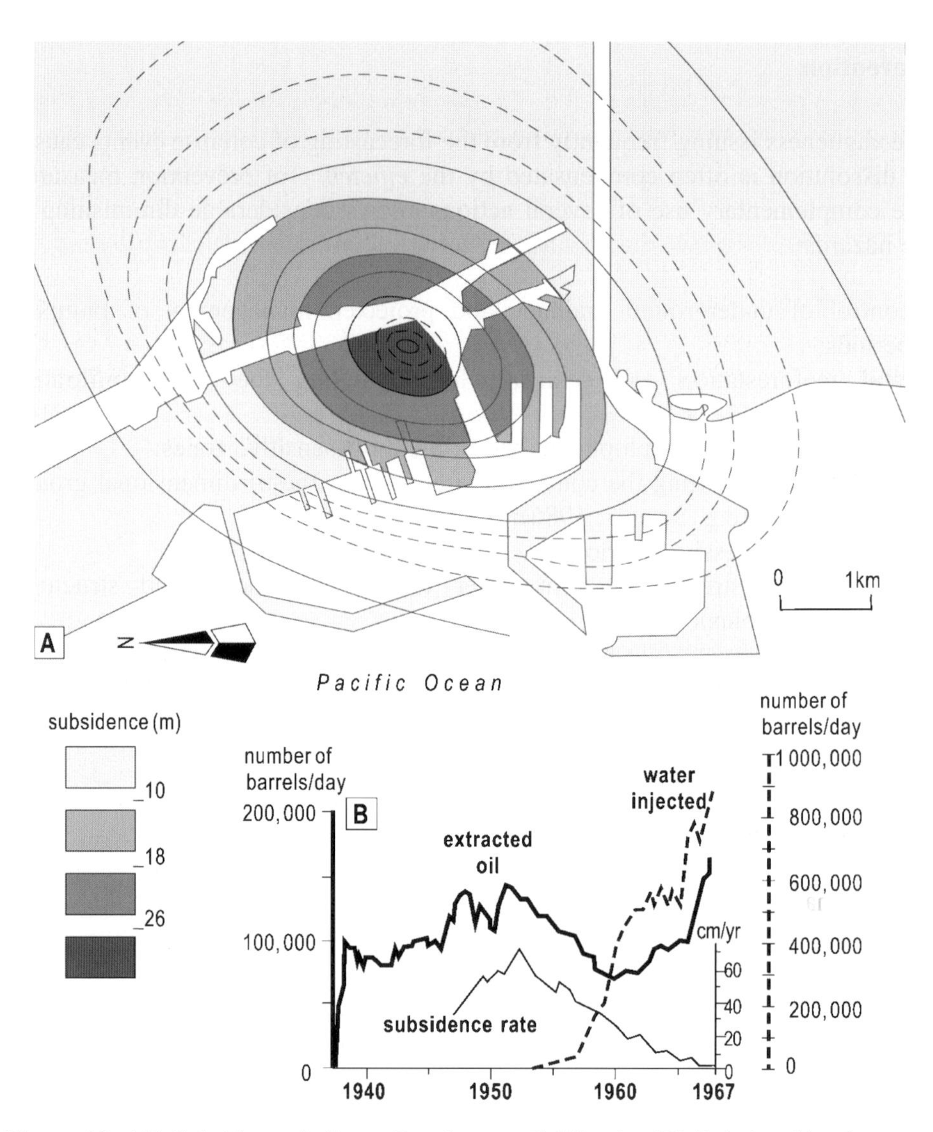

Figure 46: (A) Subsidence in Long Beach area, California. (B) Relationships between subsidence rate, oil extraction and water injection amounts (after Cooke & Doornkamp, 1990; Pipkin & Trent, 2001).

extraction. A noticeable decrease in the subsidence rate was then observed (Fig. 46, B), allowing the depressed area to be reduced over a surface of 8 km^2, and the risk to be much better controlled. The cost of damages was estimated to exceed 100 million U.S. dollars.

The Risk

The very progressive character of morphological changes due to ground subsidence movements induce little risk for human life, except when surface disorders have indirect consequences; this is, for instance, the case when hurricanes affect flat coastal areas that have been lowered by continuous hydrocarbon extraction (e.g. northwestern coast of the Gulf of Mexico), or of the dam of an artificial lake breaking due to ground morphological distortions (e.g. 1963 breaking of the Baldwin Hills dam close to Inglewood oilfield, California). Most damage concerned *human goods and developments*: road cracking and fissuring, cable and pipe breaking, buildings becoming uninhabitable, deterioration of irrigation networks, flooding of lowered areas, etc. Subsidence phenomena often extend over a wide surface, the hazards decreasing towards peripheral zones.

Basic Mechanism

Most subsidence phenomena result from the *progressive compaction* of soil and subsoil terrains, leading to the diminution of ground thickness, which is reflected up to the surface and propagates laterally. The ground compression extends especially within under-compacted rocks, where the internal pressure decreases due to interstitial fluid removal. The compression power is weak in mildly porous and water-poor eruptive or metamorphic rocks, but may be important for some sedimentary rocks comprising abundant clay or organic matter. The ground compaction is responsible for an increasing risk when the phenomena becomes irreversible, which happens when materials lose their elasticity and porosity properties. In the Tucson basin (Arizona, USA) the rapid passage during the 1990s from a 3- to a 24-mm/yr subsidence rate was attributed to the loss of elasticity locally by terrains subject to excessive underground water extraction.

Causes

Natural causes The natural factors responsible for subsidence on a global scale are related to the plate tectonics, and work especially at continent–ocean transition zones (Table 7). Usually lower than 10 mm/yr the *tectonic subsidence* may significantly increase due to extension processes in seismic or volcanic regions (Chapters 1 and 2), or to a sediment or ice overload. Remember that Earth crust subsidence movements are often compensated for by positive isostatic movements, responsible for slow landmass *raising* and relative sea-level drop. Such compensatory movements are amplified in high-latitude regions where the ice cover has melted due to the climate warming that followed the

Table 7: Main factors responsible for ground subsidence (after Cooke & Doornkamp, 1990).

	Causes
Natural causes	tectonic movements volcanic eruption seismic vibrations sedimentary burden clay desiccation organic matter oxidation dissolution of carbonate or saline rocks
Anthropogenic causes	extraction of water or hydrocarbons mining surface drainage soil soaking hydrocompaction artificial overloading (dams, buidings)

glacial stages. The Scandinavian and Baltic Sea regions, where the Fennoscandian icecap has progressively disappeared over a period of 12,000 years, have experienced an uplift ranging from 25 to 255 m, the rise of the land having locally exceeded 1 m/century and continuing presently at a slower rate.

Other natural causes are more local:

- local subsidence of surface deposits and correlative formation of endoreic depressions, due to the post-glacial melting of mixed ice and rock accumulations (e.g. New England kettles);
- continuation of coastal delta subsidence effects after the sedimentary supply has been cut due to the upstream capture of a river course, the formation of a dam, or vegetation renewal;
- decrease in the internal volume of geological formations because of a particle rearrangement caused by earthquake vibrations;
- organic-matter oxidation through the action of running and seeping water, leading to a material loss and densification (e.g. peat, plant accumulation);
- desiccation of clayey material responsible for physical retraction and cracks formation.

Anthropogenic causes

(a) *Water and oil or natural gas extraction* represents the most common and important subsidence factor due to human activities (Table 7). The reduction of fluid pressure in water and hydrocarbon reservoirs favours the increase in interparticle pressure and induces compacting effects (Chapter 6.4.2). The risks determined by both types of fluid extraction have considerably risen during the 20th century, first subsidence movements having been identified above some Texas oil fields in 1925. This is of course due to the exponential demographic augmentation and to the always-higher industrial demand (see Introduction). Differences between the two types of risks are related to the usually shallower location of aquifer reservoirs (ground pressure lower than 13 atmospheres) relative to that of hydrocarbon fields (pressure reaching up to 275 atmospheres). In addition underground aquifers are generally more porous and involve larger ground surfaces than oilfields. As a consequence the surface *subsidence above oilfields is commonly stronger and of a lesser extent than above over-exploited aquifers.* Notice that compaction effects resulting from water extraction are especially important for *peat* formations, which may contain up to ten times more water than their own weight and may be compressed by up to 75% of their initial volume (Fig. 47). The compaction of soft sediments and soils is also potentially very high in other sedimentary *wet zones* (e.g. exposed areas of estuary and delta environments), where active surface drainage may induce the partial removal of the interstitial liquid phase.

(b) *The extraction of mining resources* may be responsible for various subsidence effects, the cause and importance of which depend on working

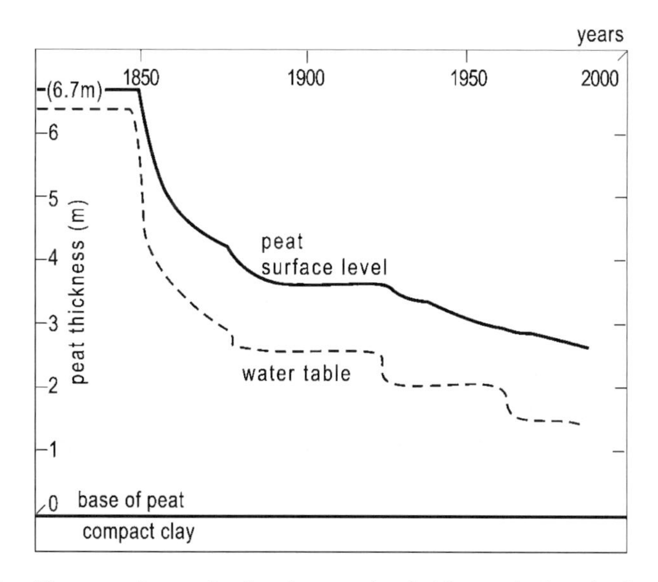

Figure 47: Aquifer pumping and related ground subsidence during the last 150 years in Holme Post peat formations, Fenlands, Great Britain (after Waltham, 1994).

depth, mine extent, nature of superimposed geological formations, extraction modes, hydrogeological conditions, etc. This question, which primarily concerns coal mines (e.g. 800,000 ha of the USA surface are affected by coal-mining subsidence), is considered in Chapter 5, which is devoted to ground mineral resources.

(c) *Hydrocompaction* results from the impregnation by irrigation drainage water of little consolidated and light surface sediments, which were initially lacking interstitial water and deposited under arid conditions. The eruption of water within interparticle voids induces the release of internal tensions, resulting in total collapse. Hydrocompaction effects concern mainly aeolian sediments (e.g. loess), as well as some alluvia deposited well above the groundwater level by temporary watercourses. Resulting damage may be widespread and of large extent, as evidenced by various dry regions in the USA and former Soviet Union: deterioration of irrigation networks, buildings and transport links.

(d) *Overloading of little-compacted terrains* by heavy buildings or other constructions determines local subsidence. The "Leaning Tower" of Pisa (Italy) started to lean over soon after its 1174 building on ancient under-compacted sediments of the Arno River. The movement increased during the 1960s due to excessive groundwater extraction from alluvial deposits, and was finally stopped by cement injections around the monument foundations. La Guardia airport (New York City) has undergone a 2-m subsidence in less than a quarter of a century, because of extensive artificial overloading.

Mixed causes *Numerous subsidence movements result from the combination of natural and anthropogenic factors.* This is especially the case for conurbations located in coastal zones (Table 8), where tectonic subsidence may be combined with the effects of a rise in sea level, sediment under-compaction, underground water extraction, and building overloading:

(a) *Venice* merits a case study. Built on marsh islands separated from the Adriatic coast by a lagoon, the town has always been vulnerable to storm waves associated with heavy rain and high tide, inducing temporary flooding (*aqua alta*). The proportion of flood periods started to increase considerably in the 1920s, following both the development of groundwater extraction and a rise in the level of the Adriatic Sea (estimate of 1.27 mm/yr between 1896 and 1967). The subsidence rate increased tenfold during the 1950s (Fig. 48), due to excessive groundwater pumping for meeting the industrial needs. The pumping was interrupted in 1969, allowing the subsidence to stop and the well hydrostatic pressure to be partly restored. A

Table 8: Examples of coastal and inland subsiding big cities (after Murck et al., 1996).

Town	Maximum subsidence (m)	Surface (km^2)
Coastal regions		
Bangkok, Thailand	1	800
Houston, USA	2.7	12,100
London, UK	0.3	295
Long Beach, USA	9	50
Nilgata, Japan	2.5	8,300
New Orleans, USA	2	175
Shanghai, China	2.6	121
Taipei, Taiwan	1.9	130
Tokyo, Japan	4.5	3,000
Venice, Italy	0.25	150
Inland regions		
Baton Rouge, USA	0.3	650
Denver, USA	0.3	320
Mexico City, Mexico	8.5	225

"rebound" was even observed in some sectors, but did not extend since the over-compacted ground formations had lost their elasticity properties.

(b) *London* has, during the past century, undergone a 30-cm isostatic subsidence combined with significant compaction induced by over-exploitation of the Cretaceous chalk aquifer. As a result the tide has undergone a 60-cm rise along the lower course of the Thames River, and numerous subway tracks and stations have become liable to flooding. This has made it necessary to build a tidal barrier across the river, downstream from the London conurbation.

(c) The *New Orleans* region close to the Mississippi river delta and mouth is subject to a natural subsidence of 12–24 cm/century, which is due to isostatic bending and sediment compaction. Man-induced subsidence mainly results from aquifer pumping and locally attains 60 cm/century. In addition extensive developments of dykes and channels all along the river course have determined quicker downstream flows and therefore caused most suspended matter to settle offshore in the Gulf of Mexico rather than in the vicinity of New Orleans; this diminishes the local sedimentation rate. As a consequence, 45% of the conurbation is now situated below or just at sea level, which necessitates important heightening and protection work.

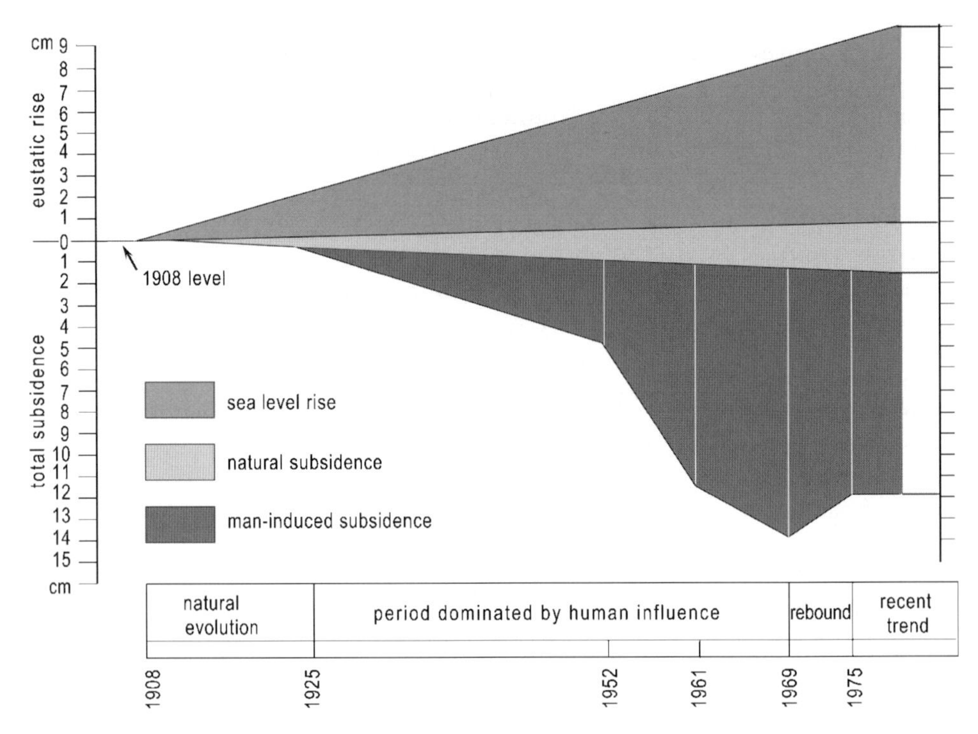

Figure 48: Rise in sea level and ground subsidence of Venice lagoon (after Gatto & Carbognin in Lumsden, 1992).

Forecasting, Prevention

Still insufficiently documented, subsidence forecasting is based on the precise *mapping* of vulnerable zones, as well as on the characterization of ground-level *physical and hydrological properties*. When subsidence is due to artificial fluid extraction, computer simulations help to anticipate and to some extent control the forthcoming evolution.

The prevention and protection against subsidence effects have generally reached a satisfactory degree of efficiency. Several *measures* may be taken, according to local ground characteristics:

- diminution of the fluid amounts extracted from underground rocks, and reservoir refill for preventing both compaction and loss of natural elasticity;
- artificial compaction in order to suppress potentially dangerous hydro-compaction or overloading effects;
- civil engineering developments for stabilizing the building foundations, communication lines, pipelines and cables;
- building localization and design taking into account the local subsidence risks.

3.3. Physicochemical Change of Surface Formations

Here are presented together the land movements not dominantly controlled by gravity processes and those which mainly result from physicochemical changes affecting outcropping formations. These movements are generally restricted to a few metres thickness at the lithosphere–atmosphere interface, and therefore pose few risks to human life. They may nevertheless determine important damage to human property, especially through secondary effects controlled by gravity (deformation, destabilization, collapse).

3.3.1. Swelling of Clayey Soils

Dark grey soils called *vertisoils*, that are characterized by *important volume changes related to seasonal humidity variations*, form in numerous arid climate regions: South Mediterranean and Sahel countries, southern Russia, Gulf of Mexico peripheral countries, South Africa, Australia, etc. (Fig. 49). During the dominant *dry season*, which lasts 7 to 10 months, active evaporation processes

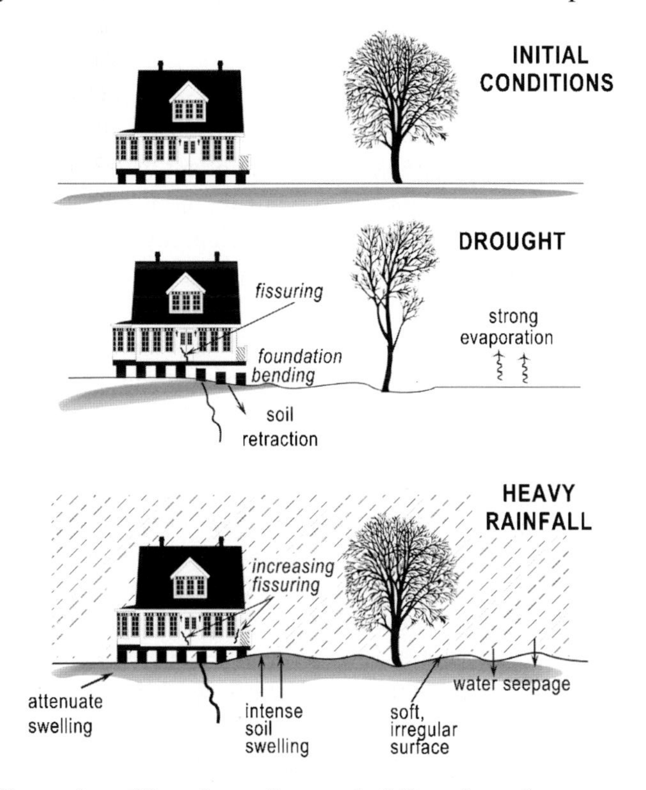

Figure 49: Effects of swelling clay soils on a building, from data commonly observed in Australia under subarid climate (after Bryant, 1991).

cause the soils to shrink and crack, more rapidly in open air and in forested areas than beneath houses and other human developments. As a result the building foundations tend to bend and fissure, and the constructions tend to be destabilized. During the short but violent *wet season*, water impregnates the soil and subsoil, preferentially in the sectors directly exposed to rain. The ground becomes plastic, unstable and distorted down to a depth of 6 m, exacerbating the damage to constructions. The seasonal repetition of such alternating drying-wetting processes causes the *deterioration of surface constructions*. The swelling pressure of the water-impregnated soil clay may reach 0.2 to 0.7 t/m^2. This natural hazard may be amplified by human activity through periodical irrigation of swelling soils or through modification of natural drainage conditions.

Similar wetting-drying cycles are observed in less arid regions, where *clay-rich rocks* of various geological age may be affected. This applies to various Cretaceous and Early Tertiary claystones from Western Europe and North America, which are susceptible to swell to more than 1.5 m thick. The resulting damage to property may be considerable. In the USA the annual repairs to buildings and other constructions attributed to swelling soil nuisance are estimated to amount to several billion dollars.

Soil or sediment swelling is principally determined by the presence in the clay fraction (<2 μm) of *abundant minerals of the smectite family*, the interlayer spaces of which are able to retain or release abundant water. This is the case for montmorillonite, beidellite, saponite, stevensite, nontronite, smectite-bearing mixed-layered minerals and bentonite (see Chamley, 1989; Parker & Rae, 1998). The water adsorption during the rainy season may mulitply the clay volume by two or even ten; its evaporation during the dry season causes the clay to retract, harden and fissure. The swelling capacity increases if clay particles are arranged in a disorganized network (*flocculated clay*; Chapter 3.3.2), the abundant voids of which are able to store water during the wet periods. The possible heterogeneous distribution of clay mineral species within soil aggregates on a microscopic scale may be of importance, namely if smectite particles are located inside the network voids and exert a strong internal pressure when swelling (i.e. "hydraulic jack" behaviour). The nature of interlayer ions compensating the cationic charge deficit of most clay minerals may also intervene; for instance, sodium smectites are much more unstable than calcium smectites. In addition,the potential for clay to swell increases when the material density is higher. Finally the stability hazard is stronger if buildings have shallow foundations, which are particularly sensitive to physicochemical changes affecting the clay-rich surface formations.

Protection against swelling-soil effects requires a clear delineation of the regional sensitivity of surface formations (e.g. Fig. 50). Data on *physical*

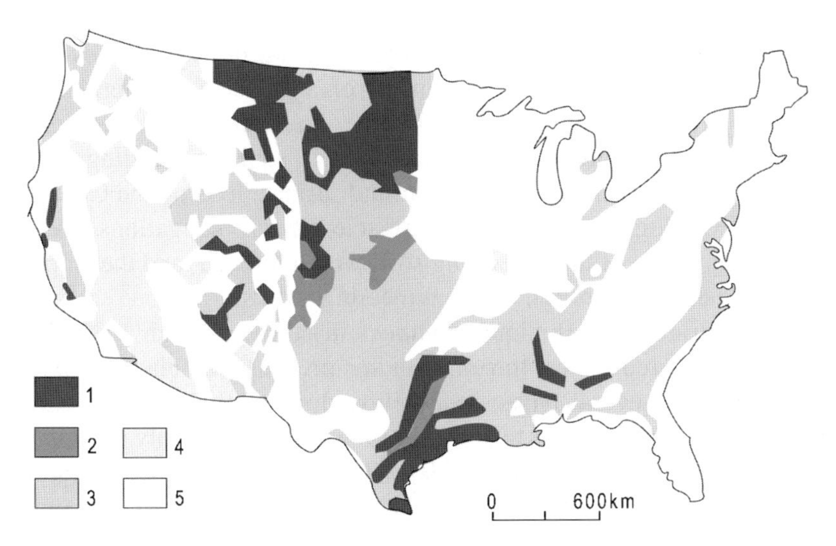

Figure 50: Swelling clay abundance in conterminous USA. 1, very high. 2, high. 3, moderate. 4, low. 5, very low to nil (after Schuster et al. in Flageollet, 1989).

properties help to characterize and quantify the hazards, namely through petrographic, mineralogical and mechanical analyses: abundance of the clay fraction, nature and estimated abundance of the different clay species and other minerals likely to swell (sodium and calcium sulphides, sulphates, etc.); microstructure observed under the scanning electron microscope; values of Atterberg plasticity and liquidity limits; swelling pressure values and other geotechnical trials. *Practical remediation* comprises various measures: removal of the swelling soils if they form a thin layer at the ground surface; water drainage; cationic exchange by replacing smectite and bentonite interlayered sodium by calcium (Chapter 3.3.2); application of heavy loads compensating the loss of swelling pressure; foundation extension deep in the ground for securing the buildings within unaffected subsoil. Generally, *the forecasting of swelling-soil hazards is quite satisfactory, and prevention tools are fairly efficient.* Unfortunately the implementation of protective measures is often poorly anticipated, costly and made as things arise.

3.3.2. Liquefaction and Textural Changes

In 1978, a peasant from Rissa, Norway, decided to build a barn close to a small lake bordering his property. By digging foundations and compacting the clayey soil, he suddenly caused the formation of a little landslide. The slide extended very rapidly and propagated upstream in the valley feeding the lake, to a distance exceeding 1 km. *The formerly hard soil instantly liquefied* over an area extending 300,000 m^2. About 6 million m^3 of argillaceous material began to move. Some small buildings carried by the fluidified mass were displaced downwards in a practically intact state, at a speed exceeding 20 km/hr. Considerable disruption resulted at the ground surface, leading the Geotechnical Institute and Geological Survey to start scientific investigations and assess further hazards. Some of the remaining soils were removed, and the slopes were restored and stabilized through civil engineering works.

Ground materials able to liquefy almost instantly through several metres thickness are referred to as the *quick clay* or sensitive clay *group*. These materials comprise several types: *quick clays* dominantly comprise clay-sized particles (< 2 μm), quick sands are coarser (> 63 μm) and mainly encountered in coastal intertidal zones, and *collapsible soils* are of an intermediate grain-size (loess, aeolian dust). *Hazards* linked to the quick clay behaviour are sometimes considerable, especially in *Northern Hemisphere regions subject to a temperate-cold climate* such as: Canada, Scandinavia, and Russia. In Norway some disastrous liquefaction events developed in 1345 at Gualdalen (500 deaths), and in 1893 at Verdal (112 deaths; area disrupted over 3 million m^2, displacement of 55 million m^3). Numerous, less dramatic but comparable movements occur commonly in these regions, where clayey deposits may be activated on slopes less than 1°. Damage to human property is often considerable: building destruction, disorganization of communication networks, and obstruction of river courses, for example.

The *mechanism* responsible for the sudden liquefaction of 'sensitive' clays and soils corresponds to an *instantaneous loss of shearing strength*. The change of the material texture is principally triggered by natural factors: attenuated seismic vibrations, earthquakes, volcanic eruptions, exceptional flooding. Sometimes it results from man-made actions such as: reshaping of natural slopes, vibrations due to mining work, heavy vehicle displacement, and digging of foundations, as at Rissa.

To understand the liquefaction mechanism it is necessary to refer to the concept of *flocculation – deflocculation*, which is characteristic of the clay mineral group. Made more or less of hydrated alumino-silicate layers, the clay minerals are frequently characterized by a deficit of cationic (i.e. positive) charges, due to the substitution in the mineral framework of tetravalent by trivalent cations (e.g. Si^{4+} by Al^{3+}) or of trivalent by divalent cations (e.g. Al^{3+} by Mg^{2+} or Fe^{2+}). The importance of the cationic deficit depends on the clay mineral family considered: nil or very low for kaolinite, moderate for illite and chlorite, and high for vermiculite, smectite and other swelling minerals as well as for fibrous clays (palygorskite, sepiolite). The deficit of positive charges in clay layers tends to be compensated by cations (K^+, Na^+, Ca^{2+}, Mg^{2+}) that are

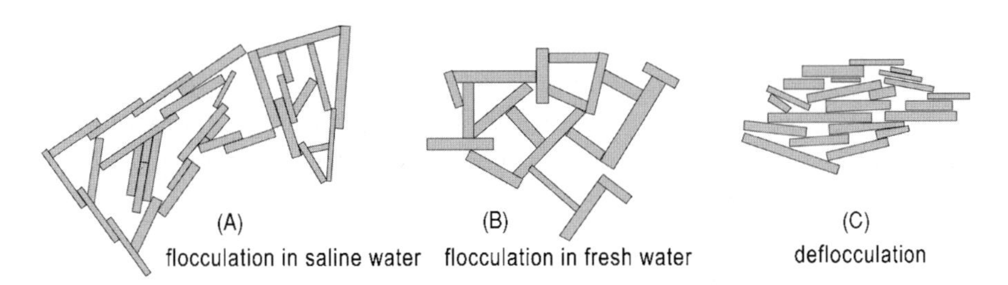

Figure 51: Sketch showing flocculation (A, B) and deflocculation (C) settling processes of clay mineral particles (after Costa & Baker, 1981).

adsorbed within the spaces surrounding and separating the sheets. The settling of clay particles in a sedimentary environment rich in dissolved cations (e.g. sea water) determines active electrochemical connections to develop, both in interlayer spaces and at layer edges; the particles get strongly connected in a disorganized way, which characterizes the *flocculation* state (Fig. 51, A); flocculation therefore corresponds to a type of coagulation that ensures a certain mechanical stability.

In fresh water the cation availability is low and results in looser flocculation, ionic bridges forming predominantly at the layer edge (Fig. 51, B).

In aqueous environments almost devoid of cations or even rich in anions (SO_4^-, NO_2^-), clay particles repel each other and tend to be deposited flat, without electrochemical connections (Fig. 51, C); this corresponds to *deflocculation*, a sort of peptisation potentially leading to sheet-on-sheet sliding. Such clay-rich sediments settling in repulsive anionic environments may have formed under various climatic conditions.

Understanding flocculation and deflocculation mechanisms in clayey materials helps to explain the sudden liquefaction process typical of the quick clay group. The most commonly quoted cause is the *deflocculation of initially flocculated clay mineral-rich sediments*. For instance, the intense land erosion that developed during and after the last glacial maximum stage (from about 20,000 to 10,000 yr BP) caused abundant clayey material to settle in the coastal marine environments bordering Canada, Scandinavia and Russia. The cation-rich seawater (Ca, K, Na) induced clay to actively flocculate; open particle networks, therefore, formed and were characterized by a balanced ionic composition. The post-glacial icecap melting determined the progressive isostatic uplift (Chapter 3.2.2) of North Hemisphere landmasses, leaving former marine coastal deposits exposed in subaerial conditions. Fresh water from repeated rainfall and soil surface or aquifer seepage (Fig. 52) progressively caused the removal of interparticle and interlayer cations. This induced the shearing strength to progressively decrease, and led clay materials to become

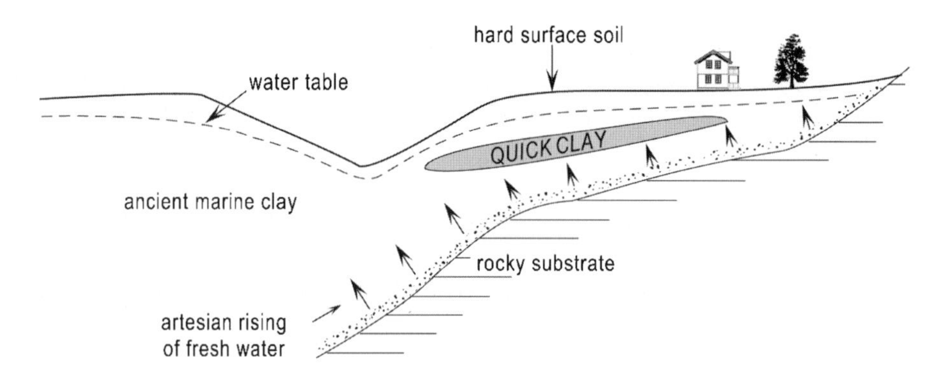

Figure 52: Example of former marine clay leaching through underground freshwater seepage, and resulting formation of quick clay (after Lumsden, 1992).

highly vulnerable from a mechanical point of view. Slight shocks or vibrations in water-soaked conditions were therefore sufficient for triggering instantaneous liquefaction and movements on very slight slopes (i.e. "house of cards" behaviour). This is what happened in 1978 in the Rissa accident in Norway, and many high-latitude regions are still under threat of similar movements.

Another explanation is sometimes provided, which takes into account the frequent rareness of clay mineral components in various types of high-latitude quick clays. This is especially the case of "rock flow" that settled during cold Quaternary stages in Scandinavian and North American glacial lakes. The lacustrine sediments are actually of a predominantly clayey size (i.e. $< 2\ \mu m$), but mainly comprise very small quartz, feldspar and mica particles, which lack electrochemical properties typical of true clay. Such minerals cannot have undergone in glacial–postglacial time the successive formation (i.e. flocculation) and destruction (i.e. deflocculation) of ionic bridges. The liquefaction is then attributed to a sudden modification of the Van der Waal's forces, which ensure a weak interparticle cohesion due to the high ratio existing between the bond energy and the very small-sized particle mass. Slight shocks or vibrations may cause the *breaking of Van der Waal's bonds*, triggering the immediate decohesion of sediment and its liquefaction.

The *forecasting* of ground liquefaction is based on laboratory analyses of the textural, physical and chemical properties of soils and sediments from the areas concerned, which are mainly located in present-day peri-glacial regions. Geotechnical tests help to anticipate the material reactivity (shear- and shock-resistance, plasticity and liquidity limits), but are rarely very discriminating because of the prominent electrochemical control. Precise mapping of hazard zones indirectly contributes to avoid misplaced liquefaction events, by leading to the implementation of perimeters where building is not permitted. As economic needs often impede such measures, various *preventive* earthwork and civil engineering developments are conducted on dangerous sectors: removal or formwork of local quick clay accumulations, water drainage and diversion, etc. An electrochemical stabilization is sometimes employed, which consists of percolating calcium-saturated water within deflocculated argillaceous terrains. The *in situ* exchange of Na by Ca induces the clay to flocculate. To summarize, the prevention of sudden liquefaction events is feasible in most cases, but remains difficult to implement, is always of punctual application and very expensive.

3.3.3. Frozen Soils Movements

A 1,300-m-long and 50-m-wide aeroplane runway was built during the 1950s at Sachs Harbour on Banks Island, Canadian North-West Territories. As the frozen ground locally displayed an uneven surface morphology, it was decided to level the soil, with sand and gravel being extracted from pits dug around the runway. The excavated zone extended over 5 ha to a depth of 2 m. In 1962 some *subsidence and undulation movements* started to form in the excavated areas, leading to the formation of several-metre-thick and -metre-wide highs and lows. The ground distortion continued to develop for one decade and was attributed to *alternating melting and freezing stages affecting the frozen soils*, which had been artificially exposed to surface temperature changes. The use of the runway became progressively dangerous, and its repair or extension was abandoned. Similar deformations of frozen soils are reported in many high latitude regions, where they may be responsible for considerable damage. For instance, in the mining zone of Vorkouta in Russia, 130 buildings among 165 deteriorated or were destroyed between 1948 and 1950, due to ground thawing as a result of domestic heating.

Soil movements in cold regions are related to the existence over almost a fifth of world landmasses of deeply frozen formations referred to as the *permafrost*. Permanently frozen soils and subsoils are widespread in the North Hemisphere (22.4 millions km^2), where they cover 80% of Alaska surface, 50% of Canada, 47% of Russia, 22% of China . . . (Fig. 53). Permafrost is continuous at the highest latitudes where its thickness averages 600 m and locally reaches 1,500 m (Siberia). It becomes discontinuous in peripheral areas, where it is generally less than 60 m thick and where it is responsible for most soil movements.

Permafrost is naturally subject to partial surface melting during the summer season. The thickness of the melted layer averages a few centimetres in coldest zones, but may reach 4 m in peripheral areas. Such seasonal melting alternates

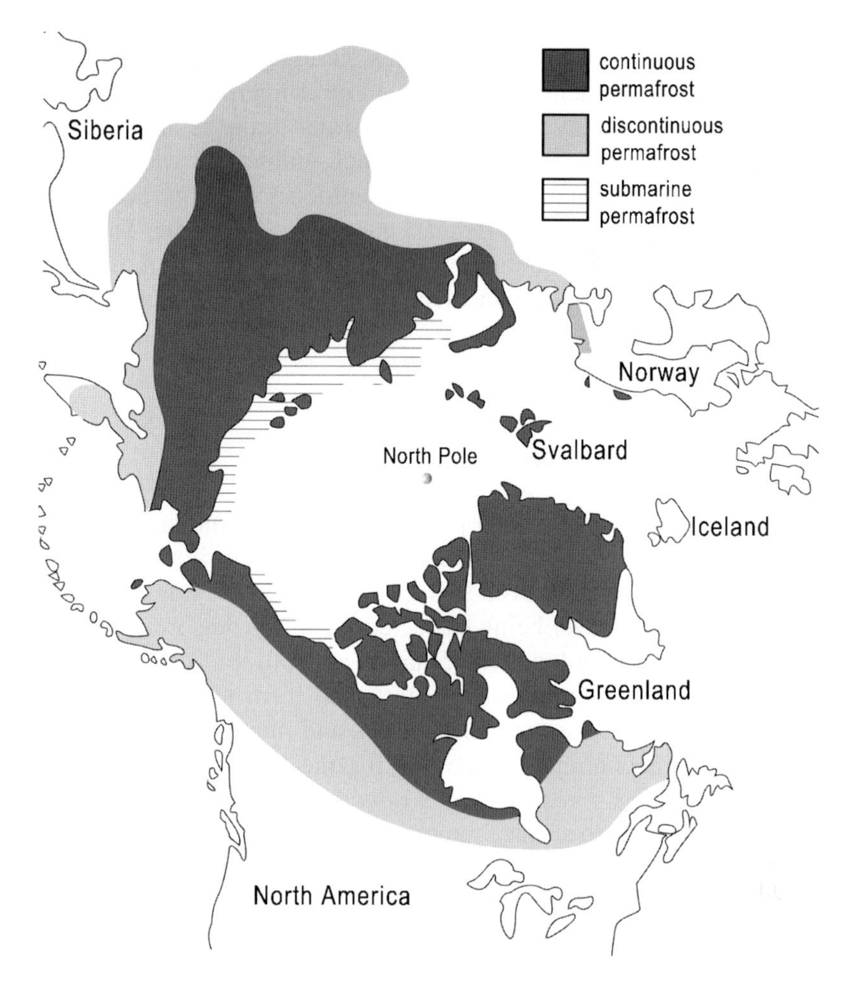

Figure 53: Distribution of permafrost in the North Hemisphere (after Perrins et al. in Bennett & Doyle, 1997).

with winter freezing for determining numerous compression and deformation effects called *thermokarsts*. The resulting damage may be important in the deterioration and destruction of roads, railway tracks, aeroplane runways, pipelines, etc. Various *anthropogenic factors* may amplify these effects: digging of excavations inducing misplaced seasonal melting (e.g. Sachs Harbour); deforestation and grass removal leading to elimination of the natural solar screen at the ground surface, and to artificially increasing the thickness of the seasonally destabilized permafrost layer (Fig. 54); ground heating for domestic, industrial and urban needs propagating heat in subsoil, and destabilising foundations; building overloading on the periodically softened and therefore distorted ground; etc.

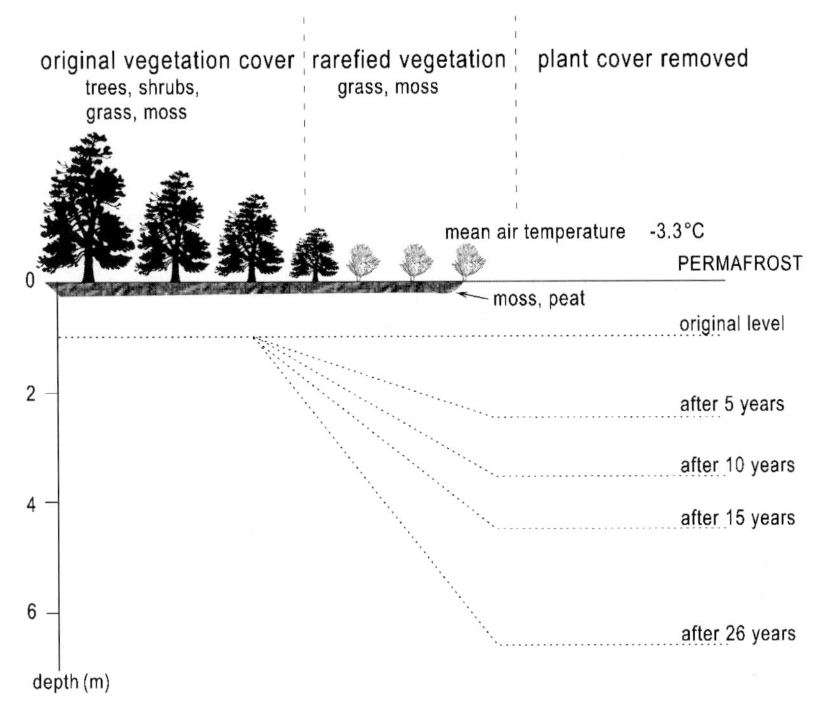

Figure 54: Influence of the plant-cover removal on the layer of melting permafrost in the Fairbanks region, Alaska (after Cooke & Doornkamp, 1990).

A possible planetary warming, which is expected to be particularly intense at high latitudes (Chapter 11), would have serious environmental consequences: widespread development of thermokarsts, increased rainfall amplifying the ice-melting, increased greenhouse effects due to methane degassing from submarine permafrost melting, etc.

Simple measures help to *prevent or limit the permafrost melting*: reforestation and returfing, special thermal isolation around the lowest floors of buildings, etc. Prevention includes the usual methods of mapping, geotechnical tests, earthwork, formwork, etc. Constructions are made of low and broad buildings with deep foundations. Civil engineering developments appropriate to permafrost regions have recently undergone big progress, because of the importance of economic issues raised by high-latitude ground resources.

A Few Landmarks, Perspectives

- Land movements at ground level and in subaquatic environments are triggered by three main types of processes: along slope displacements, vertical collapse or subsidence, and physicochemical changes. The influence of gravity is fundamental for the two former processes, accessory and working as a relay for the latter. Partially superimposed and transitory stages characterize most of these movements, leading to complex classifications.

- Land movements are widespread in Earth's surface environments. They display a broad diversity of mechanisms, materials involved, displacement rates, and hazards. The danger is potentially maximum for very fast and long-distance displacements such as debris avalanches or cliff collapse, and minimum for local and slow movements such as creep or subsidence. Human casualties are usually low or moderate, except for some huge slope movements affecting populated areas. The damage to goods is often considerable, especially when triggering factors are of an internal geodynamical origin (earthquakes, volcanic eruptions) and/or affect steep, densely urbanized regions.
- Slope movements include falling, sliding and flowing phenomena. Extending from local rock fall and solifluction to wide landslides and mudflows, they are favoured by steep gradient, water seepage, rock fracturing and lithological heterogeneity. They tend to be amplified in an increasing way by human actions, which comprise slope reshaping, raising of the artificial water table, ground overloading and deforestation.
- Ground collapse and subsidence constitute local movements that may be widely distributed, mainly in the regions subject to underground dissolution of carbonate or saline rocks (karstification), ice melting, permeable soil saturation, excessive loading, mining, and especially oil field and aquifer overpumping. These movements depend on both natural and human causes, and are responsible for numerous morphological changes on the land and sometimes on the bottom surface of the ocean.
- Some surface, little-compacted soils and sediments display diverse, sometimes imperfectly understood state changes: clay swelling, electrochemical liquefaction (quick clays), sudden collapse of deflocculated material, destabilization through ionic exchange, and permafrost seasonal thaw. These various phenomena result from natural physicochemical causes, which tend to be exacerbated in an increasing way by human activities: irrigation, drainage, deforestation, overloading, artificial vibrations, and ground heating.
- Water, which is principally responsible on the Earth's surface for both chemical weathering and physical or physicochemical dissociation, plays an essential role with gravity in triggering land movements. Another important factor consists in the abundance and mineralogical nature of the soil and sediment clay fraction, the influence of which controls most ground instabilities: wetting-drying alternations, interparticle and interlayer swelling, compressibility and permeability resistance, soil fluidification, and deflocculation.

– Traditionally controlled by natural factors, the land movements tend to depend more and more on human factors, a fact directly correlated to the recent and still-ongoing demographic explosion. The forecasting of the movement types and localization is roughly satisfactory, but remains highly aleatory as far as triggering moments are concerned. Many efficient prevention measures have been developed during the last decades, but it remains impossible to implement them on a wide scale. Most mitigation developments are restricted to densely populated and actually threatened areas.
– A significant reduction of natural hazards induced by the widespread and very diverse land movements would imply the human pressure to be decreased in some especially sensitive regions of the planet. As such an aim is presently impossible to achieve, the only way to progress consists in improving forecasting, prevention and mitigation methods. Several developments are being considered: improvement of simulations and probabilistic modelling reliability; codification of hazard levels and adequate people education; building and development techniques more appropriate to field characteristics and danger.

Further Reading

Bell F. G., 1999. Geological hazards : their assessment, avoidance and mitigation. Spon, London, 648 p.
Casale R., & Margotini C. ed., 1999. Floods and landslides. Springer, 373 p.
Cooke R. U. & Doornkamp J. C., 1990. Geomorphology in environmental management. Oxford University Press, 2nd ed., 410 p.
Flageollet J.-C., 1989. Les mouvements de terrain et leur prévention. Masson, Paris, 224 p.
Parker A., Rae J. E. ed., 1998 Environmental interactions of clays. Springer, 271 p.
Singh V. P. ed., 1996. Hydrology of disasters. Kluwer, Dordrecht, 442 p.
Voight B. ed., 1978. Rockslides and avalanches. Elsevier, Developments in Geotechnical Engineering, 14A & 14B, 833 & 850 p.

Some Websites

http://www.brgm.fr: updated review of actions conducted in France and some other countries in the natural hazards field
http://www.ngdc.noaa.gov/: past and recent information on the main types of natural hazards, established by the USA National Geophysical Data Center at NOAA (National Oceanic and Atmospheric Administration)
http://geohazards.cr.usgs.gov/: information on all main types of natural hazards, especially those developing in North America

Chapter 4

Wind and Water Hazards

4.1. Flooding

4.1.1. Summer 1993, Major Flooding of the Mississippi River

The drainage basin of the Mississippi river extends from the southern border of Canada over a large part of North America (Fig. 55). It covers more than 40% of USA conterminous states. Due to the irregular character of rainfall regime and snow seasonal melting, the river and its tributaries (Arkansas, Missouri, Ohio rivers, etc.) have for long been subjected to important civil engineering work such as water confinement, canalisation, and diversion. The first semester of the calendar year 1993 was marked by very heavy precipitation, especially in the upstream basin located to the northwest between the Missouri and Illinois

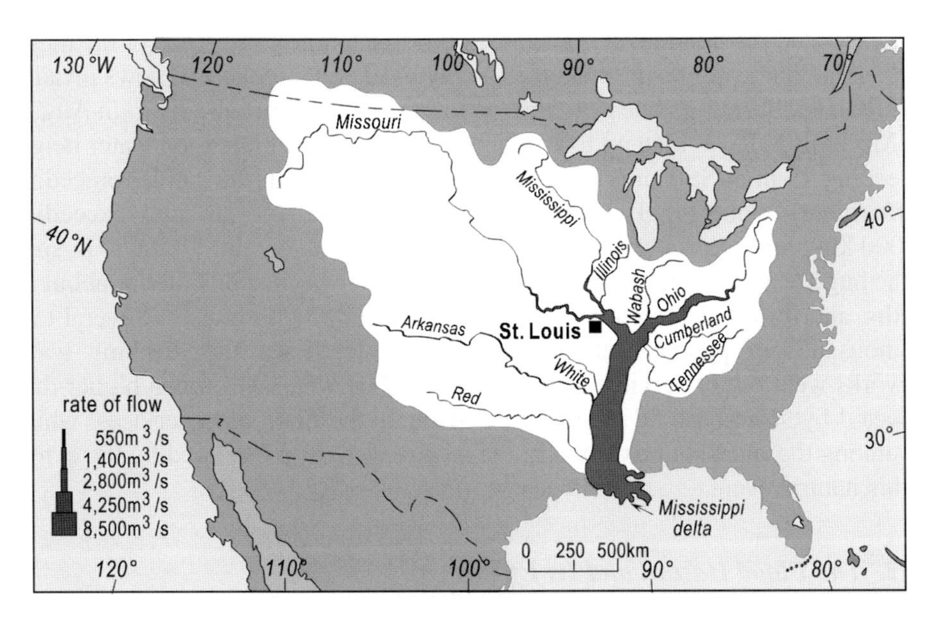

Figure 55: The Mississippi River basin and mean water outflow of its main entities (after Murck et al., 1996).

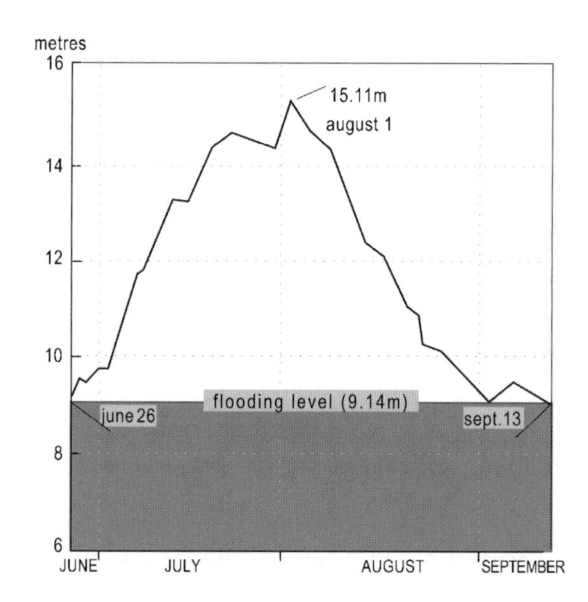

Figure 56: Variations of the mean flow rate of the Mississippi river at St. Louis, Missouri, during the major flood of summer 1993 (after Murck et al., 1996).

rivers. The amount of rainfall in North Dakota, Iowa, Kansas, etc. between January and June 1993 was twice as high as the 6-month average rainfall during the 1961–1990 period. On the 26th of June, the Mississippi River reached and then exceeded the flood level in the St. Louis area, at the confluence of the major tributaries. Water went over the dykes, many dyke walls broke and flows invaded the urban and farming zones situated on the major course of the river. In August the Mississippi River rose up to 6 m above the flood level, the total water height exceeding 15 m at St. Louis (Fig. 56), i.e. 2 m more than during the preceding extraordinary flood of 1973. The flood extended over an area exceeding 40,000 km^2, and the water level remained above the dykes for 79 days. In spite of the huge rescue resources deployed, the flooding was responsible for about 50 deaths, and for the destruction or major damage of 55,000 houses. Several tens of thousands of people were displaced; the electricity and drinking water networks went out of service. The level reached by water was much higher than expected by scientists and engineers. Obviously the flood-alert services, which are among the most sophisticated in the world, were overwhelmed by the extent of this natural event.

4.1.2. The Flood Hazard and Its Effects

Flood-induced damage is often very serious. The floods of February and March 2000 of the Zambezi and other rivers in Mozambique (southeast Africa) caused the death of 700 people and disrupted one-third of the landscape. Six months after the disaster, 250,000 inhabitants were still unsheltered and 650,000 lived in makeshift encampments. The gross national product fell down to half of its annual average value.

The countries of south and southeast continental Asia, characterized by very flat relief and submitted to intense monsoon precipitation, are particularly exposed to disastrous flooding. The number of deaths due to floods in China during one century (i.e. between 1860 and 1960) is estimated at 5 million. The 1887 Huang He River flooding caused 900,000 victims; the 1911 and 1931 Yangtze flooding determined, respectively, 100,000 and 200,000 victims. The countries situated on downstream flood plains are particularly threatened, especially in densely populated regions where the standard of living is low and where prevention and protection measures are very limited. *Bangladesh*, a country located at the confluence of three very large river basins (the Ganges, Brahmaputra and Megna) and exposed to Indian Ocean tropical storms (Chapter 4.3.2), is particularly subject to widespread flooding. Most of the 143,000 km^2 forming the country constitute a huge deltaic region, where 100 million people live. The annual floods commonly cover about half of Bangladesh (Fig. 57), causing considerable damage to inhabitants and property. In Vietnam the alluvial valleys and deltaic regions, which are densely populated and devoted to traditional farming (rice cultivation), present a flooding hazard for more than 70% of the 71 million inhabitants of the country.

The developed countries characterized by a high degree of technology and moderate demographic pressure are certainly not exempt from flooding hazards. In the USA almost two-thirds of natural hazard-induced damages registered between 1965 and 1985 resulted from flooding events. About 70% of New Zealand towns of more than 20,000 inhabitants are vulnerable to floods.

In France, river flooding constitutes the most common and widespread risk, and almost 80% of financial compensation due to natural disasters is linked to water damage. All regions are more or less concerned; most important damage occur in mountainous areas and in some poorly drained alluvial valleys (Fig. 58). In early spring 2001, exceptionally serious flooding affected the northwestern part of the country (Brittany, Normandy, Picardy) and persisted during several months due to the conjunction of very heavy rainfall, aquifer saturation and insufficient protection systems. In southern France, the Tarn River flood in 1930 caused 200 deaths and 10,000 injuries. Devastating flooding periodically affect the plains bordering the mountains in the regions submitted to a Mediterranean climate (e.g. Nîmes, 1988; Vaison-la-Romaine, 1992; Aude valley, 1999; Sommières, 2002).

Flooding of river plains should represent an expected hazard. It is in the nature of things that after long and heavy precipitation water extends in the

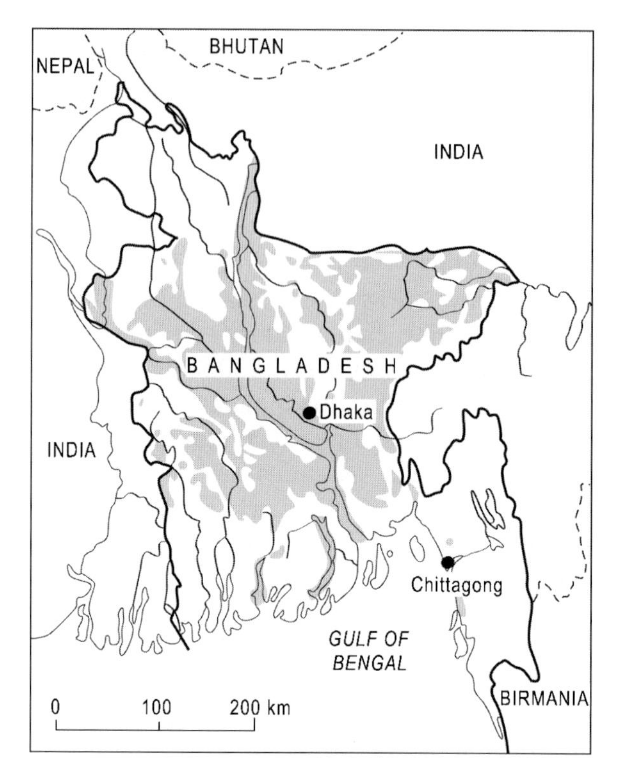

Figure 57: Map of Bangladesh showing the Ganges–Brahmaputra–Megna hydrographical system (after Smith, 1992). The shading represents the surface seasonally inundated under more than 90 cm of river water.

major stream and river courses that have been annexed by man for his activities such as communications, farming, cities, and industrial development. The damages induced by seasonal or occasional overflowing result from common and quite normal phenomena, which should be anticipated in a much more banal way than earthquakes, volcanic eruptions, or major rock collapses and landslides. The disasters due to river flooding basically result from the inappropriate occupancy by people of the ordinary routes used by nature for surface water drainage and ground water emergence.

Inundations have some beneficial consequences for people that are usually of great value. The *water supply* allows refeeding of streams, rivers, lakes, artificial reservoirs, soils and underground aquiferous rocks, as well as sustaining of irrigation, fish farming, etc. In many alluvial plains, the farm yield is based on huge water needs, especially in developing countries where food mainly consists of rice and fish (for example, Asia).

The renewed provision of fine-grained *sedimentary material* (sand, silt, clay) as well as of organic matter, *nutrients and trace elements*, favours the *agricultural quality* of delta plains, including those from arid regions where large rivers flow (e.g. the Nile in Egypt, the Tigris and Euphrates in the Middle East, the Senegal River in West Africa). Inundated zones may give way to the proliferation of freshwater microscopic algae (Chlorophyceae, Cyanophyceae), which accumulate nitrogen and contribute to enhancing the soil's fertility; the algae-produced nitrogen may reach 30 kg/ha per year in some flooded areas.

The accumulation of sand in the lower part of some river courses leads to the building of *natural barriers*, which may act as natural protection systems against coastal storm effects (e.g. Bangladesh).

The alluvial valleys, the flat topography of which predisposes to agriculture and livestock farming, shipping and communications, industrial and other *man-made developments*, are crucial for maintaining the *biological diversity* characteristic of wet zones.

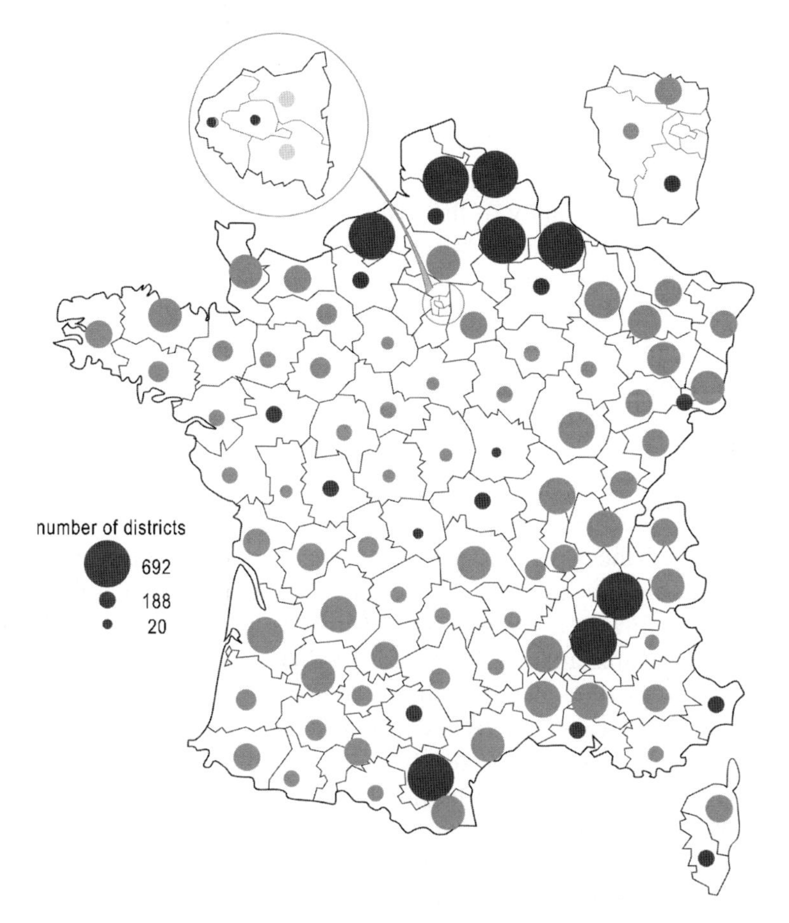

Figure 58: Distribution of French city groups officially recognized as disaster-stricken between 1984 and 1993 (after Ledoux, 1995).

Finally the plants and soils of river plains constitute efficient *natural filters* that retain polluting and toxic products. This helps to prevent the pollutants to be dispersed and therefore to both maintain the quality of downstream water resources and reduce the cost of wastewater processing.

4.1.3. Types, Mechanisms and Natural Control Factors

Two types of natural disaster may be induced by the rise of river waters: plain inundations and torrential floods.

Plain inundations develop in the middle to lower part of hydrographical basins, when heavy and continuous rain causes the rivers to overflow their usual banks (i.e. the minor course) and to occupy their major course that forms the *alluvial plain* or inundation plain. Floods generally occur after several days or weeks of uninterrupted rain. The number of human victims is sometimes considerable, as dramatically illustrated by major flooding having affected eastern and southeastern Asia in the past 150 years (Chapter 4.1.2). The water rise is progressive and develops with some delay after intense precipitation has begun, depending on the major rainfall location and amount in the upstream basin and on the physiography of the hydrographical network. The existence of large river basins submitted to heavy rain and converging at a same point determine a particularly rapid water rise, as shown in June 1993 by the Mississippi River at St. Louis (Chapter 4.1.1). The total amount of rain in a given river system is expressed by a *pluviogram*, which corresponds to the height of water from rain fallen during a given time interval. The amount of rain strictly involved in a given flood constitutes only a part of this total water amount and is called the *effective precipitation*. The starting point and progressive rise of the flood, the moment and intensity of the flooding maximum, as well as the characters of the falling-level phase, are expressed by a histogram called the *stream flow hydrogram* (Fig. 59). The duration of the interval separating the maximum rainfall in the upstream zone and the maximum rate of flow in a given downstream zone is critical for implementing the protection measures against flood effects (Chapter 4.1.4).

Torrential floods occur in mountainous regions due to fairly short but very violent rainfall. Rainwater converges in small, narrow and steep valleys, and occupies the stream courses, the levels of which rise rapidly and constitute a *flashflood*. For instance the torrential flood that affected in 1972 the Black Hills in South Dakota (USA) resulted from the accumulation of 400 mm of rainwater in less than 6 h. This amount, which represents almost the total annual precipitation in this basin, was responsible for 300 deaths, 3,000 people injured, 750 houses destroyed and 2,000 cars damaged. Torrent waters often carry abundant solid material eroded from surface soils and rocks (i.e. torrential

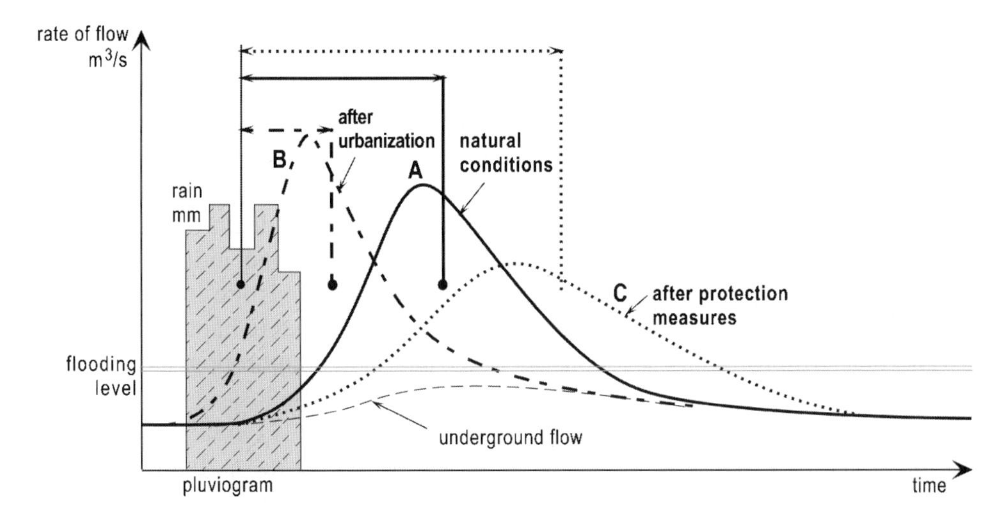

Figure 59: Stream flow showing the time necessary for river water to concentrate and rise after heavy rainfall in upstream zones (after Leopold; see Keller, 1999). Comparison of situations under natural conditions (A), after urbanization (B), and after the implementation of protection measures (C).

lavas), which accumulates at valley mouths and constitutes large sedimentary wedges.

Floods, which represent a natural phase of the water cycle, result from the heterogeneous distribution of rain location, intensity and duration in the time course. Always triggered by *heavy rainfall*, they may be amplified by some *other natural factors*:

- snow melting due to climate warming or to volcanic material rising inside magmatic chambers;
- overflowing or breaking of crater lakes;
- inability of the plant cover to ensure sufficient evaporation–transpiration processes and water seepage;
- soil and subsoil super-saturation due to the excess of previous rainfall;
- breaking of valley dams due to rock collapse, landslides, excessive sediment accumulation;
- sudden discharge of under-glacier lake water, induced by ice movements.

The flooding hazard is exacerbated in alluvial plain areas situated *close to flat coastal areas*, where the effect of rainfall on the water level may be combined with those of low atmospheric pressures, high tides and sea storms (Chapter 4.3).

4.1.4. Human Actions

Increase in the Flood Hazard

Inundations linked to artificial reservoirs are mostly determined by *dam breaking*, as was the case in south-eastern France when the Malpasset dam broke during the water-filling phase (1959) and caused the death of 423 people in the Fréjus area. Sometimes the man-built basins play an indirect role, as illustrated by the 1963 Vaiont disaster, where a major landslide leading to the lake spared the dam but produced a giant, devastating wave (Chapter 3.1.1).

The *soil waterproofing* by constructions and other anthropogenic surface developments noticeably aggravates the flooding risk. The loss of soil seepage properties is particularly important in *urban environments* (Table 9), where the various materials covering the streets, walkways, parking areas, buildings, etc. induce active surface streaming and accelerate the water rise (e.g. flashflood of Ouvèze River at Vaison-la-Romaine in 1992, or at Bal el-Oued in an Algiers suburb on November 10, 2001). Generally the stream flow hydrograms typical of rivers in urban areas are more rapid and intense as the urbanization degree becomes higher (Fig. 59, B). This is especially the case in megalopolises (Fig. 60), which comprise extensive impervious surface soils, as well as frequently clogged and overflowing pipe and channel networks. Urban developments may modify the sensitivity to floods of the different sectors of a given urbanized region (e.g. Los Angeles county, California). They may also induce an acceleration of the water-rise rhythm, for instance by the presence of civil engineering structures impeding the water evacuation (under-sized pipes, bridge piers, etc.).

In rural environments the increase in flood hazards is dominantly due to *deforestation* practices. Determined by direct or indirect causes (e.g. forest fire),

Table 9: Impermeability percentage of different types of urban surface materials (after Newson; see Smith & Ward, 1998).

Type of surface	Impermeability (%)
public and private gardens	5–25
unpaved paths, railway tracks, vacant sites	10–30
gravel roads	15–30
macadam roads	25–60
block paving with open joints	50–70
stone paving with cemented joints	75–85
asphalt	85–90
building roofs	70–95

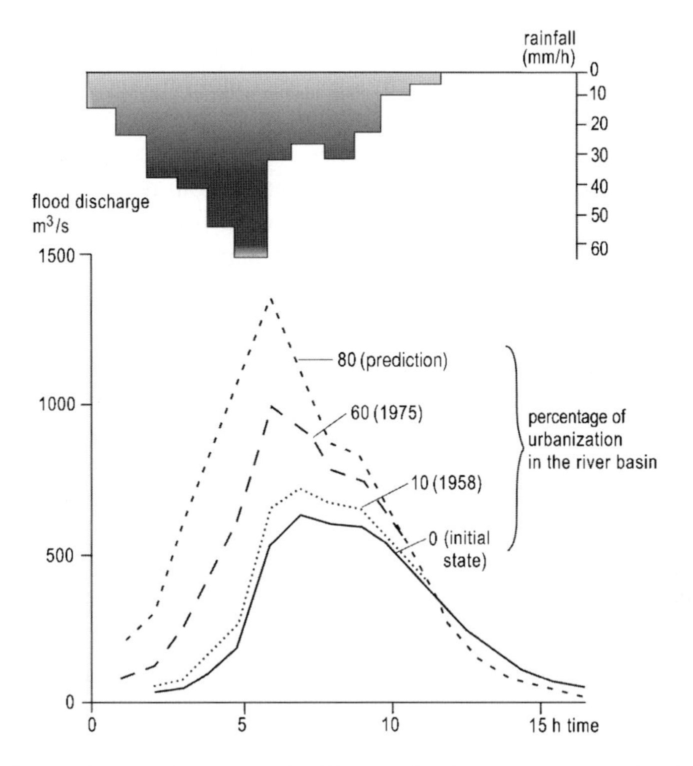

Figure 60: Stream flow hydrograms at different development stages of urbanization in the western part of Tokyo (after Yoshimoto & Suetsugi; see Smith & Ward, 1998).

the destruction of the vegetation cover diminishes the ability for water to seep within the ground, increases the surface streaming and aggravates the soil erosion potential (Chapter 8). The *agricultural practices* also increase the flooding effects, for instance, through systematic regrouping of lands, suppression of hedged farmland, downslope ploughing, and artificial implementation of seasonal fallows.

Forecasting

Flood prediction implies an accurate knowledge of threatened zones, as well as of rain and water-rise characteristics. Such information allows production of *flood hazard maps* that are based on precise geomorphological descriptions, long-term observations, aerial picture data, and satellite imagery. The maps provide data on the levels reached by water in the different sectors of a given river basin under various weather conditions, on the extent of inundated zones, on the location and lithological nature of past flood deposits, and on the intensity and duration of successive rainfall events collected in archive files.

The establishment of precise *stream-flow hydrograms* helps to estimate the time interval separating the upstream precipitation and the downstream inundation stages. Notice that past stream flow hydrograms tend to be of disputable usefulness when river flow has been modified by natural or man-made developments: cutting of meanders, canalisation, new dyke constructions and river course confinement. This was the case for the major flooding that affected the Mississippi River basin in 1993 (Chapter 4.1).

Physical and numerical modelling also contributes in an efficient way towards anticipating the flood characteristics under various geological, topographical and meteorological conditions. The computer models should be completed and refined as and when new civil engineering works are implemented along the river basins.

A second approach consists of developing *statistical investigations* that are intended to understand the water rise process and inferred flooding risks. The identification of the frequency of floods of diverse intensities is based on the record of the maximum water outflow measured each year during as many years as possible. The return time, or *recurrence interval*, predictable for a flood of a given type, is calculated by different formulae. The Weibull equation is one of them:

$$R = n + 1/m$$

where: R, mean time interval separating two floods of a given magnitude; n, number of years during which a flood of such a magnitude has effectively be registered during this time interval; m, maximum water outflow for a given number of years; $m = 1$ for the highest averaged values, $m = 2$ for the closest lower averaged values, etc.

For instance, the outflow of the Euphrates River responsible for a water discharge of 10,000 cubic feet/sec in the area of Hit, an Iraqi city located west of Baghdad, shows a statistical recurrence interval of 10 years (Fig. 61). This means that a flood determined by a water rise of similar intensity may statistically happen every 10 years, independently of the precise year of occurrence: the flood may effectively occur 10 years after the preceding one, but also after 2 or 20 years, or even twice the same year.

The grouping of data collected from observations and statistics allows implementing of *flood warning systems*, which comprise several alert levels. The predictions, the application time of which is proportional to the size of the reception basin, are more and more based on computerized communications and processing equipments: continuous recording and real-time transmission systems of data on rainfall and on the status of different river tributaries, immediate ongoing simulation of the flood wave progression and its impact, etc. The training of riverside populations to face the flood effects is crucial and

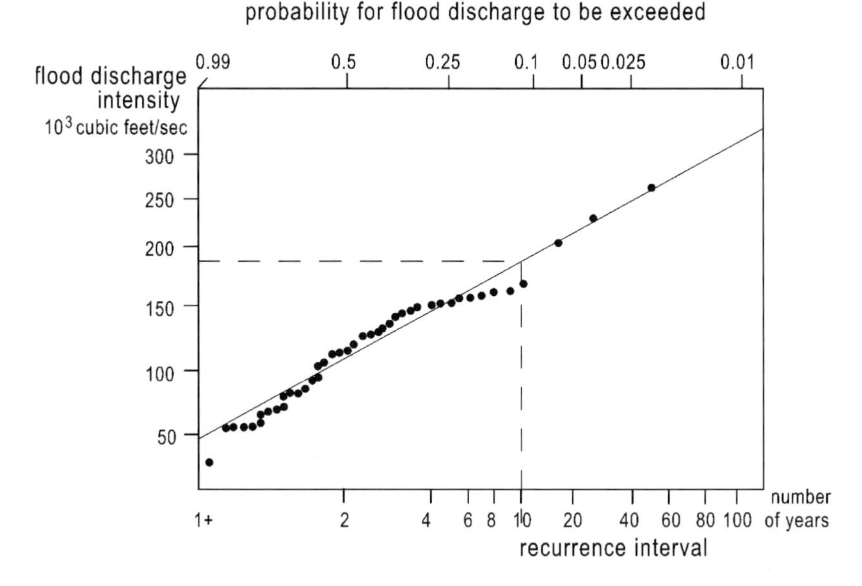

Figure 61: Statistical frequency of floods in the Euphrates River, Iraq (after Kolars & Mitchell; see Rogers & Feiss, 1998).

implies specific educational programmes. For instance in the year 2000, 55,000 pupils of 215 district schools subject to floods in the Aude department, southwestern France, received an adapted theoretical plan and practised evacuation and rescue exercises.

As a general result, *flood forecasting frequently reaches a satisfactory level of reliability in the context of alluvial plains*, especially as far as high-technology countries are concerned. It nevertheless remains of a disputable quality when the rainfall intensity and duration are particularly high (e.g. the 1993 Mississippi flooding; Chapter 4.1), when the hydrological balance and stream-flow hydrogram have been modified by natural or anthropogenic actions (e.g. shortening of the water transfer time, new channel network), or when the flood hazard is associated with serious coastal hazards (Chapter 4.3). The forecasting remains very difficult for torrential floods occurring in mountainous regions. As a consequence some prevention developments are frequently needed for complementing or relieving the forecasting tools.

Prevention

River control and management practices have been developed for several centuries in order to limit the water extent and overflowing potential during flood events. *Channelisation* consists of the straightening, deepening, widening, and often confinement of river courses, and is frequently associated with the removal of obstacles (large rock pieces, wide meander curves). River dykes, which are built with clay and gravel and are sometimes waterproofed with synthetic sheets, form the levees typical of most inhabited alluvial valleys, and often serve to support roads, railway tracks, houses, etc. Channels allow flood water to be rapidly evacuated downstream, and lead to a gain in additional farming or building land at the expense of river major beds. On the other hand channelisation works induce or favour a diminution of the surface area of wet zones, a loss of balance for the hydro- and ecosystems characterizing river environments, the concentration of pollutants, and sometimes a downstream transfer of the flooding hazard. This gives way to various conflicts of customs and interests, and sets up in an increasing way the question of the means to pursue the embankments' construction. Such a question leads to examination in a planned and integrated manner of the occupancy and management of alluvial plains that are potentially liable to flooding (Fig. 62).

The river *outflow regulation* is carried out by the mean of flood-absorbing reservoirs, of lateral inundation basins able to store the temporary water excess, of diversion canals, and by developing the irrigation networks. The risk of floods threatening people and property tends to decrease in densely cultivated, urbanized and/or industrialized alluvial valleys. This is due to increasing diversion practices of river water for human use, leading to a chronic situation of low-water levels.

The flood management of *urban areas* is more difficult to achieve than that of rural areas, for several reasons: widespread canalisation of river course inducing an artificial narrowing of transfer paths; critical interactions developing between rainwater, surface and underground water, and sewage systems; and potential pollutions. *Specific developments* help to facilitate the water flow: construction of embankments and installation of pumping stations; channel widening and enlarging; construction of diversion canals; digging of artificial basins; and, if necessary, progressive abandonment of most exposed areas. The cities newly built on alluvial flood plains tend to be designed with successive levels of occupancy planes, which correspond to different risks and uses: leisure and storage surfaces implemented at the lowest levels, houses at the highest.

Secondary Effects, Perspectives

The management of river basins for preventing flood hazards sometimes leads to considerable modification of the water and sediment budget. Amplification and reduction effects may happen according to the type of human actions performed at successive periods, and there may be interference with other human activities. The Ebro, a Spanish river flowing into the northwestern

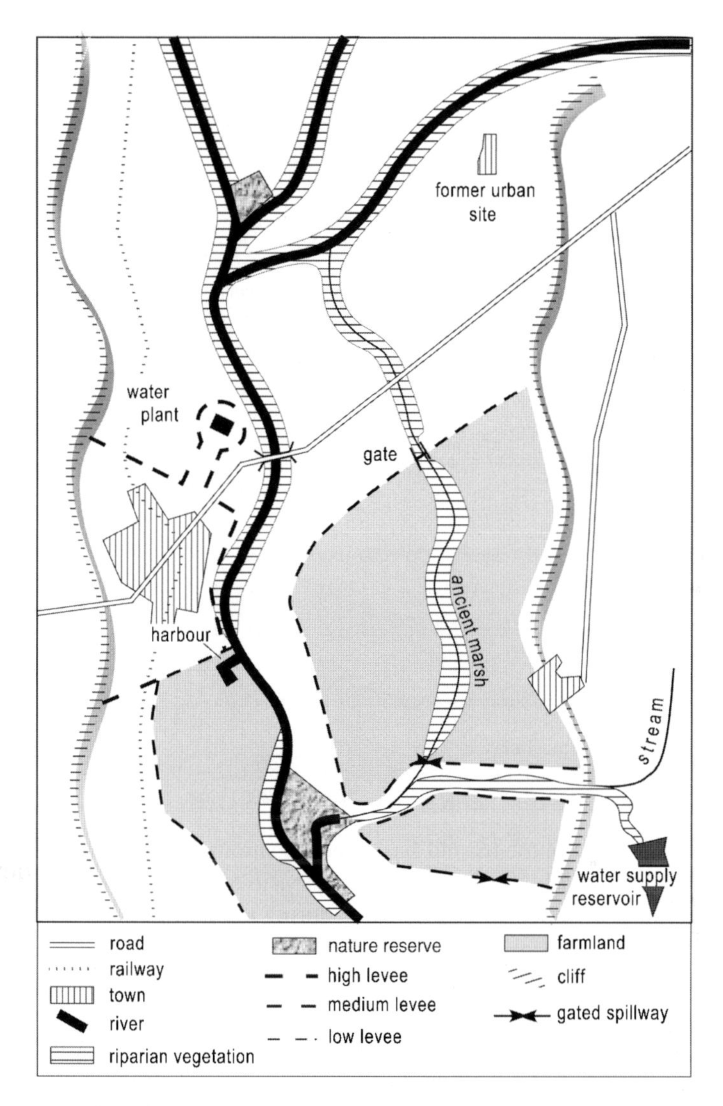

Figure 62: Example of an idealized environmental management in an alluvial plain liable to flooding (after Rasmussen; see Smith & Ward, 1998).

Mediterranean Sea, has been subject to extensive and changing human impacts during the past two millennia, which is clearly expressed by the historic evolution of the delta (Fig. 63, A and B). The building of numerous dams, especially during the last century, has resulted in the regulation of the Ebro hydrologic regime and the near-suppression of flooding hazards. The maximum water outflow was reduced from more than 3,000 m^3/sec before the reservoir building to 800 m^3/sec around the 1950s, and to 250 m^3/sec in the 1980s to

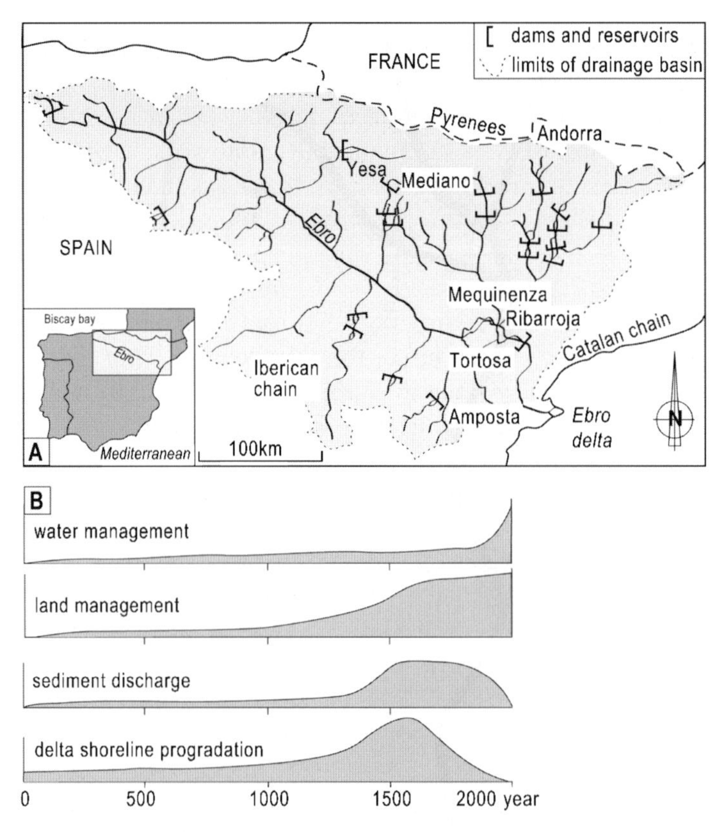

Figure 63: Human developments and basin modifications in the Ebro River and delta (after Guillen & Palanques, 1997). (A) Drainage basin and dam location. (B) Temporal variations of some delta parameters during the past two millennia.

1990s. Irrigation, agriculture and livestock farming have become more and more important, and hydroelectric resources have increased in a similar way. These man-made changes brought about some worrying consequences. For instance the sedimentary input through to the Ebro delta, which had been growing since the 15th century because of active deforestation and consequent soil erosion, progressively diminished during the 19th and especially the 20th century, due to an increasing trapping of suspended matter in artificial reservoir lakes. The delta progradation therefore progressively ceased and now tends to be replaced by a net shore retreat (Guillen & Palanques, 1997).

The evolution recorded for the Ebro River is similar to that affecting many other world rivers subject to extensive human occupancy and systematic civil engineering work: Mississippi, Nile, Rhone, for example.

In southern Poland the *Vistula* River has undergone important changes since the mid-19th century, in relation to human developments and protection works against flooding (Lajczak, 1995). First, the soil erosion induced by deforestation in the Western Carpathian Mountains was responsible for an active sedimentation on the piedmont plains. This sedimentary input in downstream regions was later amplified by meander cutting and canalisation work, which increased the river water transfer. Second, the active building of artificial reservoir lakes started during the 1960s, which progressively induced both a decrease in the flood hazard in the drainage basin and important upstream trapping of suspended matter. As a consequence the downstream deposition of sediment started to decrease. This reverse trend should be amplified and extended to the lower part of the drainage basin during the next decades, since additional dams are planned for construction along the Vistula lower course. It is likely that the land-derived sedimentary input in the Baltic Sea north of Warsaw will decrease.

The amount of suspended matter carried by the *Mississippi* River down to southern Louisiana at the end of the 20th century constituted only a fifth of the amount transported in the mid-19th century. This was the result of huge human developments deployed along the drainage basin to prevent inundations and to increase irrigation resources. The augmentation of river embankments has therefore indirectly induced an important loss of wet zones in the delta region, which progressively tends to be salinized by marine water seeping from the Gulf of Mexico. A partial, strictly controlled removal of downstream river dykes could allow the partial restoration of the freshwater humid zones, the Mississippi water being able to freely migrate again in some lower parts of the basin. Significant amounts of suspended matter carried by the river could then be distributed again within the delta, reversing the present erosion and salination trend (Kesel, 1989).

The examples of the Ebro, Vistula and Mississippi rivers illustrate the close relationships linking the different parts of a given drainage basin and the complex consequences induced by interrelated natural and human actions on the river environmental balance. Some incompatibilities arise from different uses of potentially flooded alluvial plains. Their control is crucial for reconciling the need to both overcome the flood hazards and preserve the whole river ecosystem.

4.2. Aeolian Hazards

4.2.1. End of December 1999, Storms in France

During the night of 25th to 26th December 1999, gales of an extraordinary intensity came from the Atlantic Ocean and reached the French coast between Vendée and Pas de Calais regions. The storm passed through the whole country from west to east on the morning of Sunday, December 26th, and continued over Germany, where its strength began to diminish. The anemometer on the Eiffel Tower jammed at a value of 216 km/h. The following day a second storm of similar intensity developed in a more southern position, passed again through France, sparing only the Rhone valley and central Alps regions, and headed towards Switzerland and Italy. As a consequence all French regions extending from Vendée to Franche-Comté, as well as most parts of Brittany, were very violently hit twice by westerly winds in less than 2 days (Fig. 64).

The damage resulting from the two storms, which have been called Lothar and Martin, was considerable: 130 dead and thousands of injured people; several million trees uprooted, among which 100,000 were in the Boulogne and Vincennes woods in the vicinity of Paris; 80% of the ancient forest badly damaged; road and railway networks paralysed by fallen pylons, masts, panels,

Figure 64: Trajectory of the two major storms that crossed France on 26th and 27th December 1999 (from *Le Monde*, December 29, 1999). The hatched area represents the regions hit by the two storms.

trees, buildings; 50% of telecommunications disconnected from the national network; numerous cities and firms deprived of electricity, heating, drinking water, freezing facilities, or even completely isolated. The total damage was estimated at 8 billion euros.

The rescue teams, who were operating locally under great difficulties, comprised thousands of French people helped by Belgian, British, Dutch, German, Italian, and Spanish colleagues. On this occasion, the meteorological forecasting was not adequate. Wind gusts had been announced after strong depressions had been identified above the Atlantic Ocean. But the speeds expected were of 100–120 km/h in continental France for the first storm, and of 60–80 km/h for the second. The actual intensity and power of each of them was not adequately anticipated.

The devastating character of the Lothar and Martin storms resulted from the exceptionally strong and repeated interaction of high altitude (5–10 km) jet streams and of ocean-surface eastward winds induced by atmospheric depressions forming off European coasts. Winter storms commonly develop over the Atlantic Ocean, due to the confrontation of cold polar and warm subtropical air masses. These storms generally develop above the ocean surface far away from Europe and correspond to pressure values of 1,020–1,040 hectopascals (hPa) in anti-cyclonic zones and 960–1,005 hPa in cyclonic zones, the average pressure approaching 1,013 hPa. Westerly winds frequently attain 150 km/h at the ocean surface, and 300–400 km/h at heights of several kilometres. On 24th and 25th December 1999, the *depressions formed more to the south and closer to Europe than usual.* Lothar formed 3,000 km west of the French coast (960 hPa) and Martin 2,500 km west of the Portuguese coast (965 hPa). The exceptional strength of these two storms resulted from their relative proximity to the European landmass, as well as from the *rapid descent towards the ocean surface of powerful high-altitude winds* (360 km/h), resulting in a “meteorological bomb”.

The current difficulty in forecasting such storms and especially their extraordinary violence, and therefore anticipating the possible consequences and preventive measures, results from several causes:

- strong spatial and temporal variability of low-pressure fronts and of winds forming over the ocean;
- insufficient density of measurement stations at the ocean surface, and insufficient frequency of sonde balloon releases;
- still excessive duration of the calculation time necessary for processing the billions of data accumulated in a very short time by the meteorological stations’ network;
- difficulty for computer codes to take into account some values of an abnormally high intensity.

Significant progress is expected during the forthcoming years, due to the implementation of more measurement stations, the launching of much more precise and sensitive meteorological satellites, and the augmentation of computer processing capacities.

4.2.2. Effects and Damages

Whatever their violence, the winds usually claim few human victims, except indirectly due to tree or object fall, building collapse, etc. Hurricane Andrew, which devastated the Bahamas, Florida and Louisiana in 1992 caused the death of 25 people, while the damage to property was estimated at 20 billion U.S. dollars (see Chapter 4.3.1).

The *immediate damage* resulting from wind action first lies in tree uprooting and fall, and sometimes in the destruction of entire forests. This is associated with the fall and breaking of pylons, boards, cranes, chimneys, roofs, parts of buildings, etc. *In arid or subarid*, poorly plant-covered *regions*, sand storms induce the obstruction of communications, the sapping of buildings and other man-made developments, and the partial covering of rural and even urban areas (Chapter 8). *In farming zones* the wind action determines the soil erosion, especially during drought or fallow periods. The Great Plains were subject during the 1930s to huge wind-erosion phases (Fig. 65), which were exacerbated by both a long, preceding drought and poor preservation of surface soils (Chapter 4.2.4). Several hundreds of thousands of cubic kilometres were damaged; and storm clouds were carried at altitudes reaching 4 km, the suspended matter in the air attaining locally 35,000 t/km^3. The soil erosion continued in the U.S. central plains throughout the 20th century (Fig. 66), with a variable intensity depending on the wind strength, the seasonal droughts, gravity, the growing state of cultivated fields during the storms, etc.

More diffuse and often belated damage is caused by sand and dust aeolian transport. Sometimes expressed in the long term only, they are of diverse types and their consequences may be significant (Pye, 1987):

- defoliation, and flower, seed, stem or root corrosion in cultivated fields;
- abrasion of paintings, windows and other materials;
- damage to combustion motors, accumulation of static electricity charges preventing engine ignition;
- creation of intense electric fields interfering with radio and telecommunications;
- road or air accidents due to reduced visibility.

The aeolian deflation may also induce *health problems*: breathing difficulties, allergies, gastro-intestinal troubles, absorption of minerals with a silicigenic or

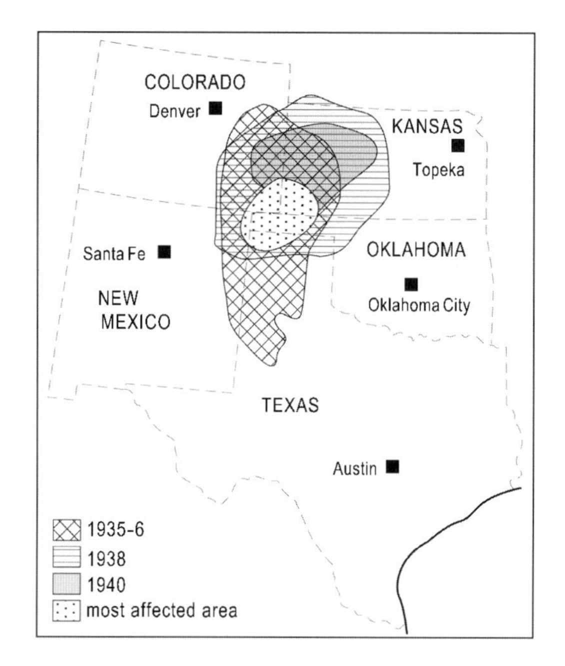

Figure 65: Wind erosion of the Great Plains ("dust bowl") during the 1930s (after Worster; see Cooke & Doornkamp, 1990).

carcinogenic potential, etc. Finally the distribution of large amounts of dust and gas by wind in the atmosphere is likely to cause *global environmental changes* (Chapter 11): air and ground cooling due to the partial absorption of solar radiation within the stratosphere; soil-surface warming due to the retention of air-reflected radiation by tropospheric aerosols; augmentation of the cloud cover

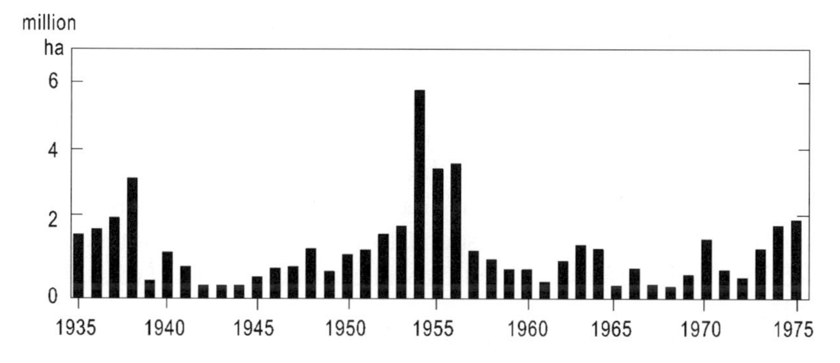

Figure 66: Extent of soil surfaces damaged during the 20th century in the Great Plains (after Pye, 1987). Values in million hectares (ha).

and of precipitation due to condensation of aerosols around aeolian suspended particles.

Winds may also have *beneficial effects.* Storms induce the mixing and accelerate the renewal of air masses at various latitudes and altitudes, and therefore participate to the *climatic regulation* provided by the fluid envelopes of our planet. In addition to their ability to *disperse the pollutants* and noxious gases that accumulate in urban and industrial zones (organo-volatile components, SO_2, NO_2, tropospheric ozone), winds may be responsible for *chemical re-balancing* of agricultural soils' composition. For instance the Sahara-derived dust provides mineral nutrients (Mg, K, Ca) and compensates for the disruptive effect of acid rain in the terrestrial ecosystems of northeastern Spain (Avila & Peñuelas, 1999).

Wind also participates in the production of non-polluting and renewable *production of electricity.* An increasing number of countries, namely those having few nuclear plants, acquire aeolian turbines of relatively large power, which are placed in coastal or continental zones exposed to frequent and high winds. During the last decades wind generators have become twice more efficient and twice less expensive than in the 1970s. Very marginal in 1980, worldwide wind-generated electricity in 1996 reached a value of 7 GW. Initially restricted to North America, this kind of renewable energy production extended during the 1990s to European countries (Fig. 67), where it continues to grow. Germany is presently the primary European producer of wind energy. Spain expects to produce 8,900 MW of wind-derived electricity in 2010, which would represent more than 4 times the 2001 production and roughly corresponds to the amount of energy produced by seven nuclear reactors.

4.2.3. Basic and Exceptional Aeolian Processes

Wind results from exchanges between air masses characterized by different pressure. The wind intensity and speed increase when the pressure difference increases, especially if relatively nearby air masses display opposite temperature and/or humidity values. The wind speed increases in the direction of low-pressure regions, as well as towards mountain zones where frictional forces diminish with the altitude and where the valley shapes induce the air flows to concentrate. The wind intensity is measured on the *Beaufort scale*, the successive degrees of which correspond to increasing speed intervals and are expressed by modifications endured by water surfaces (seas, lakes), vegetation and buildings.

Aeolian hazards develop when wind becomes very strong (speed above 90 km/h, level 10 and more on the Beaufort scale) and presents a very turbulent and cyclonic regime. *Storms* are particularly devastating in coastal and

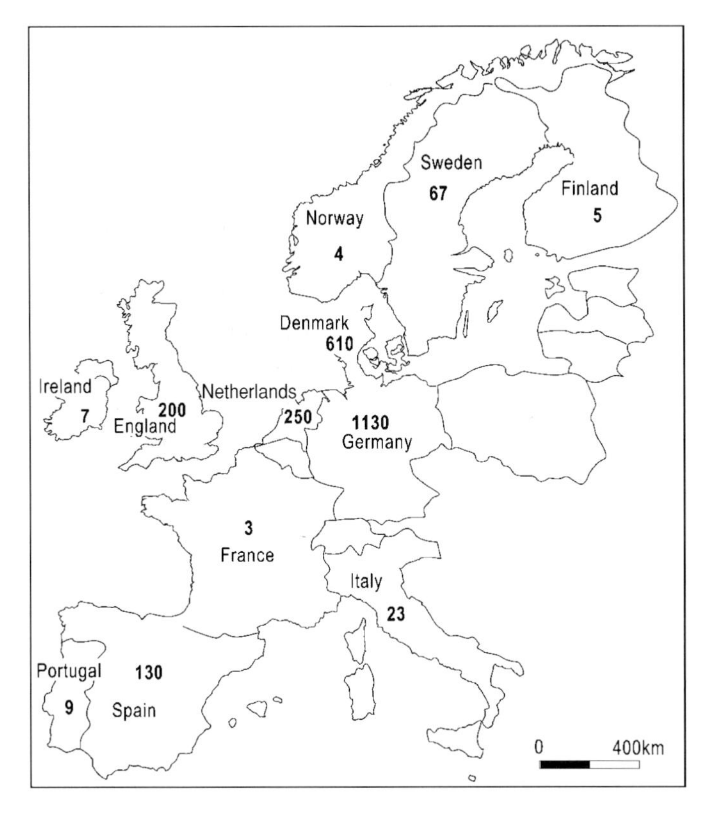

Figure 67: Electric energy capacity of 12 European countries in 1996 (after Lundgren, 1999).

mountainous zones, where the wind speed may exceed values of 250 and 300 km/h, respectively. The low-pressure air masses where storms are forming tend to constitute very large systems that spiral up. Due to the Coriolis effect, the movement is cyclonic in the Southern Hemisphere and anti-cyclonic in the Northern Hemisphere. Storms are often associated with heavy rain, which accentuates the risk of flooding hazards.

Tropical storms are called *cyclones* in the Indian and South Pacific oceans, *typhoons* in the Northwest Pacific, and *hurricanes* in the Atlantic Ocean and Gulf of Mexico. They constitute very large and active low-pressure systems. Always originating above the ocean, they move globally from east to west under the action of trade winds. Their trajectory is relatively erratic and their displacement may last several weeks. Tropical storms, the diameters of which potentially exceed 600 km, consist of spiral, highly turbulent and opaque cloud masses. The central part of the mass is clear and calm, and called the eye of the cyclone. The cloud mass forms because of the massive evaporation of warm seawater, the surface layer of which exceeds a temperature of 26.5°C and a thickness of 50 m. The atmospheric condensation of the ascending steam occurs at high altitude due to air-cooling, which determines very heavy precipitation. The cloudy column may rise up to 15 km above the ocean surface, and starts rotating due to Coriolis forces. This determines the formation of a violent *storm wave* at the base of the system. The regions of the Northern Hemisphere exposed to the displacement of such an air and water mass successively experience increasing wind and rain coming from the north, a temporary calm corresponding to the passage of the cyclone eye (diameter 30–40 km), and finally decreasing wind and rain coming from the south. A reverse succession characterizes the cyclones moving in the Southern Hemisphere.

Tropical storms, which constitute a usual phenomenon during the warm and humid season (average of 80–90 storms per year), have a *considerable destructive power* in island and coastal regions bordering the low-latitude oceanic domains (Chapter 4.3.1). The wind speed in the low-pressure moving mass may approach 350 km/h, and the depth of rainfall 10 m. The daily energy flux displaced within a cyclone may be similar to that of 400 hydrogen bombs of 20 megatons each. The danger is all the more serious if the cyclone or hurricane sweeps across a large area, moves fairly slowly, is only progressively evacuated, and is associated with violent rainfall. Hurricane Mitch, which was responsible for about 10,000 deaths and other tragic consequences in November 1998, extended over a large part of Central America, raged over 4 days, and induced the formation of huge mudflows due to torrential rain.

Tornadoes constitute cyclonic storms that are very intense but of *short duration* (i.e. a few minutes to the maximum) and of *local extent* (diameter of less than 300–400 m). Some tornadoes are practically stationary, whereas some others move at various, sometimes high speeds (up to 100 km/h) along narrow and more or less sinuous corridors. Tornadoes form in mid-latitude regions at the edge of low-pressure fronts and correspond to the confrontation of cold and warm air masses. The warm air is aspirated upwards and forms an ascending spiral or *vortex* within the core of the system, whereas the cold air flows downwards and constitutes a peripheral spiral. The tornado strength is sometimes amazingly high. It may be responsible for the moving of 80 t of railway wagons, the sliding of metal boxes over 2 km, the disintegration of light buildings, and the sucking up of soils and various materials at high altitude. Such power results from the extremely high speed of some rotating winds (up to 450 km/h), as well as from the very low pressure occurring inside the vortex core, the values of which are possibly less than 60% of that characterizing the surrounding atmosphere.

Tornadoes occur commonly in central and south-eastern regions of North America, and are also known in various European countries. Some tornadoes are

Table 10: Intensity levels of tornadoes (Fujita scale) and number of events registered in France from 1680 to 1988 (after Ledoux, 1995; Muck et al., 1996).

Intensity level (F)	**F2**	**F3**	**F4**	**F5**
wind speed (km/h)	182–253	254–332	333–419	420–513
1680–1959	23	17	8	1
1960–1988	27	27	3	1

periodically recorded in France, where their intensity may be considerable and tends to increase since a few decades (Table 10).

Sand and dust storms are typical of desert regions such as the Sahara, the Middle East, central Asia, and southwestern USA. Material swept away by wind from dune systems or loose surface soils are transported more easily further away from the sources because the particle size is small and the altitude attained is high. The corresponding hazards mainly result from reduced visibility and burying of surface man-made developments (roads, railway tracks, aeroplane runways, etc.).

4.2.4. Human Action

Accentuation of Aeolian Risk

The main way by which human activities tend to exacerbate the damage induced by storms consists in inappropriate *cultivation practices* that increase wind erosion. The case of the Great Plains is particularly relevant. Initially covered by huge natural prairies, allowing the soil to be permanently fixed, the plains have been exploited since the 19th century through intensive cultivation. Soils were ploughed at depth and seasonally planted with cereals and tubers. This led the partly disintegrated and loose soils to be temporarily exposed to wind action. Early in 1930, many cultivated soils became over-exploited, due to the economic recession that had hit the USA. Huge soil surfaces were therefore seriously exposed to wind hazards (Fig. 65). The unexpected occurrence between 1935 and 1938 of winter seasons marked by serious storms and droughts induced massive soil erosion, the loss of most fertile surface levels, the removal of huge amounts of inorganic and organic particles, and considerable damage (Chapter 4.2.2). More than half of the farmers in the stricken regions received financial compensation for the soil deterioration they had involuntarily been responsible for. Many inhabitants were forced to migrate towards the Western States. The improvement of cultivation practices later in the 20th century progressively diminished the wind-erosion risk, but did not stop it.

Other ways to artificially increase the aeolian hazards include the extensive *deforestation* of regions subject to strong winds, the hedge removal in divided farmlands, human settlement that indirectly induces over-pasturing, plant destruction, erosion and drought, and the planting of exotic species unable to retain the soils (Chapter 8).

Control and Protection Measures

The ways to *protect cultivated land* from normal wind effects are diversified and effective: planting of tree rows perpendicular to the dominant wind direction, or of semi-permeable and adequately spaced artificial windbreaks; choice of plant species adapted to the silt and sand environments characterizing the aeolian dunes; permanent keeping of soil surface humidity; artificial augmentation of the soil roughness to prevent deflation effects; etc.

Storm forecasting is based on the processing of data from meteorological stations which are inserted in numerical models immediately they are acquired. The prediction of *mid-latitude* storms is now *reliable with an accuracy of several days*, except for some events of an extraordinary intensity (Chapter 4.2.1). Nevertheless the limited density of meteorological stations and the huge amount of data necessary to be real-time processed generally do not allow the anticipation of the precise location of maximum impact zones. A diminution by one-half of the observation network would lead to an eightfold increase in computer calculation time and would therefore handicap the forecasting capability. In addition the complexity and irregularity of the turbulent wind course, which is locally controlled by the topography and air-mass location (e.g. concentrated corridors of storm flows), also hinder the efficient improvement of storm prediction.

The forecasting of tropical storms' formation and displacement is still of a *rather poor quality*, mainly because the observation and measurement network is not dense enough in low-latitude oceanic regions. The prediction of the cyclone, typhoon and hurricane trajectory is mainly based on satellite imagery (see Fig. 79). The error on the displacement path above a given region may reach 400 km 2 days before the storm passage, and 200 km the day before.

Preventive measures consist of reinforcing the buildings and other surface developments (e.g. concrete shelters built in many house basements of Florida), training the population, and evacuating the threatened zones. In the same way as for other natural hazards, *hurricane prevention has been considerably improved* during the 20th century, especially in developed countries. Two hurricanes of comparable intensity and trajectory that hit the north of the Gulf of Mexico were respectively responsible for 8,000 deaths in 1900 in Texas, and for 25 deaths in 1992 in Florida and Louisiana. By contrast the damage to people and property remains very serious and sometimes catastrophic for some over-populated, poorly equipped and insufficiently protected developing countries.

4.3. Coastal Hazards

4.3.1. Role of Hurricanes and Loss of Wet Zones in Louisiana

The Mississippi River flows into the Gulf of Mexico by forming a very large delta, the wet zones of which extend over 300 km along Louisiana coasts and constitute a 130-km-wide continental area. The wet zones are protected from marine erosion by a chain of barrier islands, but nevertheless undergo a progressive decrease in their surface, which has been especially important during the past century. The continuous loss of Mississippi Delta wet zones is attributed to two main causes. First, the extensive and continuous construction along the river basin of levees, dams and other protection devices as well as of irrigation systems has induced a diminution of the water influx and extent in downstream regions (Chapter 4.1.4). Second, the coastline tends to retreat because of several combined effects: rise in sea level associated with sediment compaction and subsidence, alongshore sediment export induced by littoral transfer, and violent erosion by hurricanes. The respective importance of each factor is determined by comparing the information gathered through geomorphological and sedimentological investigations that are conducted on different time scales.

The follow-up of the *Isle Derniere* area, located between Caillou and Terrebonne bays in the southernmost part of the Mississippi Delta (Fig. 68, A), elucidates the different erosion–sedimentation processes intervening in coastal domains. The island, naturally built since the 15th century from alluvia supplied by the Mississippi River, constituted a 50-km-long continuous barrier in the 19th century and provided a base for a fairly important city. The Isle Derniere was struck in 1856 by a major hurricane, which was responsible for the death of several hundred inhabitants and the destruction of the city. The coastal erosion developed during the 20th century, through alternating progressive and violent phases, inducing the reduction of the island size, its fragmentation and the destruction of most wet zones situated on its leeward side. In 1978 the coastal barrier was split up into five narrow and fragile islands, which were called Isles Dernieres (Fig. 68, B). In winter 1992 a new sand bridge was formed between

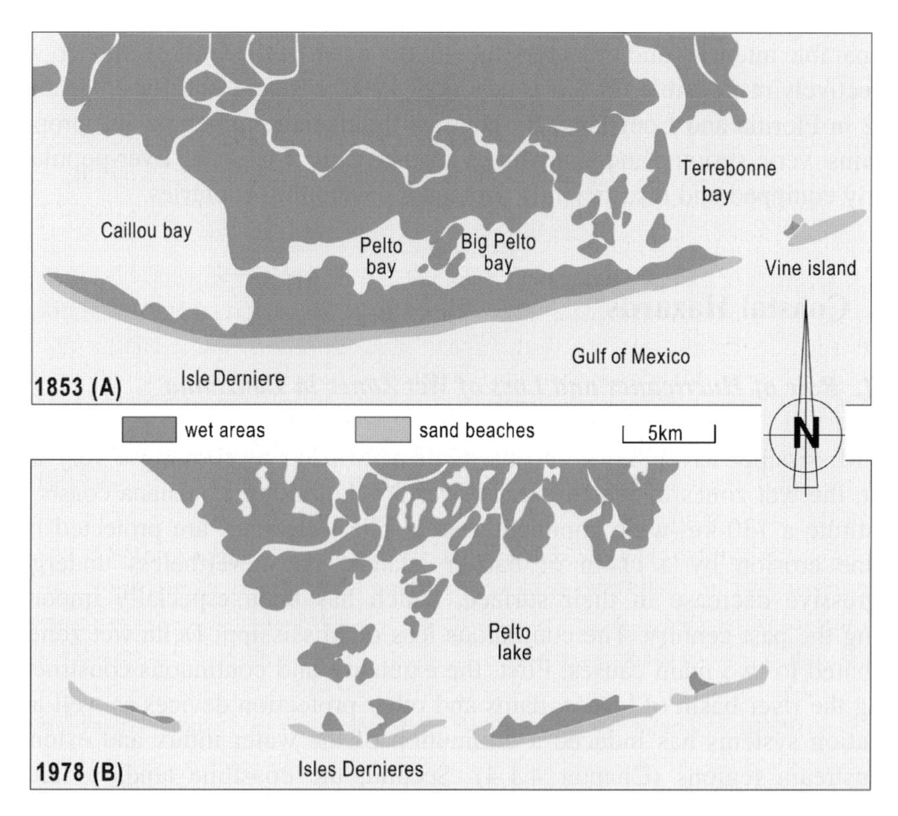

Figure 68: Morphological changes between 1854 (A) and 1978 (B) in the coastal sector of Isles Dernieres, Louisiana, before the passage in August 1992 of Hurricane Andrew (after Lundgren, 1999).

the two central islands, protecting again the leeward sectors. But in August 1992 Hurricane Andrew, which came from Florida, where it had destroyed 25,000 houses and damaged 100,000 buildings, swept across the Isles Dernieres area. Most of the sand was removed and redeposited in wet zones located behind the islands (Lake Pelto). Statistics drawn up by the Geological Survey of Louisiana report that 70% of the barrier islands bordering the Mississippi Delta were eroded in a similar way. Seventy kilometres of coastal dunes were destroyed by hurricane Andrew. The sand removed from beaches and dunes had invaded the oyster-farming zones located to the west of the delta. Re-deposited sediments locally attained a thickness of 90 cm. This example illustrates the *predominant influence of aperiodical, strong-intensity events* on coastal morphological and sedimentary changes.

4.3.2. Importance and Effects

Basic Coastal Hazards

Most world coasts are presently subject to erosion processes More than two-thirds of the marine beaches experience a shore retreat, one fifth being stable and one tenth only showing an offshore progradation trend. This phenomenon is often referred to as the "*erosive crisis*" of marine beaches.

Numerous examples are provided in the literature (Paskoff, 1994; Viles & Spencer, 1995). In West Africa the coast of Togo locally retreats by 18 m/year, those of Nigeria by 4–7 m/year. Two-thirds of Mexican, Venezuelan and Columbian beaches are marked by a loss of sediment. In the absence of extensive man-made developments, nearly all the barrier islands bordering the eastern coasts of North America, from Canada to Florida, would display a reduction of length, width and sometimes height; retreats by 0.3–2 m/year are commonly registered. By contrast most of the Pacific coast of North America is stable or moderately prograding, erosive processes occurring locally only. In Denmark the shore of Jutland is locally eroded by 5 m/year, the loss of sediment reaching 16 m^3/linear metre per year. In Italy the beaches of Emilia-Romagna, Latium and Gulf of Tarento suffer erosion of 4 m/year. The eastern coasts of England have retreated locally by 2 m/year since the mid-19th century, due to active erosion of Pleistocene glacier moraines outcropping on slightly consolidated coastal cliffs.

France, the coastlines of which extend over 5,540 km in the Hexagon and over 1,640 km in overseas territories, is also extensively subject to marine erosion (Fig. 69). Most rocky and sandy shores are retreating, the values reaching 2–7 m/year for some sand beaches directly submitted to frontal swell. The sapping, deviation and fall of blockhouse and military developments constructed during World War II in order to form the Atlantic Wall, clearly reflect this widespread shore retreat. Beach receding approaching 0.25–1 km per century has been recorded locally on the Aquitaine coast, and similar values characterize some sectors of Languedoc and Pas de Calais. Field observations in the southern part of the Gironde estuary suggest that the coastline has been moving inland by about 10 km since the 6th century, an acceleration having started in the 18th century and having extended during the 20th century (e.g. a retreat of 280 m between 1888 and 1985 at l'Amélie-les-Bains, of 700–800 m between 1881 and 1922 at Capbreton; Paskoff, 1994). Notice that some coastal sectors, which are sometimes located very close to those retreating, experience a significant sedimentary progradation. For instance the coastline of northern-most France extending along the British Channel shows successive sectors displaying alternating retreat and advance: some beaches have retreated by 20–225 m during the 1947–1977 time interval (Calais, Wissant, Hardelot, Berck, Marquenterre, etc.), whereas some others have progressed seawards by

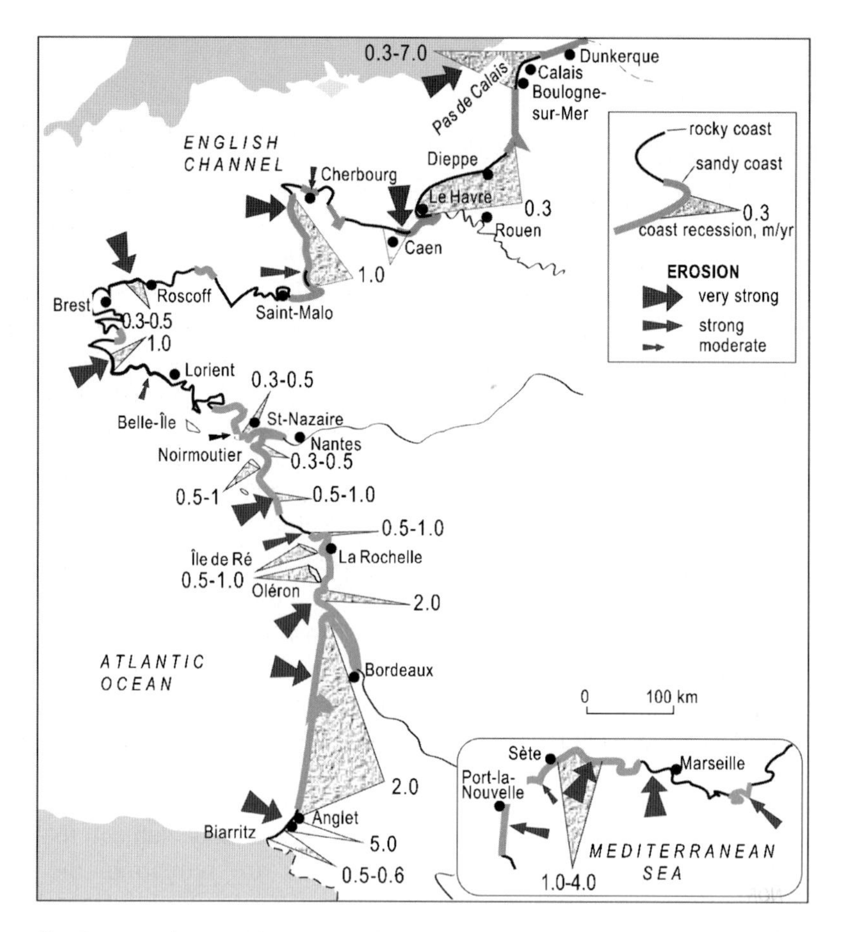

Figure 69: Present-day erosion rates of French metropolitan coasts (after Massoud & Piboubès, 1994).

20–160 m (Oye-Plage, Marck, Le Touquet, South Authie estuary, etc.). The western coast of Cotentin peninsula in Normandy has experienced during the last few years either a receding of 5–7 m/year (locally of 10 m/year), or a progradation exceeding locally 5 m/year (investigations by F. Levoy, Caen university). Nevertheless, in all these regions the average sedimentary budget is clearly negative.

On a worldwide scale, the slow and widespread shoreline retreat is potentially responsible for moderate but recurrent damages to property: deterioration of roads, buildings, urban and industrial developments, and harbour constructions, etc. This determines important and costly protection works, which need to be periodically renewed.

Coastal Storms

The effect of storm winds forming above the ocean surface and heading for landmasses exacerbates in a dramatic way the basic coastal hazard. This is the case for *mid-latitude regions storms* (Chapter 4.2.1), which determine significant sea-level rise and important damage when they are associated with spring tides and low atmospheric pressure values. The westerly storm that formed over the southern North Sea in the last days of January 1953 broke the sea wall and the seawater eruption in a large part of the polder zones of the Netherlands. The flooding resulted in the death of 1,800 people (Fig. 70), the inundation of 1,600 km^2 of land, and huge damage to property. The region situated between Rotterdam and the Scheldt estuary, where the seawater retreat remained incomplete 5 months after the disaster, was particularly affected.

In *tropical regions* the damage resulting from cyclones, hurricanes and typhoons (Chapter 4.2.3) is generally very important, since the energy of these moving air masses tends to be maximum when they reach the coastal zones. The power of tropical storms usually decreases when they progress towards inland zones. The consequences of such storms are particularly dramatic in flat deltaic regions subject to summer, ocean-originating monsoons. Sixty serious cyclones associated with powerful storm waves hit the Gulf of Bengal between 1797 and 1991. Bangladesh is usually chronically affected by such tropical storms. In

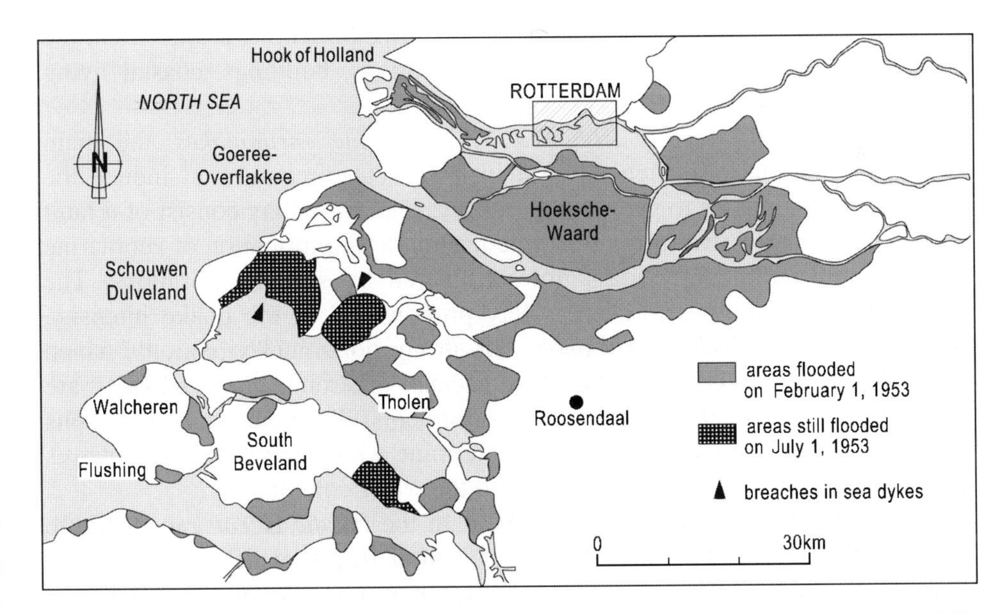

Figure 70: Southwestern Netherlands areas flooded by North Sea water during the 1953 winter storm (after Smith & Ward, 1998).

November 1970 it endured the impact of a major cyclone associated with a 6-m rise in sea level (959 hPa), which caused the deaths of 300,000 people. The loss of farming resources amounted to 63 million U.S. dollars, and 65% of the coastal fishing fleets were destroyed. Another extremely violent cyclone (940 hPa) hit Bangladesh in spring 1991, causing the deaths of 100,000 inhabitants and damage to property amounting to one billion dollars.

Tsunamis

The ocean surface is sometimes submitted to specific undulatory movements, marked by a *considerable wavelength* (successive crests 100–200 km apart), *very high speed* (400 to more than 950 km/h), *and very low height* (about 1 m). Called tsunamis (or tidal waves), these undulations constitute a major risk when approaching the coastal zones of islands and continents. At a shallower depth, the wave curve tends to deepen, and the successive crests get closer to each other. Differential friction forces determine a strong accentuation of the wave bending and height. The breaking on the shore of resulting waves, the height of which may attain 5–30 m, and sometimes more, is associated with the release of considerable energy. The height and power of the breaking wave first depends on the initial impulse strength and displacement distance, and is amplified in steep slope sectors and when submarine topography forces the wave crests to converge. The damage may be very serious in terms of people, property and the environment, e.g.: ships, harbours, communications, buildings, coastal living resources, ecosystems.

Tsunamis are principally triggered by submarine earthquakes, the main resulting vibrations of which develop in a vertical direction and induce the overlying water masses to get moving rapidly. The cause may consist of a fault rupture in a compressive tectonic regime leading to the formation of submarine scarps, or of sedimentary mass flows triggered by seismic vibrations. The tsunami intensity is therefore not solely proportional to the quake intensity. Some violent earthquakes responsible for strong horizontal vibrations, which could be particularly devastating on land because of strong shearing effects, are of a limited danger in submarine environments. By contrast, subaqueous earthquakes of moderate intensity may determine a major risk if seismic waves induce important vertical vibrations within the water column.

Most earthquake-induced tsunamis occur in the *Pacific Ocean* range, where their recurrence averages several years. Tsunami waves, which travel at high speed on the ocean surface for very variable times according to the distance to be covered (from a few minutes to about 20 h), may originate on all active margins. They pose a recurrent threat to the shores of Japanese islands, the

Hawaiian archipelago, the Kamchatka peninsula in eastern Russia, Alaska, Central and South America, and the Southwestern Pacific islands.

Numerous tsunami-induced disasters have affected the Pacific coastal zones during the past century: 76,000 deaths on Honshu island (Central Japan) in 1896; 150 deaths in Hawaii Big Island in 1946, due to a tsunami originating in the Aleutian arc west of Alaska (the wave travelling at 800 km/h for 4.5 h and reaching a height of 8 m when breaking on Hilo shore); about 2,000 Chilean and 120 Japanese dead or missing after the 1960 tsunami formed off the central Chile coast (Fig. 71); 120 deaths in California due to the tsunami triggered by the 1964 Good Friday earthquake in Alaska (Chapter 1.2); 170 victims in Nicaragua after the September 1992 tsunami; 2,100 deaths in Papua New Guinea due to a tsunami wave of 10–15 m high which travelled along the sea's surface for 10–20 minutes only.

Earthquake-induced tsunamis sometimes affect various *other regions* of the world, including Europe. The submarine earthquake that occurred in 1755 in the eastern Atlantic Ocean off Portugal induced a series of tsunamis that were responsible for the deaths of 60,000 people in Lisbon, a city of about 240,000 inhabitants. The city was badly damaged, and return waves were registered several hours after the disaster in the western Atlantic Ocean, as far as the Lesser Antilles.

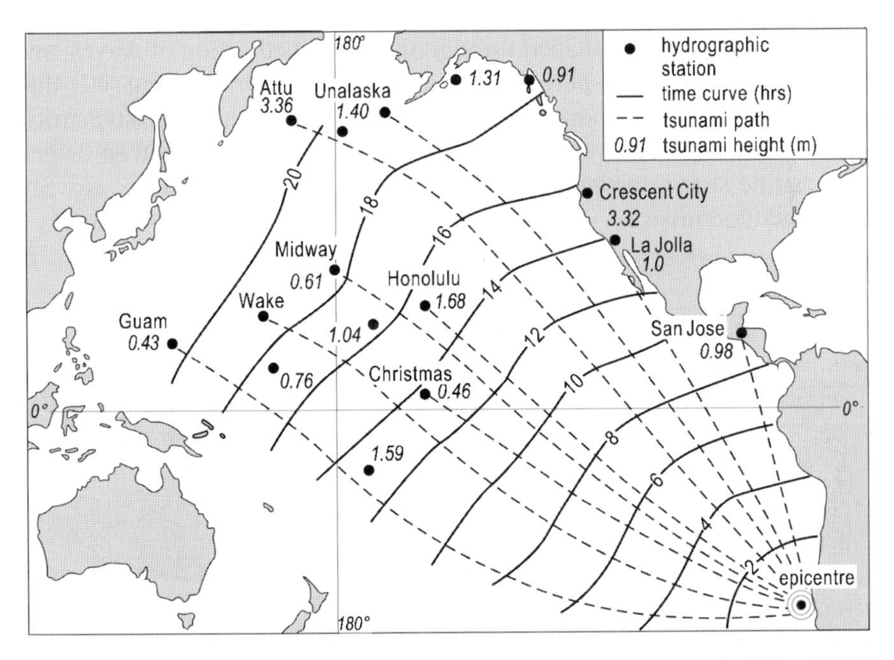

Figure 71: Propagation in the Pacific Ocean of a tsunami triggered in May 1960 by a submarine earthquake formed off central Chile (after Smith & Ward, 1998).

Tsunamis may also result from *volcanic eruptions*. Much less frequent, they are sometimes disastrous, especially if the eruption is marked by the explosion and collapse of an island volcano caldeira (e.g. Santorini in 1,600 BC, Aegean Sea; Krakatoa in 1883 AD, Indonesia; Chapter 2.2.4, 2.2.5). Other triggering factors include *submarine mass movements* (that are often induced by earthquakes or eruptions), and *man-induced submarine explosions* (e.g. the nuclear tests in Marshall archipelago during the 1940s and 1950s).

4.3.3. Types, Mechanisms and Natural Factors

Diversity and Variability of Coastal Environments

Coastal environments are subject to highly complex dynamics, the very changable characters of which depend on various factors:

- shore location and orientation;
- time scale considered;
- both cyclic and aperiodic constraints;
- both hydraulic and atmospheric interacting factors;
- either opposite or additional influence of marine and continental agents;
- variable but statistically increasing human impact.

The coastal zone, which is shaped through the combined action of waves, swell, tidal currents and wind, displays a characteristic *basic profile* (Fig. 72) that is diversely expressed in the morphology according to local characteristics: rock or sediment nature and cohesiveness, exposure to dynamic agents, steep or gentle slope, coastline shape, terrigenous input, biological activity, climate, etc. Shore environments comprise *very different types*, which include: rocky beaches and

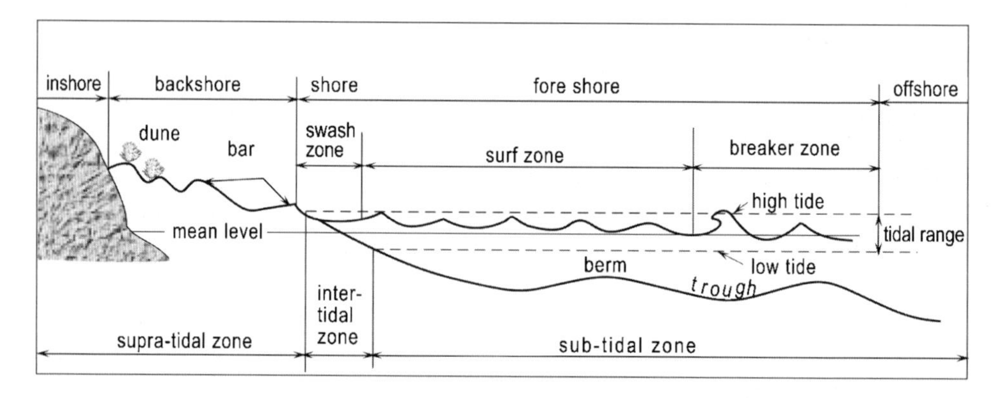

Figure 72: Reference profile of the coastal zone.

cliffs; low-relief sandy to silty-clayey beaches; backshore dunes; lagoons and marshes; estuaries and deltas; reef shoals and barriers or mangroves of intertropical regions; and seasonally ice-covered coasts of glacial and periglacial regions (Viles & Spencer, 1995).

The morphological and sedimentological changes of shore and coastal zones during tidal, seasonal and long-term cycles *depend on the local characteristics of sediment supply, erosion, transportation and deposition processes.*

The nature and amount of *sediment supplied* to coastal domains are essentially controlled by *terrigenous input*: suspended and bottom-load carried by rivers, coastal streaming, cliff erosion, and soil and dune reworking by land winds. Land-derived materials are particularly abundant near river mouths supplying suspended matter that feeds the coastal sedimentation and progradation. Sea-originating sedimentary material is usually less abundant and results from different sources:

- minerals and bioclasts brought back up from the subtidal zone by storm or tsunami waves;
- mud and clay plugging of estuaries and bays by repeated flood tidal currents;
- sea winds blowing the beach sand towards coastal dunes during ebb-tide phases.

Sedimentary particle loss in the coastal zone proceeds from two main causes: (1) the sand lateral transfer and export by waves and swell that break at an angle on the shore (see below); and (2) the particle reworking by coastal rip currents that discharge the sediment offshore, sometimes as far as within submarine canyons.

The basic coastal hazard largely depends on the respective importance of sedimentary losses relative to the gains (Chapter 4.3.2). Calculation of the *coastal sedimentary budget* is therefore essential for assessing the risk threatening a given shore zone. Most world beaches are presently retreating, since particle losses are statistically larger than gains (Figs 68, 69). In some regions the budget is stable or even positive, because the river or sea sedimentary input is important, the longshore drifting or offshore discharge is moderate, and/or both dune and intertidal systems undergo balanced quantitative exchanges (Fig. 73). Each beach or even part of a beach constitutes a sort of independent cell, which explains the occasional juxtaposition of retreating and prograding littoral sectors (e.g. the beach systems of northernmost France; Chapter 4.3.2). Each cell is characterized by its own budget of particle input, transfer and output, the algebraic sum of which determines the shore stability or instability over the course of time.

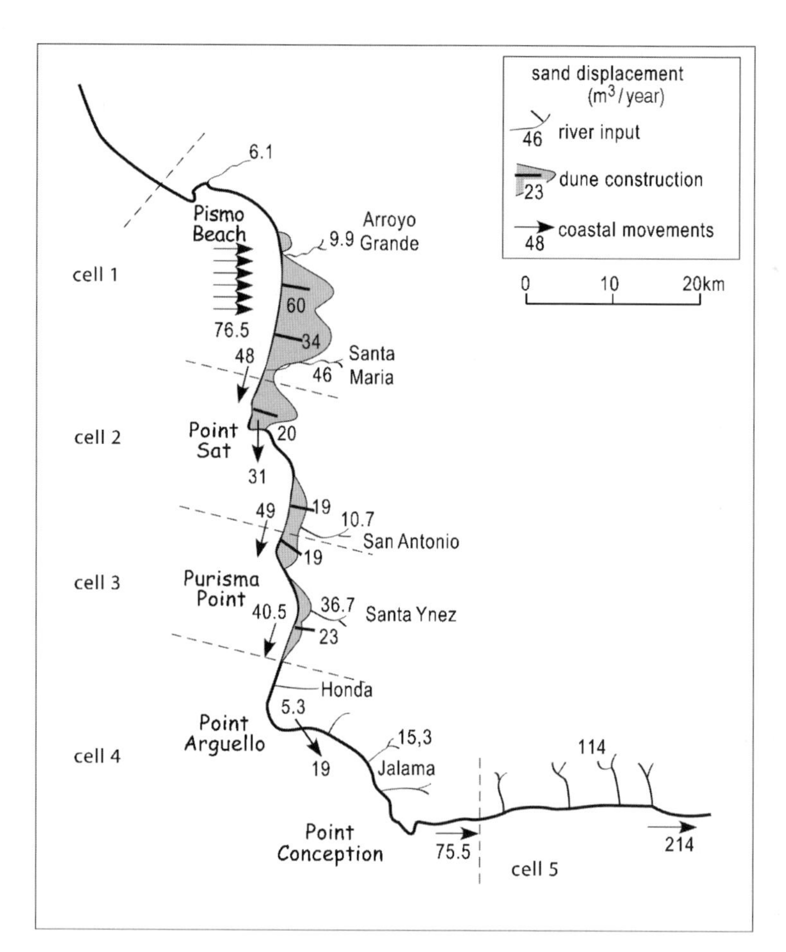

Figure 73: Example of coastal sedimentary balance: shore sand movements to the northwest of Santa Barbara, California (after Bowen & Inman; see Viles & Spencer, 1995). Several sedimentary cells are identified, which communicate with each other through longshore drift.

Coastal Erosion

The initial shaping of the coastline is principally controlled by geological and geomorphological characteristics of the land–sea transition zone (Fig. 74). The sectors essentially constituted by *soft rocks* are first actively eroded. They tend to constitute *bays*, the wide-mouthed shape of which progressively leads to disperse and diminish the wave and swell energy, to induce preferential deposition of small-sized particles (clay, silt, fine sand), and finally to protect the shore against further erosion.

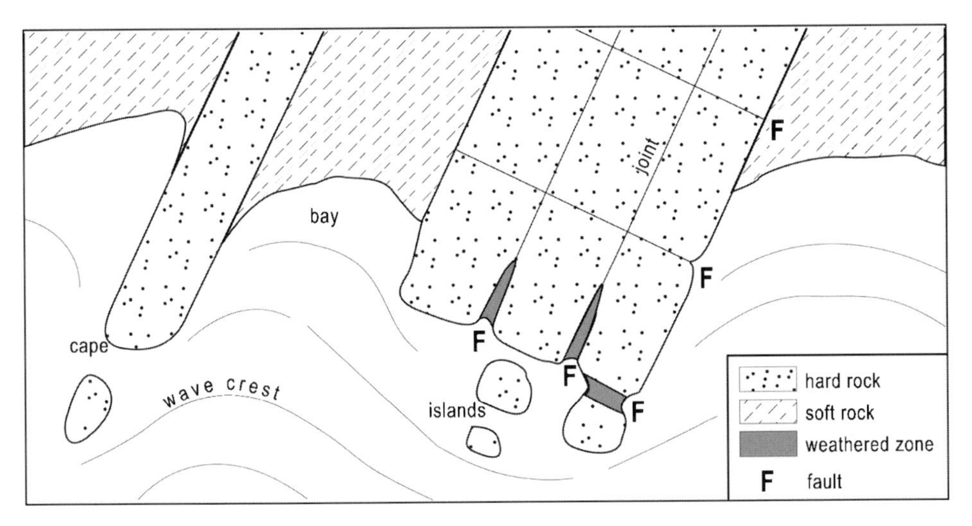

Figure 74: Geological and geomorphological control of coastal shaping (after Waltham, 1994).

By contrast the *hard rock* sectors are more slowly eroded and progressively form *capes* and salient promontories, often associated with cliff morphologies. They are submitted to high-energy converging waves and swell, and suffer continuous erosion. They constitute more or less overhanging cliffs, preceded on the offshore side by an almost flat platform situated at the mean sea level. Covered by scattered coarse sediments derived from coastal erosion (blocks, pebbles, gravel coarse sand), the platforms develop landwards at the cliff's expense, more rapidly where rocks are faulted, fissured, broken, weathered and softened.

The shore morphology formed very progressively during Quaternary times through the complementary action of sea-level changes, river inputs, and locally tectonic instability. The shape of coastlines for a given sea level is acquired all the more rapidly as all dynamic agents work in a concurrent way. For instance erosive coasts rapidly get an ad hoc morphology if the following conditions occur together: exposure to dominant winds inducing high-energy waves and swell; steep submarine slope morphology and topographic narrowing, leading waves and currents to be stronger; a slope-breaking zone, causing water energy to be violently released. In addition shore erosion processes tend to increase if breaking water carries abundant particles, namely pebbles and gravel that bombard the rocks, induce vibrations within the ground and enlarge the shore crevices.

Water set in motion by wind on the sea or lake surface is responsible for most sediment reworking and transfer processes occurring in the shore zone. By contrast the currents resulting from tide action play an accessory role in particle erosion and displacement, their influence being larger under macrotidal conditions (tide range >4 m) than under mesotidal (2–4 m) and especially microtidal (<2 m) conditions. Forming offshore beyond the coastal zone, at a distance of between a few hundred metres and several thousand kilometres from the shore, wind-induced *waves* get progressively organized and constitute a regular system called the *swell*. The energy released by the breaking waves and swell approaching the coastal zone and attaining the shore determines the water erosive force:

$$E = \rho g H^2$$

where: E, energy; ρ, water density; g, acceleration of gravity; H, wave or swell height.

The recurrent tendency for waves and swell to erode the coastal zone increases during storms, at which time several factors may combine to result in a strong, temporary sea-level rise:

- strong atmospheric pressure diminution when air masses of very different temperature get in contact (Chapter 4.2.1). A pressure diminution of 100 hPa determines a water rise of 1 m;
- violent winds pushing surface water towards the land. A wind blowing at 80 km/h causes water to rise by 1 m;
- spring-tide period. In macrotidal conditions the maximum flooding level during the spring-tide phase may be 2 m above that of the neap-tide phase;
- resonating effects due to basin narrowing and shallowing, or/and to an almost horizontal seafloor slope. This is the case, for instance, where the English Channel- or North Sea-originating swell enters the Pas de Calais between southwest England and northernmost France, or where wave and tidal currents enter the Mont St. Michel bay located between Brittany and Normandy.

The conjunction of such factors may determine the formation of *storm waves* that reach or even exceed a height of 6 m more than in normal conditions. In addition heavy rainfall is often associated with coastal storms, causing an increase in river levels and dynamics. The gathering together of all these conditions is responsible for the major phases of coastal erosion, the breaking of littoral sea walls, washover of flat coastal dunes, and flooding of backshore land depressions. Finally the storm waves that break on the shore induce violent water-retreating movements, which belong to the *rip currents* group (Fig. 75). These return currents dig erosive furrows, remove and disperse particle materials and therefore actively participate in aggravating the coastal sediment depletion.

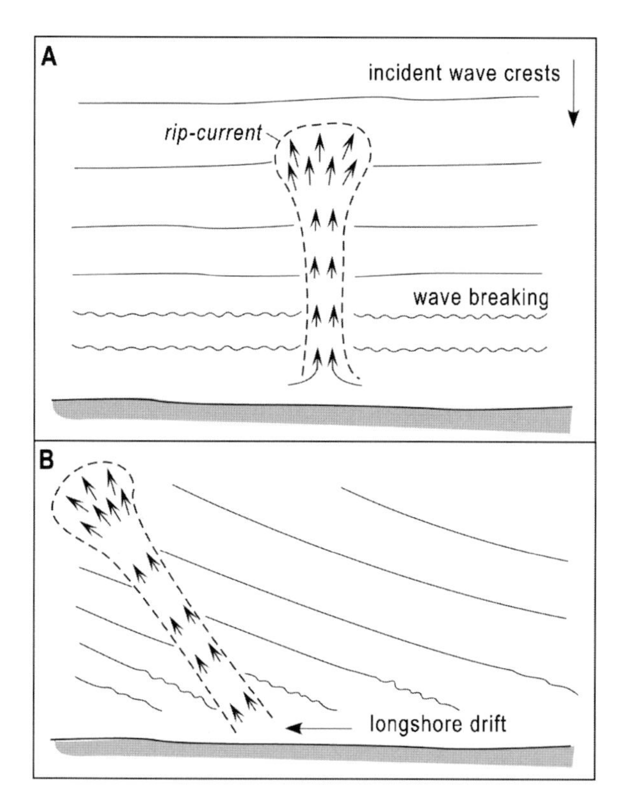

Figure 75: Formation process of rip currents through the action of waves and swell displacing parallel (A) or obliquely (B) to the shoreline. During storms such return currents are responsible for active sediment erosion and particle export towards offshore zones.

Longshore Sedimentary Drifting

The waves and swell pushed coastward by sea winds are generally oriented obliquely to the shoreline. Successive breaking and swash therefore occur at a certain angle (usually less than 10°), whereas the backswash tends to be oriented along the bottom, greater slope, i.e. at a right angle to the shoreline. The repetition of these movements at each wave-breaking and swash–backswash determines a continuous displacement of water parallel to the shore and a correlative transfer of suspended or bottom-displaced sedimentary particles, especially of sand. This transfer constitutes the *coastal drift*, which works both along the shore (*beach drift*) and at shallow water depth (*longshore drift*). The drift speed depends on the angle value of incident waves reaching the shore, on the wave period and height, and on slope and roughness values of the very shallow bottom surface. Sand displacement values attain commonly 0.3–0.6 m/

sec and may exceed 1 m/sec for sea-originating storms, when the resulting movement is due to the combination of tide and wave currents.

The longshore drift, which frequently operates continuously in the same direction, is responsible for the *lateral transfer of considerable sediment amounts*. For instance a swell characterized by an offshore bending (i.e. the swell height/wavelength ratio) of 0.02 and reaching a sandy shore (mean particle diameter of 0.3 mm) at an angle of 10°, with a periodicity of 8 sec and a breaking height of 2 m, induces the daily lateral transfer of 1,520 m^3 of sediment (Paskoff, 1994). The Aquitaine coast south of the Gironde estuary along the Bay of Biscay displays a southwards sand drift over a distance exceeding 200 km, the annual amount of displaced sediment varying between 200,000 and 630,000 m^3, depending on the beach sectors considered.

By causing the erosion of "upstream" beaches (relative to the current direction) and the sand feeding of "downstream" beaches, the *longshore drift participates in an active way to the coastal sedimentary dynamics*, as well as to sediment losses and gains affecting the successive beach cells. Nevertheless *this process modifies only slightly the regional sedimentary budget* in the coastal zone, the material subtracted from a given 'upstream' sector being restored in 'downstream' sectors. On the other hand the longshore drift is notably responsible for the *geomorphological shaping of sedimentary coasts*:

- formation of sand spits under the lee of capes (Fig. 76);
- building of sand or pebble bars facing the river and tidal currents at estuary mouths;
- construction of sand tombolos connecting an island to the continental shore;
- smoothing of coastline irregularities by isolation of backshore marshes and formation of natural polders;

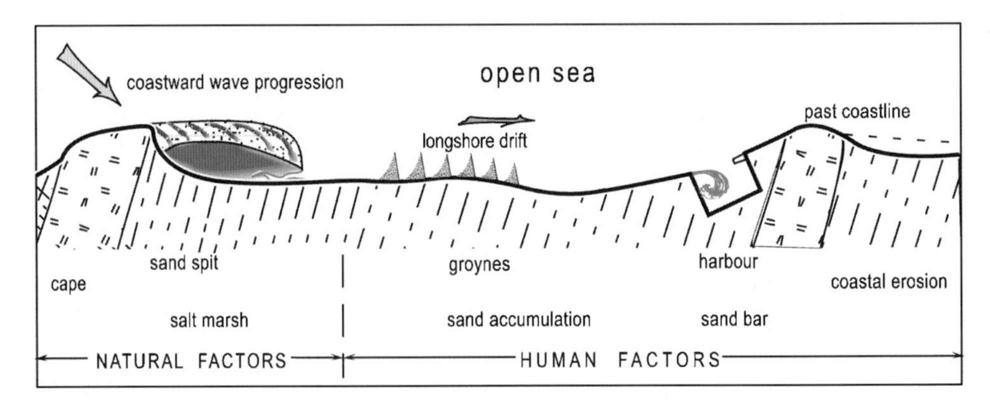

Figure 76: Sedimentary consequences of the longshore drift: natural and man-induced factors (after Waltham, 1994).

– seaward or landward migration of barrier islands and submarine bars by 'upstream' erosion – 'downstream' sedimentation processes.

The Holocene Erosive Crisis

The worldwide tendency for coastlines to retreat because of an excess of sedimentary losses relative to the gains (Chapter 4.3.2) is essentially attributed to the *natural evolution of erosion and deposition conditions* since the last glacial maximum:

- The last glacial period, the intensity of which was at a maximum about 20,000 years ago, corresponded to a very low sea level (about –130 m relative to the present level). Such a situation favoured intense continental erosion, under the combined effects of profile readjustment of river basins and abrasion by glaciers. Huge amounts of detrital sedimentary material eroded from exposed landmasses accumulated during this period on the outer part of continental shelves.
- The rise in sea level that started around 16,000 years ago, i.e. when the post-glacial warming began, determined the landward displacement of a large part of this detrital material, which was pushed by waves and swell in the direction of present-day shelves and shores.
- Due to the generalized rise of both water and sediment, the early to middle Holocene period was marked by abundant stocks of detrital material in the coastal zone. The present sea level was reached about 7,000 years ago, during a phase characterized by the stabilization in inner shelf zones of large sandbanks, the feeding of sand beaches and functioning of longshore drift, and the beginning of the sedimentary plugging of river estuaries. During this period the sedimentary stock was fully sufficient for compensating the loss due to rip currents and coastal storms.
- During the few last thousand years (late Holocene), the terrigenous detrital supply diminished drastically due to the high sea level that induced the river basin profiles to stabilize and therefore the continental erosion to decrease. The diminution of the sedimentary input concerned particularly the coarse particles (gravel, sand) that are in dynamic balance with most coastal agents (waves, swell, tidal currents). It affected most world regions except those subject to active tectonics (e.g. the Alpine, Himalayan and Andean ranges). The sediment loss due to coastal erosion is now less and less compensated by new detrital supply, which tends to aggravate the disequilibrium of the sedimentary balance. The recent and present situation therefore results from the *transition from abundance to depletion of the sediment availability within coastal zones*.

– Some observations suggest that short-term changes marked by a recent increase in storm frequency and intensity are superimposed to this late Holocene, long-term evolution. For instance the last 19th century decades and most of the 20th century have been statistically marked in the North Sea by an increase in the storm waves' height and energy. It would be premature to draw conclusions about either the possible long-lasting significance of such changes or their potential relationships with global environmental modifications (Chapter 11); but such phenomena indisputably aggravate the coastal hazards and the present-day disequilibrium of the littoral sedimentary balance.

4.3.4. Human Influences in the Coastal Zones

Aggravation of the Coastal Risk

Demographic Pressure

Man has always occupied coastal zones more densely than inland zones. This is due to the multiplicity of occupational modes offered by sea-bordering regions to mankind for developing its activities: building and communications facilities, cultivation and livestock farming, fishing and hunting, harbour and industrial facilities, sand and gravel exploitation, outdoor and leisure activities, etc. *This propensity for the privileged human occupancy of coastal zones has increased in a dramatic way parallel to the population growth characteristic of the 20th century.* At the beginning of the third millennium three-quarters of world inhabitants live less than 60 km from ocean and sea coastlines, and about 60% of them live in the immediate vicinity of shore areas. In the USA the number of people living close to sea borders doubled between 1940 and 1990, whereas that of inland inhabitants increased by one half only. Most modern megapoles are located close to marine coasts: Tokyo, Shanghai, Bombay, Jakarta, Sao Paulo, for example. Such considerable and still increasing shore occupancy determines both an accelerated deterioration of coast environmental quality and an augmentation of protection devices built against the natural hazards caused by waves, storms, hurricanes, etc. (Fig. 77). In the same way as for the densely populated hydrographical basins along major river courses, the excessive occupancy of the normal hazard zone constituted by the littoral fringe, where short-term (storms, inundations) and long-term (erosive crisis) risks combine, expose mankind to more dramatic, nature-induced disasters.

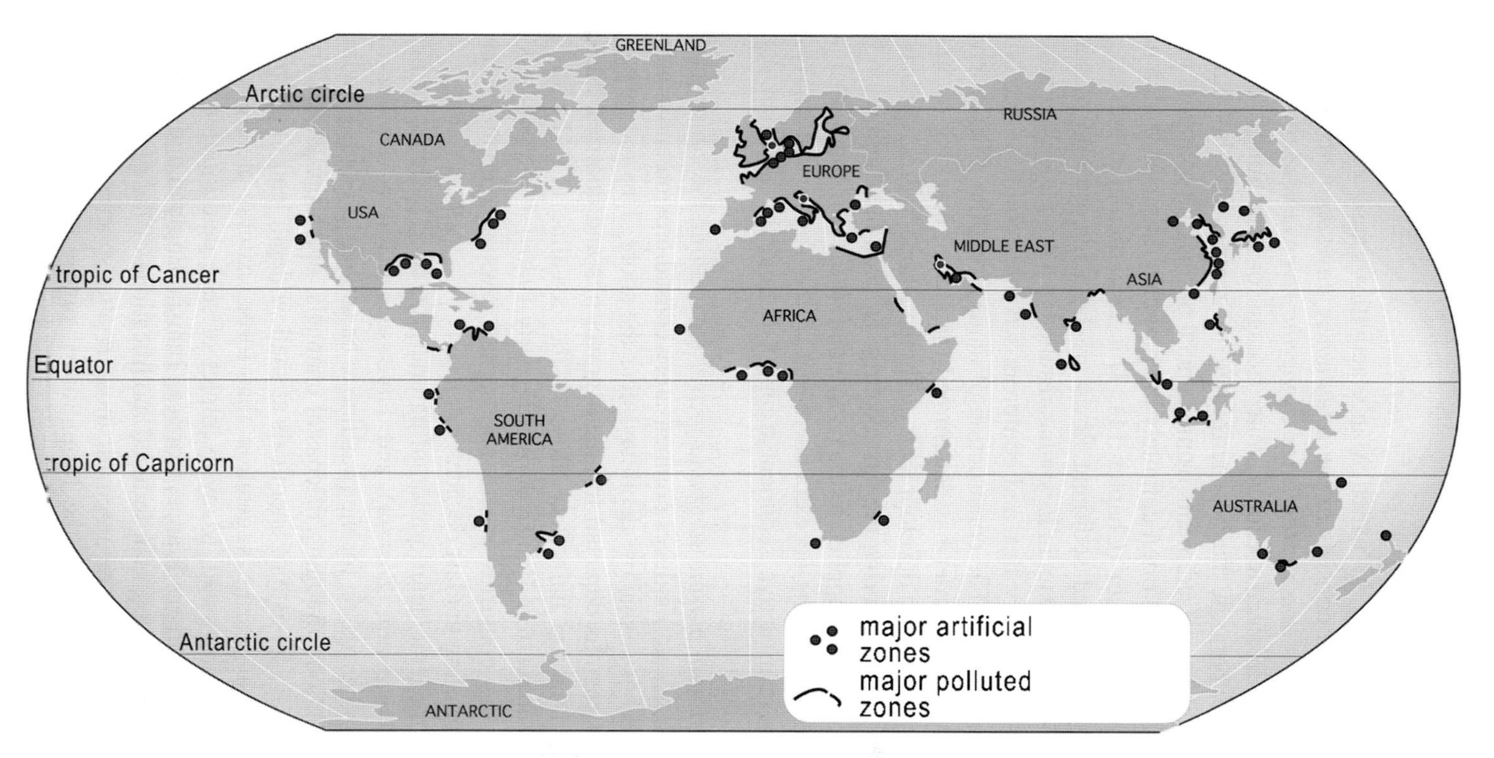

Figure 77: World distribution of main coastal protection and pollution zones (after Kelletat; see Viles & Spencer, 1995).

Acceleration of Erosion

The most serious way in which Man's activities cause accentuated coastal erosion is indirect. It consists of the *retention of huge amounts of sedimentary material behind the dams established along the river course* for various uses: production of hydroelectric energy, water storage for irrigation, flood diversion, etc. In spite of periodic flushes performed to compensate this unintentional supply and to evacuate sediments from artificial basins, the countless man-made lakes retain durably and often definitely the particles in transit, which no longer reach the marine domain (Chapter 11). This is particularly the case in arid regions where water and sediment flows are both intense and seasonal. The building of the Aswan Dam in Egypt has resulted in a dramatic diminution of the downstream sedimentary flux and compromised the stability of the Nile River delta (Chapter 11.4.2). The construction of dams along the Tijuana River close to the Mexico–USA border has determined the annual blocking of 600,000 m^3 of alluvia. Hydroelectric developments along the Rhone hydrographical basin in southeast France has, during the past century, induced the diminution by about 90% of the solid load supplied to the river mouth (i.e. from about 40 to less than 5 million tons/year). The accelerated retreat of the Togo coast in West Africa results partly from active sediment trapping behind the Ghanian dams built along the Volta River.

The extraction of marine sand and gravel for meeting the construction needs (roads, buildings, etc.) induces compensatory coastal erosion and transportation if it is carried out at shallow depth and near the shore. This risk is particularly important for regions depleted in inland gravel pits and quarries, and/or when the constructions are built very close to the coastline (e.g. building of the Atlantic Wall in western France during the early 1940s). It nevertheless tends to diminish in the countries where appropriate laws and controls have been implemented during the last decades.

The anthropogenic deterioration of natural protection systems formed by sea grass (e.g. *Posidonia*, *Zostera*) and large seaweed (e.g. *Laminaria*) fields represents another reason for the coastline to be actively eroded. Trawling on the inner-shelf seafloor, dredging of coastal channels, inappropriate boat mooring, and the disposal of waste or chemicals at shallow depth lead to the progressive destruction of submarine vegetation that forms natural breakwaters against landward waves and swell. Many examples of such deterioration are recorded off the coasts of southern France, Corsica, Tunisia, England and Denmark.

Let us also recall that man-induced modifications of the global environment are likely to determine an intensification of atmospheric and oceanographic phenomena responsible for the formation of highly erosive storms (Chapter 11).

Disruption of Natural Exchanges

The man-made developments intended to prevent beach sediment export (e.g. groyne building) or to offer safe sites for port operations (harbour and channel digging, pier construction) induce a modification and most frequently a *diminution of the longshore drift*. The trapping of particles in transit determines a down-current depletion of the sediment supply and a correlative augmentation of erosion processes (see Figs. 76 and 77).

The construction of buildings, walkways and alongshore jetties at the immediate border of sand beaches leads to prevention of cross-shore particle migrations and sedimentary exchanges between the coast and backshore dunes. Such constructions also frequently lead to strengthening of breaking waves. This is responsible for an irreversible sedimentary depletion, which increases each year and finally aggravates the beach erosion processes

By harnessing the continental discharge *through channel confinement at river mouths*, human activities may trigger an upheaval of the coastal morphology. The local diminution of particle supply may determine intense shore erosion, and active deposition being displaced either offshore or in the down-current coastal direction. For instance, the artificial channelling of the Magdalena River delta in north Columbia (South America) induced a series of spectacular changes:

- retreat of the deltaic front at a speed averaging 65 m/year since 1935;
- disappearance of several islands located near the river mouth;
- very rapid growth (230–430 m/year) of a sand spit located 15 km to the west (Puerto Columbia; Fig. 78);
- destruction of several coastal cities;
- silting-up of some harbours located up to 60 km to the southwest on the border of the Caribbean;
- other changes affecting the coast as far as in Cartagena region located 120 km west of the Magdalena river mouth.

Indirect Effects

The existence of close interactions between natural processes and the human occupancy of coastal domains is illustrated in several other ways. *Dredging* for maintaining ships' access to harbours and channel *digging* for allowing boats to cross shallow-water sand banks, lead to some modifications in the sedimentary

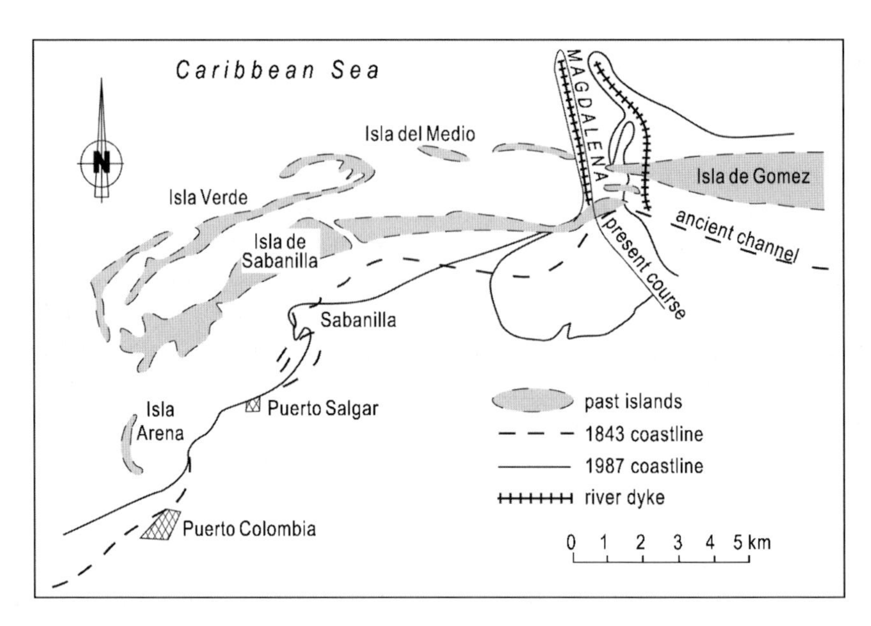

Figure 78: Modification of the coastal morphology of Columbia, west of the Magdalena River delta, between 1843 and 1987 (after Martinez et al., 1990). Most changes result from the construction of river dykes situated close to the mouth between 1925 and 1935.

balance, causing increased erosion and sand transfer, or to interruption of the sediment drift

The construction in the 1950s within Corpus Christi bay (Texas) of a navigation channel separating Mustang Island and its western end (which became Shamrock Island) induced a disruption in the alongshore sand drift, and strong erosion starting in the early 1960s. Then the drifting was displaced from the north to the south of Shamrock Island, causing, until the end of the 1990s, important morphological changes, the deterioration of maritime marshes, and a long-standing disruption of the coastal ecosystem (Williams, 1999).

Reciprocally the natural mobility of shallow-water sediments is likely either to *hamper boat navigation* through the modification by storms of the sand banks' volume, or to threaten the stability of seafloor cables and pipelines. This is particularly the case in coastal regions marked by intense maritime (or lacustrine), urban and industrial activity, as for instance in the southern North Sea and eastern English Channel (Trentesaux & Garlan, 2000). Let us also remember that excessive water or oil pumping within the ground of coastal regions may cause serious subsidence and compaction phenomena, favouring marine ingression and erosion, especially during offshore storms (Chapter 3.2.2).

Forecasting, Prevention

Beach Protection

Coastal works have been conceived and implemented for a long time to protect shores against marine erosion and storms. *Sea walls* built along the backshore parallel to beaches frequently ensure an efficient protection against waves and swell; but they are generally unsightly and, more important, they induce both a more intense wave breaking and the suppression of beach–dune sedimentary exchanges. This favours the progressive loss of the sediment stock and may end in the disappearance of sand beaches

Groynes, which are built perpendicular to the shoreline and generally arranged in series, are implemented in order to block the transit of sedimentary particles subject to longshore drift. But the diffraction of wave crests favours erosion on the groyne lee-side (Fig. 79, A). More globally, the local halt of the longshore transit is detrimental to the down-current beaches, which tend to dwindle progressively and disappear

Breakwaters constitute sea walls built directly on the seafloor at very shallow depth and close to the beach, and are intended to absorb the wave energy, as do natural barrier-islands. Breakwaters form simple bars or are T-shaped and consist of large blocks of rock, tetrapods linked to each other, or concrete masses. They are placed at various submersion levels depending on the tidal regime, and their spacing is calculated for both allowing seawater to progress landwards and breaking the wave-system energy. Breakwaters usually serve in an efficient way for protecting the beaches from erosion, but nevertheless tend to induce disequilibria of the sedimentary balance in the direction of the longshore drift (Fig. 79, B)

Jetties and piers ensure a useful protection of harbours, access channels, and river mouths, but may induce noticeable changes in the sediment transit and erosion–sedimentation processes (Fig. 79, C and D; Fig. 78). This may lead to negative effects, mainly because of the interruption of the longshore drift.

The influence of coastal defence systems on a worldwide scale is considerable (see Fig. 77). The highest density of permanent structures is probably registered in Japan, where more than 25% of the coastlines are equipped with protection walls and thousands of breakwaters, groynes and harbour constructions. Similarly, the Netherlands coast, the north Mediterranean shores and the U.S. coast of the Gulf of Mexico are also densely protected by artificial structures. In France one-tenth of the metropolitan coast length is designated to fighting against shore erosion and drifting, and on average one protection or control system is listed per kilometre. The numerous structures implemented are often

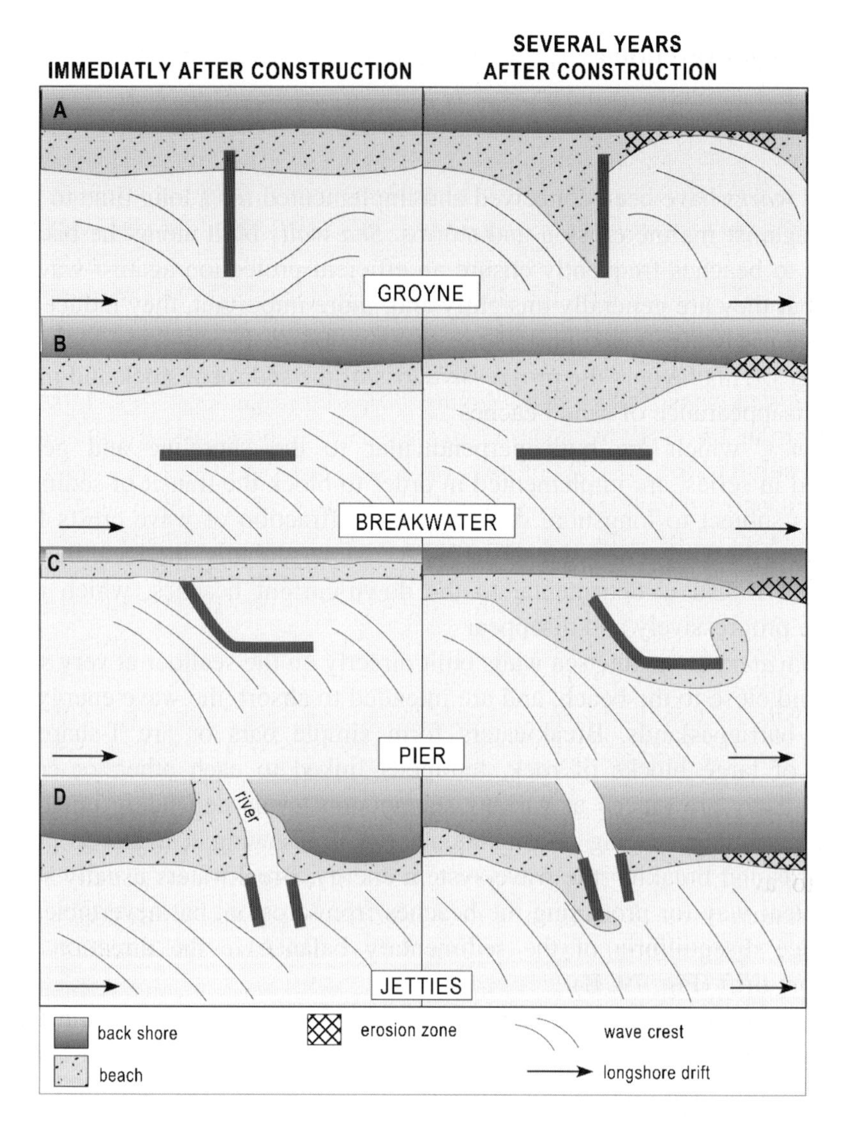

Figure 79: Sketches showing the effects on longshore drift of groynes, breakwaters, piers and jetties (adapted and completed from Keller, 2000).

of an ambivalent efficiency, cause disruption of natural processes and necessitate periodic, expensive replacements.

Beach Treatment

Alternate solutions to coastal engineering constructions have been implemented progressively during the last decades for limiting coastal erosion. They correspond to *soft, mildly disruptive treatments*, which mainly lie in *beach widening and building up by sediment replacing*. This practice consists of replacing the sediments from thinned beaches by new sediments of identical density and grain size: sand, gravel or pebble. The substitute material is either dredged or pumped in submarine (or sub-lacustrine) sectors located at a depth greater than 20 m and devoid of indirect risks, or sampled in fossil coastal dunes. The reloading methods were first developed on eastern North American coasts periodically submitted to storm and hurricane waves. The shore of Miami Beach in South Florida, which has been eroded dramatically since the 1950s, despite the construction of numerous permanent structures (groynes, sea walls, concrete walkways, etc.), has been subjected to a huge restoration programme that began in 1970. About 160,000 m^3 of sand have been pumped annually in a nearby submarine site, which has allowed re-establishing a 200-m-wide, stabilized and protective beach. The total volume of displaced sand attained 10,000,000 m^3. The sediment reloading has been very expensive (about 62 million U.S. dollars) and complementary sediment supply is periodically necessary, especially after hurricanes (e.g. in 1979, 1992). But the artificially restored beach system proves to be durable, scarcely disruptive for the environment, and aesthetically pleasing

Replacing sediment on eroded beaches has been increasingly developed since the 1980s. More than 600 km of U.S. coasts have been artificially supplied with sand. Additional sand of 600,000 m^3 is used annually to replenish the Netherlands beaches. Various Mediterranean and Atlantic beaches of France are periodically reloaded with sediment. Precautionary measures are essential in order to avoid indirect effects (e.g. shore-re-activated erosion when filling shallow-water pits, and sediment depletion in the down-current direction). In some cases the restoration work remains insufficient due to the excessive intensity of longshore drift or of rip currents.

Other soft techniques are employed locally for limiting or compensating beach erosion:

- Racking of sea water issuing from breaking waves and swash, by pumping directly inside sandy beaches. This action favours rapid water seepage, prevents the sand becoming over-saturated, and therefore tends to limit the longshore drift.
- Water spraying during low-tide phases on wide beach surfaces in arid and windy regions subject to a macrotidal regime. This helps to prevent sand desiccation and subsequent aeolian erosion.
- Establishment of by-pass systems for clearing the man-made obstacles to the natural alongshore transit: groynes, jetties, piers, and harbour constructions. The by-pass is obtained by pumping the sand to take it beyond the obstacles, or by using pressure pipelines.

– Raising of the backshore, planting of vegetation adapted to brackish conditions in coastal dunes or wet zones of barrier islands.

Beach Surveillance and Construction Rules

Prevention and protection against coastal hazards are required to better understand and therefore to anticipate the danger. This supposes that people living close to shore zones take into account the inevitably changing character of the coastline shape and if necessary accept that pressure from humans needs to be lessened, especially in the littoral zone.

Watching methods are more and more efficient at assessing and modelling the chronic risks of coastline erosion and retreat:

- precise topographic survey of beach longitudinal and transversal profiles;
- use of sampling, sonar, or air-transported laser data for drawing up, at successive time intervals, bathymetric and lithological maps of the sub-tidal to supra-tidal range;
- measurement and monitoring, by multi-spectral scanner tools and in situ automations, of the suspended matter distributed in the water column and near the bottom, and projections for anticipating the mid-term evolution of the near-shore sedimentary balance;
- combined surveillance of both exposed and submarine zones by establishing geographic information systems and processing field numerical models (e.g. Gérard, 1999).

Another approach that is developed in a growing way consists in *leaving the natural processes to interact* in little-populated coastal zones exposed to hazards, in discouraging people to settle in or to exploit most critical areas, and sometimes in abandoning or destroying the existing man-made structures and buildings. Protection efforts are then concentrated on coastal sectors that can be better controlled and are more essential for human activities. This requires an adequate education of shore-side populations, and finding a way to reconcile the opinion of user communities whose respective interests diverge a priori (e.g. fishermen versus builders, and planners versus ecologists).

Building rules and standards, specifically enacted for coastal zones and based on precise mapping and scientific data, should be implemented progressively. Such rules should allow for defining successive zones characterized by a decreasing risk when moving from shore to inland levels: e.g. immediate shore-bordering level, where destruction could possibly happen within 10 years and where building is not permitted; backshore level, where the risk is estimated at 30 years and where buildings are of light and movable construction; a further level, with bigger but non-permanent buildings for a 60-year risk periodicity; etc.

Hurricane and Tsunami Forecasting, Protection Measures

Nature-induced risks threatening the coasts with a potentially exceptional intensity require specific protection measures. *The hurricane risk*, which combines aeolian and coastal hazards (Chapter 4.2.3, 4.3.1), is much better controlled in high-technology countries than in developing regions (Chapter 4.2.4). Constructions and other developments must be designed for facing very strong wind and/or sea-level rise: underground shelters, windbreaks, water defences, various other protective structures. Information and evacuation procedures are largely based on historical data, field characteristics, experiments and simulations, and take into account the risk maps established on various scales along coastlines (e.g. Fig. 80).

The potentially considerable danger represented by *tsunamis* (Chapter 4.3.2) is generally reduced by the relatively long propagation time needed for devastating waves to cross the affected ocean domain. Forecasting measures

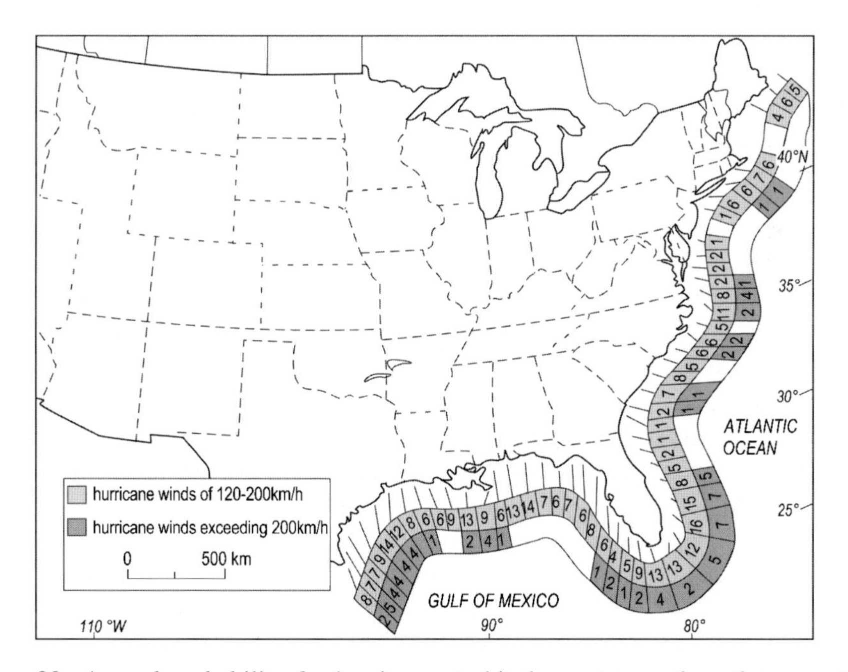

Figure 80: Annual probability for hurricanes to hit the eastern and southern coasts of USA, considered through 80-km-long segments (after Council on Environmental Quality; see Keller, 2000).

become efficient when tsunamis form at a distance that exceeds 750 km from threatened coasts. This allows for at least 1 h for alerting the shore-side populations. An *international warning system* has been implemented for most critical regions of the Pacific Ocean, where tsunamis form frequently. The system is based on a network of 30 seismograph and 78 tide stations, which deliver continuous information on seismic activity and mean sea level. The numerous data obtained are processed in real time. The time interval available for evacuating the populations living on Hawaiian islands from coastal zones varies between 5 h and more than 15 h, depending on whether earthquakes occur in Alaska or southernmost American margins (Fig. 81). The Pacific coastal regions statistically hit by the strongest giant waves (e.g. Japanese coasts) are equipped with protection and prevention structures that have risen at different levels starting from the subtidal zone: large, immersed breakwaters, re-shaped beaches absorbing the water energy, emergency coastal roads built on highly resistant stilts, buildings moved to inland zones and protected by living or artificial barriers.

The efficiency of tsunami-prevention measures is often remarkable. The regional and international alert systems are completed by local information and evacuation operations that tend to increase the quality of the whole protection

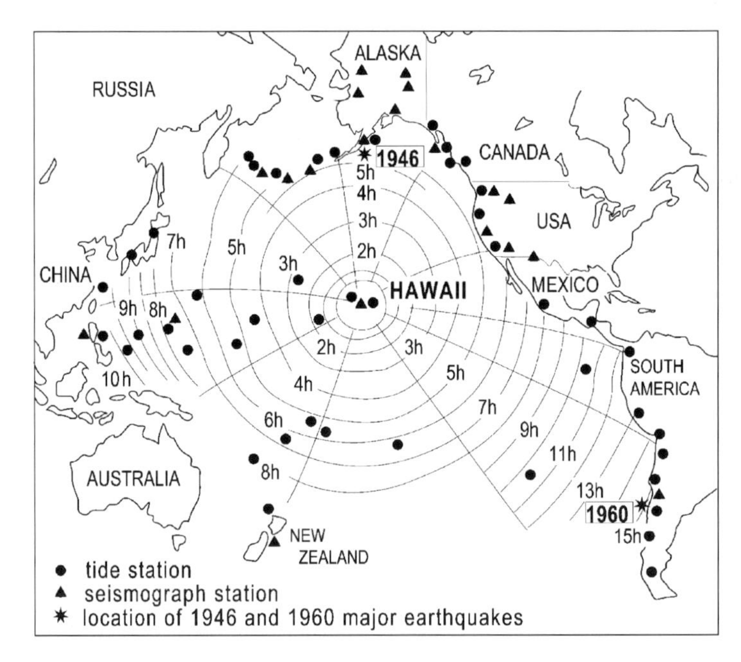

Figure 81: Tsunami alert system in the Pacific Ocean, and wave propagation time from several tsunami formation zones to the Hawaiian archipelago (after Murck et al., 1996, and NOAA; see Keller, 2000).

system. The results obtained speak for themselves. For instance more than 6,000 people died in Japan due to 14 tsunamis that hit the coasts before the implementation in 1990 of the integrated alert and protection network; the number of deaths was reduced to 215 when the network was working during the 1990s, a period marked by as many as 20 tsunamis.

Land Reclamation from the Sea

The fight against coastal hazards, which for centuries led the inhabitants from flat countries located near sea level to build numerous protection structures, has also permitted to annex large marine surfaces, which became *polders* and were progressively transformed into rural and urban continental zones. For instance, more than half of *the Netherlands* surface is situated at or slightly below the level of the North Sea. The construction of dykes for containing the river floods and storm-induced sea water started about 1,000 years ago. Water pumping by windmills for drying the land gained from the sea began as early as 700 years ago. Large and systematic extension works have been developed for more than 300 years. The building of barriers linking prominent coastal sectors has allowed, via natural longshore drift and huge construction and pumping works, the transformation by polders of 20% of the present land surface (Fig. 82). The 30-km-long Barrier Dam has been built since the 1930s for preventing sea ingressions and transforming the Zuider Zee lower estuary in a freshwater lake (Ijsselmeer) that ensures the drainage of 165,000 hectares of inhabited land.

Note that excessive polder developments may determine unforeseen and unwanted sediment accumulations, communications' blocking and landscape deterioration. This is the case in the accelerated silting-up of the Mont-St. Michel bay located in western France between Brittany and Normandy, or for the sediment filling of the Seine estuary.

A Few Landmarks, Perspectives

- Some serious damage affecting Man and his property, as well as the solid surface envelopes of the Earth, result from natural processes forming in the atmosphere, and which the fluid envelopes of our planet transmit. Heavy rain, strong winds and powerful waves are responsible for floods and inundations, storms and cyclones, coastal erosion and shore deterioration. The water and wind action is able to mobilize soft materials of various sizes that are reworked from rocks, soils and sediments (blocks, pebbles, gravel, sand, silt, clay). These dynamic factors intervene frequently in an independent way within terrestrial environments (e.g. floods, storms), whereas they are often combined in marine coastal zones (e.g. waves and swell associated with storms).

- Flooding constitutes the most common, frequent and expected natural hazard, for it is inevitable that a river periodically overflows its major course. Two main types of flooding occur, which present very different characteristics and danger. First, the alluvial plain inundations tend to cover very large surfaces, resulting from rainfall in upper basin regions and collected by upstream tributaries, potentially presenting a big danger and causing considerable damage to property. The time interval separating upstream precipitation and downstream flooding allows documenting of the alert systems and limiting of the danger to people. This is facilitated by monitoring and real-time simulation techniques showing the water advance and danger, as well as by the extension of protection devices (dykes and embankments, flood-absorbing

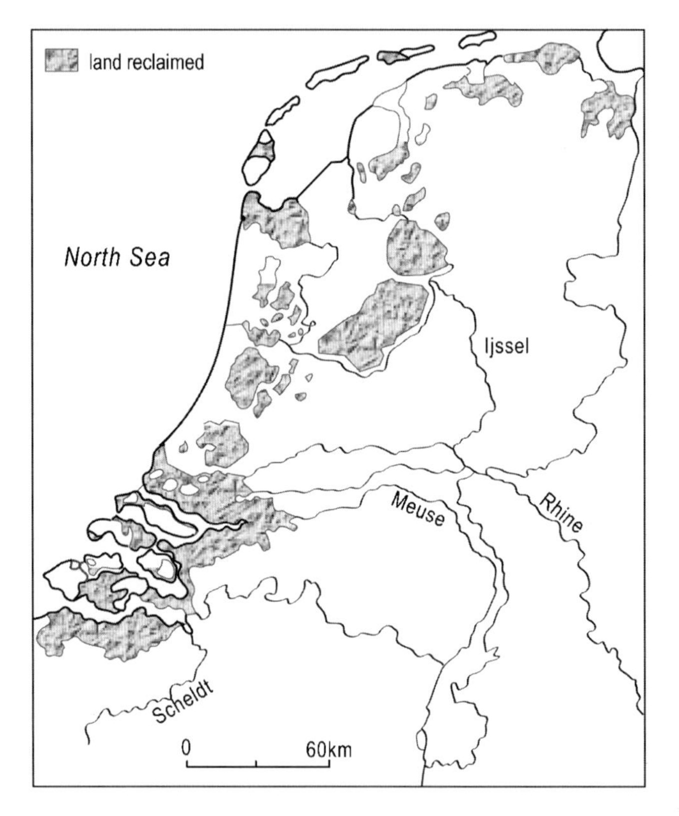

Figure 82: Regions of the Netherlands gained from the North Sea through dyke building and water pumping (after Craig et al., 1988).

and diversion basins, etc.). Progress in flood forecasting and protection is particularly efficient in high-technology countries located outside over-populated deltaic zones.

Second, torrential floods basically form in mountainous regions marked by irregular precipitation, are of local extension but of strong intensity, and are difficult to forecast and especially to prevent. The damage may be considerable for both people and goods. Flooding risk has been indirectly aggravated by recent population growth and increasing valley occupancy, which leads to artificial soil waterproofing and river-course channelling, water seepage reduction through deforestation practices, etc.

- Inundations have several beneficial consequences for human activity, which are added to the ordinary valley conveniences such as agriculture, communications, and constructions: abundant water supply, soil fertilization, and pollutants dilution, for example. On the other hand the multiplication by Man of protection structures against flooding causes damaging secondary effects: loss of wet zones, reduction of the biological diversity, disruption of delta and adjacent coastal zones, institution of a false sense of security leading to the exacerbation of the human pressure. This points to the high complexity of interrelationships between river-basin functioning and various human activities, which is responsible for various conflicts of interest. The arbitration of such conflicts necessitates an in-depth knowledge of intervening processes and of possible environmental consequences.
- Wind represents an occasionally important risk in temperate regions exposed to storms and tornadoes, and a recurrently serious risk in tropical regions subject to cyclones, hurricanes and typhoons. Wind-induced damage to people is frequently indirect: inundations associated with rain wind, storm-originating marine waves, tree and object fall, etc. The damage to property is often considerable: forest destruction, communications loss, construction collapse. Human activities may lead to amplification of the aeolian risk, particularly in arid regions: cultivation practices favouring soil erosion, planting of inappropriate species, etc. A beneficial aspect lies in the aeolian energy harnessing, the importance of which is increasing.
- Coastal hazards result from two main processes, the consequences of which to the sedimentary balance differ greatly. On the one hand, the shore erosion by waves and swell, which is accentuated by storm and cyclones, is responsible for an almost worldwide coastal retreat. This phenomenon, induced by the present inter-glacial period marked by limited terrigenous input, is locally exacerbated by human activities: active particle trapping in artificial lakes constructed along many river basins, submarine sand and gravel extraction, deterioration of natural protection structures (barrier islands, shallow submarine banks, sea-grass fields).

On the other hand the oblique arrival at the shore of most waves and swell systems determines a longshore sedimentary drift, which does not significantly modify the global balance but causes important sediment transit and morphological changes: sand spits, shore-connected ridges, tombolos, natural polders, etc. Human activities disturb these coastal dynamics by implementing numerous defence structures and other constructions such as groynes, jetties, dykes and harbours. Such structures often aggravate the down-current coastal erosion processes. Present-day protection methods tend to reduce the density of permanent coastal devices and, when possible, prioritise the sand replenishing of eroded beaches and the reshaping of the shallow submarine floor.

- Natural hazards threaten particularly the coastal zones, where various geodynamic agents may combine and be accentuated by human factors: addition of water and wind action; temporary sea-level rise due to the accumulation of rain, onshore wind, low atmospheric pressure, spring-tide and resonance effects; relative sea-level rise induced by coastal erosion, natural and artificial subsidence, and climate warming; and particularly high human density and pressure.
- Some risks of an exceptional intensity can affect the coastal domains. The tropical storms cause catastrophic damage to human property and may be disastrous for people living in flat, poorly equipped and poorly protected regions. The developed countries generally benefit from efficient forecasting and protection systems. The tsunamis acquire a considerable height and energy when approaching the coastal zone. Forming frequently in the Pacific range, they have given way to very efficient alert and protection systems, which are facilitated by the rather long propagation time of these giant waves.
- The different natural hazards induced by water and wind action display numerous common characters: (1) they are relatively frequent and widely distributed, and have been suffered by many people; (2) they are latent and inevitable, for it is quite normal for people to be subjected periodically to dynamic agents in the domains where they work naturally and which they occupy in a frequently inappropriate and excessive way; (3) they result from strong interactions between atmosphere and hydrosphere, upstream and downstream processes, offshore and coastal phenomena, natural and human influences; (4) they systematically induce severe damage to property, and more rarely and less extensively to people; (5) their prediction, which is largely based on observation and simulation data, and their protection, based on civil-engineering work and real-time monitoring, have reached a satisfactory level of efficiency, especially in high-technology countries; and (6) further progress in forecasting and prevention necessitates a concentration

of observation and data recorder networks, improvement of numerical models, and greater capacity of mathematical processing systems.

- The protection by permanent structures against the wind and continental or marine water action has been multiplied during the past centuries and is of a widespread extent in many sensitive zones (e.g. Fig. 76). But these devices have frequently attained an excessive development stage, which leads to various negative aspects: disputable efficiency when facing very high-energy events, coast and landscape deterioration, misleading appearance of security which accelerates the concentration of human occupancy and therefore the potential risk, displacement of hazards in the down-current direction, and a series of disturbances in the different ecosystem compartments. A public awareness progressively develops about the necessity to implement softer prevention and protection systems: partial destruction of some river levees for allowing a controlled meandering of water courses and restoring some wet zones; replenishing of sand beaches for compensating the shore erosion and drifting; and planting of vegetal species adapted to strong winds and soil erosion risks.
- The success of further soft protection methods against external geodynamic hazards necessitates developing an approach integrating different challenges: (1) in-depth scientific quest for understanding the mechanisms interacting at the interface between solid and fluid Earth envelopes; (2) development of qualitative and not only quantitative investigations permitting deciphering of the intimate connections linking natural and human influences; (3) accession to a simplified vocabulary allowing the different scientific disciplines to converge in a beneficial way; (4) setting up of multidisciplinary research programmes, and of PhD thesis groups focused on precise, complementary subjects and application fields; and (5) development of exchanges between scientific and non-scientific communities, as well as re-launching of wide-scope scientific journals.

Further Reading

Bodungen B. von, Turner R. K. eds, 2001. Science and intergrated coastal management, Wiley, Dahlem Series 85, 378 p.

Casale R., & Margotini C., ed., 1999. Floods and landslides. Springer, 373 p.

Cooke R. U. & Doornkamp J. C., 1990. Geomorphology in environmental management. Oxford University Press, 2nd ed., 410 p.

Gérard B. éd., 1999. Le littoral. Problèmes et pratiques de l'aménagement. BRGM, Orléans: 351 p.

Hickin E. J. ed., 1995. River geomorphology, Wiley, 255 p.

Komar P. D., 1998. Beach processes and sedimentation. Prentice Hall, 2nd ed., 544 p.

Lefèvre C. & Schneider J.-L., 2003. Risques naturels majeurs. Gordon & Breach.

Nordstrom K. F., 2000. Beaches and dunes of developed coasts. Cambridge University Press, 352 p.
Paskoff R., 1994. Les littoraux. Impact des aménagements sur leur évolution. Masson, Paris, 256 p.
Pye K., 1987. Aeolian dust and dust deposits. Academic Press, 334 p.
Smith K. & Ward R. C., 1998. Floods: physical processes and human impacts. Wiley, 408 p.
Viles H. & Spencer T., 1995. Coastal problems. Geomorphology, ecology and society at the coast. Edward Arnold, 350 p.

Some Websites

http://www.irn.org/: grouping by the International River Network of scientific, practical and sometime protest information on river, their types and regime, flooding hazards, dams, etc.
http://wcatwc.gov./main.htm: site devoted to tsunamis and built under the auspices of the West Coast & Alaska tsunami warning center: mechanisms, functioning and examples, risks and effects, forecasting
http://www.aoml.noaa.gov/hrd/tcfaq/tcfaqHED.html: grouping of worldwide observations and information on tropical storms: hurricanes, cyclones, typhoons
http://www.ngdc.noaa.gov/: past and recent information about most natural hazards that are gathered by the National Geophysical Data Center of NOAA (National Oceanic and Atmospheric Administration, USA)
http://www.meteo.fr/meteonet/services/ser.htm: extensive information on climate and meteorology, from French sources
http://www.cemagref.fr/: information on French research and technology in the field of water and erosion
http://www.brgm.fr: actions conducted in France and in various other countries in natural hazard, underground resources and hydrogeology fields
http://geohazards.cr.usgs.gov/: information on the diverse natural risks, especially those concerning North America

Part II

Exploiting Geological Resources

Chapter 5

Earth's Materials and Ores

5.1. A Strong Demand for Non-Renewable Ground Resources

5.1.1. Consumption

The Earth's mineral resources used by Man cover all facets of human activity and are of an extraordinary diversity: common and rare metals, building materials, single minerals and complex mineral compounds employed in the chemical industry, fossil and nuclear fuels, plastics and synthetic matter, etc. (see Fig. 7). *Most of these materials* buried underground or outcropping at ground level *are increasingly exploited.* This trend results from both the demographic explosion that started in the 19th century and the growing need for goods and conveniences typical of modern developed societies. Let us recall that a U.S. citizen at the dawn of the third millennium statistically requires each year more than 6 t of stone and aggregates, 6 t of fossil fuels, 800 kg of iron (for example, Introduction, 4). The daily use of building materials by each inhabitant of the planet currently represents an average of 20 kg. In 1995 U.S. inhabitants, who represent 5% of the world population, had consumed 30% of the oil on the international market, and Western Europeans (i.e. 7% of the world population) about 20%. In the same year all other countries, where almost 90% of world inhabitants live, consumed only 50% of the available oil. Close to the end of the 20th century, the annual energy consumption in developed countries reached about 40 million kilocalories (Mkcal) per person, that of developing countries only 5 Mkcal (Valdiya, 1987).

For a few decades, developing countries have nevertheless been experiencing a strong recovery. The living standards in Central Africa, India or China at the end of the 20th century were approximately 10 times lower than in West European countries; but simulations show that it should strongly increase during the next few decades. The total energy consumption in developing countries is expected to exceed that of industrial countries in 2050 (Barnier, 1992).

The world production of Earth's resources has attained considerable values during the last decade (Table 11). The most abundant materials extracted include gravel, aggregates and cements used for civil engineering works (8,870

Table 11: World production in 1994 of various materials, minerals and metals (after Evans, 1997). Values in million tons, or megatons (Mt).

Mineral resource	Megatons
aggregates	7,500
coal	4,450
crude oil	3,120
calcareous cement	1,370
iron ore	975
clay	200
sodium salt	185
silica sand	110
phosphate rock	105
gypsum	95
sulphide	50
kaolin	23
potash	23
manganese ore	22
aluminium	19
chromium ore	10
copper	9.4
bentonite	8.2
talc	8.0
zinc	6.7
feldspar	6.5
lead	5.1
borates	4.7
barytes	4.5
titanium	3.9
asbestos	2.6
perlite	1.8
diatomite	1.5
graphite	1.0
zirconium minerals	0.92
nickel	0.88
vermiculite	0.50
magnesium	0.27
mica	0.23
tin	0.18
strontium minerals	0.11

Table 12: Western Europe production and estimated reserves of some metals in 1989, and corresponding percentages relative to world values (after British Geological Survey; see Lumsden, 1992).

Metal	European production (Mt)	World production (%)	European reserves (Mt)	World reserves (%)
iron (ore)	45.000	4.6	5,200	3.5
aluminium (ore)	3.855	3.6	600	2.8
chromium (ore)	1.480	11.7	21	2.0
zinc	0.922	12.6	11	6.5
titanium (ore)	0.900	10.3	32	11.2
copper	0.274	3.0	?	?
lead	0.263	7.7	6.0	6.3
nickel	0.028	3.1	0.588	1.1
tin	0.006	2.7	0.160	3.7
tungsten	0.003	6.8	0.061	2.4
antimony	0.002	2.9	?	?
mercury	0.001373	25.0	0.079	60.8
silver	0.001035	6.9	?	?
gold	0.000025	1.2	?	?

Table 13: Western Europe production of some non-metal minerals, and corresponding percentages relative to world production (after British Geological Survey; see Lumsden, 1992).

Mineral resource	European production (Mt)	World production (%)
sodium salt	39	21.3
gypsum, anhydrite	21.6	23.8
kaolin	5.2	22.0
potash	4.8	16.3
bentonite	3.3	25.0
feldspar	2.5	45.4
talc	1.2	15.6
baryte	0.90	16.7
perlite	0.78	33.3
diatomite	0.53	28.6
asbestos	0.17	3.7
graphite	0.04	5.9

Table 14: Percentage variations of the world production of ores and minerals between 1970 and 1994 (after Evans, 1997). The recycled metal production is not included.

Mineral product	Variation (%)
aluminium	+163
feldspar	+158
perlite	+140
graphite	+138
platinum group	+ 98
sillimanite	+ 89
gypsum	+ 84
vanadium	+ 71
talc	+ 69
vermiculite	+ 66
kaolin	+ 62
sulphur	+ 61
copper	+ 47
silver	+ 43
gold	+ 38
nickel	+ 33
iron ore	+ 27
phosphate	+ 27
zinc	+ 24
potash	+ 22
mica	+ 21
diatomite	+ 15
tin	− 4
cobalt	− 16
lead	− 24
molybdenum	− 25

Mt/year, in 1994), as well as solid and liquid fossil fuels (coal, oil, 7,570 Mt/year). Some metal ores of major use are subject to huge extraction (iron 975 Mt/year, copper 9.4 Mt/year). Less common metals are exploited in relatively high quantities (tin 180,000 t in 1994; silver 6,900 t and gold 2,500 t in 1990). As a comparative example the annual world production of coal at the end of the 20th century was approximately the same as the annual solid discharge of the huge Gange–Brahmaputra–Megna river system (2,750 Mt). The production of iron ore was equivalent to that of the suspended matter annually discharged in the Atlantic Ocean by the giant hydrographical basin of the Amazon River (900 Mt).

The European production of underground materials is important as far as valuable construction and ornamental stone is concerned: granite, marble, some limestone and sandstones. Metals, the annually extracted amounts of which are large only for a very few types (iron 45 Mt, aluminium 3,85 Mt), represent only a moderate proportion of the world production (Table 12), with the exception of mercury (25%), zinc, chromium and titanium (10–12.5%). A significant number of non-metal minerals are produced by European Union countries with a high percentage relative to the world production (Table 13); this is the case for feldspar (45%) and perlite (33%) extracted from magmatic rocks, as well as for some sedimentary products: salts (halite = sodium chloride, gypsum and anhydrite = calcium sulphates, potash = potassium chloride), clay minerals (smectite or bentonite, kaolinite, talc) and biosiliceous rocks (diatomite).

The worldwide increase in ore and mineral consumption includes almost all Earth-extracted products, with the exception of a few metals such as molybdenum, lead, cobalt and tin (Table 14). The relative augmentation is generally greater than 20% per 20 years and exceeds 100% for some materials: aluminium, feldspar, perlite, graphite, for example. The growth in consumption shows a continuous trend since 1950, with a slight slowdown following the 1974 oil crisis that forced a reduction in energy expenditure (Fig. 83). Despite this important change in the international situation the world energy requirements increased by about 50% between 1980 and 2000.

5.1.2. Reserves

The ground materials, ores and minerals used by Man occur in a finite quantity within Earth's external envelopes, and most of them are not renewable (Introduction, 4). Crucial questions therefore arise about how long extraction will still be possible for, taking into account the recent demographic growth and the escalation of both requirements and consumption. A particularly critical question concerns energy-producing materials such as fossil fuels, as well as some metals of great value (gold, silver) or of very large use (copper, zinc). *Assessments of Earth's remaining, available reserves* are periodically conducted to answer such questions. As an example based on the mean annual consumption and the importance of deposits currently recognized, the time remaining at the end of the 20th century for fossil fuel exploitation was estimated to be more than 200 years for coal, 60 years for natural gas and 40 years only for crude oil (diverse sources).

Concerning the *metal ores* with sufficiently high content for allowing profitable exploitation, estimates made more than 15 years ago predicted that world exhaustion would happen in 2004 for silver, 2012 for gold, 2024 for zinc,

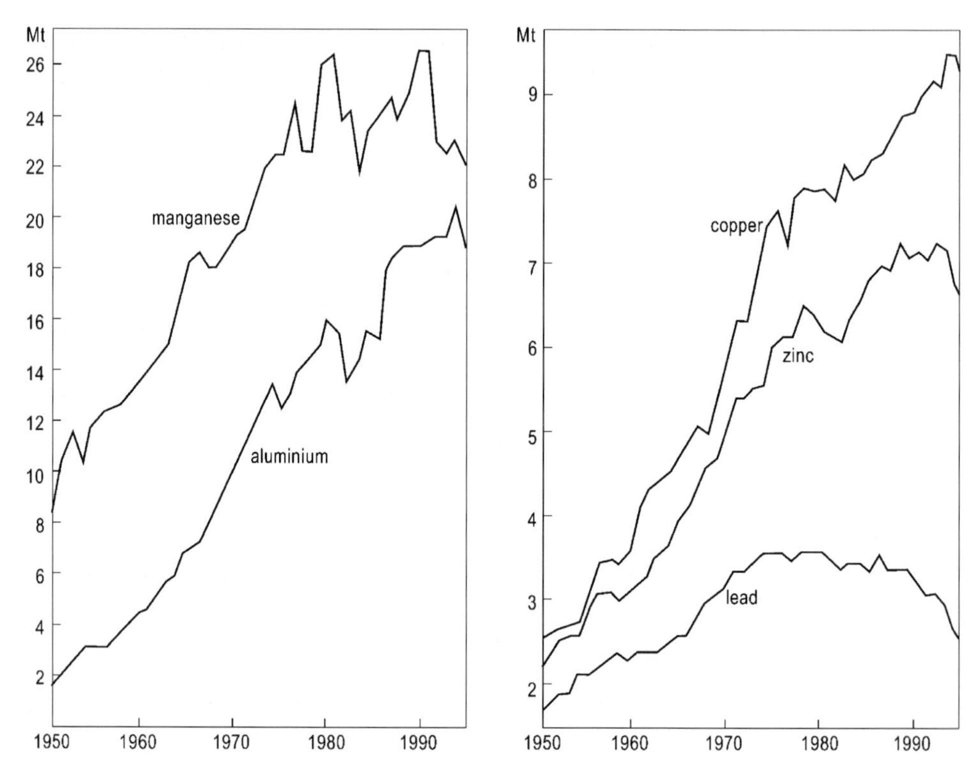

Figure 83: Variation of the world production of some metals between 1950 and 1994 (after Evans, 1997). Values in million tons (Mt).

2034 for lead, 2044 for titanium, 2045 for tungsten, 2055 for copper, 2110 for manganese and iron and 2235 for aluminium (Allègre, 1990). Other estimates based on the 1989 production of metals in Europe relative to that of the whole planet concluded different values (Lumsden, 1992): availability of zinc until 2013, of lead until 2018, of aluminium until 2190. The exhaustion of tin at that time was expected in 2010, of mercury in 2013, of iron in 2500. A few years later more optimistic dates were presented (Allègre, 1993), namely because of progress obtained in prospecting methods, discovery of new ground deposits and development of recycling techniques allowing the re-use of most metals.

The assessment of underground mineral resources and the expected duration of their exploitation are based on diverse information: amount of reserves estimated at a given time, world consumption during a given year or couple of years, simulated evolution of the needs determining the volume of extractions. Other parameters taken into account and more difficult to quantify comprise the accessibility of ore deposits and production costs, which fluctuate with technological progress, labour costs and geopolitical situation. The approximation affecting each of these various criteria explains the wide range of estimates and their changing character over time.

The future needs of mineral resources are particularly difficult to predict, for several reasons:

- The amount of an exploitable ore in a given rock first depends on the *local concentration* of the element. The concentration itself is determined by the ore deposit type and the thermo-dynamic conditions of genesis, rather than by the absolute abundance of the element in the ground. Zirconium is more abundant than copper in terrestrial formations, gallium more than lead, and vanadium more than tin, despite the fact that the former metals are much less abundant according to mining criteria. Hafnium, a metal that was discovered in 1925 only, is statistically more abundant but much more scattered in Earth rocks than boron, uranium, cadmium, iodine, mercury, silver, platinum or gold. This results from the complex, very heterogeneous distribution underground of numerous natural substances. The progress acquired in the knowledge of ore deposits, or the unexpected discovery of new mineral ores, may modify in a very noticeable way the estimate of known reserves.
- *Prospecting techniques* are based on petrographic, geochemical and geophysical investigations that become more and more efficient. The discovery of some unexpected ore deposits often results from technical progress and from the possibility of exploiting regions of difficult access and still poorly prospected: high-latitude or -altitude zones, the foot of continental margins in the oceans, deep marine basins, oceanic ridges, etc.
- The *requirements* for some underground mineral products tend to diminish due to the increasing emergence of substitute products and of synthetic and composite materials, the quality and cost price of which progressively become more competitive. The pressure on the mining resources extraction is largely controlled by the balance between needs and costs, which has determined some less-required natural products to be much less extracted during the last decades of the 20th century. This is for instance the case of iron-derived metals that were traditionally used in car manufacturing and have been partly replaced by composite substitute materials. Conversely some ore deposits such as zinc minerals have been recently subject to a rapidly increasing demand, which has determined an augmentation of exploration and extraction processes of the metal, as well as costs.
- The *recycling* of metals and other mineral products, and the *miniaturization* of equipment and numerous technical devices were considerably developed during the last decades. This has induced a slowdown of the demand and therefore a lengthening of the expected exploitation duration of various mineral resources: platinum, tungsten, copper, lead, zinc, etc.

– *Other short-term factors* may participate in hampering the predictability and induce either an augmentation of needs or costs and therefore an acceleration of prospecting investigations, or unexpected noxious effects, new legal measures and therefore a lengthening the availability of mineral resources. These factors are of a diverse nature:
  • discovery of new uses for a given ore;
  • planned limitation of production (e.g. oil, gas, metals), tension of international trade markets, political crises, embargo measures;
  • new discovery or fear of health risks (e.g. lead, asbestos);
  • difficulty in ensuring people's security (e.g. radioactive rocks and minerals) or in preserving environmental quality (e.g. some heavy metals);
  • official designation of ore-rich areas as national or international heritage zones that become protected against human activities (e.g. Antarctica, National Parks).

The unpredictable character of natural resource reserves is particularly critical for *energy-generating products*, especially if they are:

– buried deep underground and therefore difficult to locate and quantify;
– of crucial importance for the world economy;
– likely to be exhausted in the fairly near future;
– governed by confidentiality rules due to political or commercial reasons.

Such restrictive conditions apply particularly to crude oil and natural gas resources.

5.2. Main Resources and Exploitation

5.2.1. Continental Resources

The mineral resources extracted from the ground essentially comprise materials used for construction and civil engineering, ores destined for industrial use, metals marked by specific physical and chemical properties, and fossil fuels (Introduction, 4; Fig. 7).

Building materials, which are often referred to as *geomaterials*, mainly comprise igneous or sedimentary *hard rocks*. The dominant petrographic groups used in construction works include granite and other plutonic rocks, basalts and rhyolites, gneiss, schist and slates, limestone, sandstones, and flint (Table 15). The geotechnical characters and properties of these various rocks depend on their mineralogical and chemical composition, crystallization and cementation degree, grain size and texture, bedding and fracturing. Other geomaterials consist of *soft rocks* that are employed for two main uses: (1) outdoor work: more or less silty and/or calcareous clay, sand, and gravel; (2) indoor work:

Table 15: Main geomaterials: types, uses and origins (after Jackson & Jackson, 1996; Bennett & Doyle, 1997).

Material	Main uses	Main origin
magmatic and metamorphic rocks, limestone, sandstones	building material	France, Italy, Portugal
clay	bricks, tiles, pipes	UK, USA
siliceous sand	glass	Mexico, USA
limestone, clay	cement	
sand, cement	mortar	
gravel, sand, cement	concrete	
gravel, bitumen	asphalt	
anhydrite, gypsum (Ca sulphates)	plaster	Germany, Canada, France, Russia, USA
asbestos	pipes, insulation	South Africa, Canada, China Russia, Zimbabwe

plaster, amphiboles and serpentines (e.g. asbestos), and some clay minerals (e.g. vermiculite).

Ore deposits and minerals extracted for an industrial purpose are extremely diverse (Table 16). Their use strongly contributes to the efficiency and comfort of various human activities: farm production, chemical applications, civil engineering work, pharmacology, new materials, computer systems, arts, etc.

Metals are generally split into five groups:

– (1) iron group: iron, manganese, nickel, cobalt, molybdenum, chromium;
– (2) non-ferrous group: copper, lead, zinc, tin, mercury, cadmium;
– (3) light metals group: aluminium, magnesium, lithium, titanium;
– (4) precious metals group: gold, silver, platinum, palladium, uranium, etc.;
– (5) rare metals group: beryllium, caesium, gallium, zirconium, etc.

Most exploited metals comprise eight species: iron, manganese, aluminium, lead, zinc, copper, tin and nickel. Uranium constitutes another important metal that is used for producing nuclear energy (Chapter 7). These different metals *are distributed in a very heterogeneous way within Earth's rocks*. They occur mainly in old continental shields of Africa, South America, Australia, Canada, and Russia (Fig. 84). The characteristics, properties and impacts of the seven most common metals are summarized in Table 17. Iron is by far the most exploited

Table 16: Examples of the industrial use of ground mineral resources (after Bennett & Doyle, 1997; Rogers & Feiss, 1998).

Use	Ores, minerals
drilling mud	bentonite, barytes
fertilisers	phosphates, guano, potash
refractories	clay, limestone
ceramics, earthenware, porcelain	kaolin and other clay minerals
abrasives	diatomite, quartz, garnet, corundum, diamonds
pigments, fillers	iron and titanium oxides, copper and manganese salts, calcite
pharmacology	calcite, clay minerals (palygorskite, smectite)
various chemical uses	Na, K, B and Mg salts, trona kaolinite and barytes (paper-making) fluorine (metalworking); calcite (polymerisation)
jewellery	precious, semiprecious stone

metal. The annual amount of its extracted ores has dramatically increased during the last half-century: about 240 Mt in 1950, 500 Mt in 1960, 750 Mt in 1970, 857 Mt in 1980, and 975 Mt in 1994 (Evans 1997). Manganese and aluminium are also actively extracted, with respective amounts of 9 and 2 Mt in 1950, and of 22 and 19 Mt in 1994 (see Fig. 83). The underground distribution of metal ores of a strategic importance is highly unequal, most deposits being located in South Africa and Russia. Both these countries produce 99% of platinum extracted in the world, 98% of manganese, 97% of vanadium, 96% of chromium, and 21% of titanium. Europe and Japan are particularly dependent on metal production from other countries (Caristan, 2000).

Most metals in the Earth's crust are present in very low concentrations. Only aluminium and iron constitute more than 1% of host formations. On average, 1 ton of crust rocks comprises 82 kg of aluminium and 56 kg of iron, but only 75 g of nickel, 70 g of zinc, 55 g of copper and 12 g of lead. Due to extraction costs and market conditions, the amount of a given metal contained in an ore should be much larger than these average values for justifying its extraction. This is even the case for aluminium and iron. The minimum content needed for a given metal is called the *cutoff value*. The crust rocks likely to be extracted should

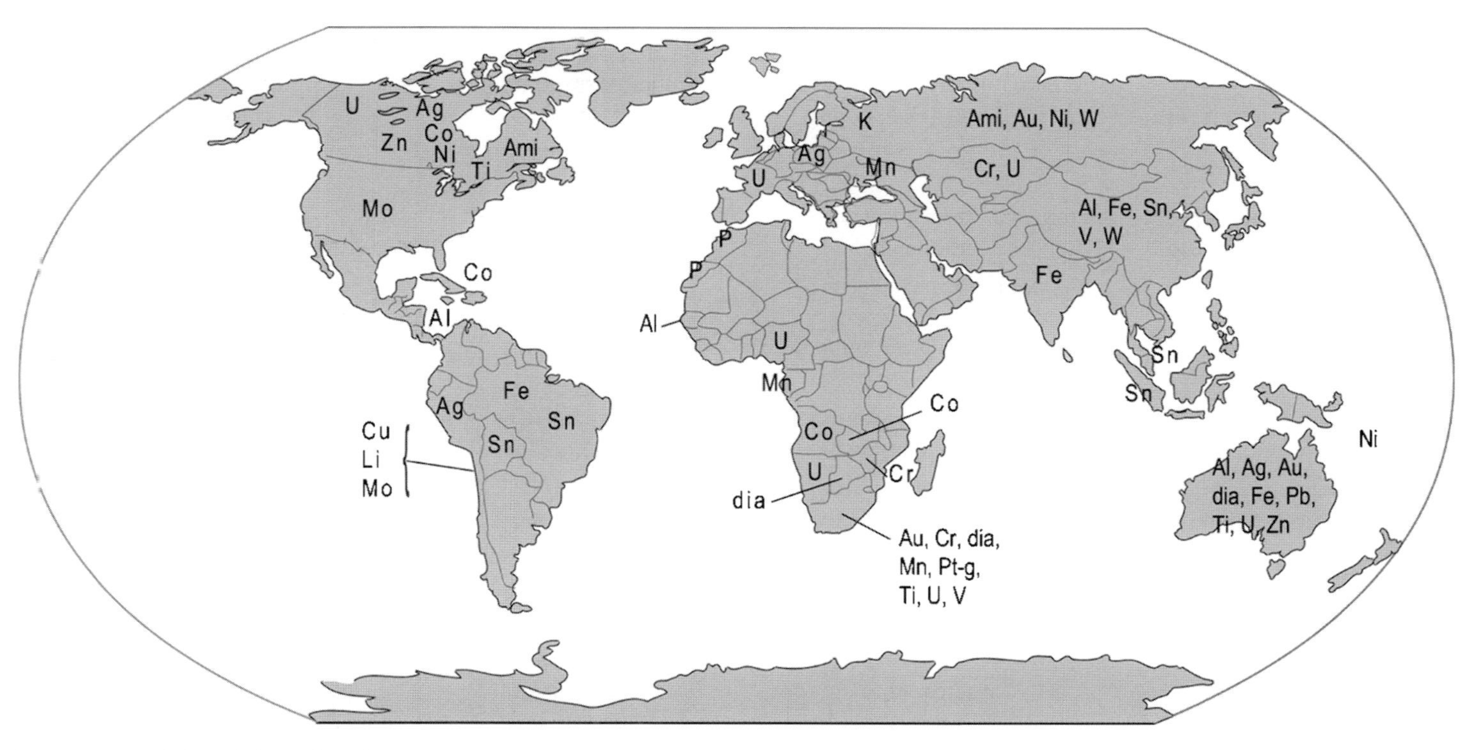

Figure 84: Main origins of various metals, diamonds, asbestos and phosphates (after Rogers & Feiss, 1998). Am, asbestos; dia, diamonds; Pt-g, platinum group.

Table 17: Main metals: shapes, uses, origins, environmental impact (after Jackson & Jackson, 1996).

Metal	Mineral type	Use	Origin	Impact
iron	oxides (magnetite, heamatite)	construction, machinery, tools	Brazil, China Australia, Russia, USA	local pollution
aluminium	oxide (bauxite)	aircraft and electrical industry	Australia, China, Senegal, Guinea-Bissau, Jamaica, Surinam	high energy and water consumption
lead	sulphide (galena)	electrodes, pigments, additive in petrol	Australia, Canada, Peru, Russia, USA	production of SO_2, acid water
zinc	sulphide (sphalerite)	metallurgy, chemical industry, die-casting	Australia, Canada, Peru, Russia, USA	production of SO_2, acid water
copper	elemental Cu, sulphide (chalcocite)	electrical wiring, cooking utensils	Chile, Peru, Canada, USA, Russia, Zaire, Zambia	production of SO_2, acid water
nickel	sulphide	steel industry, chemical industry, (dyes, catalysts, pigments)	New Caledonia, Canada, Russia, Australia	production of SO_2, acid water
tin	oxide (cassiterite)	steel plating, alloy in bronze making, additive against oxidation goldsmith's art	Indonesia, China, Bolivia, Thailand, Brazil, Nigeria	local pollution

therefore be characterized by a natural enrichment factor called the *concentration factor*, which varies between 5 and 3,000 relative to average values (Table 18).

The two major types of *fossil fuels* present a very different distribution. Coal is essentially of a Palaeozoic age and has a continental origin, whereas hydrocarbons mainly result from the evolution of Jurassic and Cretaceous sediments deposited in shelf and slope marine environments.

Coal-bearing formations are abundant, widely distributed and dominantly exploited in Canada, the USA, Australia and China. Some coal deposits contain few impurities, some others comprise non-carbonaceous elements, including sulphur dioxide (SO_2), which is rejected in the air during fuel combustion (Chapter 11.3).

Hydrocarbon deposits are particularly abundant in the Persian Gulf countries. Other oil- and gas-rich natural reservoir rocks occur in Russia, Venezuela, Mexico, Nigeria and Libya, and also in North America, Indonesia, Algeria, in

Table 18: Average proportion in the Earth crust, cutoff value and concentration factor of common metals (after Allègre, 1990; Rogers & Feiss, 1998).

Metal	**Proportion in the Earth crust (%)**	**Cutoff value (%)**	**Concentration factor**
aluminium	8.2	40	5
iron	5.6	25	4.5
titanium	0.57	1.5	25
manganese	0.095	25	260
chromium	0.019	40	2,100
vanadium	0.013 5	0.5	35
nickel	0.007 5	1	130
zinc	0.007 0	2.5	360
copper	0.005 5	0.5	90
cobalt	0.002 5	0.2	80
lead	0.001 25	3	2,400
uranium	0.000 27	0.01	40
tin	0.000 17	0.5	2,950
molybdenum	0.000 15	0.1	660
tungsten	0.000 15	0.3	2,000
silver	0.000 008	0.005	625
platinum	0.000 000 5	0.000 2	400
gold	0.000 000 4	0.000 1	250

North Sea bottom formations as well as in Iran, Iraq and China. In 1995 the world production of oil amounted to 22.4 billion barrels (1 barrel = 159 l). One-third was extracted from Middle East countries bordering the Persian Gulf. Russia, the USA and European countries exploiting North Sea fields produced 30% (i.e. 10% each), and Mexico, Venezuela plus China 15% (5% each). The rest (about one-quarter of world production) came from all other producing countries (Rogers & Feiss, 1998). Considerable amounts of heavy hydrocarbons are present in bituminous sand and schist of Canada, Venezuela, etc., but their potential exploitation is presently hampered by technical and environmental difficulties (e.g. heavy air pollution).

To summarize, the most important fossil fuel fields presently recognized are located in the Middle East for oil, in Russia and the Middle East for natural gas, and in Asia–Oceania, Russia and North America for coal (Fig. 85). The world reserves identified or declared in 2000 ensure an availability of a few decades for oil, of more than half a century for natural gas, and of more than two centuries for coal. Such estimates imply that present-day conditions remain unchanged, namely the level of consumption, the various energy sources employed (including nuclear energy) and the political equilibrium.

Notice that modern techniques allow temporary *storage of some natural resources in underground formations*. This is particularly the case for *natural gas*, the need of which considerably varies with the season (heating, energy production), which often must be transported over great distances by pipelines and can hardly be stocked in large amounts under surface conditions. More than 500 underground reservoirs of natural gas are operational in the world.

In France the Gaz de France Company exploits about 15 reservoirs located in sedimentary formations buried between 400 and 1,500 m below the ground surface. During cold winter periods more than 50% of gas used in the country come from artificial underground stocks. Most sites, one of which has a 7 billion m^3 capacity (i.e. Chémery), correspond to natural aquiferous rocks that are located in the Paris and Aquitaine basins. Gas is injected within porous rocks, especially *sandstones*, where it progressively takes the place of water. Reservoirs are naturally covered by impermeable formations (e.g. clay deposits topping an anticline structure). Continuous monitoring ensures the control of storage, racking and safety conditions. Other French sites of natural gas stocking consist in artificial cavities dug in Mesozoic and Cenozoic saline rocks in the east of the country. The *salt* was dissolved under echometric control through successive freshwater injections. Salt brines were pumped up to the ground surface and progressively replaced by gas. The cavities are of moderate size (100,000–500,000 m^3), mainly for safety reasons. The gas stored in salt cavities can be extracted quickly, which ensures a regular supply if consumption rises rapidly.

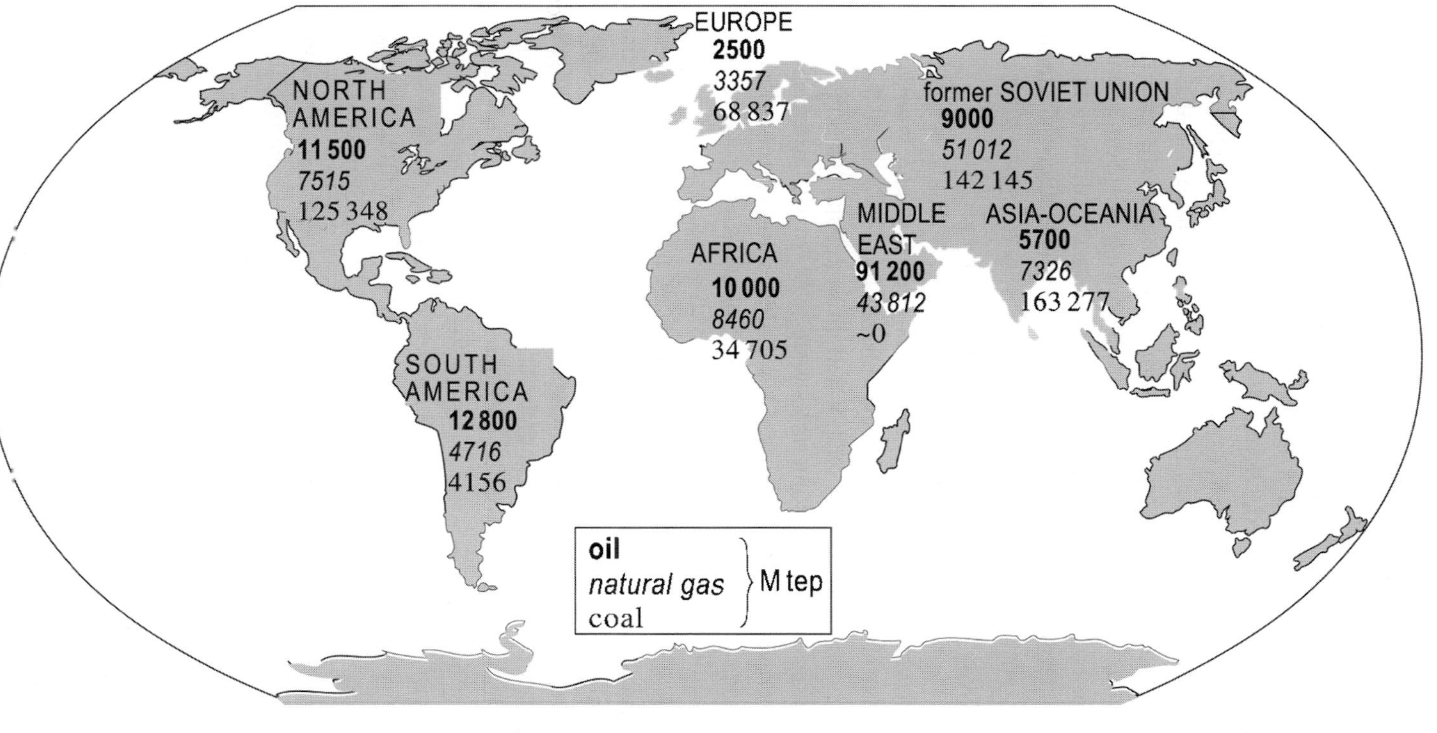

Figure 85: Oil, gas and coal reserves established in winter 1998 (after Caristan, 2000). Values in oil-equivalent million tons (Mtep).

Let us remember that *soils* forming at the ground–atmosphere interface represent an essential indirect resource because they host and influence most agricultural and livestock activities (Chapter 8.1). Soils result from the physical and especially chemical weathering of rocks under the combined action of water, temperature, dissolved chemical elements and organic activity. Their distribution is largely controlled by climate, slope and drainage conditions: brown soils of temperate regions, laterites of warm-humid regions, vertisoils of subarid domains, podzols in cool and humid countries, siliceous rankers and calcareous rendzines of mid-altitude mountains, etc. Some ancient soils preserved from erosion constitute specific ore deposits: aluminium in bauxites, kaolin in old laterites, bentonite and smectite in various clay-rich *palaeosoils*.

5.2.2. Submarine Resources

Coastal environments Various *sodium, magnesium and bromine salts* are traditionally extracted from littoral crystallizers formed by salt flats and pans, where sea water is progressively evaporated through solar heating.

Shallow water sediments sometimes give way to the extraction of *sand, gravel and aggregates*, which are used for building and civil engineering work. Such extractions are more and more controlled since pits resulting from submarine dragging are liable to cause various disturbances: unexpected erosion of nearby shores, modifications in the sedimentary transit, damage to submarine cables and pipelines. In addition active dredging activities may disrupt the biological cycle of benthic or even of plankton or fish species (e.g. loss of benthic larval stages, excess of suspended-matter discharge impeding biological growth). These diverse coastal resources do not lead to any significant renewal problem, so long as extraction is adequately planned and ore management correctly controlled.

Some *mineral placers* are exploited on beaches and in shallow coastal zones of various regions, where extraction is performed by traditional methods or by automatic devices (dredging, drawing up, sorting by machines):

- titanium- and zirconium-bearing minerals concentrated by high-energy currents and waves along the coasts of crystalline rock regions: ilmenite and titano-magnetite (e.g. Madagascar), monazite, rutile, zircon;
- tin contained in cassiterite deposited down to 45 m depth off Southeastern Asia, Indonesia and New Caledonia;
- diamonds from shore and marsh zones of Sierra Leone and Ghana, or from South African coasts.

Offshore environments *Hydrocarbon exploitation* expanded during the 20th century on continental shelves where oil fields constitute a noticeable proportion of world resources: the Gulf of Mexico, the Persian Gulf, the North Sea, etc. For a few years, the need to find new oil deposits associated with considerable progress in submarine technology led to the beginning of prospecting and exploitation of deep offshore sediments. Oil wells have been progressively drilled 1–2 km below sea level in the lower continental margin of different regions: Brazil, the Gulf of Guinea, the Gulf of Mexico.

The recent discovery of large amounts of *carbon hydrates* mainly made of methane (CH_4) dispersed in hemipelagic or re-sedimented deposits from deep offshore environments (continental slope and rise, abyssal plains) has led to intense research. Hydrates are generally buried at several hundred metres below the seafloor and constitute more or less diffuse, solid and unstable sedimentary components. Various scientific and technical questions about these materials need to be addressed: origin, formation and migration mechanisms within the sediment; detailed geographic and bathymetric distribution; relative place and importance in the global balance of carbon; possible concentration due to topography, sedimentology or biological–microbiological activity; age and possible renewal, potential extraction and exploitation.

Potential resources from remote deep-sea environments are either known for several decades and still very little exploited, or only suspected and currently investigated:

- *Metal oxides, sulphides and silicates* forming in brine-rich rift valley depressions (e.g. Red Sea) and within deep-sea brown to reddish oozes of the Pacific and Indian Oceans. *Polymetallic nodules* locally occur at the surface of such deep-sea clays, where they may cover large areas between 4,000 and 6,000 m-depth. Nodules, the mean size of which varies between 5 and 10 cm but may be much smaller or larger, contain abundant manganese and iron associated with variable amounts of less common metals such as cobalt, nickel and copper. Their potential exploitation depends mainly on extraction methods and costs relative to those of terrestrial metal ores.
- *Minerals, metals and organic substances forming in hydrothermal environments characteristic of oceanic ridges.* Current research is developing to investigate these zones marked by active chemical exchange between oceanic crust, deep-sea sediments and sea water, where various minerals precipitate, and where thermophile bacteria and autotrophic biological communities develop. Potential pharmacological and medical uses could be discovered from better knowledge of this uncommon biosphere compartment.

Much has still to be learned about the occurrence and formation modes of deep-sea resources, their concentration and potentially renewable character, the extraction processes and costs, the possible impact of exploitation on deep ecosystems, etc. Stimulating challenges are progressively taken up that imply sophisticated scientific research, innovative submarine engineering techniques, and complex legislative and legal debates on international economic zones.

5.2.3. Methods of Exploitation

Surface Extraction

The massive deposits that crop out at or close to the Earth's surface and extend to the depths are exploited in classic *quarries* (Fig. 86). The topsoil and other unproductive formations covering the exploited material are removed from the extraction zone as work progresses and are stored as slag heaps or tailings. Quarries are shaped to form successive terraces dug deeper and deeper into the rock, which allows work on rock faces of 10–15 m high. The stripping ratio, which corresponds to the surface of overburden formations removed over that of extracted material, depends on the deposit shape and underground extent (Fig. 86, A).

If the material to be extracted close to the surface constitutes an almost horizontal layer, the exploitation is conducted by *strip mining*. The miner spoils can then be stored within the excavation as the exploitation goes on (Fig. 86, B). This method, which is widely used for extracting shallow coal and lignite seams (e.g. Canada, Germany, USA), allows filling of the quarry again with unproductive material, and therefore limiting of subsequent subsidence movements and landscape deterioration (Chapter 5.3).

Underground Extraction

Underground mines and quarries are exploited by three main techniques (Fig. 87).

Longwall exploitation, which is particularly used for coal mining, consists of working the coalface perpendicularly to the vein by following its natural slope. Progression of the cutting face induces the formation of a fairly narrow and low gallery that is temporarily sustained by hydraulic supports (Fig. 87, A). The supports are displaced forwards as the coalface progresses along the vein, which causes the decompressed roof rocks to collapse and to fill the gallery cavities. The collapse may be reflected up to the ground surface, inducing subsidence occurring parallel to mining work. Subsidence effects are more serious when exploited veins are thick, superimposed and located fairly close to the surface (i.e. less than 100 m).

Pillar-and-stall working techniques are mainly used for extracting hard rocks such as massive limestone and sandstones, and sometimes also more fragile deposits such as coal. The mining work extends over a fairly wide area, leading

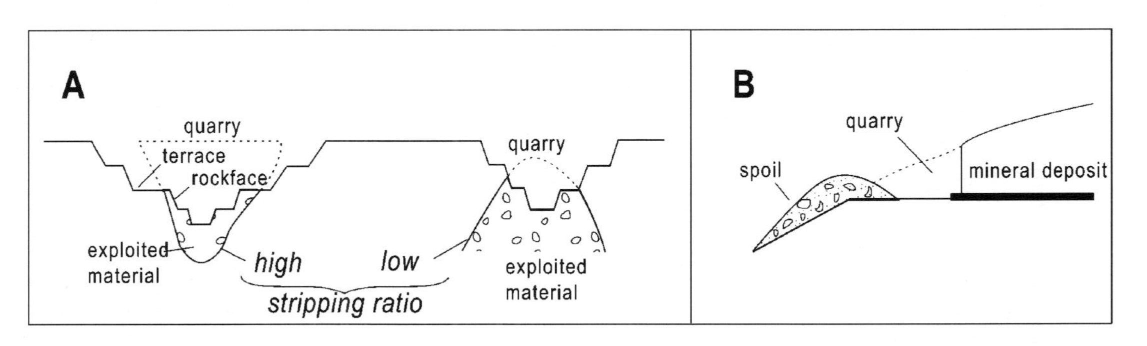

Figure 86: Examples of surface mining methods (after Bennett & Doyle, 1997). (A) Geometry of mineral resource deposits exploited in a quarry, and consequence on the amount of unproductive overlying formations to be removed. (B) Strip-mining quarrying.

to shaping an underground chamber called a “stall”, the height of which corresponds to the thickness of extracted strata. The chamber roof is supported by “pillars” made of non-extracted material (Fig. 87, B). The long-term stability of cavities depends on the mechanical properties of the pillars and roof, on the distribution of plenums and vacuums (e.g. mines with several levels of chambers, the pillars of which are superimposed or not), and on the strength of over- and underlying geological formations (Chapter 5.3).

Bell-, funnel- or bottle-shaped pits constitute smaller mining cavities, which are dug directly from the ground surface and do not exceed a lateral extension of about 10 m (Fig. 87, C). Such pits are typically worked for extracting material located close to the surface, extending horizontally, and mainly exploited for local use. The stripping ratio is generally low for these cavities, which sometimes connect with each other underground. The frequent density of such pits predisposes to a series of collapse, especially if the material extracted is little resistant. This is the case of chalk “*catiches*” in the north of the Paris Basin, of the “*minette*” iron ore in northeastern France (Lorraine region) and of shallow coal pits.

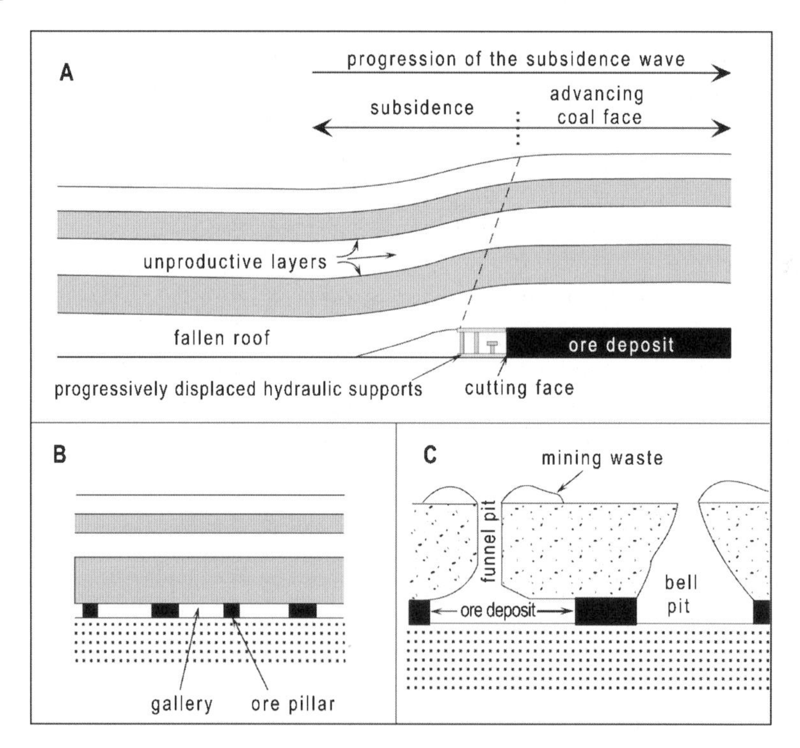

Figure 87: Underground mining methods (after Bennett & Doyle, 1997). (A) Longwall exploitation. (B) Pillar-and-stall working. (C) Exploitation by bell-, funnel- or bottle-shaped pits.

Rock and Ore Processing

Most rocky materials used for construction and civil engineering work do not need any specific preliminary treatments. By contrast mineral ores such as oxides, sulphides, carbonates and arsenides necessitate extraction out of a gangue of metals and other useful substances. In all cases, sorting, selection and purification treatments are necessary that imply a series of specific processes and lead to the production of more or less abundant spoil and waste (decantation sludge, tailings, slag heaps). The first treatment step often consists of reducing the rough material size (i.e. comminution) through crushing, grinding, underwater flaking, etc. Strong water jets are sometimes used for selecting heavy minerals and metals from little-consolidated gangues (e.g. gold, magnetite–ilmenite in fossil alluvial placers). The second step consists of extracting the metals and other products through density sorting, application of a magnetic or electric field, coagulation or chemical dissociation.

Dissolving Extraction Methods

Solvent liquids are used in several cases for extracting underground ore deposits. Fresh water constitutes an active agent for dissolving *rock salt* (halite) and other highly soluble evaporites. The liquid is introduced within the ore by vertical pipes, determining salt to dissolve and the formation of an underground cavity (Fig. 88, A). The brine is pumped to the surface and concentrated by evaporation. Some *sulphides* are dissolved by very hot water injected under pressure into fairly deep deposits (Fig. 88, B). Chemical solvents such as reactive acids are sometimes employed for extracting *metals* from deeply buried ore deposits, e.g. copper, gold and uranium (Fig. 88, C). Such methods, the impact of which is poorly known on the underground environment, allow exploitation of ores that are characterized by a metal content much lower than conventional deposits (e.g. 0.6 instead of 6 ppm for gold extraction).

Hydrocarbon Extraction

Crude oil and natural gas contained in rock pores and fractures are extracted by pumping via *wells* drilled directly on the continental surface or from artificial platforms anchored on the sea floor. During pumping the oil escape is prevented by using water and drilling mud that ensure the hole impermeability. The fluid transfer to oil refineries is made first by appropriate, often buried pipelines, and second by shipping or rail transportation. Except for a few cases of explosion due to violent gas or oil discharge (blowout) the risks induced by hydrocarbon exploitation are moderate and mainly linked to subsidence effects (Chapter

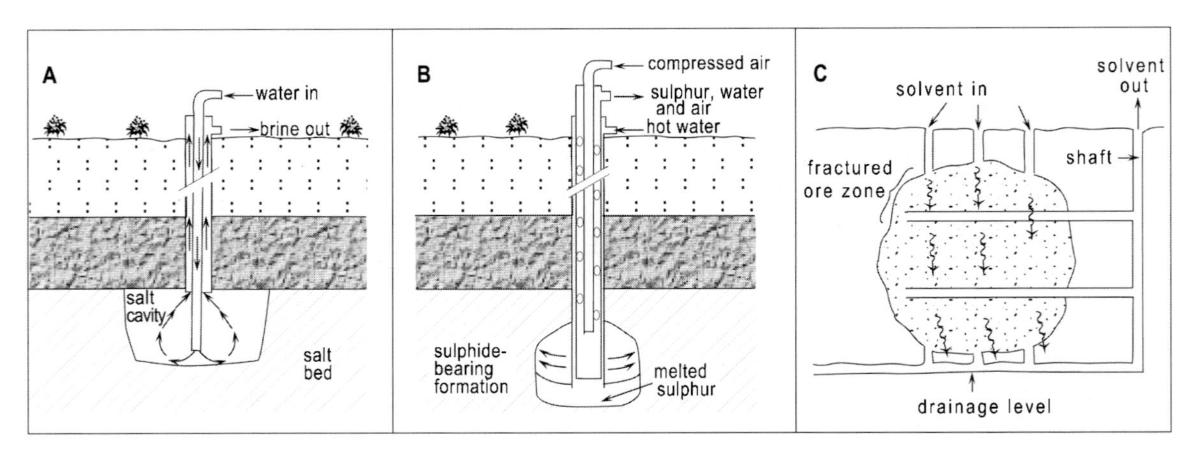

Figure 88: Solution mining techniques (after Craig et al., 1988). (A) Water injection for dissolving salt. (B) Sulphide melting by superheated water injected under pressure. (C) Acid leaching of fractured ore deposits.

3.2.2). The impact on the ground surface is low so long as the number of wells remains low.

5.3. Environmental Impact

5.3.1. A Chain of Disruptions

Consequence of the Witbank Coalfield Mining, South Africa

A large coalfield of 150 × 100 km situated in South Africa along the Transvaal–Natal border began to be mined at the end of the 19th century. The Witbank mines in this field started to be exploited in 1908 by pillar-and-stall working (Chapter 5.2.3) and gave way to intensive coal extraction until the 1940s. Numerous underground cavities were dug, which progressively induced extensive disorders at the soil surface: rock fracturing and collapse, ground subsidence, etc. Mining waste was stored as slag heaps scattered on the coalfield surface, which increased the excess load on the underground excavations.

The minefield was closed in 1947. Then spontaneous combustion started to develop in both underground cavities and surface slag heaps, leading to the local, unpredictable emission of sulphide and nitrogen toxic oxides. Underground combustion was still active at the end of the 1990s. The disused mine chambers, pillars and shafts were submitted to intense rock fracturing associated with an upward propagation of voids, which led to an increase in the rainwater seepage. The underground drainage of ancient coalfaces induced the removal of abundant suspended matter and of dissolved, partly toxic elements (sulphur and sulphates, iron, manganese, aluminium, etc.; Table 19). The drainage water progressively re-emerged at ground surface in the early 1990s, displaying very low pH values (less than 3; Table 19). This acid and toxic mine-originating water determined the destruction of vegetation in various surface sectors of the coalfield. Life disappeared in certain rivers, some of which are connected with the hydrographical network of Kruger National Park.

Faced with the serious issue of environmental damage and the difficulty of clearly identifying all those responsible, an action group made of industrial managers, users and inhabitants of the Witbank area was planned to be implemented around 2000 (Bullock & Bell, 1997), more than half a century after the mine closure.

Tailing Spill of the Aznalcollar Sulphide Mine, Southwestern Spain

Located upstream from the Guadalquivir delta the Aznalcollar mine is dug in sulphide-rich Lower Carboniferous (Visean) Series from the Western Bethic

Table 19: Chemical composition of Witbank coalfield acid mine water. Comparison with South African guidelines for drinking water quality (after Bullock & Bell, 1997). Values in milligrams per litre.

	Measured values	Recommended limit (no risk)	Permissible limit (insignificant risk)	Crisis limit (for low risk)
suspended matter	2,082–4,844			
pH	1.8–2.95	6–9	>4	<11
sulphates (SO_4)	1,610–3,250	20	600	1,200
iron	87–140	0.1	1	2
manganese	9.9–18	0.05	1.0	2.0
aluminium	86–124	0.15	0.5	1.0

Chain. New cutting faces called *Los Frailes* were opened in 1996 and were destined to produce 50 Mt of sulphur, with an annual output of 400,000 t of mineral concentrate.

At Aznalcollar the sulphides of commercial interest are isolated by chemical treatment followed by flotation. The mine waste is stored as acid mud in large reservoirs established along local rivers and blocked by artificial dams. The Aznalcollar dam was built in 1975 on the Agrio River upstream of the Doñana National Park and nature reserve (Fig. 89). The dam was enlarged on several occasions and attained a height of 25 m in 1998.

On 25th of April, 1998, a breach of about 50 m wide formed in the dam of the tailing pond. About 4 million m^3 of acid water and 2 million m^3 of toxic mud flooded along the Agrio and Guadiamar valleys, releasing various heavy metals and other noxious elements: iron, sulphur, zinc, lead, arsenic, copper, antimony, cobalt, bismuth, cadmium, mercury, etc. The toxic mud extended 40 km along the river course. About 4 hm^3 of polluted water flowed down to the Guadalquivir River. The spillage affected 4,286 ha of land, 2,656 ha of which belonged to the Doñana nature and 987 ha to the Doñana National Park. The Doñana park constitutes the largest European sanctuary for large birds and is a World Heritage Area protected by UNESCO.

The accident of Aznalcollar mine induced serious damage to the local fauna (e.g. 37 tons of dead fish were taken out of river courses) as well as to vegetation, soils and ground formations. The temporary release of toxic metals

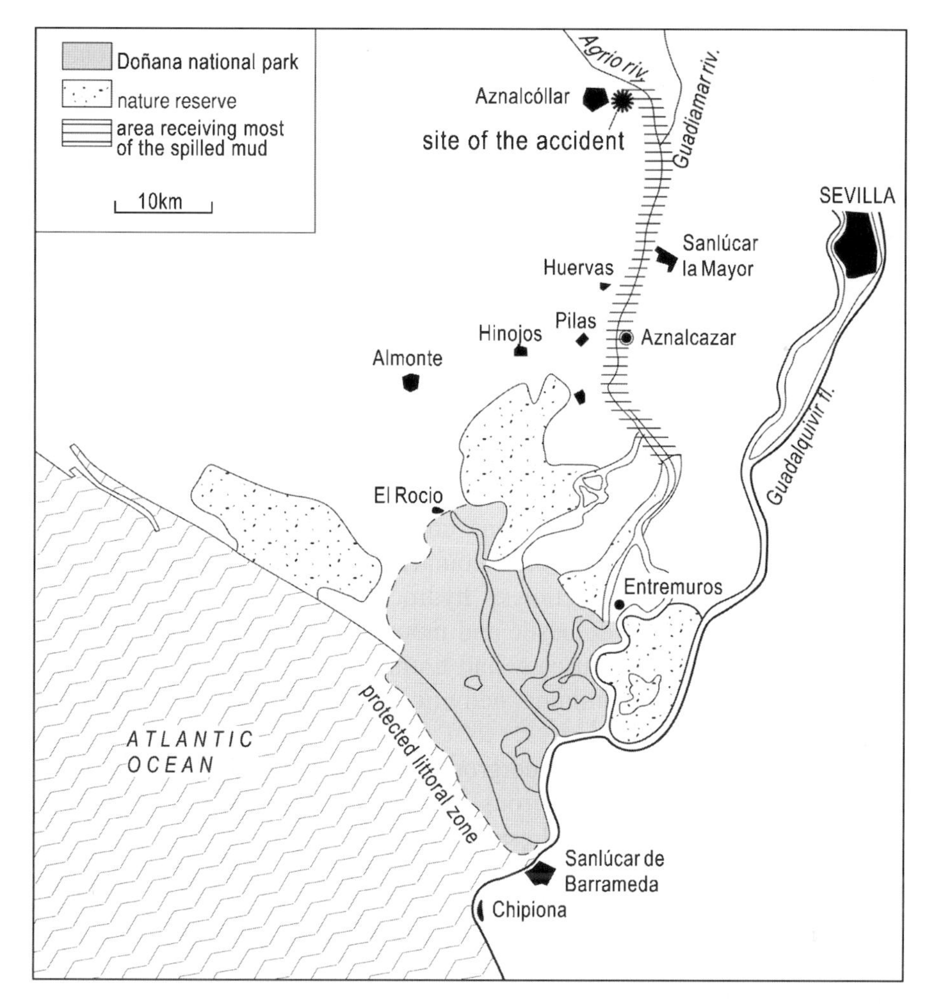

Figure 89: Region of the 1998 spilling accident having affected the Aznalcollar sulphide mine in southwestern Spain (after Grimalt et al., 1999).

and elements in the form of sulphides amounted to 16,000 t of zinc and lead, 10,000 t of arsenic, 4,000 t of copper, 50 t of cadmium and silver and 30 t of mercury. When exposed to the air, sulphides partly oxidized into sulphates that are able to release soluble toxic metals (Grimalt et al., 1999).

In order to limit the damage, temporary barriers were immediately erected along the course of the Guadiamar River. Soon after, the farming was forbidden by law and farm products were collected and destroyed. The toxic mud was removed and stored in a large open cavity of Aznalcollar mine (1,500 × 700 × 300 m), a water-reprocessing plant was built on the accident site, and an interdisciplinary research and monitoring programme was implemented to follow and control the environmental restoration phases.

5.3.2. Collapse, Subsidence

Surface Morphological Changes of Mining Areas

Regions subject to active mining very often display *progressive modifications* of their surface morphology: fissuring, fracturing, subsidence, and formation of lows alternating with bulges. Such changes are responsible for disorders in farming developments, in surface and underground water transfers, and sometimes in urban or industrial activities. *Sudden morphological changes* occur less frequently and may cause the destruction of buildings and other surface constructions, and sometimes claim human victims.

In the USA about 25% of regions concerned with coal mining display subsidence movements. In European countries ground subsidence sometimes results from the massive extraction of hydrocarbons (e.g. North Sea; Chapter 3.2.2), or water needed for domestic and industrial use (e.g. Venice), or locally for geothermal energy (e.g. Tuscany and Latium in Italy); but most subsidence movements are related to mining activities through digging or dissolution (Lumsden, 1992).

In France the impact of underground mines and quarries on surface topography is important in numerous regions that have been exploited over several centuries for extracting dressed stone, iron ores and coal. Despite the important depth of coal veins in the north of the country (400–800 m on average; extreme values of 100 and 1,300 m), multiple disruptions and inconveniences have occurred during the past two centuries:

- house and building fissuring;
- urban subsidence;
- formation of marshy lows and of ponds where groundwater flows;
- bending of railway tracks necessitating subsequent building-up (e.g. subsidence of 10 m at Lens station);
- slow land-sliding causing damage to streets and roads;
- deterioration of the water quality and flow properties;
- development of industrial wasteland.

These morphological nuisances result from the fact that most housing and industrial developments were established directly above the coal fields or/and are situated close to massive, very heavy slag heaps.

In the same region the systematic and widespread extraction of chalk in bottle-shaped cavities (*catiches*) dug close to the ground surface has induced a huge amount of local damage.

During the last centuries, about 6 million m^3 of dressed stone have been extracted underneath the city of Caen (Normandy) at 8–12 m depth in Jurassic limestone. In only 30 years, this has determined about 100 collapse or subsidence movements (Flageollet, 1989). Inestimable damage caused by underground quarries, most of which have been closed for decades, have affected or still threaten various other regions of the country such as Aquitaine, Ile-de-France, Lorraine and Touraine.

Types of Movements

Collapse Mine cavities and galleries may be subject to sudden, subvertical rock falls. *Local* collapse structures, which usually extend to a height of a few metres and attain a diameter up to 10 m, result from the progressive failure of successive rock beds forming the roof of relatively shallow underground galleries (i.e. mining depth < 30 m). The failure develops upwards and may lead to the cavity emergence at the surface and to the formation of a *crown hole* (Fig. 90, A). Local collapse may also affect pillar-and-stall quarries when exploited

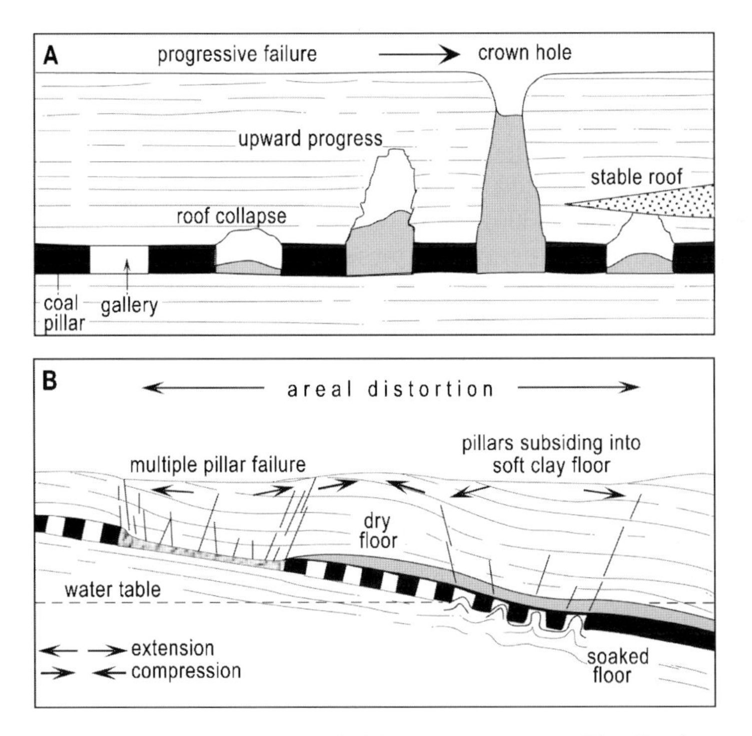

Figure 90: Local collapse (A) and subsidence movements (B) affecting underground mines (after Waltham, 1994).

structures are shallow, cavity size is important and pillars are fissured or excessively thin.

Widespread collapse affecting a large area of the mine's surface occurs much more rarely than crown holes. It generally results from the presence in the gallery roof of a hard rock slab that suddenly gives way and falls. Such brutal movements are favoured by the presence of thin pillars that break easily or tend to punch into floor beds.

Subsidence The progressive bending of rock beds overlying excavated layers results from a smooth readjustment of underground pressures and leads to a new equilibrium stage. Subsidence movements generally concern fairly deep extraction sites (1 hm and more), extend over rather large areas (1 k and more), and are metre to decametre high. Underground pressures include vertical and horizontal components, the respective importance of which depends on the lithological nature of rocks overlying the mine galleries. Subhorizontal *tension and compression movements* develop (Fig 90, B) that are responsible for both surface morphological changes and potential damage to buildings and constructions: leaning structures, fissured houses, deteriorated pipes, etc. *Modifications of hydrographic and aquiferous networks* may also result from such morphological changes. For instance the longwall mining under Burnout Creek on the Wasatch plateau in Utah (USA) has determined both geomorphic and hydrological disorders during 1992–1994: increase in length of cascades and glides; increase in pool length, number and volume; increase in the mean grain size of particle deposited within ponds; some constriction in the channel geometry (Siddle et al., 2000). Most of these changes are short-lived and potentially reversible; long-term monitoring would probably demonstrate the occurrence of more definite morphological and sedimentological modifications. Notice that subsidence movements that are subsequent to underground fluid pumping (oil, water) are considered together with all other gravity movements (Chapter 3.2.2).

Underground Processes

Pillar breaking The non-extracted parts of subterranean exploited beds constitute pillars, the progressive destruction of which may result from three causes:

(1) *crushing* through uniaxial vertical compression due to the excessive weight of roof beds. After pillar destruction the excess force is transmitted to nearby pillars that progressively give way in turn;

(2) *buckling* of thin and elongated pillars. Such degradation occurs particularly in mines comprising several superimposed working levels where pillars are randomly distributed and where one rock layer has been crushed;
(3) lateral tension and *shearing* due to the presence of a sloped surface relief directly above the mine, or to a pronounced tilting of geological beds. This is the case in coal mines where tectonic activity has induced the deformation of geological beds that comprise exploited veins.

Various indications allow assessment of the deterioration of mine pillars: fissures, edge fracturing, fallen rock blocks, rock splinters scattered on the mine floor, etc.

Slab breaking Unworkable or non-exploitable beds that form the floor and roof of underground mines and quarries break owing to three major processes:

(1) *bending*, resulting in the sudden collapse of the roofs of quadrangular chambers that frequently characterize dressed-stone quarries;
(2) pillar *shearing* owing to very slow sliding of roof beds under the force of gravity. This constitutes a common process that is responsible for numerous, sometimes very sudden mass collapses;
(3) floor *punching* by pillars that progressively subside. This phenomenon is facilitated when the gallery floor includes abundant clay and is subject to active water seepage (Fig. 90, B). The start of pillar punching into the mine floor is indicated by the formation of little bulges at the surface of the soft layer.

Aggravating Factors

Various geological, hydrological and anthropogenic influences may exacerbate the risk for mining zones to suffer significant collapse or subsidence movements.

Geological factors:

- variable thickness of exploited beds or veins and of surrounding non-exploited beds, predisposing to inhomogeneous pressure and bending effects;
- initially fractured or buckled pillars and/or roof slabs that tend to break through cavity digging;
- presence of soft or expansible layers (e.g. clay, marl, gypsum, chalk) that are prone to sliding and shearing processes;
- presence of soluble minerals such as salt that cause further subsidence,

- spontaneous combustion of coal particles or methane that participates to progressively deteriorating the pillars, fracturing the rocks and weakening the whole underground mining structure;
- occurrence of earthquakes, the transversal vibrations of which are harmful for pillar and gallery stability.

Hydrological factors:
- variation of air temperature and humidity in mining cavities leading to freezing–thawing alternations, mineral oxidation and therefore weakening of the excavated rocks' resistance;
- rise of the water table leading to ground softening and punching;
- vertical fluctuations of the water table causing local undermining, underground erosion and loss of rock stability.

Mining and post-mining activities:
- pillars excessively thin and elongated, or distributed in an aleatory manner within multi-floored excavations;
- surface overloading due to buildings and other constructions, slag heaps, housing areas;
- surface vibrations and shaking induced by working machinery, heavy-engine circulation, etc.;
- poor management of surface hydrology or areal cleaning networks, responsible for inducing unexpected water seepage and cavity filling;
- accidental connection of surface water and soluble ore deposits. This happened in 1980 in Louisiana (USA) when excessively deep drilling in the floor of Lake Seigneur caused the massive dissolution of underlying cavities dug in the Jefferson salt dome (Murck et al., 1996);
- hydraulic filling of aquifers and underground cavities due to the stopping of water pumping after mine closure;
- storage in ancient mines of corrosive or unstable chemical waste liable to accelerate the deterioration of cavity floor, walls, and roof.

Return to Normal Conditions

The cessation of collapse and subsidence movements after mine closure is more rapid as extraction sites are located closer to the ground surface. Rough estimates show that ground stabilization is acquired after a few hours to days if rocks are extracted from several tens of metres deep, and after a few decades for ores extracted at 500–800 m depth. These assessments are very approximate, since various other factors may influence the stabilization process: plasticity properties of the different rock types involved in bending and failure mechanisms, cavity ageing, effect of underground water on clay swelling, soft bed fluidization, etc. An accelerated stabilization of past mining cavities is acquired in two specific cases that may combine:

- presence of a resistant vault preventing the upward migration of roof failure and the formation of crown holes (Fig. 90, A, right side);
- self-blockage of mine chambers or galleries due to the expansion or swelling of rock pieces constantly falling from a cavity roof.

5.3.3. Mining Waste and Drainage Water

Mine Tailings and Slag Heaps

Mining and quarrying activities determine important changes of the Earth's surface morphology, as evidenced by the 37,000 km^2 newly dedicated to ground exploitation during the last quarter of the 20th century (see Chapter 11.3.4). In addition *huge amounts of non-exploited residual and waste materials are stored in or close to each quarry and mine*, as extractions go on. Each year about 27 billion tons of non-fuel ore deposits and construction materials are extracted worldwide from the Earth's crust, which results in the storage of perhaps 2 or 3 times this amount of waste.

On average the underground of industrialized countries is more widely disrupted by mining activity than that of developing countries. But this situation is currently changing, some developing countries showing even a reversed trend. At the same time a strategy develops in many countries for re-filling cavities and restoring landscape after the closure of mining concessions.

As early as 30 years ago, the annual mining waste and the total residues stored in England represented, respectively, 45 and 3,000 Mt for coal, 22 and 280 Mt for clay materials, and 1.2 and 400 Mt for slatey schist. The values in the USA amounted, respectively, to 200 and 8,000 Mt for copper ores, 22 and 80 Mt for iron ores, 20 and 400 Mt for phosphate rocks, 15 and 200 Mt for lead or zinc ores, and 7.5 and 100 Mt for gold ores (Down & Stocks, 1977). Coalfields of the densely populated region of Upper Silesia in Poland annually produce 50 Mt of mining waste. These residues are stored in 76 slag heaps and dumps covering an area of 2,000 ha and are added to 300 ancient slag heaps (Szczepanska & Twardowska, 1999).

Slag heaps are responsible for various problems of mechanical stability that increase the risks due to chemical pollution and water acidification (see below). Slag heaps consist of an accumulation of unproductive materials and mineral residues that are heterogeneous, heterometric and not very cohesive (i.e. they often correspond to an increase by 40% of the initial rock volume). They constitute cone-shaped or mass structures that are locally high (150 m and more), broad (hectometric to kilometric bottom size), and very heavy. They are often subject to slope movements: collapse, landslides, debris flows, ground punching, and peripheral bulge formation.

In 1966 the collapse of a 125-m-high coal slag heap was responsible for the deaths of 144 people and the partial destruction of the village of Aberfan in Wales (see Craig et al., 1988). The mud and debris flows in the 1980s from dumps of the Yunnan copper mine in China, amounted to a volume of 100,000 m^3 of residues, which caused the long-term destruction of more than 6.2 km^2 farmland (Aswathanarayana, 1995). Slag heap stability is established through slope and strength calculation, the results of which determine various field work: morphological re-shaping, artificial compaction, drainage, and grass or tree planting.

Slag heaps in coal measures are frequently subject to spontaneous combustion of residues that are scattered among more or less pyritized claystone, sandstone, and limestone. The combustion is triggered by the progressive oxidation of pyrite, which induces soil warming. Burning is very slow, is kept going by the air seepage within the slag mass, and may continue for decades. Combustion often propagates from the better-ventilated outer part of heaps towards their inner part. The pyrite-containing black claystones that form a large part of coal waste get progressively oxidized and constitute stable, reddish "schist", which is used for embankments and other civil engineering work.

Mine Sludge

Mud waste results mainly from metal ore processing that necessitates rock crushing, fine grinding, flotation and washing for extracting mineral products out of their gangue. Abundant mud tailings also result from the exploitation of some building or dressed stone extracted from quarries that contain clay-rich interbeds or dissolution cavities filled by fine-grained material. The removal of such "impurities" necessitates water-consuming treatments (i.e. crushing, sawing, sorting). The fine residues are stored in the form of liquid mud within *decantation basins* where they settle, accumulate and get dryer and progressively compressed. Decantation basins sometimes consist of natural depressions (e.g. ponds, small lakes) but more frequently are established in valleys where the fine particles storage is ensured by coarse slag accumulated downstream (Goguel, 1980) or by artificial dams. When the basin is filled and the mud drainage completed, the storage area can be rehabilitated, provided that buried residues are free of instability, pollution or toxicity risks. Ancient mud waste terrains may then be subject to new planting, farming and light building construction.

The threats induced by mine sludge accumulation are essentially of a chemical nature and linked to the metal ore extraction. The geological formations exploited may contain toxic heavy metals, and purification treatments may involve the use of dangerous products (arsenic, mercury, etc.). The 1998 breaking of the dam containing the residual sludge of Aznalcollar mine in Spain led to the discharge of abundant lead, arsenic, cobalt, cadmium, mercury, and selenium components down to the Guadalquivir River delta (Chapter 5.3.1). Most frequent risks are linked to the *seepage and downstream migration of toxic elements dissolved by rainwater.* Such harmful drainage may affect the whole exploitation area, from surface developments to underground shafts and galleries.

The *dumping at sea* of residual mining sludge has been envisaged and sometimes practised during the last 4–5 decades: bauxite ore residues dumped off the Provence coast in the western Mediterranean Sea; titanium-bearing mud dumped on the seafloor between Italy and Corsica; potash waste immersed in the North Sea and dispersed by tidal currents; excess calcium sulphates produced during phosphate fertilizer processing and dumped at sea off various European and African coasts, etc.

Such dumping activity poses a risk, the extent of which is still uncertain and uncontrolled, often unacceptable in the public eye and added to various other deliberate or accidental discharge events (e.g. estuary and harbour dredging, oil spills; Chapter 10). As a consequence most dumping activities at sea are now strictly controlled and regulated.

Drainage Water

Types and risks The extraction from surface formations of materials and ore deposits considerably increases the large-scale rock permeability in quarries and mines. In addition the huge amounts of mining residues stored at ground level are decompacted and directly exposed to rainwater and drainage action. As a general result, *water leaching, percolation and seepage tend to be strongly exacerbated in mining areas.*

Mine drainage water contains a large range of chemical compositions and properties that can be distributed in three main groups:

- *saline water,* from various salt ore deposits;
- *acid water*, containing variable amounts of dissolved sulphates and heavy metals, and resulting largely from oxidation of pre-existing sulphides;
- *alkaline water*, containing few heavy metals but abundant hydrogen sulphides that result from sulphate reduction and other buffering reactions.

These drainage water types pose various *environmental and health risks*, the importance of which depends on the saline, acidic or basic character of ore deposits, and also on the physico-chemical nature of secondary host formation (i.e. the soils and rocks where water migrates). Various geochemical trapping and reactions may happen during the underground water transfer. Banks et al. (1997) show from numerous examples documented in England and Scandinavia that drainage water from lead mines may be either perfectly drinkable or dangerous to health, depending on chemical reactions developing during seepage and underground migration. Drainage water issuing from copper-, zinc-, tin- or coal-mining fields may or may not present an accentuated acid and/or toxic character.

Numerous investigations conducted in or downstream from exploitation areas show the *diversity of potential danger*, which is particularly induced by water containing dissolved *heavy metals*. For instance, the lead and zinc mines exploited for more than a century in Sardinia (Italy) at Masua (Cambrian limestone) and Montevecchio (granite borders) produce residual sludge that contains various heavy metals such as lead, zinc, cadmium and copper. The tailing-draining water is very acid (pH close to 3) and oxidizing (e.g. formation of rust-coloured coatings), which determines the sediment and soil contamination in the downstream Montevecchio valley (Di Gregori & Massoli-Novelli, 1992).

The exploitation from 1860 to 1945 of Goldenwille *gold* mines in Nova Scotia (Canada) has led to the in situ storage of 3 Mt of waste mud impregnated with toxic metals (Cd, Pb, Hg, As, Tl), which resulted from ore extraction and processing. The toxic mud has been actively drained by rain and seepage water, which has caused the contamination of downstream water, sediments and fauna as far as the Atlantic Ocean shore located 10 km to the southeast. The heavy metal diffusion from decantation basins still continues despite the fact that the mines have been closed for more than half a century. Phyto-remediation treatments are envisaged (see Chapter 10.1.3) using the properties of some horsetail species (*Equisetum*) that prove to be tolerant to arsenic and mercury (Wong et al., 1999).

In the Cœur d'Alene river valley north of Idaho (USA), the extraction since 1884 of *silver* (31,000 t), *lead* (>8,000 Mt) and *zinc* (>3,000 Mt) has caused the downstream accumulation in significant proportions of lead (3.8% Pb), zinc (3.4% Zn) aluminium (340 mg/Al kg), cadmium (120 mg/Cd kg) and mercury (7 mg/Hg kg), especially in sediments of 11 lakes bordering the valley. Toxic elements are drained from both the mine network and the surface waste tailings. Despite the mines being closed several decades ago, chemical analyses of cored sediments show that lake contamination is still active, especially during heavy rainfall and flooding periods. Toxic metals represent potential danger to aquatic life and human health. Possible solutions include the coverage by unpolluted sediments of contaminated sediments and soils, and the maintenance of a high water level and anoxic conditions in the lakes (Sprenke et al., 2000).

The distribution of *arsenic* (As), a metalloid commonly released during metal ore extraction and treatment, has been investigated in 34 mine fields from Southeast Asia, Africa and Latin America, where gold has been extracted during the past century, and also antimony, silver, copper, mercury, lead and zinc. The As content in drainage water ranges between 0.005 and 75 mg/l, and exceeds the safety level for drinking in the USA at more than two-thirds (25) of the mining sites. A serious intoxication risk nevertheless threatens only one site, thanks to the mostly pentavalent form of arsenides; four other mining sites necessitate carefully watching and monitoring. The arsenic release in drainage water depends mainly on pH and Eh conditions, but other factors also intervene that complicate the understanding of the metalloid diffusion and its possible transformation into toxic forms: hydrochemical behaviour of iron, arsenic–sulphide exchanges, presence of cyanide components and other minerals, geological and petrographic nature of local formations, climate and mine-management conditions (Williams, 2001).

In some regions the *environmental contamination by mine drainage water started several centuries or even millennia ago*. This is the case in southwestern Spain, where the Rio Tinto River has carried down to the Gulf of Cadiz various anthropogenic toxic metals since the Copper and Bronze Ages. Investigations of river and marine cored sediments reveal that pollutants such as copper, iron, lead, zinc, titanium, barium, chromium, vanadium and cobalt were first released from punctual extraction sites, and then progressively from more and more widely distributed mines (Davis et al., 2000).

Acid water Mine drainage water marked by a strong acidity (pH close to 3) constitutes the *most frequent type* and is also the *most aggressive* in the environment. Acid water seeping or flowing from past or present mine fields induces the oxidation and deterioration of downstream soils and sediments, and threatens both the flora and fauna. For instance, about 20,000 km of watercourses are concerned in the USA with potential mine-derived acidification. Acid water mainly results from the drainage of disused mines where coal, copper, nickel, lead or zinc ores have been extracted (see Table 17). The acid character is caused by the common occurrence among ore deposits of sulphides (pyrite and marcassite FeS_2, pyrrhotite $Fe_{1-x}S$) that have been oxidized when put in contact with humid air during extraction, treatment and storage processes. Sulphide oxidation is strongly facilitated by the presence of bacteria such as *Thiobacillus ferrooxidans* that play an important catalytic and acidifying role. Oxidation gives way to the formation of various acidic sulphate and ferrous/ferric radicals, which result from chemical reactions the general type of which is

$$4FeS_2 + 14H_2O + 15O_2 = 4Fe(OH)_3 + 8SO_4^{2-} + 16H^+_{(aq)}$$

The main factors responsible for the acid character of drainage water comprise a low initial pH, fairly high temperature, significant amount of air- and water-dissolved oxygen, contact with humid air of sulphide-containing porous material, and active Fe^{3+} ions generating H^+ in the presence of water. *Additional factors* are related to the intensity of bacterial activity: population density, activation energy, growing rate, nutrient availability (nitrogen, phosphorus, carbon), and absence of inhibiting agents (Sengupta, 1993).

The maintenance of the acid character and the continuous downstream flowing of drainage water mainly depend on the abundance of rainwater soaking the surface ores, slag heaps, dumps and decantation basins, and also feeding the aquifers crossing underground deposits and waste. The water recharge of mine networks through seepage after extraction and pumping cessation is often responsible for renewed leaching and drainage of oxidized and acidified underground rocks (Fig. 91). This largely explains the frequent increase in the

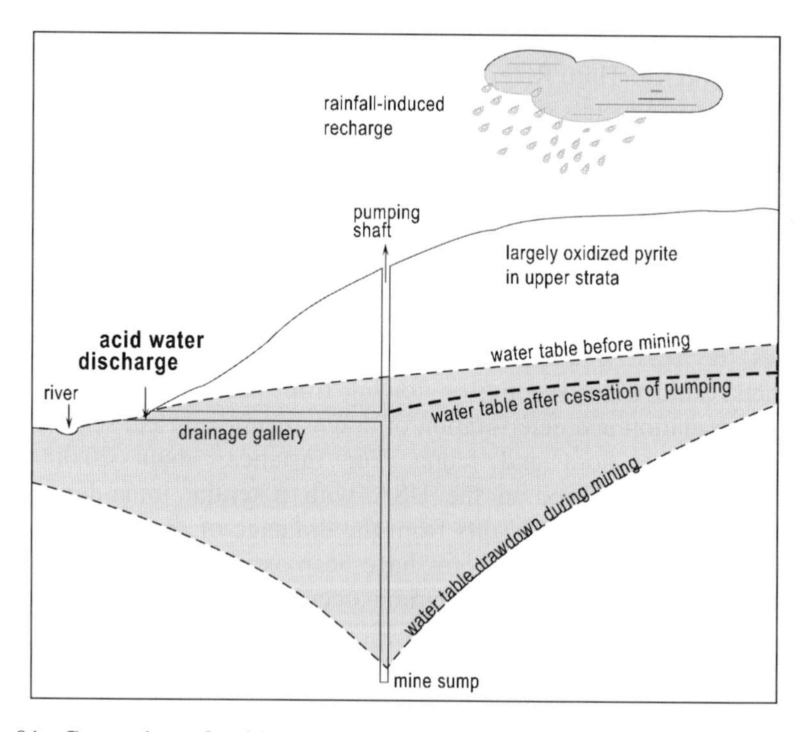

Figure 91: Generation of acid water due to mine dewatering followed by closure of exploitation (after Evans, 1997).

abundance and acid character of the water that drains mine concessions after abandonment.

5.3.4. Indirect Impact

Impact of Surface Exploitations on River Environment

The sand, gravel and pebble extraction in river courses and alluvial valleys causes local disruptions that can be anticipated and prevented through appropriate environmental impact assessments. Extraction in the active course (i.e. minor bed) of a river is a priori barely disruptive if material removed is replaced by new upstream supply as the extraction goes on. In fact, working in a river course is technically not easy if flow is strong and, more importantly, repetitive extraction determines local widening of the river section, lowering of the water level, and increasing of the river-bottom slope. Such morphological changes induce regressive erosion that is responsible for upstream meander abandonment and river-bottom undermining. Downstream from the extraction site, the water gets shallower, which favours the formation of new meanders (Fig. 92).

Most fluvial sediment extractions are conducted in *alluvial plains*, i.e. in river major beds. Pits dug in alluvial valleys tend to cause the deepening of areas liable to flooding, the preferential settling of fine particles and a subsequent loss of permeability. In addition the water table may be excessively drained upstream from the extraction site and inappropriately fed in the downstream direction (Campy & Macaire, 1989).

Excavations carried out in *alluvial aquifers* bordering the valleys often necessitate active water pumping (Chapter 6). This may induce a lowering of the water table and various disruptions in the catchments field, the vegetation growth, etc. Sometimes the gravel and pebble extractions are performed on valley borders characterized by *superimposed aquifers*, the pumping of which leads to soil removal, disruption of surface water flows and accentuated vulnerability to seeping pollution.

Impact on the Atmosphere

In addition to physical, chemical and biological disruptions affecting most regions subject to underground exploitation, mining activities and subsequent treatments may modify the air properties and quality, and therefore the local or even global environment (Chapter 11.2–11.4). The widespread use of wood struts in numerous mines has led to the progressive *deforestation* of large areas (Introduction, 4), inducing a diminution of the CO_2-fixation capacity on land.

For instance a mine producing annually 400,000 t of coal necessitates consolidating the shafts and galleries with 9,000–12,000 m^3 of wood that usually comes from tree felling in neighbouring forests.

Blasting in quarries often results in abundant *dust and gas discharge*. The use of 200–300 t of explosives may determine the emission into the atmosphere of 20–23 Mm^3 of dust, and of 8,000–15,000 m^3 of nitrogen oxides (NO_2), ammonia (NH_3), etc.

The burning of mine waste, especially from coal measures, as well as the chemical treatments needed for ore processing, induce important discharges of various gases that increase the natural *greenhouse effect* or *rainwater acidity*: carbon monoxide and dioxide (CO, CO_2), methane (CH_4), nitrogen oxides

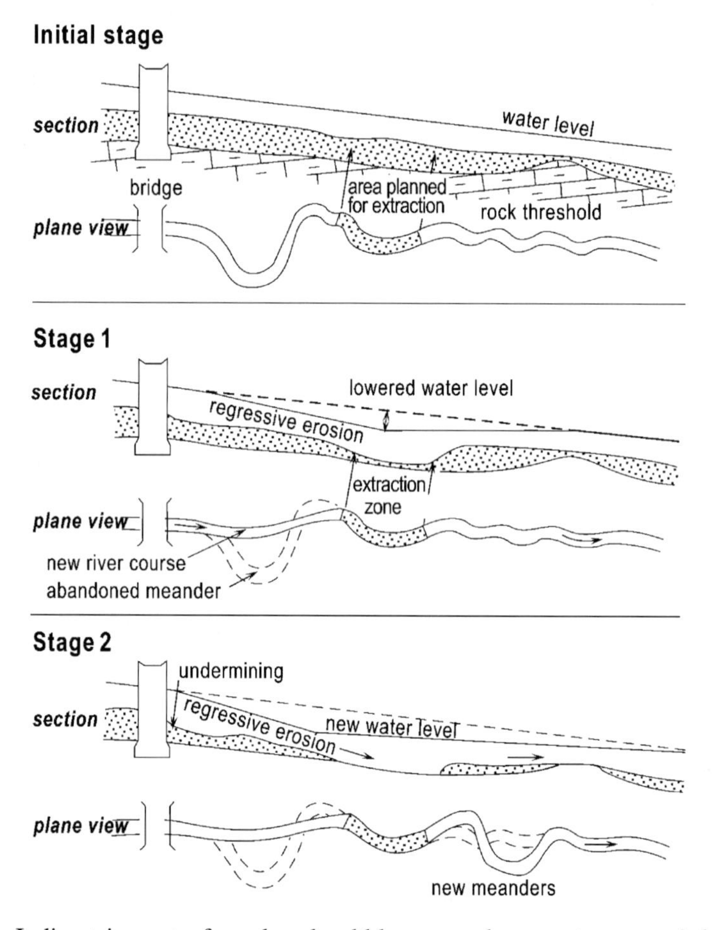

Figure 92: Indirect impact of sand and pebble removal on upstream and downstream zones of an extraction site located in an active river course (after LCPC; see Campy & Macaire, 1989).

(NO_X), sulphur gas (SO_2, H_2S), etc. Worldwide estimations report that 175 Mt of gas resulting from coal-mine exploitation are annually ejected into the atmosphere (Aswathanarayana, 1995).

Impact on Human Health

Noxious products Dust produced by ground ore extraction is responsible for *respiratory diseases* that have been known for a long time. Miner's silicosis is mostly due to fine quartz and other mineral inhalation by pit workers. Silicosis has been and is still responsible in various mining regions for numerous deaths and the shortening of mean life expectancy. The frequency of breathing diseases in the coal district of the Ruhr, Germany, has been estimated to be 60% higher than the national average.

Some chemical elements such as mercury and cyanides, which are commonly used for extracting metals out of their gangue, present a high degree of toxicity. *Mercury*, which has long been used in auriferous extraction for amalgamating the gold particles, is extremely harmful to the central nervous system. Easily transported as very fine drops in water and air, this liquid metal displays fluxes that have been multiplied by 4 towards the ocean and by 275 towards the atmosphere, since the beginning of the industrial era. Mercury is characterized by a short residence time in water (i.e. a few days to about one year) and is therefore rapidly absorbed by living organisms. The metal is accumulated and concentrated in the food chain, which determines a *biological risk amplification*. Populations from various gold-mining regions of Brazil, New Guinea, USA, Venezuela, etc. have been presented with or still suffer from serious diseases due to the ingestion of mercury-enriched water, fish, etc. (Merrits et al., 1997).

Asbestos Asbestos extraction, treatment and handling of asbestos minerals cause an increased risk of cancer (e.g. lung, pleural and peritoneal cancer) and sometimes the deterioration of lung tissues, fibrous (i.e. asbestosis). Asbestos, 150 Mt of which was extracted from geological formations between 1870 and 1992, consists of very fine mineral needles. Asbestos mineral species dominantly comprise chrysotile, which is a whitish serpentine (90% of asbestos extracted), and accessorily two amphiboles, the bluish crocidolite and the brownish amosite (5%). Other mineral species referred to as asbestos are of secondary industrial interest; they comprise amphiboles called actinolite, anthophyllite and tremolite.

Asbestos minerals are flexible, chemically inert and characterized by high fireproofing and soundproofing properties. They were extensively employed until the 1970s, and are still used to a much lesser degree, for various purposes: insulation materials, cements, fillers, tiles, paints, tyres and brakes, containers and filters, paper, clothes and curtains, etc. Lung cancer affects mainly people who have handled chrysotile and are tobacco smokers. Cancer of pleural abdominal walls is particularly due to crocidolite and sometimes amosite ingestion. The carcinogenic risk affects mainly the asbestos miners (Canada, Russia, South Africa, Australia), as well as people involved in the *close handling* of asbestos-rich materials (i.e. various construction and industrial activities). Very minor but proven health risks may appear when the average air content in asbestos needles exceeds 1 fibre/l.

Due to various pressure groups whose action was progressively amplified by the media and public opinion, the *possibility of a carcinogenic risk* similar to that threatening asbestos workers *has been presented for people occupying places or using materials that contain asbestos.* This reaction, which sometimes has been called the "asbestos panic", was born in Europe in the late 1970s and rapidly extended to North American countries. The fear of asbestos-induced cancers gave way to huge programmes destined to remove from public and private constructions the materials made of, impregnated with or coated with the fibrous mineral. More than 100 billion U.S. dollars were spent to remove asbestos from American schools and public buildings; the removal of asbestos from the big Jussieu university campus in Paris was decided in 1997 and is destined to last until 2004 or 2005.

Few *concrete arguments can be found to demonstrate any real or even potential carcinogenic risk in public, business or private premises made of asbestos-containing material.* Multiple investigations show that the 'asbestos panic' appears essentially irrational, which has led to describing the asbestos-removal strategy as a "public policy debacle" (Ross, 1995; Rogers & Feiss, 1998). For instance counting the number of asbestos fibres in 1,400 air samples from 219 North American schools has revealed the average presence of 0.000 22 fibres/ml, a value that is 7 orders of magnitude lower than the minimum risk threshold of 1 fibre/l. The potential lifetime risk for pupils who have spent 6 h/day and 5 days/week at school for 14 years would be of one additional cancer per one million pupils. In comparison with this value of one death due to asbestos, the risk of these pupils dying during their whole life is estimated to be 35 due to lightning strike, 350 to radioactive emission from brick-made buildings, 700 to the daily ingestion of one saccharin-containing drink, 1,000 to the normal use of air transportation, 8,400 to domestic accidents, and 21,000 to tobacco addiction (Ross, 1995).

Everyday life in asbestos-mining regions does not objectively include a higher carcinogenic risk than in other regions. The pleural and abdominal cancer risk is the same for non-mining inhabitants of Thetford (Canada) which is situated in one of the world's biggest chrysotile extraction sites, as for people living in Zurich canton (Switzerland), where almost no asbestos fibres have been detected. The female residents who are not or only slightly involved in mining activity of the two Canadian towns of Thetford and Asbestos since 1970 are exposed to an average concentration of 0.000 05 to 0.005 asbestos fibres/l; the rate of carcinogenic risk is the same for these women and for those living outside asbestos-producing regions. Numerous analyses and monitoring work concerning people working in the asbestos industry (e.g. cement plants, friction-material manufacturers) in Canada, England, Sweden and the USA show that 15–35 years of exposure to air containing 0.000 15 to 0.015 fibre/l induce a nil to extremely low augmentation of the carcinogenic risk. These values are nevertheless 2,000–68,000 times higher than the values of asbestos contained in the air of public schools and other buildings (Ross, 1995). By contrast the carcinogenic risk effectively increases during asbestos-removal processes, during which noticeable amounts of micro-needles are released into the atmosphere. The very expensive asbestos-removal operations are essentially of psychological benefit.

5.4. The Future of Mining Sites

5.4.1. Underground Mines and Quarries

Mining activity generally has a considerable influence on the daily life of regions subject to exploitation. Non-mining activities, especially farming, are sometimes perceived as accessory or even annoying by people directly involved with underground extraction. In fact *ground exploitation is generally of a limited duration*, due to either the exhaustion of mineral resources or modifications in mineral or material needs. Mining and quarrying work is destined to be replaced by other, often more diversified activities. The redevelopment and restructuring of mine regions is therefore ineluctable.

Mining-site restoration should if possible *be anticipated* a long time before the closure of concessions. The assimilation of such an obvious fact has often arisen too late during the 20th century. Productivity constraints and under-consideration of environmental consequences have frequently led to managing the post-mining periods as necessary and without hindsight. People's mentality is now progressively changing. The redevelopment of underground exploitation sites tends more and more often to give way to forecasting actions that include both local and regional aspects (i.e. from micro- to meso-scale).

Collapse and subsidence movements consecutive to underground extraction are generally stabilized through *natural evolution* more rapidly when mining sites are situated close to the soil surface (Chapter 5.3.2). An exception concerns the shallow dressed-stone mines that tend to deteriorate very progressively through weathering and fissuring of chamber roofs and pillars. The very slow dilapidation of such cavities threatens the stability of cities and other surface developments for many decades after the cessation of stone extraction. It is therefore necessary to plan *stabilization work*:

- pillar reinforcement by masonry or strapping work;
- building of new pillars;
- roof strengthening by bolted plates and projection of concrete;
- partial or even total filling by mining residues or other non-reactive material;
- artificial start of collapse for preventing unexpected damage.

Let us recall that the cessation of water pumping in mining cavities leads to the aquifer recharge as well as the formation and transfer of acid drainage water (Chapter 5.3.3; Fig. 89). This requires careful monitoring of the quality of water released from past extraction sites.

The *re-use of ancient underground extraction sites* may be very diverse:

- *sub-surface storage* of materials and property, conversion for commercial purposes provided that hydrological and geotechnical safety is ensured;
- *cultivation* of non-chlorophyllous plants necessitating stable temperature and humidity conditions (e.g. mushroom beds);
- *long-term archiving*. The world's largest seed collection is stored in an ancient coal mine of the Spitzbergen archipelago (Svalbard) in the Arctic ocean;
- *cavity filling* with mining residues, chemical or low-activity nuclear waste, etc.;
- *injection* within ancient oil fields of more or less noxious and toxic *fluid residues*. Injection techniques mainly concern contaminated liquids but potentially also some gasses such as CO_2, the large-scale and controlled burial of which is envisaged for limiting the man-induced augmentation of the greenhouse effect (Chapter 11.2).

It is absolutely necessary to foresee and check the *future of artificially buried products*, especially in regions marked by superimposed aquifers that could be connected by accident. The injection of chemically dangerous liquids into ancient oil wells from the Frio formation (Texas) has been performed at a depth of between 1,220 and 2,130 m below the soil surface. During the 1980s, 3.8×10^6 m^3 were buried each year. Due to past hydrocarbon extractions (a total amount of 208.10^6 m^3 of crude oil had been removed), this depth interval corresponds to under-pressurized rocks relative to surrounding geological formations. The risk of pollutants migrating underground is therefore very low. The safety of the storage area is ensured by long-term monitoring as well as by isotopic analyses of pore water. Such analyses provide information on both the residence time and the interstitial migration rate (Kreitler & Akhter, 1990). The discovery of degraded hydrocarbons in the storage area suggests that microbial reactions currently occur at serious depth, which could accelerate the chemical destruction of buried liquid waste. The existence of this "dormant biosphere" in deep underground formations has given way to new research, the application of which could be useful for future decontamination developments.

5.4.2. Surface Exploitations

Landscape deterioration has extended worldwide during the last centuries due to the dramatically increasing number of often-large strip-mining excavations and waste accumulations. Various dangers arise from pressure release and gravity movements developed on the rock faces. Different techniques help to mitigate these harmful effects caused by excessive opencast mining of ground resources.

- Some extraction processes comprise the *refilling of excavations* as work goes on. Opencast operations include continuous digging, ore removal and back-filling with displaced overburden material, i.e. with soil and non-exploited rocks. The initial shape and surface conditions are progressively restored behind the extraction zone (Fig. 93).
- *Anticipated restoration* techniques have been experimented with, especially for British limestone and lime quarries dug in karstic regions. For instance in the English Peak District extraction is performed by using moderate and controlled blasting on rock slopes, which leads to shaping a landform sequence similar to that of the natural landscape. This technique, called "landform replication", helps to reduce the visual impact, newly constructed landforms being re-planted with local plant species (Gunn & Bailey, 1993).

Sometimes the filling up of abandoned quarries involves geological formations that are of an exceptional petrographic, mineralogical, stratigraphic or palaeontological interest. Maintaining access to such formations is crucial for geoscience research and teaching, or for public education. Irremediable scientific loss could happen in the absence of preliminary consultation. This is the reason why more associations are charged to *protect the geological heritage*, with the support of various learned societies and/or official institutions.

Ancient quarries and opencast mines are converted in an increasing way into *culture or leisure places*: coal, metal or volcanic ore exploitation sites transformed into open-air museums or for other educational purpose; gravel pits and mining areas newly designed to host lakes and pools, leisure centres, sports training circuits, etc.

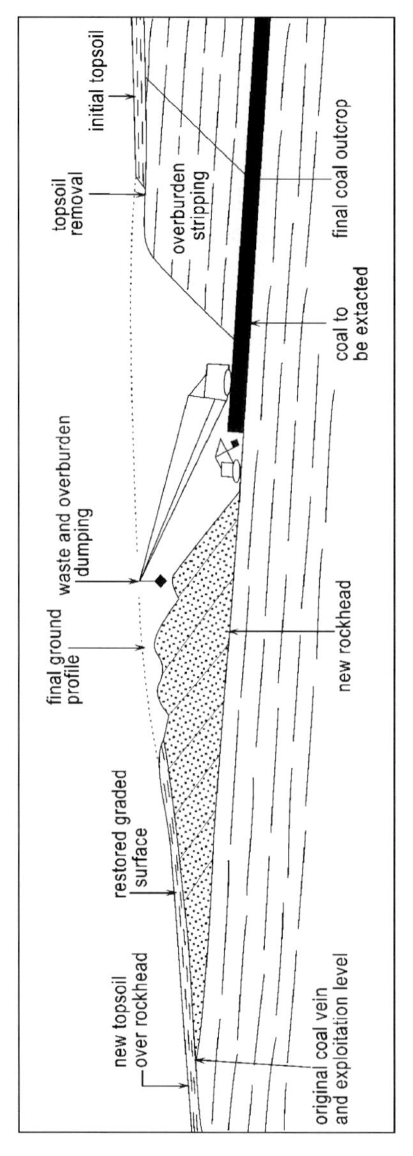

Figure 93: Opencast exploitation technique with simultaneous back-filling and restoration (after Waltham, 1994).

5.4.3. Mining Residues and Waste

The restoration of surface areas occupied by mining waste (i.e. slag heaps, tailings, decantation basins) *is generally delicate because several risks may combine*: under-compacted deposits prone to slope movements, the presence of partly toxic matter hampering the re-use of dumped materials, drainage by rainwater and river water potentially responsible for leaching of noxious dissolved elements, etc. Appropriate and rigorous control of water flows resulting from rainwater seepage is crucial for preventing downstream zones from becoming contaminated. Various technical solutions have been developed, as shown for instance by slag heap management methods implemented in the coalfields of Upper Silesia, Poland (Fig. 94).

The planting of new vegetation for recreation or farming purposes necessitates special care because of the sloped or polluted character of numerous sites occupied by mining waste. The plant colonization usually starts very slowly since steep slag heaps prevent sowing and acid soils hinder seed germination. Once the new vegetation is established, it may develop and give way to nicely landscaped areas. This is the case on metal ore mining sites around Johannesburg (South Africa), on kaolin deposits in Cornwall or on limestone in Derbyshire (U.K.), and on abandoned coalfields in the Valenciennes region (northern France).

The re-use of mining waste has become more significant during the last decades but varies considerably depending on ore deposit types concerned:

- *Coal.* Sterile formations in coal measures are often very abundant. They are partly extracted together with coal and progressively piled up. Coal waste finally constitutes vast slag heaps that include various rock types: mudstones and siltstones, slatey claystones, sandstones and limestone, coal debris, etc. The most abundant minerals are illite and kaolinite clays, quartz, micas, pyrite, as well as calcium, magnesium and iron carbonates. During storage some chemical modifications develop within slag heap masses, namely the oxidation of sulphides into sulphates and of ferrous oxides into ferric ones. In addition spontaneous combustion processes occur generating reddish 'schist' (Chapter 5.3.3), which become resistant to mechanical constraints. The re-use of coal waste commonly involves 10–20% of dumped material:
  - hard rocks and products of spontaneous combustion: road and rail ballast, and other civil engineering work;
  - more or less silty-to-sandy claystone and marl: major components of bricks and cements;
  - partial oxidation products of brown coal (i.e. lignite): soil conditioners, humus substitute;

  - coal-bearing claystone: light, cellular-shaped aggregates obtained through blowing and expansion.
- *Slates*. Difficulty in extracting slates of adequate size, shape, resistance and purity frequently results in the production of 20 times more waste than the usable material. The flat shape of slate prevents its use as aggregate or building material. Most slate debris is employed for filling the cavities after quarrying or it is abandoned as surface piles. Some schist debris is crushed and used as an additive in bitumen, rubber, paints and plastics manufacturing.
- *Phosphates*. Residual mud issuing from phosphate ore treatment tends to settle very slowly. For instance the suspended matter in decantation basins

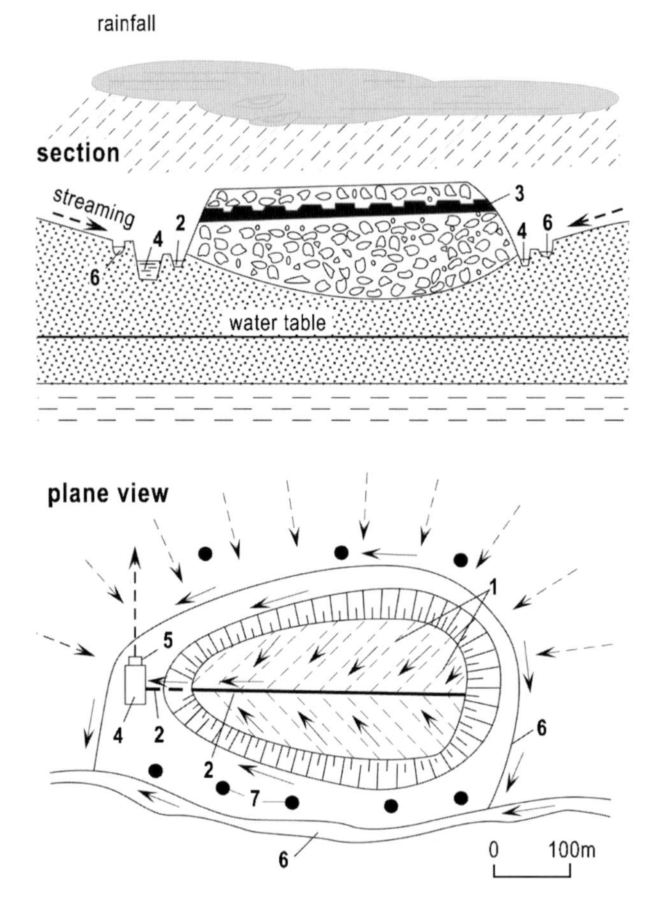

Figure 94: Drainage system of a coal-mining slag heap in the Upper Silesia field, Poland (after Szczepanska & Twardowska, 1999). 1, acid-resistant drains; 2, water collectors; 3, very little permeable base layer; 4, internal ditch and reservoir; 5, pump station; 6, external ditch; 7, monitoring wells.

from Florida phosphate plants displays settling times of up to several years. As residual mud still contains up to 40% of phosphates, most post-mining treatments are destined to additional recovery. A few other uses have proved profitable, e.g. additive for light aggregates, pipes, and bricks.

- *Kaolin*. The exploitation of clay deposits for porcelain manufacturing commonly necessitates extraction of about 9 t of waste for obtaining 1 t of kaolinite. Residues mainly consist of quartz, micas and other clay minerals, in addition to overburden material. Some waste is used for filling up the cavities again, some is sorted and used in making bricks, tiles, concrete, or artificial pozzolana.
- *Metal ores*. Metal extraction and treatment generally results in the production of voluminous, strongly polluted waste, the interest and value of which are low:
  - the exploitation of *iron* ore deposits gives way to abundant residues consisting mainly of overburden rocks and soils, the usefulness of which is aleatory: aggregates, building and roofing material, in situ precipitation of secondarily exploitable iron;
  - *copper* mining residues are generally of a huge volume. The accumulation in the USA of sterile material extracted from copper mines represented already in 1975 half of all slag heaps and tailings of the country (Down & Stocks, 1977). The possibility is low to re-use these acid and potentially toxic waste: road ballast, bitumen containers;
  - *gold* and *uranium* ore residues are also very abundant relative to sterile host formations. They may be used as road-building material, additives to concrete, or siliceous brick components (Blunden & Reddish, 1992);
  - *aluminium*-manufacturing residues resulting from bauxite exploitation are mainly composed of fine-grained, ferruginous and highly alkaline (pH close to 12) reddish mud, which settles very slowly and causes various storage or disposal problems in river courses and underground or seafloor repositories (Chapter 10). Some re-use possibilities have been tested and implemented, but nevertheless remain very accessory: thermal insulators, porosity testers, concrete additives, acid soil conditioners, pigments, etc.

5.4.4. Polluted Soils and Water

The environmental impact of mining activity and associated industrial developments extends far beyond the extraction, treatment, and waste storage sites. This is principally due to the high chemical dispersion power of surface flow, seepage and subsurface-draining water. *Water constitutes the main transportation agent* of dissolved, harmful or harmless products from underground ore deposits and surface tailings towards downstream soils, river courses and aquiferous rocks. Wind constitutes another, generally accessory agent. *The environmental restoration of mining regions necessitates considering and integrating all chemical exchanges likely to occur from upstream to downstream areas.*

An illustration of the complex interrelations acting in the successive compartments of a given mine ecosystem is provided by the old copper-mining region of Gulf Creek, in northeastern New South Wales, Australia (Lottermoser et al., 1999). Sulphide extraction and processing (pyrite, chalcopyrite, sphalerite) first led to the discharge of very acidic water (pH 2.2 to 3.4), which progressively caused the soil deterioration and plant destruction. The dispersal by drainage water of chemical elements contained in stored ores and abandoned waste was responsible for the physical accumulation in river sediments and related soils of various metals and metalloids: copper, silver, arsenic, cadmium, iron, lead and zinc. Plants and animals living in soils and river courses further caused either a bio-accumulation or a bio-rejection of some heavy metals (e.g. copper, zinc), which locally induced high concentrations of potentially toxic chemical complexes. This series of multiple inter-reactions needs to be considered globally when planning pertinent measures and long-term restoration.

The less-disruptive treatments for the environment consist in soil self-purification and trapping of plant pollutants (i.e. phytoremediation; Chapter 10.1.3). Various-acting mechanisms may occur:

- physical retention of colloids and particles during water seepage inside soils;
- contaminant trapping by plant roots (i.e. phytostabilization) and leaves (i.e. phytoextraction);
- contaminant fixation or exchange through adsorption-desorption processes involving both organic matter and clay minerals;
- precipitation and neutralization of toxic elements by bacterial strains, which may generate ammonia (NH_3) and carbon hydroxides (HCO_3^-);
- precipitation of dissolved metals through oxidation-reduction processes catalysed by bacterial activity (Sengupta, 1993).

Various limited experiments have been accomplished and are currently performed to test these various mechanisms, but almost no large scale treatments exist so far that could realistically be applied to a whole mining site and be of an industrial interest. *Wet zones* potentially constitute especially suitable environments for phytoremediation processes since they host various plant species that react with dissolved metals:

- freshwater algae (Cyanophyceae, Chlorophyceae) able to trap zinc, nickel, copper or lead, or to use iron and manganese as micro-nutrients;

– moss (e.g. *Sphagnum*) and reed family (e.g. *Typha*), the usefulness of which for extracting some metals (Fe, Mn) from water has been demonstrated.

The addition of natural substances characterized by strong adsorption properties (e.g. zeolites) is locally envisaged for cleaning up polluted industrial wastelands related to closed mining sites.

Physical and mechanical treatments help in an efficient way to prevent downstream contamination and pollution, but tend to accentuate the artificial character of the regions concerned. The *sealing* of polluted sectors with impervious material such as clay, polyvinyl, or polythene, and their artificial compacting help to prevent both the downward water seepage and the pollutant diffusion towards ground surface. But safety conditions are not guaranteed long-term; while the soil remains contaminated, lateral or downwards escape may happen, and unexpected re-use or disruption may arise.

Some investigations recommend building dams to *flood* old mines. Such a global treatment, which has been envisaged for the Shamokin coalfield in Pennsylvania (USA), would deeply modify the landscape but have several beneficial consequences: sealing of the polluted area by progressive sedimentary filling, buffering of acid drainage water, disposal of a moderately good quality water that could serve recreational needs (Rahn, 1992). Such modifications would nevertheless be irreversible.

The collection and diversion of drainage water that flows either upstream from mining areas, at the surface of mining sites (see Fig. 94), or in land areas situated downstream from ore fields, represents an efficient means to prevent the dispersal of dissolved chemical elements.

The chemical or electrolytic treatment of drainage water unfit for agricultural, domestic or industrial use aims to restore or balance the composition that has been altered during seepage and migration within mining fields. Water can then be re-injected in local aquiferous rocks. Most treatments are destined to *dilute the acid character* of drainage water issued from coal and metal ore exploitations, and therefore to restore some alkalinity. Chemical actions may be diverse: neutralization by lime, addition of ground limestone or soda, ionic exchange, passage of water through limestone drains that continuously release alkaline cations, etc. (Sengupta, 1993; Barton & Karathanasis, 1999). Neutralization effects may sometimes occur naturally, for instance when acid water released from coal tailings seeps through aquiferous carbonate formations. This role is played by Cretaceous chalk that underlies the coalfield slag heaps scattered over the coalfields of northern France.

An example of efficient electrolytic treatment is provided by the copper and gold exploitation complex of Bingham Canyon, Utah (USA), where the biggest artificial excavation in the world is located (length 4 km, depth 0.8 km). Downstream from the open-cast mines, a dense network of long, thin barriers has been implemented in deep aquifers, for blocking and diverting drainage water from its natural underground course. Water is conducted through deeply buried electrokinetic lines made of alternating positive and negative electrodes that retain the polluting heavy metals. Metals are then pumped up to the ground surface and depolluted water is re-injected in regional aquiferous rocks (Evans, 1997).

A Few Landmarks, Perspectives

- Ground mineral resources are used in all fields of human activity: construction and surface developments, urbanization and agriculture, industry and transport, manufacturing and art, feeding and health care. Either untreated or after being sorted, purified, or transformed, Earth's material and ore deposits, metals and pure or alloy minerals, fossil fuels and radioactive elements are part of our core daily needs. The world consumption of underground natural resources is gigantic, e.g.: almost 10 billion t/year of building materials or of fossil fuel, one billion tons of iron ore, and 2,500 tons of gold. These values still tend to increase rapidly for most mineral products, because of both the demographic explosion and rise in living standards. Our total energy needs increased by about 50% between 1980 and 2000, and the extraction of most metals by at least 20%. The demand of developing countries is still small but is growing rapidly.
- World distribution and accessibility of soil and sub-soil resources are very heterogeneous, as is their consumption in different countries. This results from the geological complexity of our planet for the former, and from different industrialization and technology development levels for the latter. Metal ores and radioactive or precious substances mainly occur in old Proterozoic and Palaeozoic continental shields, whereas coal and hydrocarbon products predominate either in Carboniferous freshwater deposits or in Mesozoic marine margin sediments. Among the developed countries, Western Europe and Japan are depleted in most underground resources of high commercial or industrial value. Submarine resources are partly still poorly known (hydrothermal deposits, methane hydrates and other hydrocarbons), which leads to to reactivated research as well as to debates on their ownership in the case of exploitation.
- Earth's resources basically exist in finite quantity. The question of the reserves still available is particularly crucial for energetic products, precious metals and substances of strategic interest. Reserve assessments are periodically updated but remain somewhat aleatory for various reasons: difficulty in locating some deposits and appreciating their abundance, non-

linear progress in prospecting and recycling techniques, emergence of substitute products, evolution of the world market and geopolitical context.

- Underground extraction techniques comprise: longwall exploitation mainly used for coal-vein mining; pillar-and-stall working, preferentially applied in mining the more resistant construction materials; and small-sized pits dug into subsurface deposits. Surface extraction is performed in classic quarries, the areal hold of which depends on the deposit shape, by strip mining, allowing exploitation of sub-horizontal, shallow veins, and by river-settled sand and gravel pits. The extraction is sometimes preceded by water or acid dissolution (salt, some sulphide and metal ores) or by drilling and pumping (hydrocarbons). Surface treatments comprise crushing, grinding, sorting, and washing processes that considerably increase the impact of the exploitation. Industrial plant and living areas in the past have often been implemented above mining fields, which has accentuated and multiplied the risks and nuisances.
- Environmental impacts of mining activities are widespread and very diversified. Many mining fields and basins have progressively become "anthroposystems", where natural and human factors closely interact. Physical and chemical nuisances spread at the soil surface and underground far beyond the designated area of exploitation, and generally continue a long time after concessions have been closed. Physical damage that propagates from depth to surface mainly consists of collapse and subsidence movements in and around the artificial cavities subject to gravity, water and air influences. At ground level, slag heaps, tailings and decantation basins are also responsible for gravity movements, as well as for landscape deterioration and sometimes spontaneous combustion (e.g. coal dumps). Chemical risks due to the presence of noxious or toxic elements are amplified by dissolution and transportation through rain, seepage and drainage water: oxidation, acidification or alkalinization, which may be exacerbated by bacterial activity; downstream transfer of metals, metalloids, or toxic organic compounds.
- Potential damage to the biosphere mainly results from the formation of acid drainage water, which frequently characterizes coal and metal ore mining fields (Cu,Ni, Pb, Zn) and tends to become more active after the mine closure. Acid water is carried by underground aquifers or at the ground surface and may impregnate the downstream soils and feed river courses threatening both flora and fauna. Additional risk for human health may result from the bioaccumulation in the food chain of toxic elements (mercury, lead, zinc). Other sanitary dangers are induced by dust or gas inhalation during mining or ore processing: silicosis, asbestosis, cancers. As far as asbestos fibres are concerned, the carcinogenic risk, which has been demonstrated for asbestos miners and workers, appears to be nil or insignificant in all other cases,

including for people living in asbestos-bearing premises or in asbestos-mining regions. The massive operations for asbestos removal that have been conducted since the 1980s are largely based on a psycho-dramatic perception of the risk, and tend to increase the potential danger through fibre release in open air.

– As the closure of most mining sites is ineluctable because of exhaustion of resources or a change in needs, it is highly desirable to accompany and, if possible, to anticipate their conversion. This means conducting an integrated, multi-scale study that takes into account all local interests and issues in a global restoration. Such an effort, which is based on socio-economic in addition to industrial and commercial considerations, has rarely been pursued in the past. This has often led to restoration of the mining zones as problems arise and with disputable consistency. Challenges are in fact very difficult to achieve and strongly interconnected. They concern safety questions, economical promotion and living environment: conversion of shallow mines and quarries, protection and development of the geological heritage, stabilization of both underground and dump areas, promotion of mining residue, dispersal control and chemical neutralization of drainage pollutants or toxic water. Numerous techniques for restoring the quality and attractiveness of past mining sites are currently investigated and implemented: morphological re-shaping, re-planting, self-purification, phytoremediation, chemical or electrokinetic water treatment, for example. They *should be integrated in an approach that is as natural as possible.*

Further Reading

Barnes J. W., 1998. Ores and minerals. Introducing economic geology. Open University Press, 1988, 181 p.

Blunden J., Reddish A., 1991. Energy, resources and environment. Hodder & Stoughton, The Open University, 339 p.

Down C. G. & Stocks J., 1977. Environmental aspects of mining. Applied Science Publishers, London, 371 p.

Evans A. M., 1997. An introduction to economic geology and its environmental impact. Blackwell, 364 p.

Goguel J., 1980. Géologie de l'environnement. Masson, Paris, 193 p.

Hester R. E. & Harrison R. M., 1994. Mining and its environmental impact. Royal Society of Chemistry, Cambridge, 164 p.

Panizza M., 1996. Environmental geomorphology. Developments in Earth Surface Processes 4, Elsevier, 268 p.

Rogers J. J. W. & Feiss P. G., 1998. People and the Earth. Cambridge University Press, 338 p.

Sengupta M., 1993. Environmental impacts of mining: monitoring, restoration, and control. CRC Press Lewis Publishers, 494 p.

Some Websites

http://geology.usgs.gov/index.shtml: official U.S. Government site devoted to underground exploitation and management

http://www.riotinto.com/: information site of a mining company that is a leader on a worldwide scale

http://www.idiscome/aime/: diffusion of knowledge in scientific and technological domains concerned by the production of Earth's materials, mineral ores, metals and energy. This site is maintained by the American Institute of Mining, Metallurgical, and Petroleum Engineers

http://www.brgm.fr: actions conducted in France and in various other countries in natural hazard, underground resources and hydrogeology fields

Chapter 6

Underground Water

6.1. A Strong Demand for Limited Resources

6.1.1. Over-exploitation of Underground Water in Two Large Countries

The High Plains Aquifer, also called the Ogallala Aquifer, stretches over eight states of the USA. It constitutes a very large underground water reserve that extends over a length exceeding 1,100 km from Wyoming and South Dakota to Texas and New Mexico (Fig. 95). The High Plains Aquifer is actively exploited by pumping, 95% of the extracted water being used for irrigating the huge cultivated surfaces of the Central Plains. As the recharge by rainwater is insufficient for compensating the amount of groundwater pumped to the surface, the level of groundwater progressively fell during the 20th century. The drop in water level was particularly serious during the 1960s and 1970s, especially in the southern States (Texas, south of Kansas), where precipitation is scarce, evaporation strong and irrigation demand important. The water table, which is the upper level of water within aquiferous rocks, was locally 30 m lower in 1980 than during the pre-development period of the 19th century (i.e. 100 ft). In some U.S. regions, the High Plains Aquifer, the water reserves of which have slowly accumulated during the past hundred thousand years, risks becoming exhausted in a few decades.

The reduction in water quantity has progressively induced a decrease in its quality due to the relative concentration of dissolved or particulate chemical elements: natural sodium and other salts removed from ground rocks, farming-derived fertilizers (nitrates, phosphates) and pesticides, oil and gas released from wells or dumping. This has locally determined a loss of safe drinking water. Since the 1980s, strict control measures have been implemented by both State authorities and the Federal Agency for Environmental Protection. These regulations have not reduced the rate of water depletion and deterioration but have helped to prevent if from increasing further.

The Great Artesian Basin constitutes a deep aquifer system occupying almost a quarter (22%) of the Australian continent. Located in the east of the continent, it extends over more than 1,500 km from west to east, as well as from north to

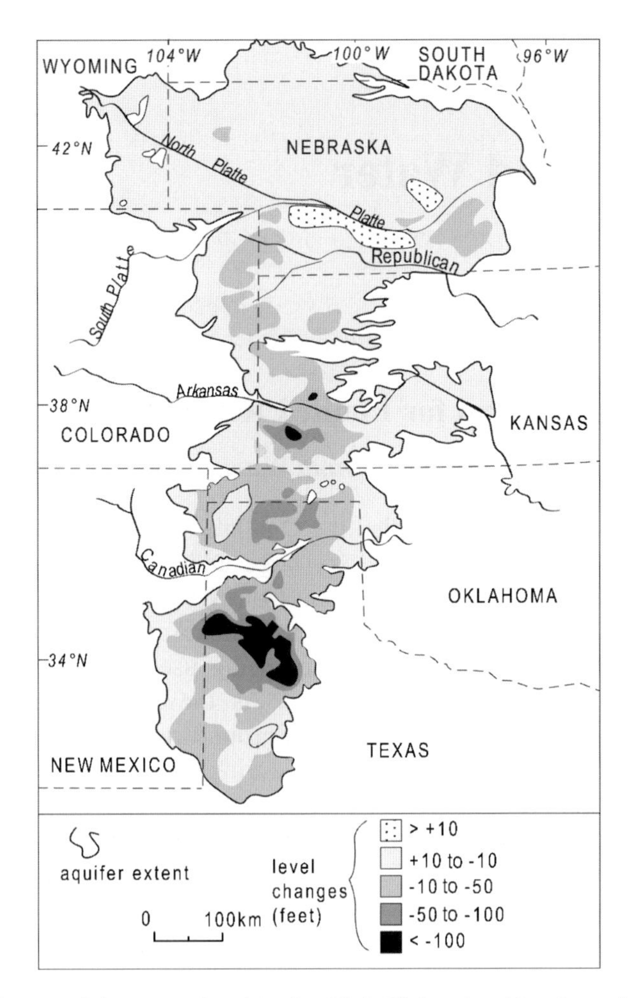

Figure 95: Change of the water level in the High Plains Aquifer since the 19th century farming developments (after U.S. Geological Survey; see Turner II et al., 1990).

south (South Australia, Queensland, Northern Territory, New South Wales). The Great Artesian Basin is exploited by innumerable wells and springs and represents the main water resource for arid to semi-arid regions of central Australia. The development in the 20th century of domestic, agricultural and industrial activities by colonists from Europe has progressively determined the dwindling of spring outflow. The situation rapidly deteriorated in the 1990s after the discovery and exploitation of the giant metal ore deposits of Olympic Dam. Located outside the Great Artesian Basin in South Australia, the Olympic Dam copper–uranium mine requires very abundant water for ore extraction and processing. This water supply is derived from boreholes located 120 and 200 km to the north in the Artesian Basin, which has induced an acceleration of water depletion within aquiferous rocks.

The Olympic Dam ore extractions in 1997 amounted to 85,000 t of copper, 1,600 t of uranium oxide, 13 t of silver and 0.85 t of gold. The mine, which is presently the largest in the world for uranium extraction, is planned to eventually produce 350,000 t of copper and other metals per year. The large ore reserves identified have a production plan for at least 50–100 years. This evolution has nevertheless recently experienced a slowdown because of the threat posed to water abundance in central and Southern Australia, as well as related industrial and environmental risks. Extensive investigations have been implemented to better understand the effects of the Olympic Dam boreholes on the mound springs from the Great Artesian Basin, and to present recommendations for sustainable development (Mudd, 2000). Meanwhile the mine's production has been limited to a maximum of 200,000 t/year of copper, 4,630 t/year of uranium oxide (U_3O_8), 23 t/year of silver, and 2 t/year of gold.

6.1.2. Water Consumption

At the dawn of the third millennium *people use annually more than 5,000 km^3 of water per year*, a large part of which is discharged into rivers or seas, evaporates, or is too polluted to be easily reused. This extraordinary quantity is predominantly used for farming (almost two-thirds of world consumption on average) and industry (about one-fifth), and secondarily for domestic needs. The percentages vary greatly depending on the country. For instance agricultural practices determine the consumption of 94% of water resources in India and 32% in the USA. The rates are of 4 and 45% for industrial use, and of 2 and 23% for domestic purposes, respectively (Barnier, 1992). The French consumption averages a proportion of three-fifths for industry, one-fifth for agriculture, and one-fifth for domestic use. China, which is the most populated country in the world, constitutes the leading water consumer, but is closely followed by the USA, India and the former Soviet Union, three countries where population is much lower. On average, one USA person uses 300 times more water than a Ghanaian (West Africa), and a European person 70 times more.

The water consumption in the world at the end of 20th century was 40 times higher than that estimated at the end of 17th century (Turner et al., 1993). *The increase registered for 1950–1980 is similar to that estimated during the last 300 years*, which represents an augmentation of the order of 1 (Fig. 96). This trend has not been synchronous in all continents. A steady augmentation of water consumption started in the 19th century in Europe, North America and Australia, and in the 20th century in most other countries. The use of water for farming practices is presently still dominant, but depending on the country it has

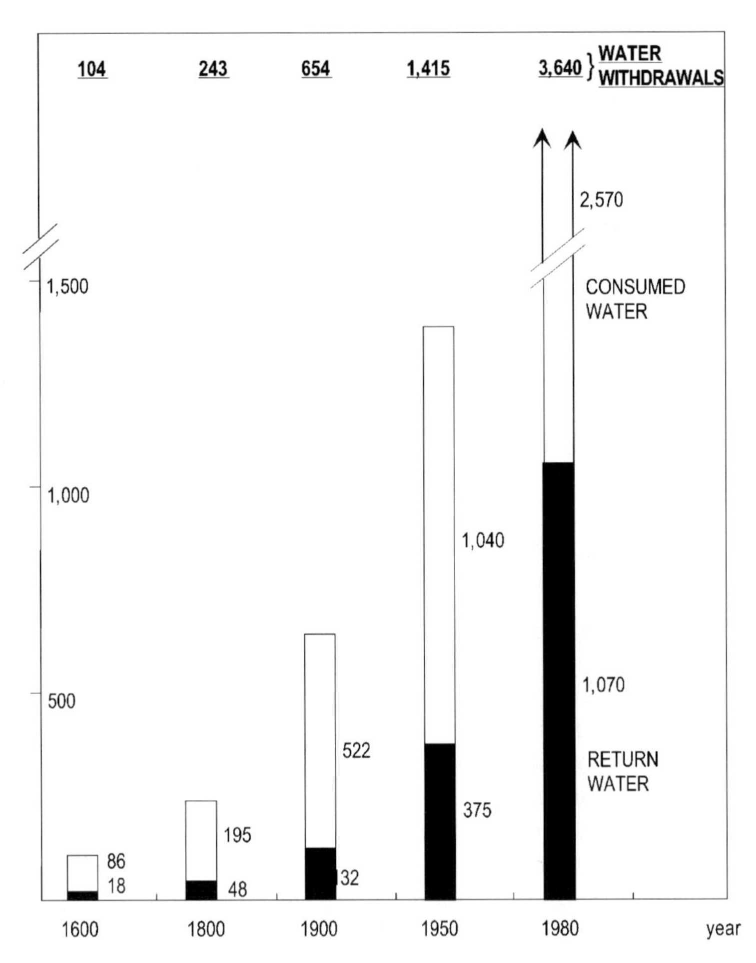

Figure 96: Estimates of the total water withdrawn, consumed and returned for all human activities in 1680, 1800, 1900, 1950, and 1989 (after Turner II et al., 1993). Values in km^3/yr.

decreased by 90%–75% relative to other uses during the last three centuries. In all countries the augmentation of consumed, non-reusable water, has been considerable during the last century due to the development of polluting practices in farm, urban and industrial activities.

6.1.3. *Water Availability and Reserves*

Earth contains about 1.4 billion km^3 of water (1,400,000 km^3), 97.5% of which are salty (oceans, seas, saline lakes) or not available (rock-bound water, deep aquiferous rocks). Fresh water constitutes only 2.53% (i.e. about 35,000 km^3) of the total amount. Two-thirds of fresh water is immobilized in a solid state in icecaps and glaciers, permanent snow and permafrost (i.e. frozen soils; Chapter 3.3.3), and almost one-third is contained in underground aquiferous rocks. Surface water filling the lakes, rivers and marshes constitutes only 0.3% of total fresh water (Fig. 97; see also Table 20).

Liquid fresh water represents less than 1% of Earth's water (i.e. about 11,000 km^3), 99% of which is contained in ground rocks. Most underground water is located too deep below the soil surface to be extracted. As a consequence, *water, which is reusable provided that it is adequately managed, is in fact only available in limited amounts for human use.* Desalination of seawater and ice-melting constitute well-controlled techniques that could increase freshwater availability (Chapter 6.5); but both the complexity of the processes and the cost of production and transportation currently prevent application of these techniques on a large scale.

The renewal of surface and underground freshwater reserves depends on the amount of rainfall, on temperature which controls evaporation, and on drainage and seepage. *Water availability therefore depends on climate, geomorphology and geology characteristics.* Some *industrialized countries*, which are predominantly situated in temperate humid regions, have abundant water resources: Canada, New Zealand, Norway, and to a lesser degree other northernmost European countries, USA, Russia, and Australia. Some others such as Japan and

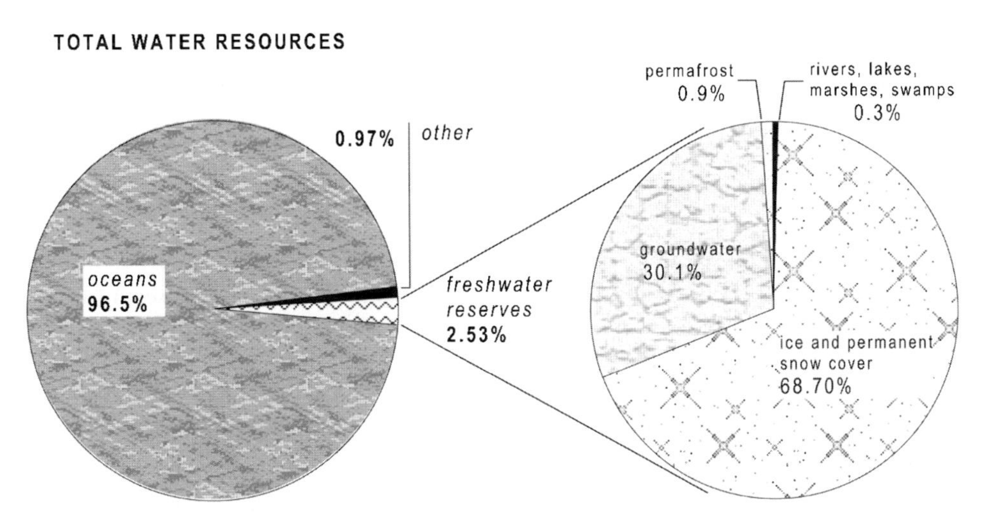

Figure 97: Proportion and distribution of freshwater resources relative to total water on Earth (after World Resources Institute; see Pickering & Owen, 1997).

Table 20: Relative abundance and renewal time of water in Earth's main reservoirs (after Owen et al., 1998). Bold characters concern easily accessible and usable water for human purposes.

Location and nature of reservoir	**Total water supply %**	**Renewal time (years)**
on the land		
ice caps	2.225	16,000
glaciers	0.015	16,000
freshwater lakes	**0.009**	**10 to 100**
saltwater lakes	0.007	10 to 100
rivers	**0.001**	**0.003 to 0.05** (12 to 20 days)
soil and underground		
soil moisture	0.003	0,75 (280 days)
fairly shallow aquifers (≤ 1 km)	**0.303**	**300**
deep aquifers (>1 km)	0.303	4,600
atmosphere		
gaseous, liquid, solid water	0.0001	0.2 to 0.03 (9 to 12 days)
oceans		
saline water	97.134	maximum 37,000
	total 100.000%	

various countries from southwestern Europe have rather limited reserves owing to their high population density. Certain other countries are encountering more difficulty in meeting their water needs: Germany, Benelux, central Europe from Poland to Bulgaria.

Most *developing countries* situated in humid intertropical zones have globally enough freshwater, provided that contamination and pollution remain low. This is the case in Central and South America, the Guinea Gulf-bordering countries, Indonesia, India, and China. Other developing countries, especially those located in arid climatic zones of Africa and the Middle East, dramatically lack fresh water.

It is generally admitted that *average water needs are comfortably satisfied when the amount available for each inhabitant exceeds 1,600 m*^{3}*/year*, and that they become acceptable for values exceeding 1,000 m^3/year (Foley, 1999). When the annual runoff is less than 1,000 m^3/year per person, chronic shortages occur, values below 500 m^3/year corresponding to strong scarcity inducing serious health problems. The world average consumption of water amounts 700 m^3/year. It exceeds 1,600 m^3/year in most developed countries, and is less than 500 m^3/year in numerous developing countries with desert or arid climates. In addition the water *quality* is often worse when a low standard of living results in ill-considered and almost irreversible pollution. The deterioration of water quality helps to reduce the resource availability, even in some developing countries with a humid climate (e.g. some regions of India).

6.2. Water Resources

6.2.1. Place in the Hydrological Cycle

The dynamics of freshwater exploited from surface rivers and lakes and from underground aquifers constitutes one phase of a global cycle called the *hydrological cycle*. The hydrological cycle involves liquid water from oceans, continents and air (rain), gaseous atmospheric water and solid water (glaciers, sea ice, cloud ice). It is dominated by precipitation (rain, snow) and evaporation (all open-air water surfaces, plant activity), which are linked by both the surface and underground gravity-driven flows and the wind-driven cloud movements (Fig. 98, left side).

Vertical water exchanges developing between the atmosphere (air, clouds) and surface hydrosphere (seas, lakes, rivers, ice, soils) balance each other under natural conditions, rainfall and evaporation flux reaching annually 500,000 km^3. *Horizontal exchanges* between terrestrial and marine domains tend also to be balanced; the fluxes amount to 35 mkm^3 for liquid water flowing from rivers and aquifers to the sea, as well as for water vapour resulting from marine evaporation and falling as rain on continents. Surface fluxes towards the ocean are much larger than underground fluxes, since water in aquifers moves much more slowly than open-air running water.

The different hydrological cycle phases, especially those involving surface land water and shallow aquifers, are diversely disrupted and quantitatively modified by human activities: land use, farming, urbanization, industrialization, recreational developments, etc. (Fig. 98, right side).

Different types of Earth's reservoirs that are subject to water exchanges *are characterized by very different volumetric capacities* (Table 20; see also Chapter

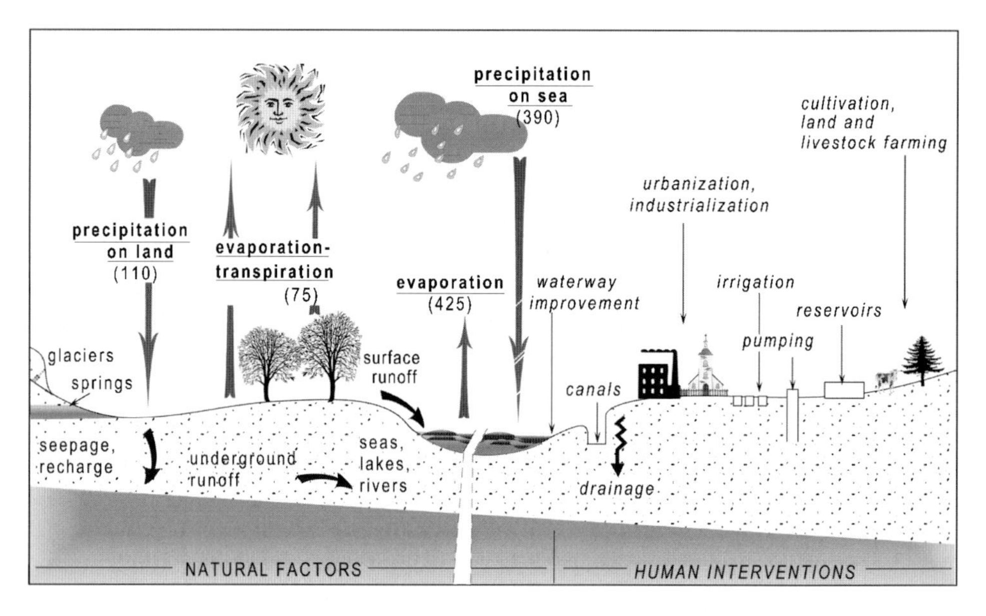

Figure 98: Natural and anthropogenic factors intervening in the water cycle (after Turner et al., 1993; Price, 1996). Figures in brackets indicate the water fluxes in thousands cubic kilometres per year (mkm^3/yr).

6.1.3, and Fig. 98). Easily usable reserves are provided by surface and subsurface liquid freshwater, which represents about 0.12 million km^3 (Mkm^3) for rivers and lakes, and 4 Mkm^3 for accessible underground stocks. These two types of reservoir contain 0.01% (rivers: 0.001; lakes: 0.009) and 0.303% of Earth's water, respectively. An overwhelming majority of water reserves are located principally in the oceans (1,370 Mkm^3, i.e. 97.134% of the planet's water), and secondarily in permanent icecaps (about 30 Mkm^3, i.e. 2.24% of total amount) and in inaccessible or saline deep rocks (about 4 Mkm^3, i.e. 0.303% of total water).

The residence time of water types commonly accessible to man is fairly short relative to that of seas, icecaps or deep aquifers. River water is statistically renewed every 2–3 weeks, and accessible aquifer water has a residence time of less than 300 years. By contrast the average renewal time may attain 47,000 years for seawater, 16,000 years for glacier and icecap ice, and 4,600 years for deep aquifer water. As a consequence the water reserves used by Man can be replaced in a relatively short time. *The renewal of accessible underground water is nevertheless 5,500–9,000 times slower than that of surface running water*, which determines much difficulty for replenishing over-exploited underground aquifers (Chapter 6.2.2, 6.4), and also for cleaning them up (Chapter 10).

6.2.2. Underground Water Reservoirs

Types of Water Contained in Rocks

Water in rocks is either retained more or less firmly within geological formations, or free and mobile in rock pores.

Retained water belongs to three different types:

(1) *hygroscopic water* is firmly adsorbed in rock pores and cannot be extracted except through calcination. For instance the mass of granite rocks contains 1% in weight of hygroscopic water;
(2) *film water* is retained on the surface of rock grains by molecular attraction forces. It is not subject to gravity and is partly extractible by plant roots in surface soils. Film water may constitute 5–10% of sand mass, and up to 50% of some clay-rich rocks, especially those containing abundant smectite minerals;
(3) *capillary water* is kept by surface tension on rock grains, is not subject to gravity but may evaporate. Its abundance increases proportionally to the pore size. Capillary water occupies a very thin layer at sand and gravel surfaces, but may extend over a thickness of 10 m in some highly absorbent clay materials (e.g. smectite-rich soil, clay or claystone).

Free water circulates within porous rocks and comes from either rainwater seepage, river or lake water infiltration, and/or slow migration from upstream geological formations. Free water is displaced by gravity forces and is therefore labelled "*gravity water*". Its displacement is expressed in the vertical flow, which depends on the rock permeability k (i.e. its hydraulic conductivity), on the thickness H of rock soaked in water, and on the thickness h of water possibly present above the ground surface (e.g. river or lake). The corresponding formula constitutes *Darcy's law*, which represents the proportionality linking the underground water flow and the hydraulic gradient:

$$Q = kS(H + h)/H$$

with S, surface of the rock section where water circulates.

Aquifers and Karst

Free water migrating within rock pores, cracks and fissures is called *groundwater*. Groundwater constitutes the largest source of water used by Man for drinking and for many other purposes. It circulates in porous rocks called *aquifers*. This is the case of the High Plains Aquifer in the USA, of the Great Artesian Basin in Australia (Chapter 6.1.1), as well as of the chalk aquifer that ensures most water resources for several northwestern European countries (Fig.

Figure 99: Extent of the chalk aquifer in northwestern Europe (after Ziegler; see Lumsden, 1992).

99). Innumerable water-rich aquifers occur worldwide, the size, depth and thickness of which greatly differ.

Groundwater migrating within rock voids via gravity forces may reach the ground surface through topographic lows where it constitutes *springs*. It may also be directly attained within underground formations by man-made *wells*.

Groundwater that circulates in more or less soluble aquifers such as limestone or Ca, K or Na salt often causes rock to partly dissolve. Underground dissolution may induce the formation of a subterranean network called karst. Karstic formations are of various size and extent, and often display morphological features resembling surface river systems: development of river courses and lakes, tributaries, cascades, seeping and re-emergence zones, etc.

Aquifer Characteristics

Groundwater migrates in geological formations that are water-saturated and constitute the *phreatic zone*. The phreatic zone is fed from an overlying non-saturated zone called the *vadose zone*, the rock pores of which contain film and/or capillary water, free gravity water seeping downwards, and air. The non-saturated zone is generally located close to the ground surface. It may sometimes also be situated deeper underground, namely when a porous rock layer is overlain by an impervious, clay-rich "aquiclude" that hampers active seepage (Fig. 100, left side).

Groundwater physical properties are especially characterized by the shape and level of the upper surface, which is called the *water table* or piezometric surface. The water table morphology reproduces with some variation that of the ground surface (Fig. 100). The slope of the water table is roughly parallel to the valley sides and its area underground is the extension of open-air rivers, lakes, or reservoirs. The water-table slope constitutes the *hydraulic gradient*, which results from the water pressure necessary for compensating the frictional forces and for allowing free water to migrate within rock pores.

Groundwater that migrates freely upwards when rainfall and seepage are abundant is referred to as *phreatic groundwater.* Corresponding aquiferous rocks constitute an *unconfined aquifer.* The water table rises or falls according to precipitation and seeping intensity, which determines its level to "beat". Phreatic groundwater connected with a river constitutes *alluvial groundwater* that beats according to surface-water flooding and fall. In the same way the water table connected with a lake or an artificial reservoir beats according to the basin filling level (Fig. 100, right side).

Perched groundwater corresponds to phreatic water that is retained above the valley level due to the presence beneath the aquifer of an impervious rock formation (aquiclude). It appears at ground level in the form of springs (Fig. 100, left side, top) or by seeping.

Underground water, the upward extent of which is limited by an impervious rock layer (e.g. clay, marl) migrates within a *confined aquifer* and is referred to as *captive groundwater* (Fig. 100, left side, bottom). The water table is then situated above the upper limit of the water-saturated porous rock forming the aquifer. Groundwater is therefore submitted to high internal pressure and may gush out of the rock when a well is dug within the aquifer. Such a well is called an *artesian* well. Captive groundwater may become artesian and give way to artesian springs in topographic sectors where the water table is situated above ground level.

Aquifer Properties

Underground rocks liable to constitute important aquifers mainly consist of porous or fissured *sedimentary rocks*: sand, sandstone, ash and lapilli, gravel and pebble deposits, crumbly limestone such as chalk, etc. Magmatic rocks of granite and basalt groups generally constitute aquifers of little extent and capacity, as do most metamorphic rocks such as gneiss and micaschist.

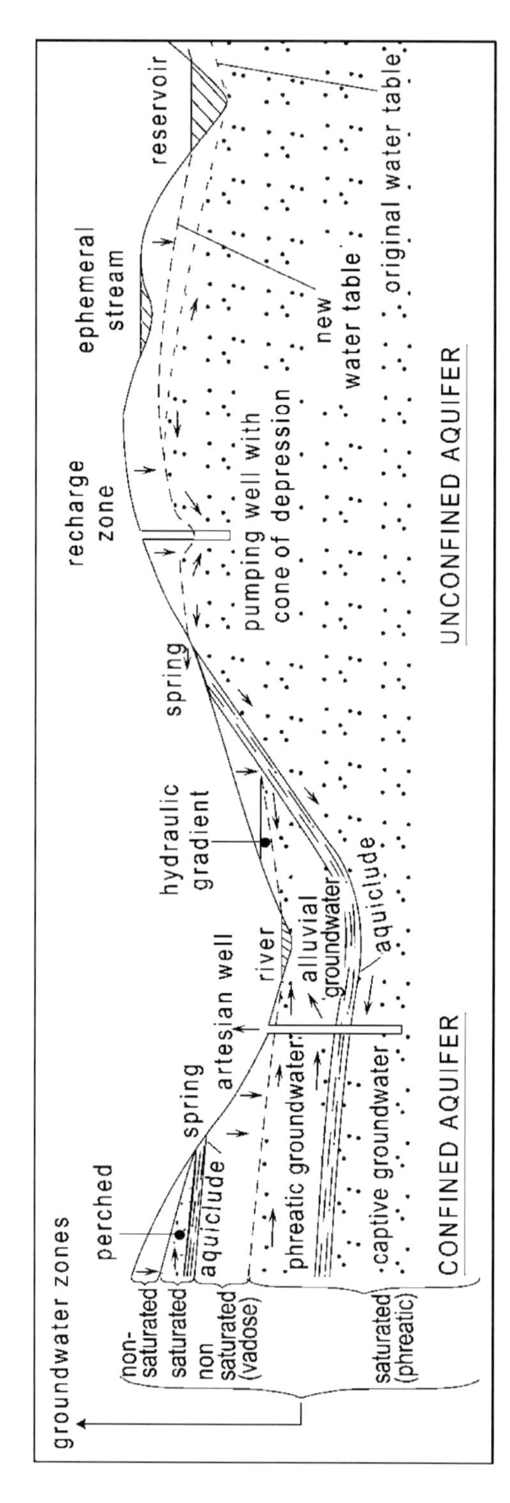

Figure 100: Groundwater and aquifer types (after Waltham, 1994). Arrows indicate the directions of water movement.

Exceptions concern some highly weathered or fractured eruptive rocks, as well as porous or vesicular volcanic products (pumice stone, pozzuolana).

Aquifers display a *highly variable lithological nature and horizontal continuity*. Some aquiferous rocks are homogeneous on a regional scale (e.g. the chalk aquifer in northwestern Europe) but heterogeneous on rock and especially thin-section scales. Other aquifers such as alluvial plain or alluvial fan deposits are very heterogeneous, heterometric and discontinuous. Their formation results from highly changing hydrodynamic and gravity processes that have induced the deposition of coarse to fine-grained, often lenticular and very local sedimentary bodies. In these environments the lateral migration of river channels and the varying force of water streaming are responsible for unsorted detrital sedimentation (Fig. 101).

The more or less heterogeneous character of aquifer lithology determines the importance and mobility of groundwater. It is crucial for allowing groundwater to be exploited, aquifers to be recharged, and remediation to be performed in cases of underground pollution.

Permeability constitutes an essential hydrological parameter that expresses the flow rate of underground water within rock pores, fissures and fractures. It contributes in an essential way to estimating the possibility for groundwater to

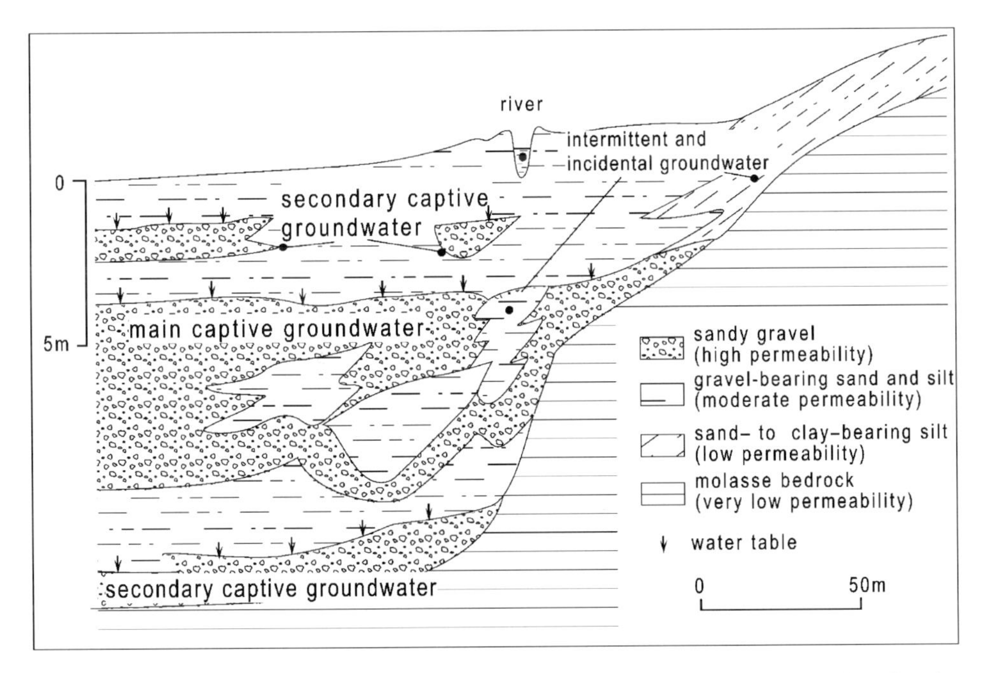

Figure 101: Aquifer heterogeneity. Example of a detrital sedimentary fan in the Arbogne alluvial plain, Switzerland (after Parriaux; see Campy & Macaire, 1989).

Table 21: Mean permeability and porosity of some common rocks (after Campy & Macaire, 1989; Waltham, 1994). Water permeability is given in metres per day (m/day).

Rock	**Permeability m/day**	**Effective porosity %**	**Total porosity %**
gravel	300	40	45
sand	20	30	40
chalk	20	4	20
fractured sandstone	5	8	15
clay	0,0002	3	47
slate	0,0001	–	3
granite	0,0001	–	1
cavernous limestone	very heterogeneous	4	5
fracture zone	50	–	10

be extracted and renewed. Permeability is measured in millimetres per second or metres per day or year. A rock is considered potentially to constitute an aquifer if the water flow rate exceeds 1 m/day. An impervious rock corresponds to values that are less than 0.01 m/day. In some rocks water moves more slowly than 1 m/year (Table 21). Common aquifers show permeability values of a few metres per day. In some highly permeable rocks, the flow speed may attain 20–300 m/day. Open karsts display values of a few kilometres per day and behave like active river systems.

Total porosity corresponds to the void volume to rock volume ratio. It is of a limited reliability for evaluating the pore water mobility, since very fine-grained rocks such as clay are highly porous but strongly retain water by electrochemical absorption, and therefore behave as impervious rocks. As clay prevents water being expelled it is both impermeable and uncompressible; its microporosity is very high but its permeability is very low, and similar to that of granite (Table 21). *Effective porosity*, which corresponds to the ratio of the total gravity water volume over the rock volume, is much more representative of the water flow potential. The actual celerity of water *Vr* is given by Darcy law to effective porosity ratio:

$$Vr = \frac{kS(H+h)/H}{\text{effective porosity}}$$

The natural water *runoff* in aquiferous rocks mainly depends on the rock permeability, the amount and importance of springs, and the level of connected rivers and lakes. Natural *groundwater recharge* essentially depends on rainfall intensity, on the possibility for aquifers to be fed by water seepage from both surface soils and vadose zone, and on connections with upstream aquifers. An aquifer situated deep in the underground of arid climate regions is practically very difficult and very slow to recharge naturally. This is the major problem instigated by exploited sandstone aquifers of arid regions from the Middle East, Arabian Peninsula or North Africa, where precipitation does not exceed a few centimetres per year. Underground water in these countries has very slowly accumulated over a hundred thousand years; if extracted it cannot be renewed at a fast enough rate to benefit mankind (Chapter 6.4).

6.2.3. Physicochemical Properties of Underground Water

The average *temperature* of groundwater that flows less than 10 m below ground surface fluctuates according to the season. Fluctuations are nevertheless attenuate and delayed relative to air and soil surface temperature. By contrast the groundwater temperature below 10–15 m depth is very stable throughout the year and is similar to that of fairly deep caves. Water located deeper than a few hundred metres displays a progressive downward temperature increase that is driven and controlled by terrestrial heat flow.

The *chemical composition* of groundwater varies considerably depending on host rock type, climate, flow duration, etc. Precursory rainwater typically contains 10–20 mg/l of dissolved material: air-derived oxygen and carbon dioxide, sea-originating salt, various nature-, city- or industry-derived aerosol components (sulphides or sulphates, nitrogen, metals, etc.). Rainwater penetrating the underground is progressively enriched by various chemical elements through seepage from surface soils and migration within porous rocks. Flowing water tends to subtract mineral ions and organic components from host formations, and to ensure their downstream transfer. *Underground water therefore bears a chemical signature that is typical of aquifer composition*, the origin of which can be identified by chemical analysis in surface springs, wells, etc. (e.g. Fig. 102).

The chemical characteristic of groundwater is usually labelled: (1) by the nature of dominant anions: bicarbonates (HCO_3^{2-}), sulphates (SO_4^-), chlorides (Cl^-); (2) by the nature of dominant cations: calcium (Ca^{2+}), magnesium (Mg^{2+}), sodium (Na^+), potassium (K^+).

Groundwater is more mineralised the deeper and longer its migration, and the more marked it is by exchanges with hotter and more soluble host rocks. Water that contains abundant dissolved elements tends to release some of them when approaching the ground surface or merging as springs. Such precipitation results in numerous and various concretions that decorate caves and karst networks

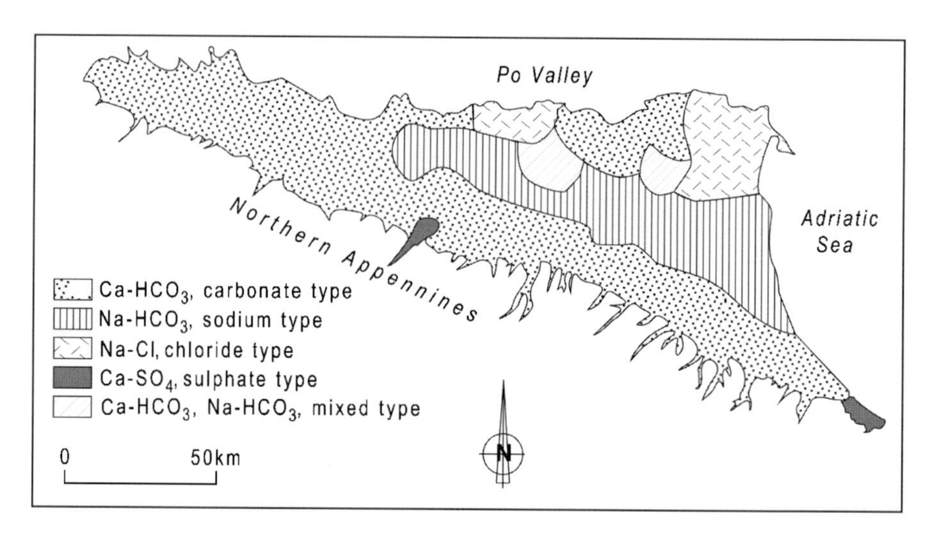

Figure 102: Chemical types of underground water in Emilia-Romagna region, Po Valley, Northern Italy (after Martinelli et al., 1998).

(stalagmites, stalactites, pisolites, etc.), in calcareous travertine and tufa at spring exits, in iron and/or manganese oxide crusts, etc. The nature and amount of chemical elements dissolved in a mineral water determine its properties and potential uses (Table 22; Chapter 6.3.1; see also Chapter 10).

The groundwater mineralisation in host rocks changes during migration depending on the precipitation regime. In *humid regions*, where aquifers are regularly recharged through rainfall, seeping water is generally ion-depleted. During the first migration stages, it becomes enriched in calcium (Ca) and sometimes magnesium (Mg), which are extracted from carbonate rocks, and tends to get "hard". This chemical change develops more rapidly when soil and subsoil are rich in carbon dioxide (CO_2) resulting from plant activity and organic-matter decomposition. Deeper underground, Ca and Mg tend to be progressively adsorbed by clay mineral particles that in turn release sodium (Na) and sometimes potassium (K). Pore waters become of a sodium type and gradually get "soft". If migration persists deeper, the water movement slows down; some less-soluble chemical elements such as sulphur and chlorine may be dissolved, giving way to the formation of sulphated or chlorinated water.

In *arid regions*, evaporation processes usually exceed precipitation. Shallow underground water tends to rise through capillary forces towards the soil surface and to evaporate, mainly if outcropping rocks are notably permeable (e.g. coarse-grained sandstone). This induces dissolved elements to concentrate in pore water, which prevents host rock elements being dissolved. The daily repetition of evaporation determines a progressive *salination* of groundwater that tends to loose its drinkable properties. This is the case for numerous local

Table 22: Chemical quality of natural water according to European Union recommendations (after Price, 1996). Comparison with data from some Spanish thermal water springs used for therapeutic purpose (after Cuchi-Oterino et al., 2000). Values in mg/l.

Chemical components	**European recommended values for drinking water**	**Maximum values admitted in Europe**	**Values of some Spanish thermal waters**
pH	6.5–8.5	–	5.7–9.2
calcium	100	50	688
magnesium	30	50	180
sodium	20	150	3,010
potassium	10	12	118
nitrates	25	50	40
chlorides	25	–	5,147
sulphates	25	250	5,182
aluminium	0.05	0.2	–
iron	0.05	0.2	5.78
lead	–	0.05	–
arsenic	–	0.05	–
mercury	–	0.001	–
cyanides	–	0.05	–
fluorides	–	1.5	16
pesticides	–	0.000 5	–

sabkha or oasis groundwater, for aquifers bordering Lake Chad in Central Africa, the Aral Sea south of Russia (Chapter 6.4.3), the Great Salt Lake in Utah (USA), etc. Under very arid conditions the capillary rise of saltwater may start as deep as 5 m in the ground. It is responsible for the precipitation of salt and gypsum crusts that induce a decrease in soil permeability and may affect plant growth.

6.2.4. Types of Exploitation

The worldwide increase in water requirements and the necessity to exploit deeper and deeper aquifers in addition to traditional springs and wells has determined the development of *complex water-extraction engineering systems* involving drilling, pumping, automatic catchment devices, gates and non-return valves, and distribution networks (Price, 1996; Fig. 103).

The current increase in underground pollution risks, the difficulty in decontaminating polluted aquifers in an efficient and rapid way, and the

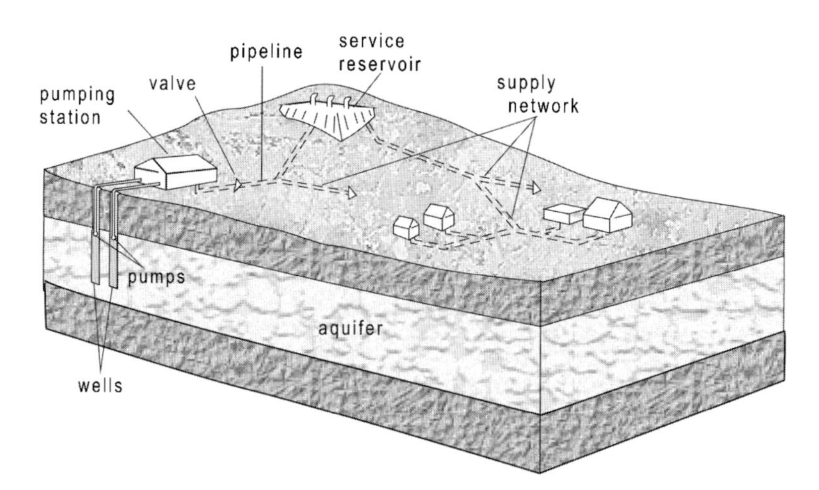

Figure 103: Example of an integrated system for groundwater extraction, exploitation and supply (after Price, 1996).

progressive awareness of hygiene and safety problems linked to water management, have led to a drastic extension of the *systems and rules designed to guarantee water quality*: definition of protection zones, specific water catchment techniques (Fig. 104), building and development rules in the vicinity of catchment fields, monitoring of groundwater level and quality (Chapter 10.2).

Quality criteria largely depend on the use for which a given groundwater is extracted. Water fitting irrigation needs and destined to be filtered by both soils and plants may be unsuitable for drinking and feeding, or even for bathing and other domestic uses. Some water types employed for treating external diseases may be unfit for being regularly ingested, etc. (Chapter 6.5.1).

6.3. Specific Uses

6.3.1. Groundwater and Health

Therapeutic properties of mineralised, gaseous and/or hot water have been known since ancient times and have given way to numerous catchment systems and chronicles: Roman and Arab thermal baths of Spain and the Mediterranean area; sulphate-, sodium- and chloride-rich thermal sources of Aachen gushing out from Devonian limestone in the Rhineland Massif (Germany) and used by Roman legions and Charlemagne's court; Karlovy Vary (i.e. Karlsbad) springs in the Czech Republic visited by Peter the Great and having generated in the late 17th century famous precursory Russian investigations into thermal waters;

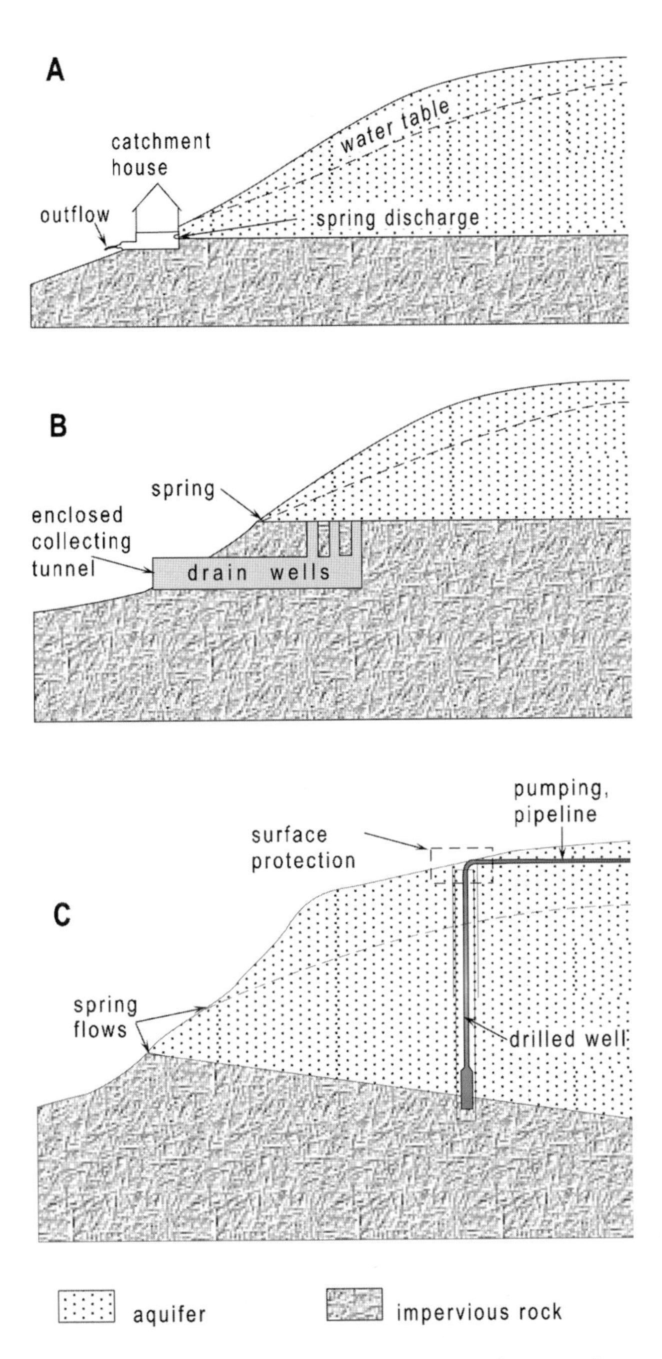

Figure 104: Examples of protective point-of-discharge interception methods (after Robertson & Edberg, 1993). (A) enclosed post-spring interception with partial excavation. (B) Underdrain interception and diversion system. (C) Well drilling and water pumping.

traditional thermal springs of the French Massif Central (Vichy, La Bourboule) and eastern Pyrenees (Amélie-les-Bains); etc. (see for instance *Environmental Geology*, Springer, volume 39/5, 2000). Thermal baths, hydrotherapeutic establishments and curative water types are countless and have given way during the last decades to spectacular developments.

Mineralised waters destined to external treatments (e.g. skin diseases) generally contain *high amounts of specific dissolved elements* and are therefore mostly unfit for drinking. Similar restrictions apply to certain "thermal" waters used for treating circulation, breathing, digestive or rheumatic diseases. They may contain chemical elements such as Cl, SO_4, NO_3, SiO_2, Mg, Na, K, Fe, Li, F or B, the amounts of which are much higher than standard values recommended for drinking or sanitary use (see Table 22, right side).

Men and animals in certain regions nevertheless regularly ingest such "naturally contaminated" waters, which induce chronic diseases. For instance some spring waters from Bolgatanga and Bongo districts in Ghana (West Africa) that drain hornblende- and fluorine-rich syenite and granite formations contain up to 4.6 mg/l F; this value is 3 times higher than the maximum content tolerated (1.5 mg/l F). The local population drinks this water, which causes numerous cases of fluorosis, a disease characterized by bone and breathing disorders. Other springs in the same region are fluorine-depleted ($F < 0.5$ mg/l) and predispose to dental decay (Apambire et al., 1997).

In Nalgonda region, India, chronic fluorosis results from the consumption of water containing up to 20 mg/l F, the dissolution of which in an acid granite environment is facilitated by Ca and Mg depletion (Rao et al., 1993). In other regions excess mercury and arsenic are recorded within spring waters, as for instance those draining the fossil hydrothermal system of Northland in New Zealand (Craw et al., 1999).

Bottled water is used increasingly for drinking or even cooking. This tendency is exacerbated by a growing suspicion of tap-water quality and media-spread dietary or aesthetic concerns. Bottled water belongs to *three categories*, the distinction of which is sometimes confusing due to the frequent use of the word "spring" on bottle labelling.

- *Still mineral water* consists of groundwater directly bottled at an original discharge point that consists of a real spring, a well or a drill site. Mineral water is characteristic of a given geographic sector that has been officially labelled. It has been neither physically nor chemically treated except through filtration and/or carbonation or CO_2 degassing. It is of a certified composition and free of pollutants. This water type is the most clearly defined.
- *Spring water* comes equally from one or several springs. It may have been treated before bottling which may or may not be performed on the catchment

site. Transportation and packaging operations are especially important for guaranteeing the product quality.
- *Table water* conforms to lesser rigorous constraints but has a guaranteed drinkable character. It corresponds either to water extracted from non-regulated springs or to aerated tap water, and may have been physically and chemically cleared of suspended matter or impurities.

Noticeable progress has been made to ensure the permanent characteristics and quality of drinking water (Robertson & Edberg, 1993):

- regular, long-term checking of the origin and chemical composition of water extracted from intermediate and deep aquifers;
- understanding of aquifer recharge modes;
- appropriate choice of catchment devices and systems (see Fig. 104), as well as of bottling procedure;
- adequate prevention of chemical and microbial contamination risks.

6.3.2 Geothermal Energy

Origin

The use of naturally heated water or steam for meeting domestic or industrial needs started about one century ago in Tuscany (Lardello) and Campania (Naples) in Italy. Geothermal energy became significant in New Zealand in the course of the 20th century, and spread to several other countries during the last decades: Iceland, Spain, France, Hungary, Japan, Mexico, Russia, California and Hawaii in the USA. Icelandic thermal waters ensure the domestic heating of whole cities, developing greenhouse cultivation, and producing electricity. In France hot water rising to ground surface is exploited at Chaudes-Aigues (Cantal region) where its temperature attains 82°C, at Amélie-les-Bains (eastern Pyrenees), where its mineralization also induces curative applications, etc.

Most geothermal sites occur in active *volcanic and seismic zones* that correspond to oceanic accretion of subduction regions (e.g. Pacific active margins), to hotspots (Hawaii archipelago), to intraplate volcanism (Russia) or to plate collision (North of Algeria). Exploited geothermal zones are practically all located in subaerial sites (Fig. 105) and of a moderate extent. The production of easily accessible hot water in volcanic regions is linked to the local occurrence of important heat flow resulting from *strong geothermal gradients* (i.e. temperature increase in several tens of degrees Celsius for 100 m of burial).

Numerous *hydrothermal vents* also exist in submarine volcanic domains, mainly at the axes of oceanic ridges where basaltic rocks continuously form

Figure 105: Main geothermal zones on exposed landmasses (after Owen et al., 1998).

through seafloor spreading. Seawater circulating within basalts fractures gets hot and induces various chemical elements to dissolve (Fe, Mn, S, etc.). Water ensures the transfer of these elements up to the seafloor where they may precipitate as oxides, sulphides or silicates.

In non-volcanic regions average heat flow values are much lower (about 30°C/1,000 m). Significantly hot water occurs only in deep aquifers that are located between 1,500 and 2,000 m below ground surface. Exploitation of the thermal energy carried by such deep groundwater necessitates strong pumping capacity and heavy, costly extraction systems. Appropriate devices are nevertheless implemented in several countries, namely in Europe: Aquitaine, Alsace and Paris basins in France; deep limestone and granite areas of Castilla and Mediterranean borders in Spain (Cuchi-Oterino et al., 2000).

Types

Geothermal energy is exploited in two ways depending on the fluid nature and temperature:

- *Convective hydrothermal systems* consist of permeable rock layers the pore water of which is heated by underlying volcanic magma submitted to convective movements (Fig. 106). Water tends to be heated at a super-critical temperature (200°C and more) and reaches ground surface in the form of

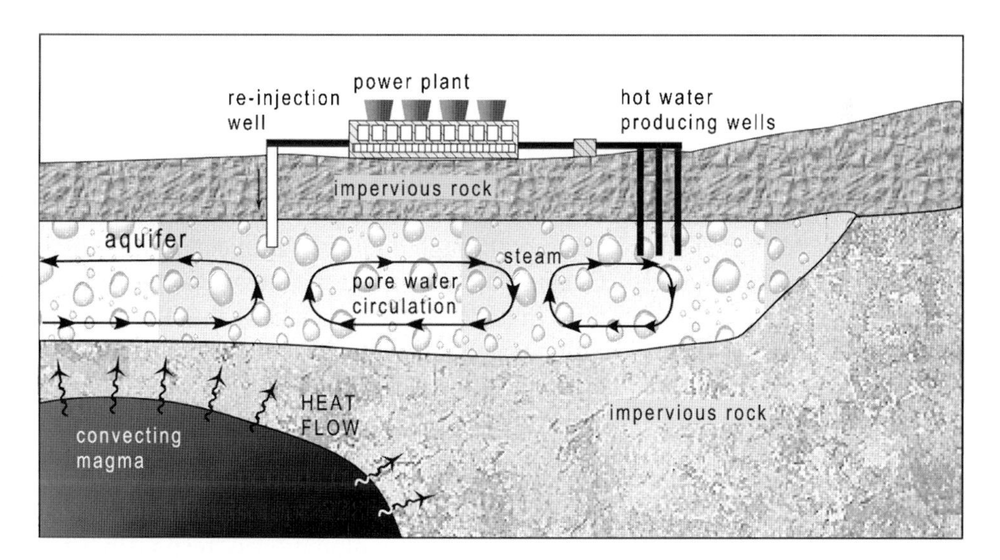

Figure 106: Example of an energy-producing hydrothermal system (after Keller, 2000).

steam (e.g. *geysers* of Iceland, New Zealand, or Yellowstone park in northwestern USA). It may also reach surface as a very hot liquid (more than 100 or even 150°C) and constitute *thermal springs* potentially able to feed hydraulic turbines and domestic heating systems. After being used and cooled, the water may be reinjected underground, where it is heated again and can be re-used for producing energy. Some recent investigations lead to the idea of injecting cold surface water within initially non-aquiferous underground rocks that are naturally heated by magma convection and could generate geothermal energy.

- *Geothermal systems* have been developed fairly recently. They are based on the temperature difference existing under normal heat flow conditions between surface air and groundwater. Their functioning corresponds to the heat pump principle. At about 100 m depth the aquifer temperature under normal heat flow and temperate climatic conditions reaches about 15°C. This fairly low value is nevertheless constant and sufficiently different form air temperature for allowing buildings to be heated in winter and cooled in summer. The water temperature and potential energy are higher when the pumping level is situated deeper underground.

Drawbacks and Advantages, Perspectives

Geothermal energy has been used in a significant way for more than three decades but remains of very *accessory importance* since it is expensive, implies the presence of fairly abundant and renewable hot groundwater, and necessitates implementing important equipments. Warm water cannot be transported over large distances by pipeline, forcing its use in the vicinity of extraction sites. Hot groundwater is often highly mineralised, which favours pipe corrosion or obstruction. In addition the groundwater extraction predisposes host rocks to subside (Chapter 3.2.2) and to contract through cooling. Re-injected water may induce artificial earthquakes (Chapter 1.4). Finally, the most energy-bearing waters are usually located very deep in underground formations that are strongly compressed and of low permeability, water capacity and renewal potential.

Several *advantages* are nevertheless associated to geothermal energy production:

- groundwater use induces almost no polluting effects;
- no or only very small amounts of gas are emitted that could participate to increase the greenhouse effect. Groundwater exploitation is responsible for producing less that 1% of NO_2 and less than 5% of CO_2 rejected by coal combustion;

- geothermal energy neither necessitates complex refining facilities nor induces the rejection of abundant and/or toxic waste. Safety problems are usually easy to control, except for a few cases of radon emission (Chapter 7).

The theoretical potential of energy production by geothermal activity is considerable. Estimates show that Earth's geothermal reservoirs, despite an efficiency of only 1%, should be able to produce 8.10^{19} J, i.e. an amount of energy amount similar to that emitted by the combustion of 13 billion barrels of oil. It is likely that geothermal energy will be of an increasing importance during future decades, due to the major changes that are likely to occur in economic and environmental situations.

6.4. Impact of Water Exploitation

6.4.1. Surface Flows

Noticeable man-induced changes in surface terrestrial water circulation first improved river *transportation* and *irrigation.* These ancient developments consisted of channel deepening, bank stabilization, meander cutting, water-course regularization, canal and lock construction, water diversion and catchment. Widespread *canalisation* work started as early as the 7th century in China, developed in the 17th century in Europe, and continued in the 19th century in North America. At the dawn of the 20th century, more than 20,000 km of canals were already in use. In 1985 the length of world waterways exceeded 500,000 km (Turner II et al., 1993).

The widespread construction of large, dammed *artificial reservoirs* along numerous river valleys started during the last quarter of the 19th century. These man-made basins induced extensive changes in the surface water budget:

- retention and stocking of huge water masses disrupting the naturally climate-controlled freshwater fluxes;
- local concentration and exacerbation of water seepage in soil and rock underground;
- local trapping and deposition of abundant suspended and bottom-tracted matter that did not feed the downstream river, lake and ocean sedimentation any longer;
- morphological change of river networks and of surrounding and downstream landscapes;
- major ecosystem changes and development of largely artificial "anthroposystems".

These large-scale morphological changes have been responsible for a *dramatic decrease in world river outflows, associated with a strong increase in*

Table 23: Estimates of anthropogenic modifications of river runoff during the three last centuries (1680–1980), and associated changes in continental evaporation (after Turner II et al., 1993). Values in km^3/yr. Estimated changes are induced by river water use for agriculture, livestock farming, industry, urbanization and domestic needs; they do not take into account soil and vegetation anthropogenic modifications.

	Total river runoff			**Evaporation from land**		
Continent	**1680**	**1980**	**difference**	**1680**	**1980**	**difference**
Europe	3,240	3,040	–200	3,925	4,125	+ 200
Asia	14,550	12,810	–1,740	18,140	19,880	+ 1,740
Africa	4,320	4,180	–140	16,450	16,600	+ 150
North America	6,200	5,880	–320	7,710	8,030	+ 320
South America	10,420	10,360	–60	18,940	18,995	+ 55
Australia, Oceania	1,970	1,960	–10	4,435	4,445	+ 10

evaporation (Table 23). Such modifications in hydrological balance tend to deteriorate conditions for plant growth, to accentuate the tendency for soils to become arid (Chapter 8) and to drastically reduce the area of wet zones. In northern and western Europe, 60% of humid areas listed in the late 19th century had disappeared in 2000. Due to demography and development, the area of water planes destined for farming, urban, industrial or hydroelectric use was extended in a spectacular way during the last decades. The number of artificial basins larger than 100 Mm^3 has been multiplied by 3 since the mid-20th century, relative to all previous periods. The total area of man-made water planes approaches 600,000 km^2.

Artificial reservoirs constitute new sedimentation basins in which the particles in transit within river courses are trapped and progressively buried. Such trapping actions have two major consequences:

– Particle accumulation determines progressive *sedimentary sealing* of artificial basins, the diminution of water storage capacity and finally difficulty in ensuring irrigation water supply or flooding control. This sediment sequestration, characterizing anthropogenic basin implementation along river watercourses, is estimated to amount to 30% for the whole planet and 50% for Europe only. Investigations performed worldwide on 633 artificial reservoirs larger than 0.5 km^3 show that the impact mainly affects North America, Europe, Africa and Oceania countries (Vörösmarty et al., 1997). Lake Nasser, which was established in 1964 for blocking Nile River water behind the Aswan dam, should be totally sealed around 2500, provided that the hydrological regime variations and sedimentary compacting occur normally, and that artificial emptying, cleaning up and flushing operations are not regularly carried out.

- Particle trapping behind artificial dams induces *strong depletion of down-stream particle supply*, which precipitates various environmental disorders: increase in marine erosional processes in river mouth areas, seawater incursion and unexpected salination in downstream alluvial valleys, concentration of contaminants, deterioration of coastal ecosystems. The Colorado water flowing downstream of the major dams implemented along the river course is practically devoid of suspended matter. In the Nile River delta, the 50-km-wide Manzala lagoon has suffered, during the last decades, major ecological changes due to particle trapping in Lake Nasser as well as to other human actions: conversion of wet zones into farmland, excessive irrigation, industrial constructions, etc. The current evolution should determine the loss of 30% of the total lagoon surface before 2050 (Randazzo et al., 1998).

6.4.2. Underground Flows

Groundwater pumping leads to lowering of the water table around catchment sites, the level being the lower upright-discharging wells. The water table surface tends to form a *depression cone* centred on the pumping point (Fig. 107). Lowering may extend laterally if numerous pumping sites are implemented close to each other within the same aquifer, or if the amount of total extracted water exceeds the natural recharge capacity of aquiferous rock. Artificial water table-lowering is deliberately practised for facilitating underground engineering work: access to mine shafts and galleries, digging of building foundations, etc. On the other hand, *active groundwater pumping may seriously affect and disrupt surface water circulation and flows*:

- loss of the artesian character of aquifers due to excessive water table lowering. The Albian "Green Sand" aquifer in the Paris Basin has been artificially lowered by about 100 m in less than one century, causing regional artesian wells to dry up;
- surface level-lowering of rivers or lakes connected to over-exploited aquiferous rocks;
- inversion of subsurface groundwater flow directions due to water draught from over-exploited aquifers;
- transformation of surface water planes (rivers, lakes, land-locked seas) from a drainage to a recharge status. This is presently the case of the Aral Sea (Chapter 6.4.3).

Change in the groundwater flow direction may determine unexpected and almost irreversible *aquifer pollution*, mainly in dense industrial and farming areas or in hazardous dump sectors (Fig. 107, right side). The pollution risk is amplified in regions subject to arid climate or to lengthy drought periods, during which chemical elements tend to concentrate in barely rechargeable aquifers. In the Glafkos basin, western Greece (Patras region), the combination during the 1980s of extended dryness and excessive pumping determined a strong diminution of groundwater quality and caused various problems for domestic, agricultural and urban activities (Lambrakis et al., 1997).

Ground subsidence due to particle compaction within over-exploited aquiferous rocks may be responsible for various disorders that propagate up to the soil surface. Such risks particularly affect the large conurbations built on under-compacted formations and facing considerable water requirements. This is the case in regions located on peat or soft sedimentary deposits (e.g. Mexico City; see also Fig. 47), and of coastal subsiding regions (e.g. London, New Orleans, Venice; Chapter 3.2.2). Corresponding disruptions may be suppressed or at least reduced by artificial water recharging, provided that compaction is

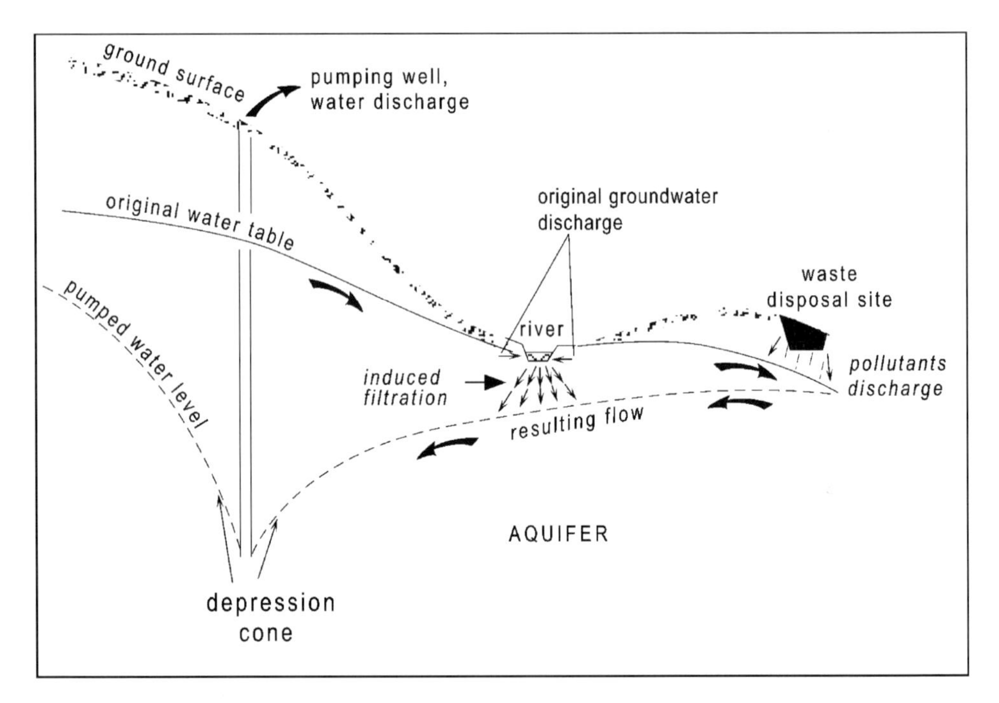

Figure 107: Consequence of groundwater excessive pumping: lowering, formation of depression cone, inversion of flow direction, and subsequent chemical pollution (after Bell, 1999).

still reversible. In all cases best remediation measures consist of diminishing the amounts of water extracted and allowing aquifers to be naturally recharged.

6.4.3 Salination

Over-exploited Coastal Aquifers

A sloped interface exists within coastal aquifers between overlying freshwater flowing seawards and underlying, denser saltwater that enters porous rocks and migrates by dipping landwards (Fig. 108, A). Exploiting coastal aquiferous rocks by pumping modifies this natural balance: the water table tends to form a depression cone, the freshwater–saltwater interface straightens up due to

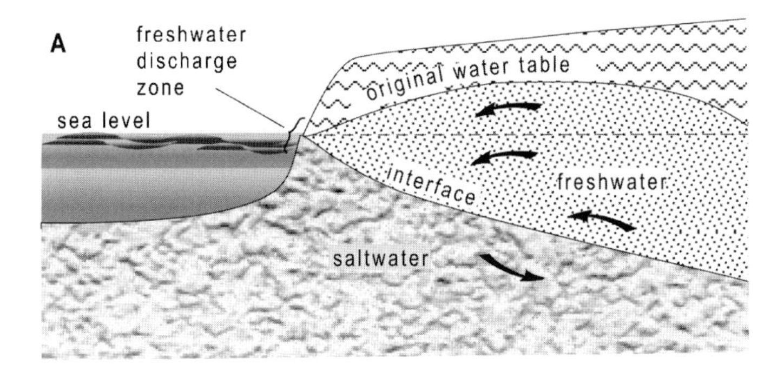

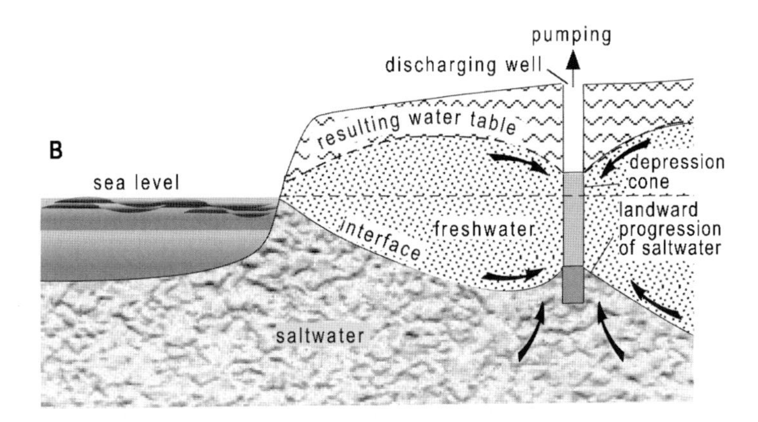

Figure 108: Hydrostatic relationships between freshwater and saltwater in a coastal aquifer. (A) Natural situation. (B) Situation induced by excessive freshwater pumping (after Price, 1996; Bell, 1999).

decrease in freshwater pressure, and saltwater progresses inland. This constitutes a *saline intrusion*. Saltwater may invade a large part of the aquifer and eventually be pumped up to the ground surface (Fig. 108, B). Extracted water may then be unfit for irrigation, domestic, urban and sometimes industrial use. This situation is common in many regions where coastal aquifers are excessively exploited: sandy coasts of the Netherlands, France and Spain; Southern Californian shores; atolls in the Pacific range, etc.

Large-scale Over-exploitation of Aquifers

The *Aral Sea* in western central Asia belongs to Kazakhstan and Ouzbekistan, two countries of the former Soviet Union. The sea is fed from the east by two rivers called Syr Darya and Amou Darya, which also drain the Kirghizistan, Tadjikistan and Turkmenistan regions. This land-locked sea, the initial surface of which amounted 70,000 km^2 and the salt content, was traditionally exploited for maritime transportation and fishing. Large-scale intensive agriculture was developed during the 20th century in all surrounding regions, which necessitated huge amounts of irrigation water. Very active extraction of regional groundwater was performed over several decades, which progressively led the Aral Sea to be drained and its waters to feed the over-exploited aquifers. The Syr Darya and Amou Darya rivers, a noticeable part of which were diverted to meet local irrigation needs, rapidly became unable to supply enough fresh water to balance the hydrological budget. The Aral Sea progressively became smaller and more saline. By the end of the 20th century the sea displayed a loss by about five-sixths of its volume, and its salt content was twice as high as that of ocean water (Fig. 109).

The progressive *drying up, emptying and salination* of the Aral Sea have considerably reduced shipping and fishing practices. *Massive salt deposits* have formed on basin borders. Regional aridification developed and led to active *wind erosion* of both soils and solid salt. Mineral and salt dust is periodically reworked by wind and may constitute aeolian plumes that rise up to 4 km high and propagate as far as 500 km from the sea. Disease and death rate have progressively increased. The infant mortality rate reached 10% in the early 1990s, i.e. 10 times higher than average values in world developed countries. The dismantling of the Soviet Union has given this problem an international dimension.

The case of the Aral Sea emphasizes the impact of groundwater over-exploitation in arid regions characterized by a fragile and unstable water balance. The groundwater extraction determines compensatory lowering of surface water planes, landscape aridification, rise of evaporation, and active *salt crystallization* (Chapter 6.2.3). Such disruptions are exacerbated in coastal

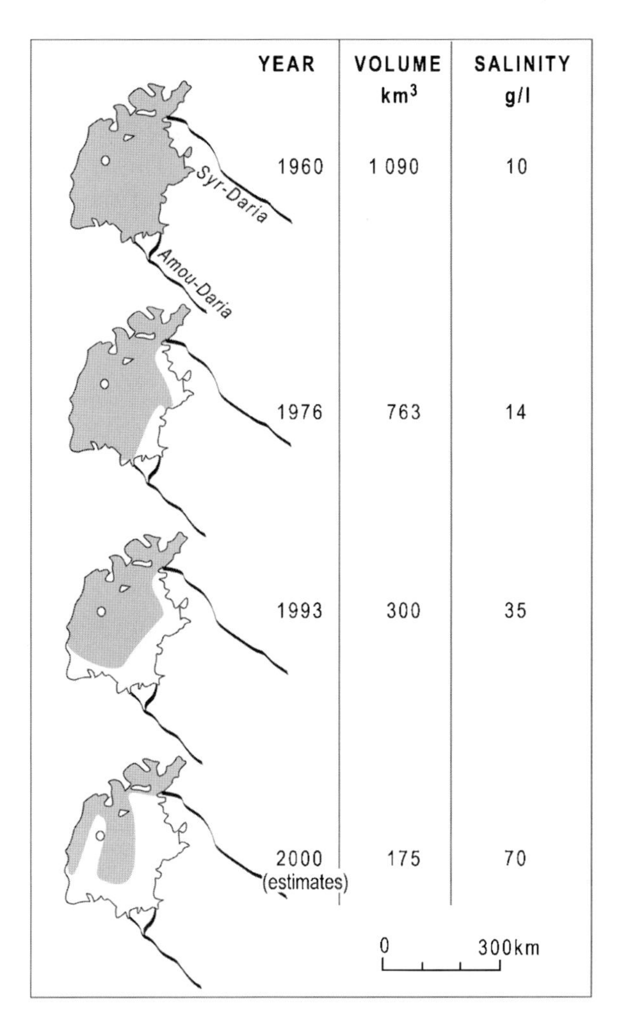

Figure 109: Modification of Aral Sea volume and salt content during past decades (after Micklin; see Rogers & Feiss, 1998).

domains where saline water intrusion and capillary evaporation amplify salt precipitation on ground surface. This is the case of the semi-arid region of Almeria in southeastern Spain, where, in addition, continuously recycled and boron-enriched pore water migrates within old evaporitic rocks. Capillary rise of evaporating water induces dissolved rock salt to accumulate at ground surface, which threatens land cultivation (Pulido Bosch et al., 1992).

More than 40 million hectares of irrigated soils in the world are considered as excessively rich in salt. One third of this huge surface results from groundwater over-pumping and other human disruptions implemented during the four last decades. The regions mainly affected by such *anthropogenic salination* comprise the Aral Sea hydrological system and numerous other basins marked by variable, sometimes high rainfall, and systematically subject to excessive water exploitation, damming, drainage or diversion. This is the case of deltas of major Chinese rivers (Hoang-Ho, Yang-Tse-Kiang), of the Ganges, Indus and Nile alluvial plains, of the Rio Grande and Colorado river valleys, etc.

6.5. Perspectives of Water Management

6.5.1. Abundance and Quality of Water Resources

Fresh water constitutes a critical resource, the usable amounts of which are essentially stored in sub-surface geological formations. This dominant location at the lithosphere–atmosphere–biosphere interface makes water especially vulnerable to chemical deterioration and anthropogenic wasting. About 12% of annual precipitation (i.e. 6,000 km^3) is used by mankind. This proportion could barely be exceeded, since reserves and requirements are unequally distributed in Earth's surface envelopes and some water resources are difficult to access. A *major challenge* for better management of water is posed for Earth's inhabitants, who should make some choices among multiple possibilities:

- economy of water use;
- water reuse after purification and recycling;
- treatment and exploitation of unavailable or non-fresh water (glaciers, icecaps, seas);
- search for the discovery of new, productive, perhaps very deep aquifers;
- transportation or displacement from low to high water-consuming regions;
- solving of contradictory interests such as water versus aquiferous rock extraction (e.g. gravel, pebbles), or water seepage versus aquifer artificial waterproofing.

A first concern exists in better delineating the *optimum adequacy existing between water quality and use*. Establishing reliable estimates of the amount of water types precisely adapted to diverse uses is crucial for better managing the resource:

– water used for irrigation can often not be drunk but may be adapted for cooking;
– water coloured by non-toxic metal oxides (iron, manganese) or clouded by suspended matter is generally rejected for feeding or washing, even if its chemical composition is suitable for such uses;
– water used for washing and bathing may contain various mineral or organic dissolved or suspended elements, but should be free of coliform bacteria and

of worm eggs responsible for bilharziasis (Trematodes) or filariasis (Nematodes);
- water destined for aquaculture must primarily contain sufficient amounts of dissolved oxygen and nutrients (nitrogen, phosphorus);
- water utilised for cooling industrial facilities or energy-producing plants can be saline and even slightly polluted, but should be free of toxic or corrosive elements and be of a relatively low temperature.

A second major concern is *improving the quality of domestic water.* This objective is more urgent to achieve than preventing and fighting agriculture- or industry-induced pollution, even if all these threats are more or less interrelated. At the dawn of the third millennium, only one world inhabitant among four has easy access to drinkable water. Two billion people lack sanitary facilities. Such lacks are particularly widespread in poor or developing countries, where 80% of diseases and 30% of deaths result from contaminated water ingestion (Barnier, 1992): cholera, hepatitis, malaria, parasitosis, etc. Water-borne diseases constitute the major lethal risk in developing countries. In addition these countries comprise an increasing number of megalopolises that are strongly prone to water contamination and related epidemics (Chapter 10).

6.5.2. Example of Large City Water Management

The *Los Angeles conurbation* in Southern California comprises more than ten million inhabitants who are large water consumers. Los Angeles County extends in an alluvial plain sloping gently towards the Pacific Ocean and subject to warm, sub-arid climatic conditions. Evaporation–transpiration (760 mm/year) exceeds precipitation (230 mm/year), which induces chronic water depletion. Groundwater resources are supplied from a porous aquifer that overlies impervious clay-rich layers. Local extraction covers about half of human needs. Intensive pumping close to the coast tends to induce saline intrusions within aquiferous rocks, which becomes critical during the driest years. In the course of the 20th century, these specific geographic, climatic and hydrological conditions progressively provided major difficulties in ensuring the supply of enough good quality water to a dramatically increasing population.

A huge programme of water management has been implemented on a regional scale during the last decades. The main developments performed are the following, which have permitted the achievement of a *fragile but globally satisfactory balance* between water supply and consumption (Fig. 110):

- freshwater import by aqueducts and canals from southern foothills of Sierra Nevada mountains. Lake Mono and Colorado River are the two major water sources, which are located 200 km to the north and to the east, respectively;

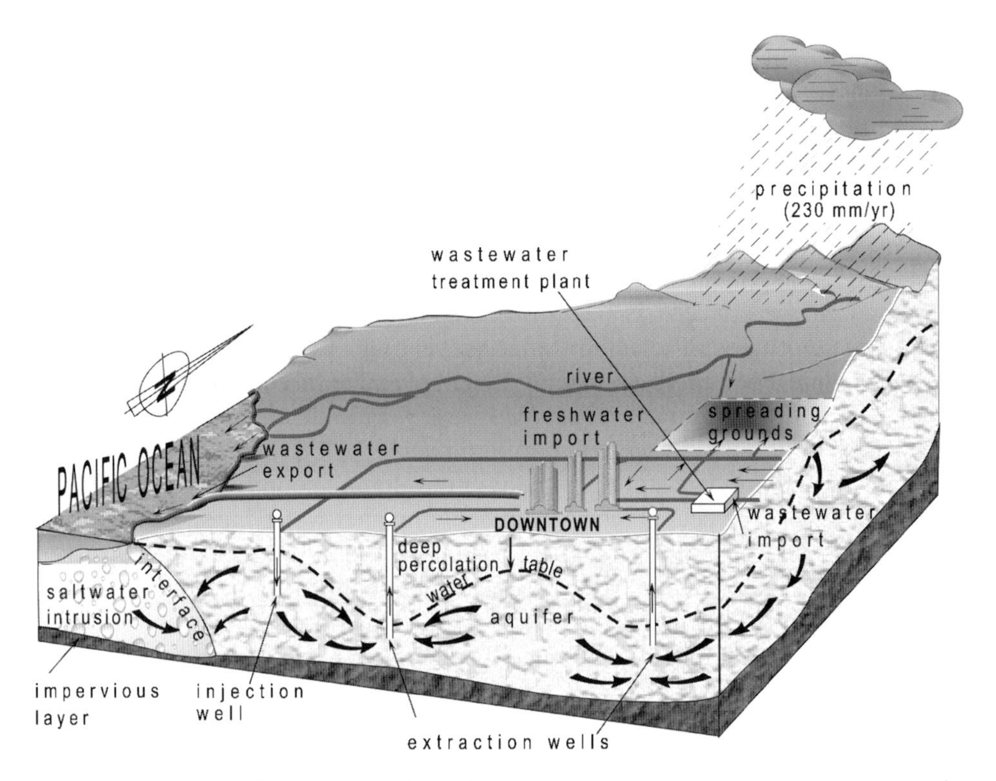

Figure 110: Schematic representation of water resource management programme in Los Angeles region, Southern California (after Merritts et al., 1997).

- implementation of large pipeline and basin networks that collect and distribute either naturally fresh or artificially purified water;
- construction of plants for treating and regenerating large amounts of wastewater;
- drilling in the coastal range of wells, where freshwater is injected for preventing saline intrusion;
- establishment of surface-spreading basins where water seepage participates to progressively recharge the aquifer located beneath the Los Angeles conurbation;
- population incentive to increase the use of bottled or containerised water for drinking, feeding and cooking.

Most large cities have tended or tend to implement integrated systems of water import, export and treatment that reconcile local climatic conditions, surface and ground water availability and exchanges, demographic and environmental pressure, and budgetary constraints. Challenges are particularly critical for some Asian, African and Latin American megalopolises where water resources are very limited, population growth is exponential, and contamination and pollution factors are innumerable (Chapters 9, 10).

6.5.3. Potential Development of Water Resources

Exploitation of new surface water and groundwater reservoirs is being investigated in various ways. Most techniques recently experimented with or developed are nevertheless still of a limited efficiency and very expensive. They can rarely be employed for significantly increasing the water supply currently needed by most developing countries. Some exploration methods are listed below:

- *Exhaustive reconnaissance* of the size and extent of groundwater reservoir rocks is developed in numerous world regions. New methods, among which geophysical prospecting is of growing importance, are implemented for delineating the actual aquifer shape and volume.
- *Exploitation of very deep aquifers* is conducted in some countries, mainly those located in arid climatic regions (Middle East, North Africa). Water extracted from rocks deep underground is often very ancient and is not renewable on a human scale. In addition, water quantity and extraction potential are limited due to the high degree of compaction of deep aquiferous rocks.
- *Seawater desalination* processes through distilling, osmosis or ion-exchange operations are put into practice in some regions where freshwater is exceptionally rare and standards of life are high (e.g. Persian Gulf countries, Malta in the central Mediterranean). Desalination treatments are currently very expensive. It is nevertheless likely that both growing needs for fresh water and technical progress will determine in the very near future an increasing tendency for further exploiting the huge and easily accessible seawater reserves.
- Large *iceberg transportation* from southern high latitudes (Antarctica range) to arid low-latitude regions is envisaged, but poses difficult melting, cost, and safety problems.

Improvement of groundwater exploitation constitutes another challenge that seems to be more easily taken up. Several types of actions may be developed.

- Certain developed countries have abundant and renewable groundwater *reserves that are inefficiently exploited* and even somewhat squandered. The water supply could be ameliorated by means of thorough investigation including detailed aquifer stock-taking, calculation of water renewal rate, estimates of consumption-averaged fluctuations and rationalization of extraction techniques. Such progress could be obtained in France, where the renewable groundwater resource is about 100 km^3/year. The total French groundwater consumption is only about half this value (i.e. 50 km^3/year), and nevertheless a few regions experience some difficulty in ensuring their water supply.

- The *recycling of used water* instead of its rejection in river and marine basins constitutes another way of contributing to the hydrological balance. Water treatment techniques have generally become efficient and may be developed on various scales, from small house groups (e.g. micro-filtration processes) to big cities (large treatment plants).
- The *separation of water supply networks destined for distinct uses* may also better save the resource. Such a goal can be reached when building new moderate-sized cities, for instance by separating the private and public or industrial water supply systems. It is much more costly, complicated and difficult to achieve in ancient and large cities.
- Didactic and repeated *information* on the limited character of water resources, especially drinking and cooking water, may help populations to effectively participate in fighting excessive consumption and potential wasting (e.g. construction of private swimming pools and artificial basins in arid regions).
- The implementation of *multidisciplinary research programmes* applied to specific river basins that constitute case studies may greatly help to better manage and anticipate the water reserve fluctuations. Combining different scientific disciplines such as geology, hydrogeology, hydrology, climatology, economy and sociology, leads to integration of the diverse factors controlling the fate of Earth's water, which participates in an increasing way to the balance of anthroposystems.

Special attention should be paid to the currently developing methods referred to as *water active management*. These methods are based on the artificial reuse of surface- or ground-originating water that has been purified and is subsequently injected or re-injected within the underground.

An example is provided by progress performed in the Netherlands, where considerable water needs for coastal cities (Amsterdam, Den Haag, Leyden, etc.) and widespread risk of saline intrusion have induced particularly efficient developments. Massive water injection and stocking are performed down to 200 m depth in Quaternary dune bodies bordering the North Sea. For instance, in 1991, 176 Mm^3 of waste water had been injected in the underground sand. This huge amount represented 15% of the total water supplied in 1 year in the whole country. Injected water was essentially of surface origin (Rhine River 35%, Maas River 39%, Lake Ijssel 17%, other sources 9%). Ninty-six percent of injected water had been treated through soft processes (coagulation, decantation, filtration) and were saved from chemical cleaning (chlorination, ozonization).

Eighty-nine percent of this actively managed water had been transported by pipeline over distances exceeding 50 km (Detay, 1997).

A Few Landmarks, Perspectives

- Water constitutes a very abundant component of the Earth that amounts to 1.4 billion cubic kilometres. Indefinitely renewable and exchangeable between its liquid, solid and gaseous states, it constitutes a priori an almost infinite resource. In fact its availability is limited for human consumption and use. More than 99% of water masses occur as saltwater in the oceans and seas, as ice within high-latitude icecaps, as molecules strongly retained within rock minerals or as pore water stored in deep and inaccessible aquifers. Water distribution in surface and underground reservoirs depends on climatic and geological conditions. It is therefore very heterogeneous and does not adequately correlate with human needs, which have become considerable due to recent population growth and concentration. The water consumption has been multiplied by 40 times during the past three centuries, and its chemical quality has often deteriorated. As a consequence water has become a natural resource of limited quantity and sometimes of disputable quality, which tends to be less easily recyclable for meeting human needs.
- Initially used for domestic consumption, fishing and transportation, fresh water has for long been exploited from surface reservoirs (lakes, rivers, springs, wells). Increasing needs for irrigation, farming and fluvial transportation have progressively induced reshaping river courses and establishing artificial dam lakes. Such morphological changes have led to strong modification of seepage and evaporation, as well as of plant and ecosystem development. Major landscape changes first arose in Europe, and progressively extended worldwide during the 19th and especially the 20th century. Meanwhile underground water reserves located in aquiferous rocks have been exploited more and more through drilling and pumping. Groundwater reserves depend on rainfall amounts and are renewed from the soil surface through seepage and migration, at a rate that is several thousand times slower than river flows. Numerous, sometimes giant aquifers therefore become over-exploited, which lowers the level of some connected rivers or lakes, as well as the capillary rise and surface precipitation of salt minerals.
- Underground water reservoirs are called aquifers and consist of permeable sedimentary rocks (limestone, sand and sandstones, gravel and pebbles) or of various fissured, fractured or partly dissolved (i.e. karstic) rocks. Groundwater migrates following gravity forces. Its basal part lies above an impervious surface along which it slowly moves downstream. Its surface or water table follows mildly the ground surface morphology and rises or drops

according to the rainfall regime. Groundwater constantly impregnates the lower part of aquiferous rocks, which is called the phreatic zone. It is overlain by a water-seeping porous horizon called the vadose zone. The vadose zone disappears if aquiferous rock is water-saturated. Free groundwater discharges at the ground surface by feeding rivers and lakes or by forming springs on valley sides. Captive groundwater migrates under pressure between two impervious rock layers and spouts up through artesian springs or wells. The possibility for underground rocks to serve as an aquifer essentially depends on their permeability. An aquifer is considered as exploitable if the groundwater flow rate exceeds 1 m/day.

– Groundwater physicochemical properties are less variable than those of surface water but nevertheless present a large diversity. The groundwater temperature is roughly stable seasonally, which may be used for heating buildings in winter and cooling them in summer. Buried water becomes warmer at increasing ground depth according to the Earth heat flow. Heated pore water from some volcanic zones or deep aquifers may be used for producing geothermal energy. The nature of chemical elements dissolved in groundwater depends on host rock characteristics, as well as on migration duration and depth. Under humid climate conditions, seeping rainwater tends to successively become calcium- and magnesium-rich, and then sodium-, sulphate- or even chloride-enriched. Underground chemical exchanges between host rocks and migrating pore water give way to the formation of various mineralised, gaseous or gas-free waters, the exploitation of which is currently developing for both drinking and sanitary uses, and poses several quality and normalization problems. In arid climate conditions capillary evaporation processes, which are often associated with aquifer over-exploitation, induce water to become chemically concentrated. This leads to groundwater salination, to water planes drying through underground drainage, and to salt crust deposition threatening both eco- and anthroposystems.
– The environmental impact of increasing groundwater exploitation is multiple and particularly critical since it specifically affects the lithosphere–atmosphere interface where life concentrates. Widespread, man-induced river course regulation and artificial basin construction have modified the balance between surface and underground water relationships, and led to increased aridification risks (Chapter 8). Huge wet zone areas have disappeared during the last decades. Continuous trapping in artificial reservoirs of river-carried suspended matter determines the depletion of downstream sedimentation. Chronic water-table lowering induces surface flow disruptions and favours both saline intrusion in coastal aquifers and surface salination of arid and/or over-exploited regions. Subsidence and excess compacting may also result from aquifer over-exploitation (Chapter 3). Groundwater is frequently subject

to chemical pollution or sanitary deterioration, the rectification of which is difficult and lengthy (Chapter 1).

- Water resources, the status of which has evolved from a local and domestic to a widespread economic importance for mankind, progressively has become a life issue. Several initiatives are being developed and should be reinforced for preventing drifts towards major confrontations in society: (1) search for new freshwater deposits: deep aquifers, desalinised seawater, etc.; (2) better management of the resource: more efficient extraction techniques, flow renewal assessments, controlled exchanges between natural reservoirs, extension of recycling and temporary stocking through active aquifer management, fight against water wasting; (3) development of integrated, multidisciplinary research programmes on continental hydrosystems and their economical and social constraints, intensification of information practices and consumers' sense of responsibility.

Further Reading

Castany G., 1982. Principes et méthodes de l'hydrogéologie. Dunod, Paris, 256 p.

Detay M., 1997. La gestion active des aquifères. Masson, Paris, 416 p.

Domenico P. A. & Schwartz F. W., 1998. Physical and chemical hydrogeology. Wiley, 2nd ed., 506 p.

Ingebritsen S. E. & Sanford W. E., Groundwater in geologic processes. Cambridge University Press, 365 p.

Marsily Gh. De, 1981. Hydrogéologie quantitative. Masson, Paris, 215 p.

Price M., 1996. Introducing groundwater. Chapman & Hall, 2nd ed., 278 p.

Some Websites

http://www.us.net/adept/links.html: inter-sites connections in hydrology, hydrogeology and geo-environnemental sciences established by Consulting Hydrogeology & Environmental Software Services

http://www.brgm.fr: actions conducted in France and various other countries in natural hazard, underground resources, and hydrogeology fields

http://water.usgs.gov/: extensive information on surface and underground water resources, quality, exploitation and management in the USA

Chapter 7

Radioactivity

7.1. Natural Radioactivity

7.1.1. An Inert, Odourless, Radioactive Gas

In December 1984 Stanley Watras, an engineer working in the building department of Pottstown nuclear plant, Pennsylvania (USA), drew the security service's attention that during his passage through the recently implemented radioactivity detectors he was systematically triggering the alarm system. Measurements performed immediately revealed that Mr Watras was emitting radioactivity of 2,700 pCi/l, an extraordinarily high value that was 700 times higher than normal. This result was all the more unexpected because the Pottstown nuclear reactor had not yet been activated, Mr Watras was not involved in the department responsible for radioactive affairs and he had stayed permanently outside any potentially dangerous area.

After multiple unsuccessful investigations on the nuclear power plant site it appeared that Mr Watras's body and clothes were impregnated with *radon-222*, which is an inert, colourless and odourless radioactive gas derived from the natural disintegration of uranium. Radon had irradiated both Mr Watras and his family, who lived in a nearby house built on geological formations made of granite. The slow, natural disintegration of uranium contained in low amounts within plutonic ground rocks had induced the emission of radon, which was continuously diffusing through the foundations of the house and the basement.

Numerous investigations were performed after this discovery, which had justifiably given cause for concern to many people living in granitic regions. Excessive and repeated inhalation of radon is known for both affecting the epithelium of the bronchial tubes and possibly inducing lung cancer. Such risks were suspected as early as the 16th century in Central Europe, where Georgia Agricola in 1526 had noticed that numerous silver miners working in granitic regions were dying from unknown lung diseases (*De re metallica*).

At the end of the 20th century some estimates concluded that 8,000–30,000 U.S. citizens potentially could die each year due to the fixation in human tissues of radon naturally emitted from uranium-containing rocks and subsequently accumulated in private houses and other buildings. Radioactivity measurements showed that several thousand dwellings constructed in Pennsylvania on granite substrate displayed abnormally high indoor radon contents; these houses were evacuated.

In the region of Albuquerque, New Mexico, the current radioactivity is between 330 and 530 mRem/year (*milli-röntgen equivalent man per year*; Chapter 7.1.4). These values are 1.5–3 times higher than the country's average values (200 mRem/year) and are mainly attributed to radon gas release from uranium-bearing granite. Radon radioactive natural emissions are two- to threefold higher than radiogenic values due to human activities: weapons, 3 mRem/year; smoke detectors 1 mRem/year; building materials, a few tens of mRem/year; electric lighting, 6 mRem/year; natural gas burning, 6–9 mRem/year; coal plant smoke, 3–4 mRem/year; nuclear plant smoke, <0.1 mRem/year (Brooking, 1992). The indoor radon content in Albuquerque region is directly correlated to the distance separating the contaminated houses from Sandia Mountains' granite and is locally amplified where porous ground facilitates the radioactive gas diffusion up to basement and lower floors.

The actual effects of indoor radon on human health are nevertheless still poorly known, quantified and even demonstrated. The low amounts of radon gas in the air renders analyses very difficult to perform and almost no long-term observations have been performed that are reliable. Preliminary tests were carried out in China during the 1970s on two groups of 70,000 sedentary inhabitants of the Guangdong province. One group was living in a valley protected from granite influence, and the other on a granitic plateau containing radioactive monazite (300 mRem/year). No significant difference was registered between the two groups after 5 years of measurements (PRC; see Brooking, 1992). However, the experiments were not specifically devoted to the indoor radon levels, the potential risk of which was still unpredictable 30 years ago.

7.1.2. Basic Mechanisms

Nuclear Fission

Earth's natural radioactivity essentially corresponds to the progressive disintegration of some unstable chemical elements that were incorporated in our planet underground during its early formation. Disintegration basically affects the nucleus of these elements and is therefore called nuclear fission. The chemical elements concerned comprise essentially heavy metals of uranium and thorium groups:

- *uranium* (U), density (d) = 18.7, atomic number 92, atomic mass of 238, 235 or 234 depending on isotope nature;

– *thorium* (Th), $d = 12.1$, atomic number 90, atomic mass 232.

Some lighter elements such as *potassium-40* are also radioactive but occur much less frequently and/or abundantly than uranium and thorium.

The natural decay of heavy isotopes is very slow. For instance the loss of half the mass of uranium-238 (^{238}U) occurs in 4.5 10^9 years, which corresponds to the age of Earth. The time interval necessary for disintegrating half the atoms of an element is called the half-life or period t. The half-life of ^{235}U is equal to 7×10^8 years, and that of ^{232}Th to 1.4×10^{10} years. The radioactive isotope disintegration is associated with an *important release of energy*, which is largely responsible for Earth's heat flow (i.e. the geothermal gradient), allows life to develop at our planet surface and has led to multiple human applications including the production of nuclear energy (Chapter 7.3). The heat produced by disintegration of 1 g of ^{235}U is equivalent to that produced by the combustion of 13.7 barrels of petroleum.

The mass loss of uranium, thorium and other unstable isotopes is characterized by the emission of alpha (α) particles or beta (β) rays out of the element's nuclei, which induces the formation of new chemical elements that progressively replace the original ones. The emission of one α particle, i.e. the nucleus of one atom of helium (He), the atomic mass of which is equal to 4 and the charge positive and equal to 2 ($^4He^{2+}$), determines an equivalent diminution of the atomic mass and number of the disintegrating element. This reaction characterizes for instance the disintegration of radium into radon:

$$^{226}Ra_{88} \rightarrow {}^{222}Rn_{86}$$

where the superscript is the atomic mass and the subscript the atomic number.

The emission of one β ray (i.e. an electron with a mass of zero and a negative charge equal to -1; β^{-1}) does not induce any modification of the atomic mass but an increase of 1 in the atomic number. This reaction characterizes for instance the disintegration of rubidium-87 (^{87}Rb) into strontium-87:

$$^{87}Rb_{37} \rightarrow {}^{87}Sr_{38}$$

The continuous emission during geological times of α particles and β rays has led and still leads to initial radioactive and relatively heavy elements (called *parent elements*) to be progressively replaced by lighter elements called *daughter elements*. The chain of successive daughter elements is systematically the same for a given unstable parent element. The final product consists of a stable chemical element of the lead family, the atomic mass of which is 207 or 208 (^{207}Pb, ^{208}Pb). An example is provided in Fig. 111 which represents a chain of 14 isotope elements derived from disintegration of uranium-238, a parent element that in natural conditions is far more abundant ($^{238}U = 99.3\%$) than uranium-235 ($^{235}U = 0.7\%$) and uranium-234 ($^{234}U = 0.005\%$). Radon-222 (^{222}Rn), a radioactive gas potentially responsible for nature-induced cancer

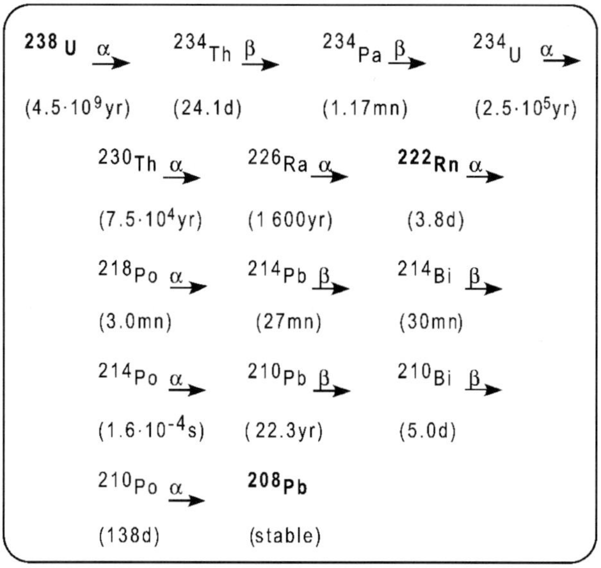

^{238}U $\xrightarrow{\alpha}$ ^{234}Th $\xrightarrow{\beta}$ ^{234}Pa $\xrightarrow{\beta}$ ^{234}U $\xrightarrow{\alpha}$

(4.5·10^{9}yr) (24.1d) (1.17mn) (2.5·10^{5}yr)

^{230}Th $\xrightarrow{\alpha}$ ^{226}Ra $\xrightarrow{\alpha}$ ^{222}Rn $\xrightarrow{\alpha}$

(7.5·10^{4}yr) (1 600yr) (3.8d)

^{218}Po $\xrightarrow{\alpha}$ ^{214}Pb $\xrightarrow{\beta}$ ^{214}Bi $\xrightarrow{\beta}$

(3.0mn) (27mn) (30mn)

^{214}Po $\xrightarrow{\alpha}$ ^{210}Pb $\xrightarrow{\beta}$ ^{210}Bi $\xrightarrow{\beta}$

(1.6·10^{-4}s) (22.3yr) (5.0d)

^{210}Po $\xrightarrow{\alpha}$ ^{208}Pb

(138d) (stable)

Figure 111: Disintegration of uranium-238 (^{238}U) in a chain of daughter-elements characterized by decreasing atomic masses (after MacKenzie 2000). U, uranium; Th, thorium; Pa, protactinium; Ra, radium; Rn, radon; Po, polonium; Pb, lead; Bi, bismuth. The emission of an alpha (α) particle or beta (β) ray determines a loss or not of the atomic mass. Each element half-life or period (*t*) is indicated in brackets. yr, years; d, days; ms, minutes; s, seconds.

(Chapter 7.1.1), constitutes one among the daughter elements of this chain, the end member of which is lead-208.

Nuclear Fusion

In contrast to fission, nuclear fusion is the *chemical combination of the nuclei of light elements synthesising new, heavier chemical elements.* This type of synthesis requires extremely high temperatures, the value of which ranges between several millions and several hundreds of millions of degrees Celsius. In natural conditions, such temperatures occur in the core of the sun and other stars, and are associated with the *production of huge amounts of energy.*

The energy produced by the sun's nuclear reactions allows heating of the atmosphere and surface of solar planets, and mainly of Earth. Reactions consist of the fusion of four atoms of hydrogen (atomic mass and number equal to 1) that combine to form one atom of helium (atomic mass = 4, atomic number = 2):

$$4\,{}^{1}H_{1} \rightarrow {}^{4}He_{2}$$

This reaction works in the sun's core, where temperatures are about 14 million degrees and where enormous amounts of hydrogen are continuously melted.

In the case of much older stars than the sun, such as red giant stars and their evolving forms, the core temperature has reached, through gravitational contraction, values as high as several hundred million degrees and more. Nuclear fusion in these stars has determined the synthesis of successive, heavier and heavier elements, from carbon-6 ($3^4He_2 \rightarrow {}^{12}C_6$) to thorium-90, protactinium-91 and finally uranium-92. All these newly synthesized chemical elements have been expelled in interstellar space through explosion of the oldest stars in supernovae. Some of these elements have been involved in planet aggregation during the early stages of the solar system formation. The heaviest, end members of the nuclear fusion chain are responsible for most of Earth's natural radioactivity.

Nuclear fusion is responsible for the production of huge amounts of heat that result from the synthesis of very common and easily available chemical elements such as hydrogen or its heavy isotope deuterium (${}^2D_1$). Nuclear fusion therefore constitutes theoretically an inexhaustible source of energy. In addition fusion reactions are free of contamination risks, since helium born from hydrogen melting represents an inert and non-toxic gas, and resulting radioactive waste is insignificant.

Unfortunately it is impossible to envisage reaching temperature values that are close to those of the sun's core in industrial applications. Such extremely high temperatures are essential for melting the different hydrogen nuclei that *a priori* repel each other. Nuclear fusion at very high temperatures has been realized on Earth only in man-produced thermonuclear bombs. As far as "cold" nuclear fusion is concerned, it still comes close to being a myth.

Cosmos-originating Nuclear Reactions

Some radioactive chemical elements are synthesized in the upper terrestrial atmosphere due to interactions between cosmic rays and gas atomic nuclei. Newly formed radionuclides comprise hydrogen-3 (period $t = 12.3$ years), beryllium-10 ($t = 1.6\ 10^6$ years), carbon-14 ($t = 5.75\ 10^3$ years), silicon-32 ($t = 178$ years), and chlorine-36 ($t = 3\ 10^5$ years).

Once formed, these cosmos-derived radioactive elements are incorporated into the air together with non-radioactive gasses (nitrogen, oxygen, carbon dioxide, water vapour, etc.). They may then be transported by atmospheric winds down to the Earth's surface, where they are susceptible to inclusion in soils, sediments, underground water, and plant, animal or human tissues.

These atmosphere-forming isotopes are responsible for a noticeable part of the Earth's natural radioactive background (see Fig. 121). Their disintegration is associated with the release of very little heat, which cannot be used for producing energy. Some cosmos-derived radionuclides such as ^{14}C (called radiocarbon) and ^{32}Si constitute precious tools for precisely dating some past events. Some other items help to trace the functioning of terrestrial environments and palaeo-environments: exchange processes between atmosphere, hydrosphere and lithosphere; time-related variations and causes of atmospheric composition and climate characteristics; etc.

7.1.3. Geological Control

Rock Radioactivity

The radioactive elements in geological formations mainly depend on petrographic parameters. The mean uranium concentration in surface rocks ranges from 0.003 ppm (parts per million), i.e. 0.37 becquerels per kilogram (Bq/kg; see Chapter 7.1.4) in meteorites, to more than 120 ppm (1,500 Bq/kg) in phosphates (Eisenbud & Gesell, 1997). The uranium concentration averages 7.3 Bq/kg in basalts, 49 in *granites*, 15–16 in limestone and most other sedimentary rocks, 610–980 in *bituminous shale* (e.g. Tennessee ores, USA), and 240–1,500 in *phosphates* (e.g. Florida, Morocco).

The relative abundance of uranium in phosphate and bituminous rocks is directly related to that of organic matter, which has contributed in fixing radionuclides on sedimentary particles during deposition processes. The link existing between organic matter and uranium is illustrated by the radionuclide distribution in Late Devonian Ohio Shale (Fig. 112, A). The intensity of uranium disintegration in shale determines the amount of radon released (Fig. 112, B).

In a similar way the recent peat and organic sandy mud of Carson valley and Lake Tahoe (California and Nevada) contain up to 0.6% (6,000 ppm) of uranium that disintegrates by forming several daughter elements (Otton et al., 1989). The average uranium content ranges between 300 and 500 ppm. The radioactive metal diffuses in surface water used for public and private purpose (up to 177 parts per billion, ppb). It is principally stored in alluvial deposits, where its concentration within two local marsh ponds is estimated to reach 24,000 and 15,000 kg, respectively.

The variability of total radionuclide concentration in most common terrestrial rocks and superimposed soils is fairly low, as is the absolute amount of the three most abundant radioisotopes (^{238}U, ^{232}Th, ^{40}K). Radioactivity of the two heaviest isotopes varies between 7 and 70 Bq/kg (Table 24). Values are lower and more diverse for potassium-40, the disintegration energy of which is weak.

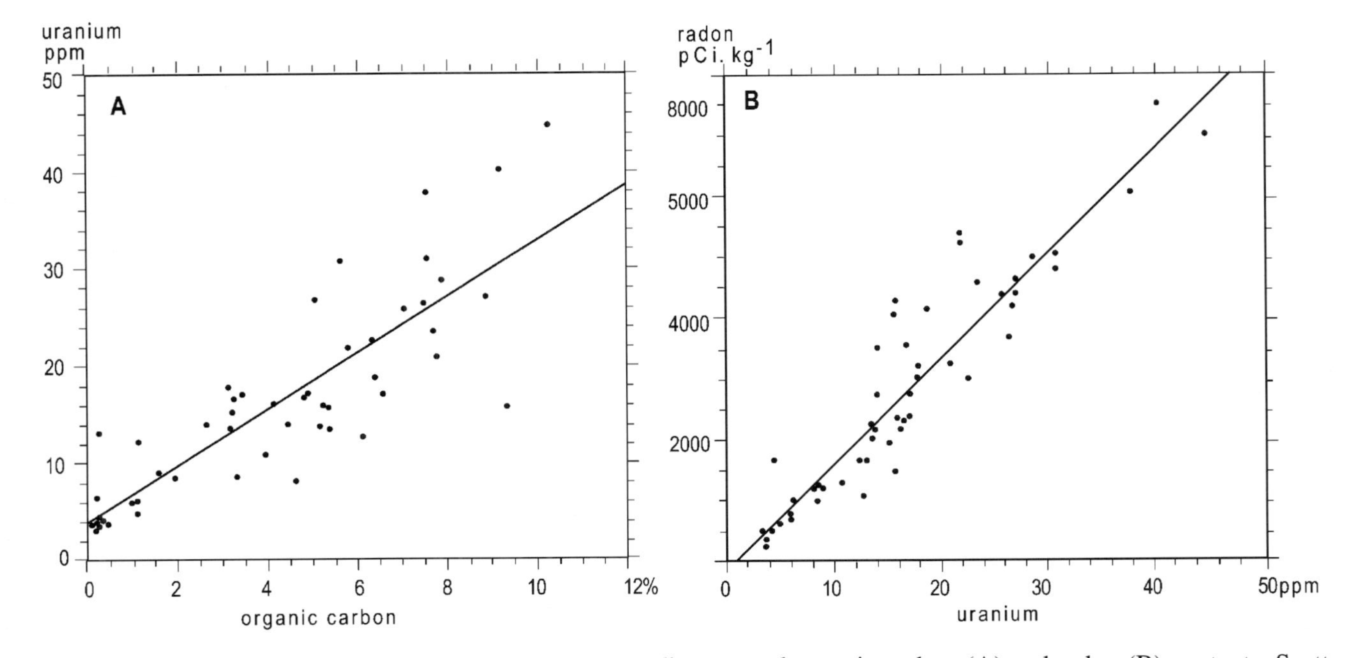

Figure 112: Uranium distributions in Ohio Shale (USA) according to total organic carbon (A) and radon (B) contents. Scatter plot and linear regression curves (after Harrell et al., 1991).

Table 24: Average concentration of the three most common radioactive metals in various rocks and soils (after Eisenbud & Gesell, 1997). Values in becquerels per kilogram (Bq/kg or $Bq \cdot kg^{-1}$).

	Uranium-238	**Thorium-232**	**Potassium-40**
basalt	7–10	10–15	300
granite	40	70	> 1,000
sandstone, claystone	40	50	800
beach sand	40	25	< 300
limestone	25	8	70
continental crust	36	44	850
soils	22	37	400

Radon

Radon (Rn) is a radioactive gas resulting from disintegration of uranium and thorium chains. It comprises three isotopes that proceed from radium through an α-particle loss and give way to polonium through another α particle emission (André-Jehan & Féraud 2001):

- ^{222}Rn belongs to ^{238}U chain and has a period t of 3.83 days;
- ^{220}Rn (thoron) belongs to ^{232}Th chain, $t = 55.3$ sec;
- ^{219}Rn (actinon) belongs to ^{235} chain, $t = 3.92$ sec.

Lesions due to radon inhalation are a priori induced by *radon-222* only; the activity duration of both other isotopes is too short, although their instantaneously released energy is higher. ^{222}Rn is emitted by various rocks containing significant amounts or ^{238}U: granite and gneiss, shale containing organic matter, etc. Its abundance varies parallel to that of uranium in parent rocks (Fig. 112; B). The radioactive gas may also be contained in recent sedimentary or pedological formations such as glacial deposits or soils resulting from erosion or weathering of uraniferous parent rocks.

The diffusion of radon up to the ground surface is facilitated by the presence in host formations of faults, fractures and cracks. This is illustrated by the region of Döttingen in Eiffel Massif, Germany, where ^{222}Rn concentration in surface soils increases close to faults crossing both Late Devonian sandstones and siltstones, and Tertiary volcanic units (Fig. 113). On the other hand the radioactive gas may be released in a preferential way from permeable and highly-specific surface formations subject to landslides (Chapter 3). Investigations performed on giant slides in Austria (Koefels, Tyrol) and Nepal (Langtang Himal) show that fairly large amounts of radon (> 1,000 Bq/m^3) are released in

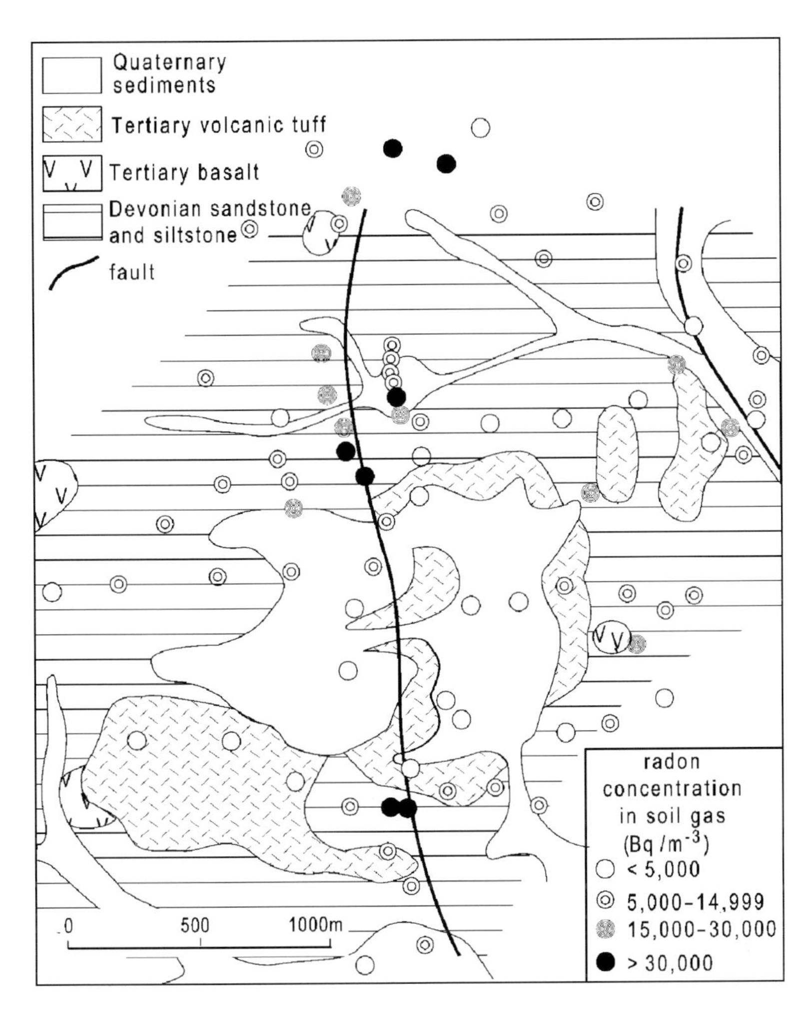

Figure 113: Distribution of radon in Döttingen village soils covering Devonian sandstones and Tertiary volcanics in the Eiffel Massif, Germany (after Keller et al., 1992). Values in becquerels per cubic metre (Bq/M^3).

the atmosphere when gravity movements occur (Purtscheller et al., 1995). Conversely, radon remains stored in the ground when host formations are sealed by impervious deposits, namely clay.

7.1.4. Risks and Prevention

Units of Measurement

Risks potentially induced by nuclear irradiation have to be quantified to be evaluated appropriately. Such a need arose shortly after Antoine-Henri

Becquerel had discovered in 1896 that uranium ore exposed a photographic plate in darkness, and after Marie Curie had isolated radium particles in 1908. Very diverse units of radioactivity have been used since these early times for measuring ionising radiation and their effects, and most of them are still used.

The oldest units comprise three types:

- *curie* (Ci). 1 Ci $= 3.7 \times 10^{10}$ spontaneous nuclear disintegrations per second, or 3.7×10^{10} Bq (see below). This unit is divided into *picocuries* (1 pCi $= 10^{-12}$ Ci $=$ 0.037 disintegrations/sec), as well as in pCi/l $=$ pCi/l^{-1} $= 37$ disintegrations/sec per litre, or 37 Bq/m^3. A radioactive dose of 100 pCi/1 ($= 1.3 \times 10^5$ MeV) is called "*working level*" (WL);
- *rad* (from radium) is a unit expressing the amount of energy released by radioactivity. 1 Rad $= 0.01$ J/kg. This highly-divided unit was later converted into grays (Gy; see below), which are twofold bigger (1 Rad $= 0.01$ Gy $= 10^{-2}$ Gy;
- *rem* (from "*röntgen equivalent man*"). This unit expresses the biological effect of ionising radiation and belongs to the same scale as the rad.

The more recent units are internationally used and also comprise three types:

- *becquerel* (Bq): activity of a certain amount of radionuclide that undergoes one spontaneous nuclear disintegration per second : 1 Bq $= 37$ pCi. Indoor or outdoor radioactivity is often expressed in becquerels per cubic meter (Bq/m^3 or Bq $\cdot$ m^{-3}). 37 Bq/m^3 $= 1$ pCi/l;
- *gray* (Gy). 1 Gy $= 1$ J/kg, which represents the dose released by 1 kg of a radioactive element producing a uniform energy of 1 J. 1 Gy $= 1.1 \times 10^{14}$ erg/g $= 100$ rad;
- *sievert* (Sv). This unit is used for measuring the biological effects of a radioactive dose absorbed by humans, animals or plants.

To sum up this rather complex terminology, becquerels and curies allow measuring of ionising activity, grays and rads the energy released, and sieverts and rems the biological effect of equivalent doses.

Impact on Man

The Earth's biosphere environment is characterized by a *basic natural radioactivity* that impregnates all living organisms. The value of this fundamental ionising radiation is close to 3 μSv $= 0.003$ Sv $= 0.3$ rem $= 300$ m-Rem). Such a value is low and normal; it does not induce any risk for human health and is even necessary. Potential danger arises when the amount of ionising radiation is much higher. Chronic diseases appear when instantaneous doses reach values of 0.1 Sv.

About two-thirds of Earth's radioactive background (2 μSv = 200 mRem) *is due to the presence in the atmosphere of radon-222* emitted by the disintegration of uranium-238 contained in geological formations. The alpha (α) particles ejected by radon radioactivity attack human cells, but their relatively large size and high electric charge prevent them from easily penetrating both clothes and skin. In addition the relatively long half-life of ^{222}Rn ($t = 3.8$ days) is responsible for a very moderate production of α particles at each air inhalation.

By contrast the daughter elements resulting from *radon decay* are characterized by a very short half-life: ^{218}Po, $t = 3.05$ min; ^{214}Pb, $t = 25.8$ min; ^{214}BI, $t = 19.7$ min (see Fig. 111). This leads to active release of α particles at each inhalation. The resulting lung cell irradiation increases when respiration goes on. This is the reason why the presence of indoor radon in poorly ventilated mines and buildings potentially constitutes a significant danger (Chapter 7.1.1).

The ways by which radon-222 penetrates houses and other constructions are diverse:

- upward diffusion in basement floors from soil and underground;
- slow, lateral penetration across fissured and gas-permeable walls;
- indoor release from contaminated water used for drinking, cooking and washing.

The few statistical investigations available so far show that the world average radon content in private dwellings is about 55 Bq/m^3 (Eisenbud & Gesell, 1997). Indoor gas principally results from ground diffusion (up to 55 Bq/m^3), secondarily from outdoor air penetration (up to 10 Bq/m^3), and very accessorily from construction material and water release (2 and 0.4 Bq/m^3, respectively).

Nevertheless, the ^{222}Rn indoor content in some regions is much higher than world average values and could pose a carcinogenic risk. The mean radon amount in Austrian houses reaches 400 Bq/m^3. In Tyrol province some systematically conducted measurements inside Umhausen city dwellings have revealed excess values relative to the country's average in 71% of cases during the winter season, and in 33% of cases during the summer. Maximum values locally attained 274,000 Bq/m^3 (Purtscheller, 1995).

Radon emissions potentially reach also a critical level in some underground mines where uranium ores and other metal ores are extracted. The radioactive gas continuously released within galleries may constitute a serious risk to miners' health.

Various studies are currently under way for better characterizing and quantifying the actual nature, importance, and long-term consequences of health risks induced by radon inhalation. The *permeability of surface and subsurface soils and sediments* situated between uranium-containing geological substrates (granite, shale, etc.) and man made constructions appears to play a critical role.

Three groups of preventive action may help to mitigate radon-induced potential risks:

- Geological and pedological survey associated with measurements of ^{226}Ra and ^{222}Rn concentrations, with construction type and quality assessments, and with statistical modelling. These investigations lead to the establishment of *potential risk maps*. This approach has been applied to 259 towns and villages of New Hampshire state (NE USA). A predictive model has been established that should be tested at several space and time scales (Apte et al., 1999). The French Geological Survey (BRGM) in 2000 conducted a predictive mapping of radon risk in Corsica, from field investigations based on ground uranium content, outcropping and hidden rock faulting and fracturing, fracture influence extent, and geomorphological characteristics. The whole island was divided into areas of three different types: areas necessitating rapid corrective measures (average annual radioactivity greater than 1,000 Bq/m^3), areas where preventive actions could be envisaged (400–1,000 Bq/m^3), and areas devoid of any potential risk ($<$500 Bq/m^3).
- *Selection of construction materials* that are free of radon and impervious to gas penetration; *sealing* and gas-proofing of basements.
- Indoor installation of *radon detectors*, of absorbing *filters* (e.g. activated charcoal), and of *ventilation* systems diluting the radioactive gas.

Beneficial Aspects of Natural Radioactivity

Research on the geochemistry of unstable, radioactive isotopes helps strongly in characterizing and understanding the *chemical exchanges occurring between Earth's natural reservoirs* of atmosphere, hydrosphere, biosphere and lithosphere. Such investigations also allow improving the chronology of palaeoenvironmental evolution (MacKenzie, 2000).

Natural radioactivity may also have direct *positive effects on human health.* For instance the radon-containing warm water of Badgastein mineral springs in the Austrian Alps helps efficiently to fight rheumatisms and polyarthritis (Zötl, 1995). The therapeutic action of radon on various inflammatory and skin diseases is used in several clinics from this region. In 1992 the annual consumption of Gastein mineral water reached thirty million litres.

7.2. Exploitation of Nuclear Energy

7.2.1. Requirements, Consumption, Reserves

Human harnessing of radioactivity for producing domestic energy dates back to the beginning of the mid-20th century. The first demonstration of man-induced nuclear fission on a potentially industrial scale was performed in 1942 at Chicago University (USA). This fundamental experience occurred in on international context marked by two special features of crucial importance for further development and perception of nuclear energy.

- On one hand, the extraordinary world population growth characterizing the 20th century and especially its second half induced an *exponential increase in energy needs*, and particularly of the demand for electricity. These needs had to be met by operating nuclear reactors.
- On the other hand the continuation in Asia of World War II conflicts accelerated the search for *military applications of nuclear energy*. The first full-scale tests were performed in July 1945 at Alamogordo in New Mexico (USA). They preceded the release of nuclear bombs on Hiroshima and Nagasaki cities in August of the same year.

The contemporaneous occurrence of these two very different applications of nuclear energy has led mankind to pursue simultaneously, and often to lump together, the development of public and military nuclear industries. This potential confusion partly endured until the end of the 20th century.

The construction of nuclear power plants began in 1954. The number of plants built strongly increased between the mid-1960s and the mid-1970s, and then tended to diminish (Fig. 114). The electricity-generating capacity of the world's nuclear plants was moderate until the 1970s, and then increased dramatically until the early-1990s (Fig. 115). The total capacity was of 340 gigawatts (340×109 W) in 1995 and was produced by 432 operating plants.

Since the last decade of the 20th century, the nuclear plant implementation curve has become stable. This results both from political decisions linked to crucial questions concerning either nuclear plants' safety (Chapter 7.2.3) or nuclear waste disposal (Chapter 7.3), and from some countries' decision to activate plant building. In 1995 nuclear electricity amounted to 12×10^5 W/h (12,000 TWh), which corresponded to 17% of the world's energy production. This proportion is roughly the same at the beginning of the 21st century, which results from a balance between the countries abandoning the nuclear reactor system and those developing active nuclear plant construction programmes.

The proportion of nuclear energy produced by diverse world countries greatly differs and largely depends on national policies. Most industrial countries developed active nuclear programmes until the end of the 1980s. Some of these countries, especially France, have considerably invested in nuclear research and industry (Fig. 116). This strategy was largely driven by the wish for technological control and energy independence.

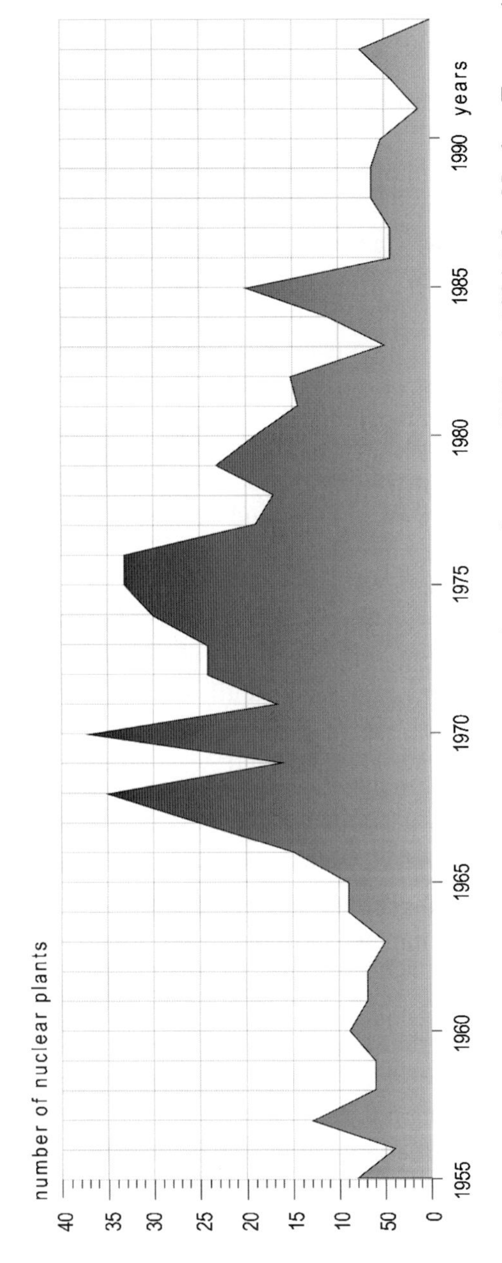

Figure 114: Number of world nuclear power plants construction starts between 1955 and 1994 (after Nuclear Energy, 1995).

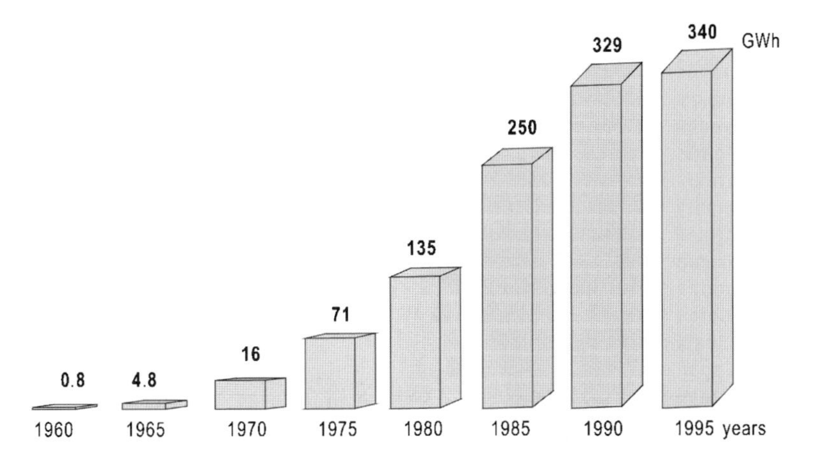

Figure 115: Energy-generating capacity of world nuclear plants between 1960 and 1995 (after Worldwatch Institute; see Nuclear Energy, 1995).

Since the end of the 1990s, demands for nuclear plant openings have come increasingly from developing countries that are facing crucial demographic or economic problems. These countries are mainly located in Asia, despite some of them having suffered major nuclear plant accidents (e.g. Ukraine) or being potentially subject to major earthquakes (e.g. Armenia).

The uranium-ore reserves currently identified are important and are principally located in Australia, South African Namibia, Niger, Kazakhstan and Canada (see Fig. 84). They are sufficient for ensuring the production of nuclear energy during at least the whole 21st century, provided that annual consumption remains similar to that of the last decade (e.g. 32,000 t of uranium extracted in 1993).

7.2.2. Production of Nuclear Electricity

Mechanism

The nuclear energy utilised by human technology and destined to produce electricity results from *fission* processes that resemble those working through Earth's natural radioactivity (Chapter 7.1.2). Man-induced nuclear fission essentially concerns *uranium*, and especially its *isotope-235* (^{235}U). Due to a relatively short period ($t = 710$ millions years) ^{235}U is much more easily fissionable than isotope-238 (^{238}U period = 4,510 million years). This property has also led ^{235}U to be much more actively disintegrated than ^{238}U since the Earth was formed 4.5 billion years ago. During geological history the ^{235}U proportion has decreased from 22% to 0.7% of uranium, whereas that of ^{238}U has increased

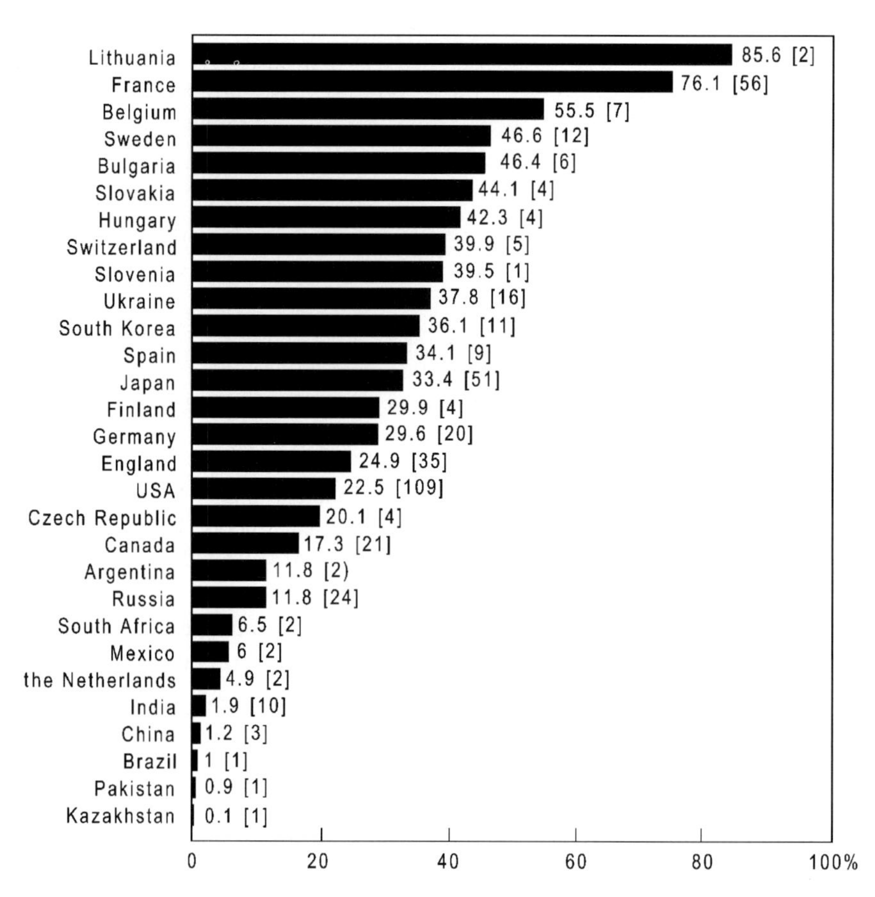

Figure 116: Percentage of nuclear-fission generated energy in different countries in 1995 (after International Atomic Energy Agency; see Montgomery, 2000). Values in brackets indicate the number of nuclear reactors per country.

from 78% to 99.3%. The use of uranium for producing electricity on an industrial scale therefore necessitates that the isotope-235 proportion is artificially increased relative to isotope-238, so that fission reactions can produce enough instantaneous energy. *^{235}U enrichment* constitutes a preliminary treatment that is conducted after uranium ore has been extracted, and before the unstable metal is employed as a fuel in nuclear reactors. Enrichment operations include the isolation of the metal out of its gangue (production of *yellow cake*), its purification and its conversion into uranium hexafluoride.

In most plant reactors, a minimum content 3.5% of ^{235}U is necessary for allowing the nuclear fuel to be used. Fission of the ^{235}U–^{238}U unstable nuclei mixture determines through atomic bombardment a disintegration chain similar to that acting usually in the nature: formation of new nuclei marked by lower atomic mass and/or number, emission of neutrons, and release of considerable energy (Chapter 7.2.1). Uranium-235 disintegrates to form various successive daughter elements: thorium-231 (^{231}Th), protactinium-231 (^{231}Pa), actinium-227 (^{227}Ac), thorium-227 (^{227}Th), radium-223 (^{223}Ra), etc. Uranium decay and subsequent element formation are accompanied by the emission of α particles and β rays, and with an energy release of several hundred million electron volts. Energy is collected in the form of steam under pressure, which is conducted in turbines for producing electricity (Fig. 117).

The neutron flux released through fission processes and used for disintegrating new uranium nuclei is regulated for preventing excessive energy production and reactor deterioration. This is why atomic piles made of nuclear fuel rods (i.e. uranium dioxide) include two types of control systems:

- *control rods* preventing the excess emission of neutrons. Some neutrons are absorbed by boron or cadmium that is fixed on rods;
- *nuclear moderators*, in the form of pressurised water, heavy water or graphite, which permits neutron speed to be slowed down.

The nuclear fuel combustion progressively induces both an *impoverishment in fissionable uranium* and the *production of waste* made of variously radioactive daughter elements. Nuclear reactors must therefore periodically be re-fed with

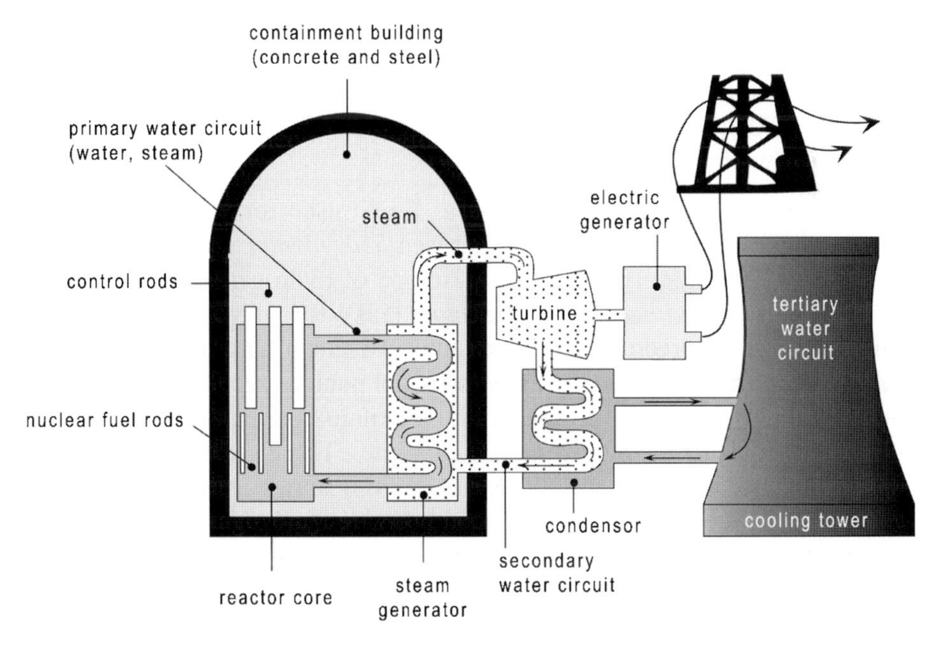

Figure 117: Schematic representation of a nuclear power plant (after Turk & Thompson, 1995).

nuclear fuel rods, i.e. approximately annually. Impoverished fuel must be removed and stored as nuclear waste or reprocessed.

Nuclear reactors are usually designed to operate for about 40 years, after which they should be downgraded. This step constitutes the last phase of the nuclear cycle proper (Fig. 118). Reactors are subsequently dismantled, nuclear power plants decommissioned and destroyed, and all contaminated materials processed or stored (Chapter 7.3).

Reactors, Fast-breeder Reactors, Reprocessing

Reactors Three-quarters of reactors operating in nuclear power plants and producing 85% of all nuclear-generated electricity are cooled with ordinary water called *light water.* These *light water reactors* (LWRs) comprise three independent water circuits (Fig. 117):

- a primary water circuit is located in the reactor core. It is used for controlling and regulating the flux of heat produced by nuclear combustion. Water is maintained either under liquid form at a high pressure (i.e. pressurised water reactor, PWR), or as a mixture of gas and liquid (boiling-water reactor, BWR);
- a secondary water circuit is destined to store heat issued from the reactor core via the primary circuit. Water consists of steam that activates turbines for generating electricity;
- a tertiary circuit of liquid water ensures the cooling of the whole power plant. It is totally disconnected from the two first circuits. The water is usually moderately warm.

Some reactors designed during World War II and used for industrial purposes mainly in Canada are cooled with pressurised *heavy water* consisting of water enriched with deuterium (i.e. the heavy isotope of hydrogen; atomic mass = 2, $^{2}D_{1}$). These *pressurised heavy water reactors* (PHWRs) may work with natural uranium comprising predominantly the slowly disintegrating isotope 238. This saves the ^{235}U enrichment phase of the industrial nuclear cycle.

Other reactors, mainly developed in Great Britain, use carbon dioxide as the coolant, the nuclear moderator being graphite. These *gas-cooled reactors* (GCRs) and advanced gas-cooled reactors (AGRs) may be working with ^{238}U-rich natural uranium, as do PHWRs, or with nuclear fuel only slightly enriched in ^{235}U.

Finally some hybrid-type reactors use ordinary water as a coolant and graphite as a nuclear moderator. These light-water graphite reactors (LWGRs) were designed in Russia and largely used in the former Soviet Union. Reactors of this type (called RBMK in Russia) were operating at Chernobyl, and one of

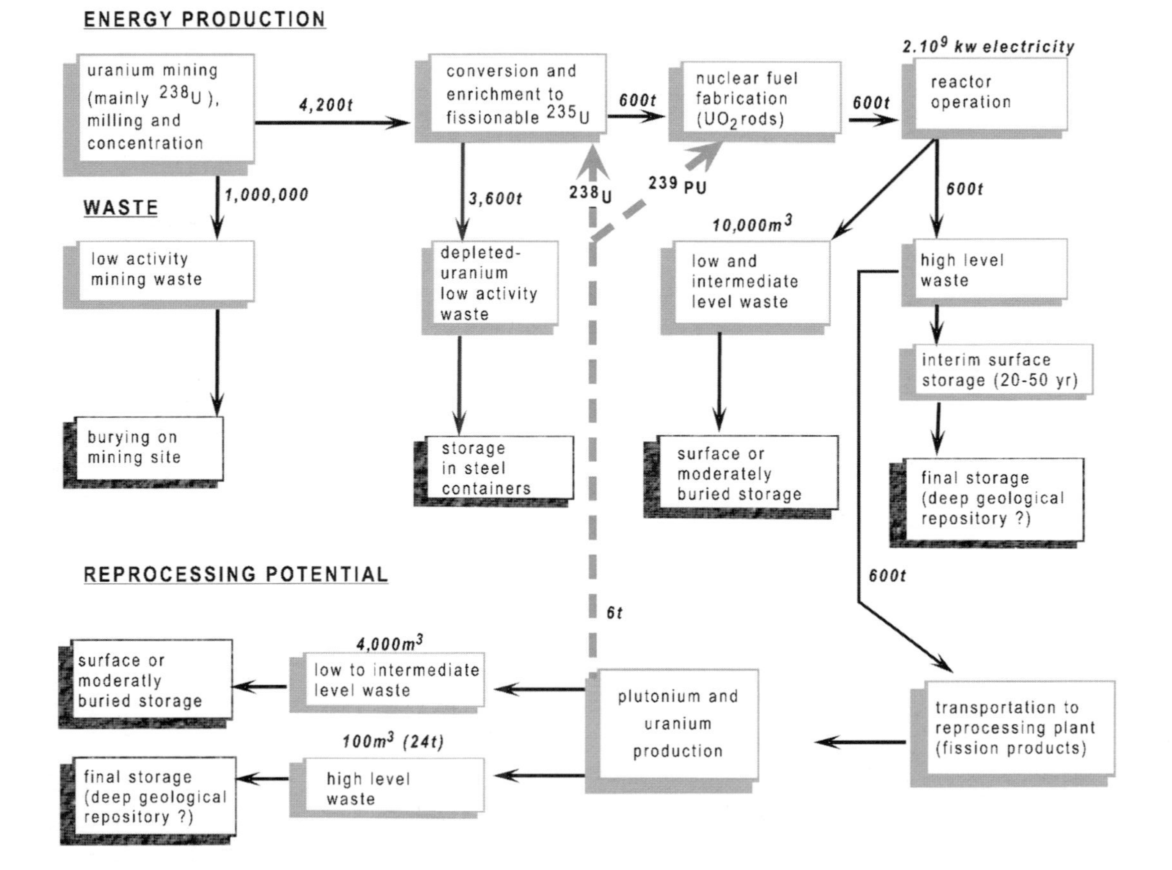

Figure 118: A typical nuclear fuel cycle on an industrial scale (after Nuclear Energy, 1995). Numbers constitute estimates for a light water reactor producing 2 billion kWh and working for 30 years.

them was responsible for a terrible nuclear accident in 1986 (Chapter 7.2.3). Notice that the world's oldest working reactor, at Obninsk in Russia, is also of the same type. This reactor has never suffered any serious incident since it started producing electricity in 1954.

Fast-breeder reactors In a fast-breeder reactor (FBR) the nuclear fuel consists of a *mixture of uranium oxides* (^{238}U and ^{235}U isotopes) *and plutonium* (Pu), and no moderator is used. Only a very small amount of ^{235}U is necessary for starting the nuclear combustion process. Plutonium, the atomic number of which is 94, constitutes a mildly radioactive element formed by uranium neutronic irradiation in the core of nuclear reactors. This metal, of which the most common isotope has an atomic mass of 239 and a half-life of 24,000 years, is therefore a *residue of the nuclear industry.* The combustion of ^{239}Pu in fast-breeder reactors allows us to partly disintegrate this nuclear waste and therefore to diminish its quantity. The nuclear fission of the U–Pu mixture nevertheless generates the formation of new plutonium nuclei. This determines an almost endless electricity-producing cycle, the energy efficiency of which is 50–100 times higher than that of a nuclear reactor operating solely with uranium.

The main problem induced by FBR functioning results from the need to use *liquid sodium* as a coolant in the primary water circuit. This fluid is a heat conductor that is much more efficient than water but is potentially very dangerous. Liquid sodium is highly corrosive and catches fire spontaneously when put in contact with the air. Another problem is that plutonium constitutes the major fuel for nuclear weapons (Chapter 7.2.4). Despite the fact that this element cannot be used for bomb-making without sophisticated techniques, its formation through fast-breeding reactions and stocking in large amount could amplify the risk of misappropriation or terrorist actions.

Fast-breeder reactors have been designed mainly in France (e.g. Phénix at Marcoule, Superphénix at Creys-Malville), in the United Kingdom and in Japan. Their development has barely exceeded the experimental stage.

Reprocessing Nuclear fuel reprocessing techniques were envisaged as early as the 1960s, when active development of nuclear power plants and atomic weapons was causing concern about uranium-ore shortage around the end of 20th century. Nuclear reprocessing consists of the removal of combustion residues from the reactor core, nuclear waste transportation in special containers and the *recovery of a large fraction of residual uranium and plutonium for reuse* in the normal nuclear cycle (Fig. 118).

Reprocessing operations imply delicate handling of highly radioactive solid, liquid and gaseous substances. Eighty-five percent of world's nuclear reprocessing capacity is located: (1) in France, at Marcoule in the Gard department, and at La Hague on the Cotentin peninsula headland : and (2) in Great Britain, at

Sellafield on the border of the Irish Sea. Other reprocessing units are mainly located in India, Japan, and Russia. The need to develop reprocessing activities has become much less important during the last decades, for several geological and political reasons: discovery of important new uranium ores, establishment and application of the Nuclear Non-proliferation Treaty (Chapter 7.2.4), the decision taken by some countries to abandon the reactor system, etc.

7.2.3. The Nuclear Risk

April 1986, Explosion of a Nuclear Reactor at Chernobyl

At 01:24 in the night of April 26, 1986, reactor number 4 of the nuclear power plant at Chernobyl, in the Ukraine, underwent *violent superheating*. This unexpected event happened during a control test designed to check the capacity of the reactor to automatically stop working in case of a general electric power failure. The energy released by the suddenly uncontrolled fission of nuclear fuel rods instantaneously caused water to vaporize in the primary cooling circuit, the reactor to explode, and the containment-building roof to be seriously damaged.

Twenty minutes later the mass of hydrogen released by water disintegration triggered a second explosion that destroyed the whole containment building and caused the ejection of nuclear fuel debris, reactor pieces, and concrete blocks around the site. The graphite rods destined to moderate the neutron flux became suddenly exposed to air and caught fire. This induced a *mini-eruption* marked by the emission of a mixture of graphite, water steam, melted nuclear fuel and fission products.

A large *radioactive cloud* formed which first moved to the northwest and then was progressively wind-transported in the direction of both northwestern Europe and Asia (Fig. 119). The contamination determined by the Chernobyl catastrophe was more than 10 times higher than that caused by the nuclear bomb explosion in Hiroshima in 1945.

During the 2 weeks following the accident, 31 people died. Some of the victims were fire-fighters who had been excessively irradiated at the site of the fire. Almost 200,000 inhabitants were evacuated from a 30-km wide leeward side zone situated to the northwest of the Chernobyl plant, essentially in Byelorussia. The ground in this zone was severely contaminated; flora and fauna were destroyed or damaged.

Two months after the Chernobyl disaster, measurements showed that several hundred thousand people were presenting with excessive radioactive iodine concentrations in their blood, and a fear of cancers, leukaemia and other serious diseases developed in the whole region. The proportion of thyroid cancers increased considerably. Despite these various risks and the damage, few concrete and indisputable proofs exist, *15 years* after the accident, of serious and

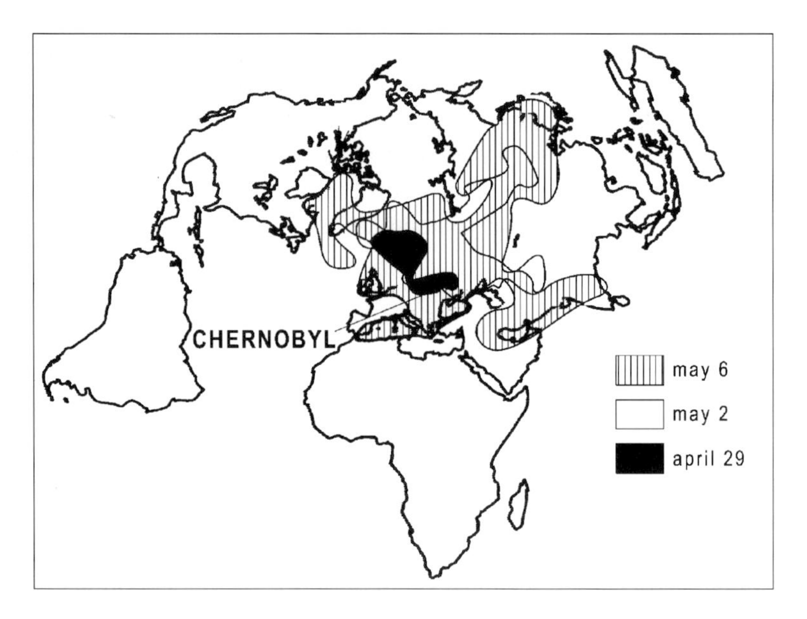

Figure 119: Extension of radioactivity in the atmosphere after the nuclear reactor accident at Chernobyl on April 26, 1986 (after Barnier, 1992).

long-standing consequences on human health. Numerous reports and journal articles nevertheless exist that point to prenatal malformations and miscarriages, tumours and post-traumatic stress. Very long-term, disastrous effects could still happen potentially, but this probability becomes less and less.

Despite the various and relatively abundant radioactive elements ejected into the atmosphere after the accident (Table 25) the soil pollution has not been serious outside the 30-km-wide area situated around and to the leeward side of the Chernobyl nuclear plant. Numerous precautions were taken by all countries potentially threatened by radioactive fallout (Fig. 119), especially as far as milk and farm produce were concerned.

On site, reactor number 4 was locked and sealed inside a steel and concrete *sarcophagus*, which nevertheless displayed some local fluid leaks during the last decade. In spite of further potential risk, the two other nuclear fuel reactors at Chernobyl have been working continuously since 1986 to satisfy Ukraine electricity needs. Many inhabitants of the severely contaminated zone who had been displaced have returned to their homeland and practise farming again. The cessation of the Chernobyl nuclear power plant was planned for 2001.

The Chernobyl accident is by far the most serious that has happened since the production of domestic nuclear power started. Its *causes*, which have become known progressively and belatedly, are multiple, complex, and due to both *technical and human* failures (Bourgeois et al., 1996).

Table 25: Nature and abundance of main radionuclides emitted in the atmosphere due: (1) to Chernobyl accident in April 1986, (2) to Sellafield nuclear reprocessing plant (U.K.) since the 1980s, and (3) to military nuclear tests performed between 1952 and 1997 (diverse sources; see MacKenzie, 2000). Values in exabecquerels (1 EBq = 10^{18} Bq).

Radionuclide	Period (d, days; yr, years)	Chernobyl 26/4/1986	Sellafield	Military tests
^{3}H	12.32 yr		0.060	240
^{14}C	5.730 yr			0.22
^{85}Kr	10.7 yr	0.033		
^{89}Sr	50.55 d	0.094		91.4
^{90}Sr	28.6 yr	0.0081	0.0063	0.604
^{95}Zn	64.0 d	0.016	0.024	143
^{103}Ru	39.3 d	0.14		238
^{106}Ru	371.6 d	0.059	0.034	11.8
^{125}Sb	2.73 yr			0.524
^{131}I	8.0 d	0.67		651
^{134}Cs	2.07 yr	0.019	0.058	
^{137}Cs	30 yr	0.037	0.041	0.912
^{140}Ba	12.8 d	0.28		732
^{141}Ce	32.5 d	0.13		254
^{144}Ce	284.9 d	0.088	0.062	29.6
^{238}Pu	87.7 yr	$3.0 \cdot 10^{-5}$	0.000 125	
^{239}Pu	$2.41 \cdot 10^{4}$ yr	$2.6 \cdot 10^{-5}$	0.000 61	0.006 52
^{240}Pu	$6.56 \cdot 10^{3}$ yr	$3.7 \cdot 10^{-5}$	0.000 61	0.004 35
^{241}Pu	14.4 yr		0.022	0.142
^{241}Am	432.7 yr			0.003 8

On the one hand, the light-water graphite reactor type used had initially been designed for predominantly military purposes, necessitating very powerful functioning. The void ratio was therefore extremely important when reactors were working at low power, as necessary for industrial use, and this encouraged strong thermal instabilities. This weakness was partly corrected after the accident by appropriate modifications performed on all Russian reactors of the same RBMK type. Improvements have mainly consisted of changing nuclear fuel-enrichment techniques. This major defect was associated with various other technical shortcomings: lack of a global containment system allowing complete isolation of the faulty reactor; relatively slow falling time of urgent stopping rods (i.e. 20 sec instead of 0.5 sec in Western nuclear power plants); low number of automatic safety systems; practical inconveniences for applying emergency procedures.

On the other hand, the implementation of the control test took place in a context of under-preparation, contradictory information, lack of discipline, overconfidence and finally under-estimation of actual risks.

Discharge in the Environment

In the course of the industrial nuclear cycle the *emission of radionuclides* in the natural environment *may principally happen during mining extraction, ore purification and potential reprocessing phases*, and secondarily during fuel fabrication and combustion phases.

Ore extraction and preliminary treatments Open-air and underground uranium-ore exploitation is associated with the *release in the atmosphere of variable amounts of radon-222*. Radon discharge also characterizes sorting and purification processes. In addition mining waste stored in extraction cavities or under specific packaging may release radon long-term scale through the ^{238}U–^{230}Th disintegration chain (Fig. 111). The risk from radon release may affect both miners and then workers on extraction sites. Its importance is of very local extent and still poorly defined (Chapters 7.1.1 and 7.4).

Radioactive ores rich in uranium (i.e. more than 10–15%) are responsible for important *gamma (γ) radiation*, which is emitted by various daughter elements of the heavy metal. γ rays potentially induce strong health risks for miners. For getting round this problem, cutting faces are mined by automatic devices.

The release into the environment of radioactive isotopes contained in uranium mining waste is facilitated by various factors, some of which may be controlled by appropriate treatment. For instance, the discharge of radium-226 in rock pore water may be facilitated and therefore managed thanks to sulphate- or iron-reducing bacteria that are put in contact with radioactive waste (Landa & Gray, 1995).

^{235}U enrichment phase and reactor functioning The potential emission of radioactive elements is low during the nuclear fuel preparation phase, because most elements other than uranium have already been removed through preliminary treatments. Small amounts of thorium-234 (period $t=24.1$ days) that directly result from ^{238}U fission may nevertheless be released and temporarily incorporated into nearby materials. This ^{234}Th discharge was noticed in Ribble River sediments deposited downstream from the Springfield factory in northwestern Britain when ^{235}U enrichment processes were conducted.

During the nuclear fuel combustion phase the safety conditions are remarkably severe in all modern nuclear power plants. The risk of radioactive elements escaping is almost nil, and could anyway be effectively controlled.

Reproceessing phase The potential discharge from reprocessing factories of *gaseous radionuclides* principally concerns *krypton-85*, *hydrogen-3* and *carbon-14*. These elements are released in very small amounts and do not induce any appreciable risk.

Liquid discharge potentially concerns *plutonium* (^{239}Pu, ^{240}Pu), *americium* (^{241}Am), and *cesium* (^{137}Cs), which may accumulate in sediments deposited downstream from reprocessing plants. This contamination has been noticed in some estuaries, coastal lagoons and shallow seafloor areas of the Irish Sea, where cooling water from the Sellafield reprocessing plant is discharged. Fortunately the activity level of these nuclear contaminants is very low. On the other hand the potential risk is very difficult to assess for several reasons: with which ease radionuclides are either adsorbed on sedimentary particles or released into water; complex inter-reactions with some non-radioactive contaminants discharged from other factories; nuclear disintegration controlled by environmental chemistry; etc. (see MacKenzie, 2000).

In the same way as for other industrial processes comprising cooling phases, *the functioning of nuclear power plants determines the emission of heat.* Tertiary-circuit issuing water is rejected outside the plant from cooling towers (Fig. 117) or by pumping. Water evaporated from cooling towers determines very local atmosphere heating, whereas water discharged in a liquid state close to the plant currently determines a temperature increase of 3°C in river courses and of 0.5°C in sea water. The tertiary-cooling circuit is totally independent of primary- and secondary-cooling circuits, and is theoretically free of any radioactive contamination. Its water may therefore be used for various *secondary applications*: domestic heating, aquaculture, etc.

Nuclear Accidents

Numerous accidents and problems have been identified since nuclear power began to be exploited in the mid-20th century. Some reports have listed the occurrence of 150 accidents between 1945 and 1996 (Sonderson et al.; see MacKenzie, 2000). Most accidents have been of very minor importance and have neither threatened people nor damaged property and environment. In fact less than ten serious nuclear accidents have happened so far worldwide.

Industrial power plants Important accidents have officially affected five nuclear power plants since exploitation of nuclear energy first started in 1954.

They are linked to either civil or military industrial activities, and result from either technical weaknesses or human errors and deficiencies (MacKenzie, 2000):

- September 1957, Kysthym, Soviet Union: chemical explosion of a military container of high-level radioactive waste. Emission of 0.75 EBq ($=0.75 \times 10^{18}$ Bq), absorption of 1,200 Sv.
- October 1957, Windscale (now "Sellafield"), an English, military plant producing plutonium and polonium: fire within an air-cooled reactor, the moderator of which was graphite. Emission of 1,774 TBq ($=1{,}774 \times 10^{12}$ Bq), absorption of 2,000 Sv.
- March 1979, Three Mile Island, USA: loss of coolant inducing the melting of 50% of the nuclear fuel in a pressurised water reactor. Large radionuclide release within the containment building. Emission of 370 PBq ($=370 \times 10^{15}$ Bq) of the noble gas xenon-133, and of 550 GBqs ($=550 \times 10^{8}$ Bq) of iodine-131; absorption of 20 Sv. The Three Mile Island accident did not cause any noticeable damage to people and the environment, thanks to the presence of a strong global containment system that is similar to those operating in Western European plants.
- April 1986, Chernobyl, Ukraine, Soviet Union: explosion of reactor number 4 (see above). Emission of 2 EBq (see Table 25), absorption of 60,000 Sv.
- September 1999, Tokai-mura, Japan: explosion and uncontrolled nuclear reactions for 20 h during enriched-uranium processing. One hundred and sixty-one people were evacuated from a 350-m-wide area around the factory, and 310,000 people living in a perimeter of 10 km were confined to home for 18 h. Measurements taken both immediately after the accident and subsequently have revealed the absence of serious environmental contamination by fission products, and the abundant emission of neutrons responsible for low physical and radiological risks (*Journal of Environmental Radioactivity*, 2000).

Other human activities Nuclear accidents have affected non-industrial activities in three domains:

- *Military* accidents: nuclear bomb release from American bombers at Palomares, Spain (January, 1966), and at Thule, Greenland (January, 1968); bomb disintegration on impact.
- *Space* accidents: burning in the stratosphere over the Indian ocean of an American satellite containing a plutonium-238 power source (April, 1964). Crash in Northwest Territory, Canada, of a Russian Cosmos satellite containing a nuclear reactor (January, 1978); radionuclide dispersal over an area of 50×800 km.

- *Medical* accidents: diverse nuclear accelerator faults, unexpected release of cobalt-60, cesium-123, etc.

Risk, Prevention, and Control

Preventing potential irradiation danger is essential at each step of the industrial nuclear fuel cycle (Fig. 118). *The most serious risks are caused by nuclear power plants themselves.* Considerable progress has been accomplished during the two last decades for preventing reactors from melting and fission products escaping. Newly implemented systems include:

- the duplication of cooling circuits;
- the possibility for reactors to be instantaneously and automatically isolated and protected by thick steel and concrete containment barriers and surrounding walls;
- the increasing use of materials characterized by high thermal inertia;
- the augmentation of cooling surface.

The possibility of human failure, which potentially induces the least-expected risks, is anticipated as accurately as possible, mainly by developing automatic control and safety devices. Recent calculations made on a worldwide scale show that the likelihood of a major nuclear fuel fusion happening in a given reactor is of 1 in 20,000 years. This means that one among all presently operating nuclear reactors statistically could melt every 200 years.

Somewhat barely quantifiable risks remain for certain *old nuclear power plants*, as for instance some of those built during the 1960s and 1970s in the former Soviet Union. The opening up since the 1990s of the Soviet bloc has allowed the introduction of the most recent safety standards and rules, and implementation of an international assessment and action programme. Some potentially unsafe power plants have been closed and their nuclear fuel has been removed. Functioning and cooling systems have sometimes been modified in depth, safety and alarm systems have been added, and various similar operations still go on. Refurbishing and ensuring the present safety standards in all old nuclear power plants, or dismantling the less reliable of them, would nevertheless necessitate much larger technical and especially financial support than presently available.

Drastically limiting gaseous and liquid radioactive discharge in the environment constitutes another major objective. Remarkably this has been achieved as far as most dangerous isotopes are concerned, especially in developed countries. For instance the physical and chemical treatment of cooling water discharged from the Sellafield nuclear plant into the Irish Sea has, since the mid-1980s, allowed removal of all cesium and plutonium components

(Fig. 120). Slightly contaminated areas remain very locally, which result from old discharges conducted through less-efficient treatment procedures.

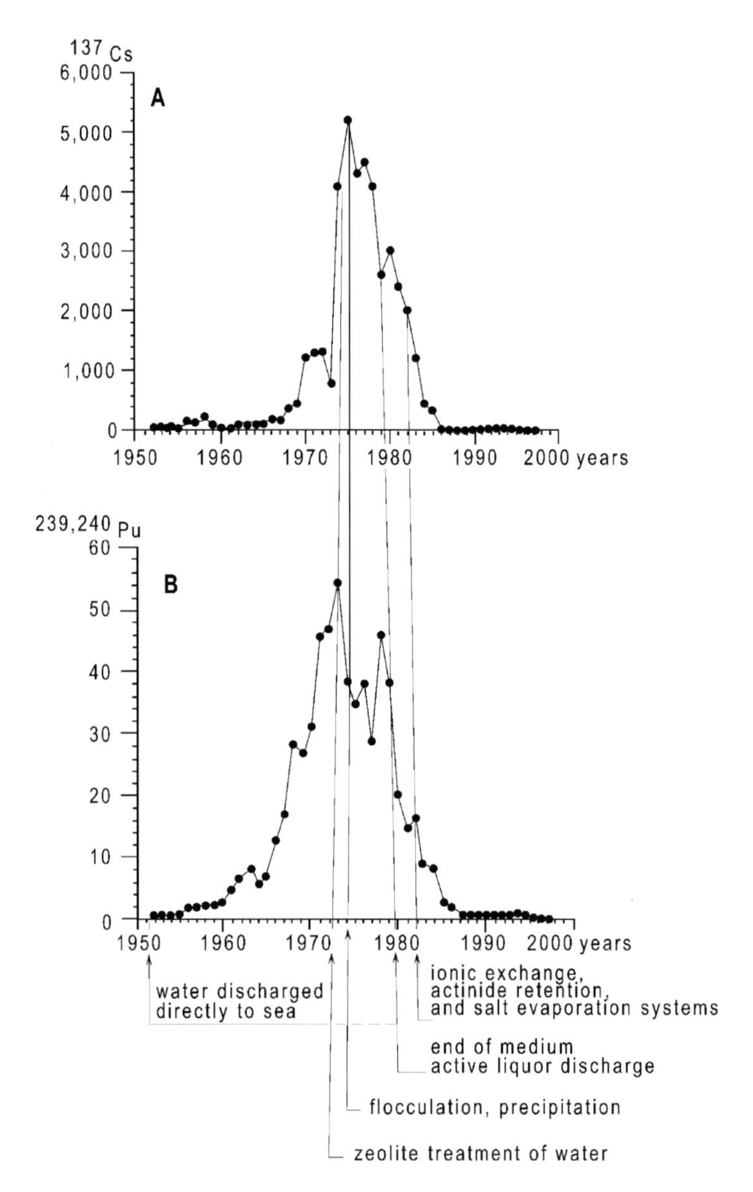

Figure 120: Radiogenic activity between 1952 and 1997 of caesium-137 (A) and of plutonium-239, -240 (B) within Sellafield low-level liquid effluents discharged in the Irish Sea, and major changes to waste treatment procedures (diverse sources; see MacKenzie, 2000). Radioactivity in terabecquerels (TBq; 1 TBq = 10^{12} Bq).

In the Chernobyl disaster area, the radioactive soils have been decontaminated through phyto-remediation processes (Chapter 9.3.3) and specific agricultural measures including fertilization, soil enrichment, and controlled ploughing. This has allowed reducing the ^{137}Cs accumulation in arable crops by a factor of 2.3–2.8, and planning to produce forage and drinkable milk again (Alexakhin, 1993).

On a worldwide scale the discharge into the natural environment of industry-, military- and accident-produced radionuclides was diminished from about 10^{22} becquerels (= 10,000 EBq) in the 1950s and 1960s to about 1 EBq during the two following decades. The decrease continued during the 1990s (MacKenzie, 2000).

Advantages and Disadvantages of Nuclear Power

Nuclear energy, the production of which increased extraordinarily between 1965 and 1985, has undergone a noticeable slowing down during the last two decades. This is not due to any growing shortage of the resource, but principally to a growing perception of major potential risks to health. In fact the use of nuclear power for industrial and domestic purposes presents various objective advantages and drawbacks, the respective weight of which is modified in our minds by many more subjective considerations: temptation to lump together civil and military industry, newly developing reluctance about some scientific approaches and ethics, fear of long-term unexpected effects, amplification by the media, etc.

Most obvious gains and advantages of nuclear energy are the following:

- production in very large amounts of electric power. A nuclear fuel reactor provides an average instantaneous energy of about 700 MW;
- uninterrupted reactor functioning during several decades, ensuring regular, reliable and long-term electricity production;
- need for small amounts of fuel that are extracted from large reserves and may be augmented through fast-breeder reactions and reprocessing;
- fuel extraction, pre-treatment, and transportation phases marked by very little impact on the environment;
- fuel costs having progressively decreased and being 35%–50% cheaper than that of the usual fossil fuels;
- moderate amounts of waste produced by modern nuclear power plants. Very low discharge of gas responsible for increasing the greenhouse effect (CO_2, NO_2) or acid rain (SO_2). Absence of flying ash emission;
- accident rate much lower than that induced by most heavy chemical industry activities. Very high-level, technical expertise ensuring plant safety and waste

management. Expertise applicable to several other activity fields: medicine, agriculture, agro-industry, physics, environmental control, etc.

Disadvantages and limitations of nuclear power production are mainly the following:

- high cost of nuclear power plant construction. The cost is about twice that of coal thermal plants, and further considerably increases when decommissioned plants have to be dismantled;
- serious consumption of cooling water, and potential risk of environmental pollution;
- complex and costly management of nuclear wastes, especially those of a high activity level and long half-life (Chapter 7.3);
- know-how usable for military or even terrorist purposes. Often inappropriate practices of confidentiality.

Whatever the uncertainties and apprehensions in relation to radioactivity exploitation, *the nuclear industry-induced environmental risks are basically extremely low* (Fig. 121). The artificial nuclear cycle functioning statistically determines only 0.1% of Earth's total radioactivity. More than four-fifths (82%) of soil, water, air, and human radioactivity are of natural origin, a noticeable part of which (11%) is specific to the human body.

Let us remember that coal combustion determines more abundant emission of radioactive substances than that of nuclear fuel, for an identical amount of energy produced. In addition some coal-released substances are known for bearing carcinogenic properties. Current human exposure to normal radioactive consumption products is 30 times higher than that determined by the Man-made nuclear cycle: television receivers (< 10 μSv for U.S. citizens), video equipment (< 10 μSv), optical glasses (< 1 μSv), incandescence sleeves (0.4 μSv), welding systems (0.2 μSv), dental prostheses (0.14 μSv), luminous watches (0.05 μSv), electronic tubes (0.04 μSv), smoke detectors (0.03 μSv), airport control systems (0.002 μSv), etc.

The challenge for people in charge of nuclear power production and to political decision-makers is therefore clear. *If safety conditions were met everywhere, as is presently the case for most nuclear plants, and if nuclear waste disposal was satisfactorily and precisely controlled, the production of radioactive energy would offer one of the best safeguards for man, the biosphere and Earth's environment.*

Even the examination of nuclear-accident consequences, which can be established objectively only a long time after events have occurred, almost systematically demonstrates their very low impact. Serious consequences were nevertheless effectively recorded after the nuclear accident at Chernobyl, where they essentially affected people and environment in the region surrounding the

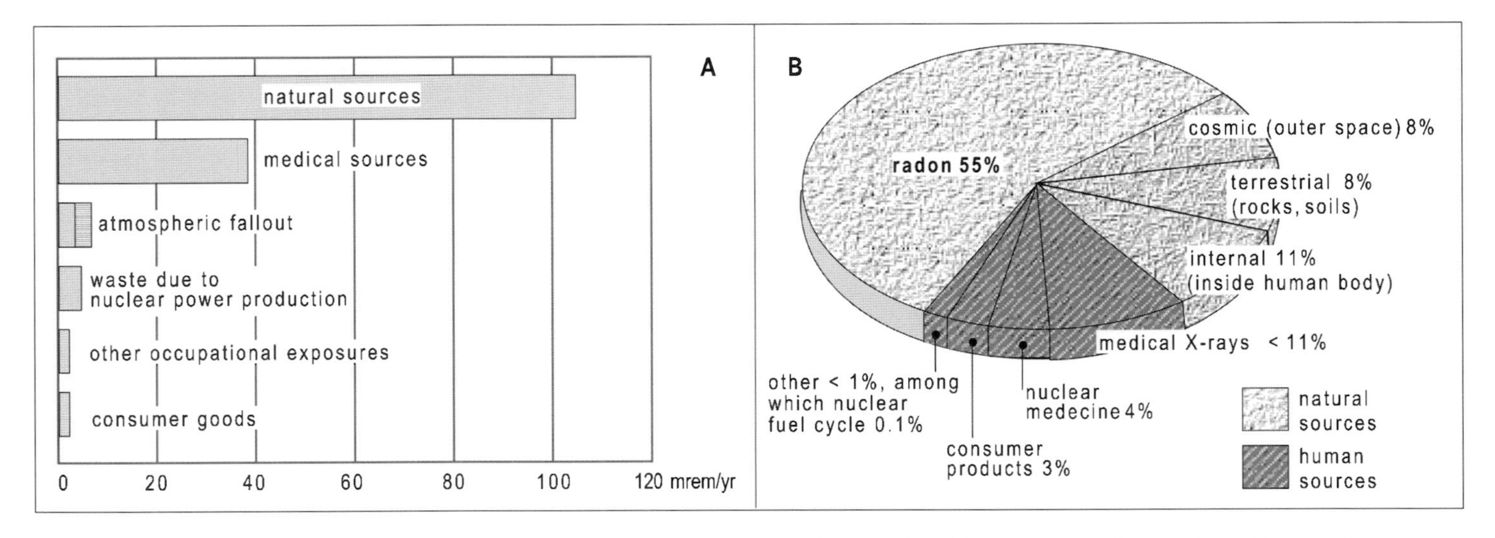

Figure 121: Importance of natural and anthropogenic sources responsible for Earth's average radioactivity affecting the human environment. (A) Doses in mRem/yr (after Murck et al., 1997). (B) Exposure percentage to different radioactive sources by a North American inhabitant (after Turk & Thompson, 1995).

nuclear plant. Chernobyl's long-distance radioactive emissions, which are often presented as having been very dangerous, were in fact considerably lower than that of natural radioactive background (Table 26). Despite such objective data, the effects of the Chernobyl radioactive cloud on health, water, and soils is still recurrently amplified by the media in an excessive and even irrational manner.

7.2.4. Military Nuclear Weapons and Tests

Historical Aspects

Following the first nuclear explosion tests in the USA in 1945 and the release of atomic bombs over Japan, most military-induced radioactive emissions have resulted from experiments destined to develop and perfect new weapons, and exceptionally from accidents (Palomares, 1966; Thule, 1968; Chapter 7.2.3).

Atmospheric tests have generated the emission in the air of abundant ionizing substances, which are transported partly locally (12%), partly in the regional troposphere (10%), and principally in the stratosphere (78%), where they diffuse widely along latitudinal trajectories. Atmospheric tests were mainly conducted between 1952 and 1962. From 1963 rather fewer open-air explosions were still practised by China, France and India. Such tests were stopped in 1980.

Table 26: Additional doses of radioactivity received by several European countries following the Chernobyl nuclear accident in April 1986 (after Bourgeois et al., 1996). Numbers are added up over a period of 50 years for taking into account the release of long-life radioactive elements (caesium, etc.).

	Values (mRems)	
natural radiation = basic radioactivity	7,000 to 14,000	i.e. 100% to 200%
		Relative increase
medical radiation	2,000 to 3,500	+30 to +40%
radiation due to Chernobyl accident		
– Greece	60	+0.7%
– Germany, Italy	40	+0.5%
– Belgium, France	9	+0.1%
– Spain, Portugal	1	+0.01%

Underground tests have been conducted almost 3 times more abundantly than atmospheric experiments (1,352 as opposed to 520). They were actively conducted between 1960 and 1990, and then progressively diminished. Underground tests do not occur any longer, mainly as a result of the 1968 Nuclear Non-proliferation Treaty (NPT), which was revised in 1995 and signed by 178 countries.

Radioactive Mechanisms and Products

In a *nuclear bomb* the explosion is produced by *fission* of uranium-235 or plutonium-239. In a *thermonuclear bomb* uranium fission is used to induce the *fusion* at very high temperatures of hydrogen (^{1}H) and deuterium (^{2}D). This reaction generates both the production of large amounts of tritium (^{3}Tr) and a frequent secondary fission due to uranium-238 and neutron inter-reactions.

After bombs have exploded, *additional radionuclides* are generated in the atmosphere. They result from neutron capture by atomic nuclei that are either present in the air or derived from bomb metal constituents. The radioactive elements produced by nuclear explosions are of very diverse nature and half-life (see Table 25). They comprise about 25 isotopes, ranging from hydrogen-3 to plutonium-241 and americium-241. They include different strontium, ruthenium, cesium, cerium and plutonium isotopes, and also some *carbon-14* issuing from nitrogen-14 neutron bombardment. The atmospheric concentration of ^{14}C almost increased twofold during the 1960s due to active military nuclear tests.

Risks and Monitoring

Numerous fission and fusion products are of a low concentration that does not threaten the environment, or have been disintegrated since the cessation of military nuclear tests. Some radioactive elements nevertheless remain stored in soils and porous rocks, which are still continuously irradiated. Residual isotopes that could potentially damage health include mainly carbon-14 (^{14}C; $t = 5{,}730$ yr), strontium 90 (^{90}Sr; $t = 28.6$ yr), caesium-137 (^{137}Cs; $t = 30$ yr), and the plutonium group (^{238}Pu, ^{239}Pu, ^{240}Pu, ^{241}Pu; period t between 14.4 and 24,000 yr). Americium (^{241}Am; $t = 432.7$ yr) also constitutes a potential danger in water and soil; this artificial radioisotope is derived from the disintegration of ^{241}Pu released by radioactive fallout.

$^{239,240}Pu$ and ^{137}Cs concentrations were recorded between 1990 and 1993 on soils from 29 atolls of the Marshall archipelago. The Marshall Islands are located between 4° and 11.5° of latitude in the northwestern Pacific and had been submitted to the effect of numerous atmospheric nuclear tests conducted between 1946 and 1958 by the U.S. army. Soils contain plutonium ranging from world average values (0.2–0.4 Bq/kg) to values 20,000 times higher. Concentration increases between 9° and 11.5°N in the direction of *Bikini* and *Enewetak* atolls where explosions took place. Topsoil contamination remains severe in ancient shooting areas, where $^{239,240}Pu$ levels locally exceed 10^3 Bq/kg. The radioisotope concentration decreases rapidly towards lower latitudes (Fig. 122). Due to isotope fractioning processes, the Pu to Cs ratio diminishes by a factor of 133 from firing zone soils to those situated at a distance of 1,000 km.

The cessation of both atmospheric and underground military nuclear tests, together with the signature of the Nuclear Non-proliferation Treaty by most countries (except India, Israel and Pakistan) and the nuclear disarmament policy decided by the five main countries officially owning atomic weapons (China, France, Russia, U.K., USA) have *considerably reduced the military- nuclear risk. Other threats* nevertheless exist, which are all the more worrying, as they remain imprecise and disconnected from international politics and diplomacy:

- probable secret development of nuclear weapons in some countries (e.g. Iran, Iraq, North Korea, Libya);

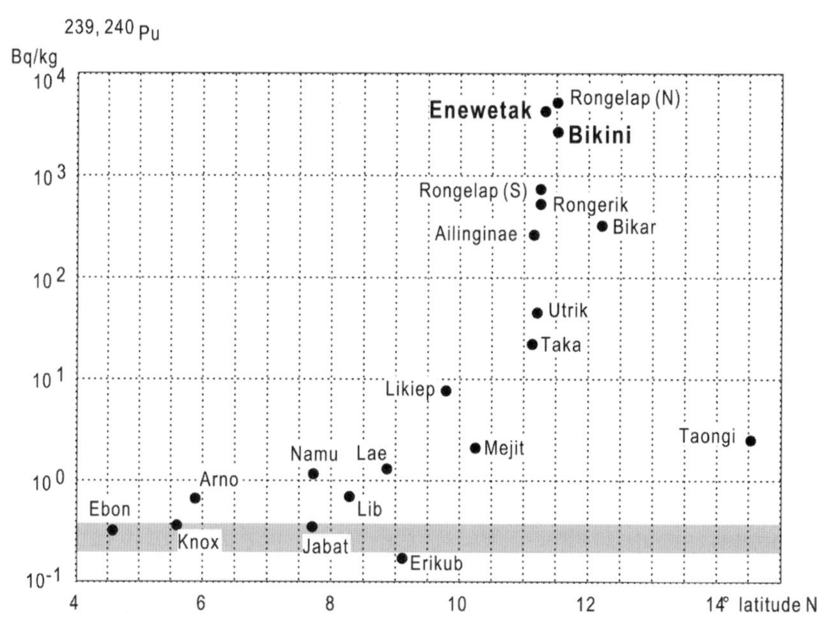

Figure 122: Maximum concentration of plutonium ($^{239,240}Pu$) in topsoils (0–5 cm depth) of Marshall Islands as a function of latitude (after Simon et al., 1999). Some atoll names are indicated, those where nuclear tests were carried out being in bold characters. World average values are indicated by horizontal strip. Values in becquerels per kilogram (Bq/kg).

- possible long-term deterioration of downgraded or excess weapons, the radioactive components of which may be released in the environment. This is the case for old Russian nuclear-powered submarines that were sunk in the Laptev and Kara Seas;
- fraudulent resale or misappropriation of plutonium stocked in downgraded nuclear power plants of the former Soviet Union, and possible use for terrorist actions. For manufacturing a nuclear bomb, 25 kg of plutonium or of slightly enriched uranium is sufficient. Handling and transporting such radioactive elements does not necessitate any specific precautions. This risk is somewhat attenuated by the fact that industry-derived plutonium needs very sophisticated, dangerous and costly techniques to be used as an explosive. By contrast, weapons-originating plutonium may be converted rather easily into nuclear fuel.

7.3. Nuclear Waste

7.3.1. Types and Amounts

Residues from nuclear energy exploitation represent *crucial environmental concerns* for human societies, since they contain some long-life fission products that potentially constitute serious health threats (Table 27). Similar problems are raised by some other industrial and medical activities implying the residual formation of active and/or concentrated radioisotopes. Highly radioactive waste is rare but nevertheless poses very difficult safety problems.

Four categories of nuclear waste are defined according to the radioactivity level:

- *Very low level waste* (VLLW) consists mainly of concrete, plaster, scrap metals and other debris that result from dismantling of diverse installations reserved for nuclear-related activities but having not been in appreciable contact with radioactive materials: mine equipment and buildings, ore-treatment and -reprocessing factories, nuclear power plant outer buildings and related constructions, research laboratories and centres, etc.
- *Low-level waste* (LLW, or A-category waste) constitutes 90% of radioactive residues but represent only 1% of industry-induced radioactivity: clothes, filters, tools, metal containers free of nuclear fuel, and other items of common use that have not been in contact with the reactors. These materials come from nuclear power plants, hospitals, research centres employing radioisotopes, etc. The resulting radioactivity principally results from beta (β) particles and gamma (γ) electromagnetic rays emission; it is of low intensity and differs very little from ambient natural values. The half-life of radioisotopes

Table 27: Nature and major anthropogenic sources of main radioactive isotopes needing specific disposal systems (after Rogers & Feiss, 1998). Period (= half-life) indicated in years (yr), days (d), or hours (h). Am, americium; Au, gold; I, iodine; Ir, iridium; Np, neptunium; Pu, plutonium; Tc, technecium.

Element	Period
fission products of nuclear plants	
^{90}Sr	28 yr
^{99}Tc	$2.1 \cdot 10^5$ yr
^{137}Cs	30 yr
^{129}I	$1.6 \cdot 10^7$ yr
^{239}Pu	$2.4 \cdot 10^4$ yr
isotopes formed in nuclear waste	
^{210}Pb	22.3 yr
^{226}Ra	1,600 yr
^{234}U	$2.5 \cdot 10^5$ yr
^{237}Np	$2.1 \cdot 10^6$ yr
^{242}Pu	$3.8 \cdot 10^5$ yr
^{243}Am	7,400 yr
isotopes of industrial and medical waste	
^{24}Na	15 h
^{32}P	14 d
^{60}Co	5.3 yr
^{82}Br	36 h
^{85}Kr	10.7 yr
^{125}I	60 d
^{192}Ir	74 d
^{198}Au	2.7 d

concerned does not exceed 30 years. Residues generally do not need to be containerised. After compacting or incineration, they may be stored in natural or artificial subsurface cavities. Low-level waste from nuclear power plants themselves constitute only 1%–2% of this category. Major LLW comes from sites where uranium ore has been mined, extracted and isolated.

- *Intermediate-level waste* (ILW, or B category) represents about 7% of nuclear wastes' volume and 4% of total radioactivity. It comprises various materials: resins, decantation sludge, fuel containers from operating plants, equipment and ordinary residues from downgraded and dismantled plants. ILW is responsible for the emission of alpha (α) particles, the period of which is between 30 and several hundred years. Some β and γ rays are also released.

This waste needs to be locked in concrete or bitumen containers as long as irradiating energy remains important (i.e. up to hundreds of years).

- *High-level waste* (HLW, or C category) occupies only 3% of nuclear waste volume but is responsible for the emission of 95% of residual radioactivity. It consists of material directly related to nuclear fission reactions: nuclear combustion ash, uranium-impoverished fuel rods after combustion, reprocessing-derived fluids. Irradiating emission is dominated by α particles that induce potential danger for thousands to hundreds of thousands of years. HLW releases large amounts of *heat* and needs to be cooled for several decades before being finally sealed and stored. As first nuclear power plants began operating in the 1950s and most reactors are 30–40 years old, all high-level radioactive wastes are still stored close to ground surface in water-cooling basins and/or under active ventilation.

Radioactive wastes produced annually by the European nuclear industry roughly amount to 160,000 t, 5,000 t, of which are of high activity (HLW). Although abundant, total radioactive wastes correspond only to 0.8% of the twenty billion tons (20,000,000,000 t) of highly toxic waste produced each year in Europe by the chemical industry. This rate drops to 0.025% when considering only high-level radioactive waste.

Controlling radioactive waste disposal in an efficient and long-lasting manner has become an essential condition for allowing long-term exploitation of nuclear energy. Initial development programmes planned that European countries should produce 25% of nuclear-derived electricity in 2050, relative to 16–17% in the 1990s. This implied that 400,000 t of nuclear waste should be annually managed in a few decades from now. Such a question becomes all the more critical as most presently operating nuclear reactors should be decommissioned and dismantled during the next two decades. This will determine the production of even more abundant waste. The decision to replace or not the nuclear power plants on a large scale will mainly depend on the feasibility and safety of solutions that will be adopted for high-activity waste disposal. The present period is therefore crucial from this point of view. This explains the large research effort presently developed by most countries having an important nuclear industry (e.g. ANDRA in France: French Agency for Radio-Active Waste Management).

7.3.2. Elimination

Eliminating radioactive waste through dispersal, immersion or burying was envisaged and sometimes carried out during the first phases of nuclear energy exploitation, i.e. between 1960 and 1975. All possible elimination processes have subsequently been officially abandoned, mainly because of long-term safety vagueness. The different solutions proposed and major inferred risks are as follows:

- *throwing into space*: risk of launching failure or of return on Earth; exorbitant cost for evacuating large amounts of waste;
- *immersion on the seafloor*, in a similar way to what has been done in polar seas with decommissioned Soviet submarines: risk of container corrosion; mistrust and rejection at the hands of the public;
- *burying in oceanic troughs* and "digestion" by subduction processes: risks of earthquakes, volcanic eruptions or violent heat release; very slow speed of potential incorporation within terrestrial crust; relative proximity of inhabited land masses;
- *burying in Antarctica or Greenland icecaps*: risk of drum crushing and rupture; necessity for transporting very sensitive materials over long distances; need for international authorization; likelihood of long-term rejection through glacier movements.

Impoverished fuel reprocessing constitutes a type of progressive elimination treatment. Reprocessing offers the advantage of new energy-producing fission reactions and significant reduction of highly radioactive wastes. The plutonium isotopes resulting from chemical separation processes and initially destined for fast-breeder reactors may be combined with uranium for constituting a new fuel called MOX.

In fact the technical and financial interest of reprocessing treatments and resulting nuclear products has diminished during the two last decades due to the discovery in abundance of new uranium ores (Chapter 7.2.2). The additional production of plutonium, a radioactive metal susceptible to misappropriation, poses an extra risk. Danger might also result from the transportation of impoverished fuel to specialized factories; such a risk is rarely accepted by public opinion, in spite of various real-size tests having shown the almost complete resistance of containers to all types of impact and collision. Finally, present reprocessing facilities are extremely insufficient for permitting treatment of all impoverished fuel stocked so far.

7.3.3. *Surface Repository Disposal*

As radioactive wastes cannot practically be eliminated, they should necessarily be stored somewhere. Almost all nuclear energy-producing countries have therefore developed active research for storage solutions. Disposal may be close to the terrestrial surface or deep in the crust in geological formations. Both these options are diversely documented and defended by nuclear scientists and engineers.

Storage at depths shallower than a few tens of metres is permitted and carried out for *very low, low-, and intermediate-level waste*. Intermediate-level residues need to be specifically packaged through vitrification and containerisation in steel drums, which are stored in cavities reinforced with concrete.

As far as *high-level nuclear waste* is concerned, surface-disposal supporters plead for the advantage of a direct and responsible watchdog and argue the difficulty for deep geological risks to be precisely constrained. They also insist on choosing very dry storage sites such as deserts, where corrosion would be very low and slow; this implies that deserts are situated in tectonically stable regions (e.g. West Africa, Central Australia). Opponents to surface-storage solutions stress the *risks of external geodynamic changes* during long geological periods:

- sea-level drop inducing reactivated erosion and possible nuclear waste outcropping;
- major climatic cooling causing glacial erosion and soil instability;
- exceptional flooding responsible for morphological disorders;
- earthquake hazard effects;
- potential meteorite fallout.

The understanding of such potential hazards determines the current development of various in-depth researches: investigation on geological analogues, experiments and tests, modelling, etc.

Man-induced risks due to robberies, terrorist actions, plane crashes, etc. also potentially threaten the future of surface-stored waste. In addition we should keep in mind the limited time duration of human societies and of their memory.

7.3.4. Deep Geological Disposal

Present-day preferences lean towards the solution of storing nuclear waste deep in the ground, as shown by the implementation of various research programmes, underground tests and experimental laboratories. The hypothesis of *reversible disposal at intermediate ground depth* between 300 and 800 m is being documented in an increasing way. Such a solution would allow recovering radioactive waste in the case of an unforeseen situation or of decisive technological progress for potential reuse.

Radioactive wastes themselves are planned to be isolated as much as possible from host geological formations. *Multi-barrier systems* are being developed that comprise successive hideaway packaging (Fig. 123): radioactive material solidification through vitrification, solid waste encapsulating in drums, drum overpacking, overpack vaulting, vault back-filling and sealing.

The lithological nature of geological formations envisaged for hosting high-level radioactive waste must be precisely characterized, which currently gives way to numerous measurements, experiments and modelling research. Host rock formations should be as *thick, homogeneous, dry, absorbent, watertight and cold* as possible, and the *permanent character of such conditions* should be practically guaranteed for at least several thousands years. A geological stability of 10,000 years is generally considered as appropriate. This implies that potential disposal sites are located in regions that are both devoid of appreciable seismic and tectonic activity, and remote from areas of rising volcanic magma. *The rocks* themselves should be little or non-faulted and -fissured, free of noticeable aquiferous properties, only very slightly permeable, and barely sensitive to neutron bombardment and long-term heat release. They should also be of no potential economic interest.

Various types of geological formations have been or are being envisaged and investigated for hosting nuclear waste. Their advantages and disadvantages are evaluated through multiple field and laboratory analyses, as well as through in

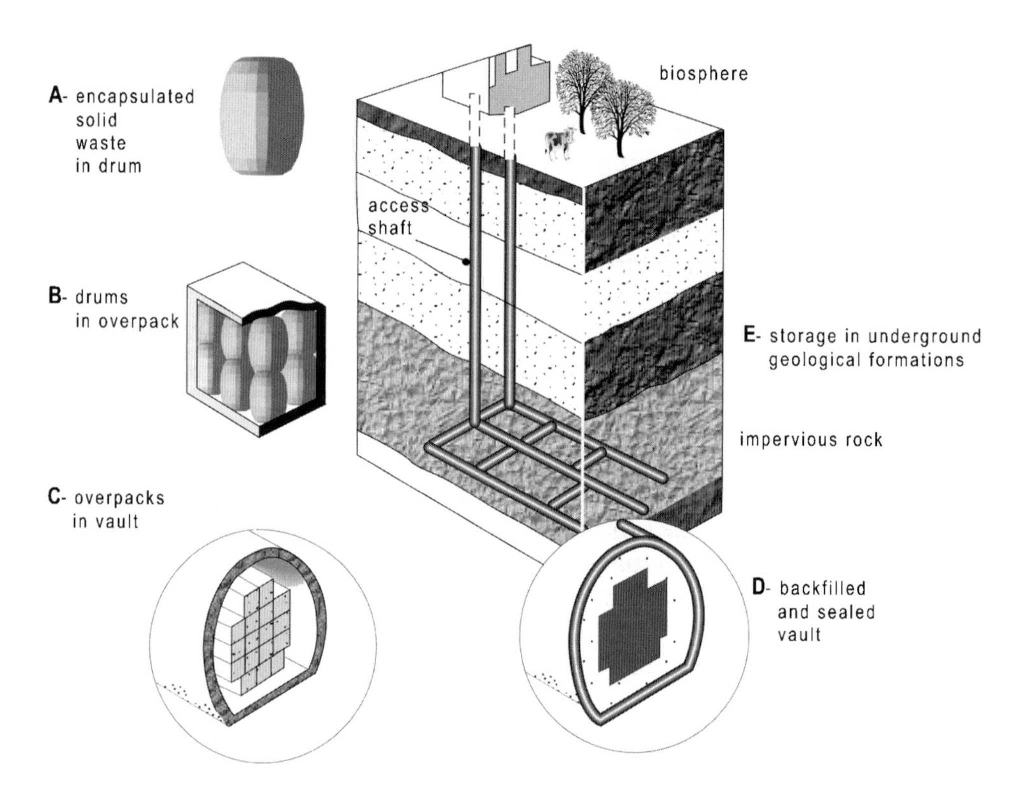

Figure 123: Schematic example of multi-barrier system planned for deep geological disposal of radioactive waste (after Bradshaw et al., 1992).

situ experiments. This is for instance the case of the French underground laboratory implemented by ANDRA between about 300 and more than 500 m deep in Late Jurassic clay formations from the northeastern Paris Basin (Meuse–Haute Marne site).

The rock types investigated by the different countries involved in these surveys are namely the following.

- *Granite* (Canada, France, Sweden, Switzerland, United Kingdom, USA). Advantage of tectonic stability, high thermal conductivity facilitating heat dispersal, high mechanical strength, wide extent in several little-populated, dry and/or cold regions formed by old and stable continental shields. Risk of rock fracturing and of unexpected intrusion of underground water.
- *Basalt* (USA). Interest of stable properties determined by homogeneous petrological and mineralogical characteristics. Important risk of fracturing due to long-lasting heat release by radioisotope disintegration; potential return to volcanic activity.
- *Volcanic tuff* (USA). Advantage of low permeability, and possible high absorption properties if zeolite minerals are present. Potential risk of significant water content, fracturing, return to volcanic activity and associated instability. The potential economic interest of volcanic tuff is incompatible with very long term waste disposal.
- *Salt rocks* (Canada, Germany, USA). Advantage of homogeneity, which is not very frequent in sedimentary formations. Salt older than several tens of million years proves the very long term absence of interstitial dissolving water. Salt is naturally plastic and therefore prevents fluid input and output. Its high thermal conductivity favours heat transfer and elimination. Potential risks arise from low mechanical properties, high corrosive power on metal-bearing parcels containing radioactive waste, potential solubility, and difficulty for storage reversibility due to self-sealing through flowing.
- *Clay, claystone, mudstone* (France, United Kingdom, USA). Advantage of impermeability, plasticity, low compaction and fracturing properties, very limited water circulation, high absorption and heat storage properties. Disadvantage of frequent lithological heterogeneity, crack formation through heat-generating desiccation, low mechanical properties and sensitivity to microfissure formation or reactivation.

The capability of clay-rich sedimentary formations to endure without significant change the very long-term effects of highly radioactive elements' disintegration may be estimated by information provided by *natural analogues such as the Oklo nuclear reactor.* Proterozoic shale as old as 1.8 billion years crop out at Oklo in southeastern Gabon, west of Central Africa. They mainly consist of carbonate- and quartz-rich claystone and siltstone that was intruded at the

Francevillian stage by highly radioactive sediments having penetrated along a fault system. Intrusive Proterozoic deposits contained up to 70% of uranium, about 3% of which were constituted by the isotope ^{235}U. This amount is similar to that presently used in nuclear power plants. *Natural fission reactions* were actively developing at Oklo for more than 500,000 years, in the presence of underground water. They continued working until the ^{235}U rate was sufficiently low to allow the neutron bombardment to stop.

Calculation of successive geochemical balances shows that most radioactive daughter elements were retained in the clayey rock during the whole activity period of the Oklo natural reactor (Brookins, 1976). This was mainly the case for plutonium, ruthenium, rare earth elements, zirconium and palladium. Alkaline chemical elements such as sodium and potassium, as well as associated elements such as calcium, were diversely migrating. But the migration process started 25 million years after cessation of reactor activity and has therefore not induced any appreciable radioisotope transportation. In addition most migrating elements have remained close to the disintegration site, except xenon and krypton gasses and perhaps iodine. During the fission phase, the environment remained anaerobic, the temperature barely exceeded 150°C and pressure was between 600 and 800 atmospheres. These conditions were similar to those presently envisaged for storing nuclear waste at intermediate depths (e.g. a few hundred metres). The case of the Oklo natural reactor demonstrates that buried clay materials are likely to constitute adequate geological formations for hosting high-level radioactive waste.

A Few Landmarks, Perspectives

- Radioactivity constitutes a natural phenomenon that is inherent to the Universe and to all Earth's environments, including the human environment. Radioactivity also represents an important energy resource that has been harnessed by Man for producing electricity, manufacturing weapons, providing medical diagnoses and treatments, and supplying information on Earth's functioning, on chemical exchanges between the different planet reservoirs, etc. As for all natural resources, radioactivity forms and uses present some advantages and some risks. As both positive and negative aspects are particularly important, they have induced both extraordinary technological developments and strong objections.
- Nuclear dissociation and synthesis characterize the whole Universe and are responsible for the original formation of all inorganic chemical elements. Nuclear reactions have induced and still induce the genesis of unstable isotopes called radionuclides, which form according to three natural

processes: (1) very high temperature fusion of atomic nuclei (e.g. hydrogen, helium) characterizes functioning of the sun and other stars. Nuclear fusion is associated with the emission of strong radiation that heats the planet and particularly Earth's surface. Fusion has been used for manufacturing hydrogen thermonuclear bombs; (2) cosmic bombardment of atmospheric gasses determines the synthesis of low-energy radionuclides (^{14}C, ^{32}Si, etc.), which serve as remarkable environmental markers and geological chronometers; (3) nuclear fission consists of the disintegration through neutron bombardment of unstable metals, which mainly comprise uranium and thorium families. Fission triggers complex chains of newly formed isotopes and of α-particle and β-ray emission. It is associated with the release of powerful energy responsible for the natural heat flow that continuously rises from Earth's solid envelopes and warms our planet's surface. This energy is also at the origin of most human applications of radioactivity. Natural radioactivity risks and advantages are illustrated by the example of radon (^{222}Rn), an inert and colourless gas forming through uranium-238 fission and diffusing towards soil, water, air and domestic basements and floors. Radon, which is responsible for more than 50% of the atmosphere's natural radioactivity, is likely to induce serious lung diseases to inhabitants of houses built on granite, organic clay or phosphate rocks. Radon is also used to treat dermatological and rheumatological diseases.

- The use of radioactivity for producing electricity is mainly based on the fission of ^{235}U, an isotope that is much more energetic (i.e. of shorter half-life) but much less abundant (0.7%) than ^{238}U (99.3%). After an ore-enrichment phase, ^{235}U disintegration is kept going by intense, continuous and regulated neutron bombardment. This releases considerable amounts of heat, which is converted into electricity via turbines. Nuclear reactor functioning progressively induces both nuclear fuel impoverishment and the synthesis of new radioactive isotopes (plutonium, iodine, technetium, etc.). Irradiated nuclear fuel rods must therefore be periodically replaced. Nuclear power plant functioning increasingly promotes the question of managing nuclear waste, which is likely to generate additional fission products (formation of americium, radium, etc.).
- Industrial exploitation of nuclear energy has started and developed since the mid-20th century, simultaneously with the world's population explosion. It is associated with potentially important risks for human health. These risks result from the extraordinarily powerful heat released through fission reactions, as well as from the concentration of high-energy radioisotopes produced by the harnessed nuclear cycle. Exemplary care and expertise have globally been brought to all nuclear-industry phases, from ore extraction to waste stockpiling. Safety guarantees are generally outstanding and much

higher than those provided by most other industrial processes. Radioactive emission in the environment from extraction sites mainly concerns radon, the main part of which is of natural origin. Emission from nuclear power plants and reprocessing factories was significant until the late 1980s but is now so low that measurement and potential effects are very difficult to assess. Environmental contamination caused by nuclear electricity production presently amounts to 0.1% of Earth's total radioactivity. There is nevertheless serious risk from accidents, which have been significant or serious in five nuclear power plants since 1957. The most disastrous event occurred in 1986 at Chernobyl, Ukraine, where 31 people died soon after the accident, and where various diseases have developed since then. Civil or military nuclear accidents have partly resulted from human errors or deficiencies. A particularly essential "safety culture" has developed widely during the two last decades and has spread to most countries, including those of the former Soviet Union. Let us remember that except for Chernobyl, the nuclear accident impacts have essentially been very localized. Anthropogenic radioactivity due to the Chernobyl catastrophe has increased the West European background radioactivity only by one ten-thousandth to one-hundredth.

- The most sensible question arising from nuclear power use concerns the future of radioactive waste. Four waste types are identified, which depend on radioactivity level and duration. High-level waste constitutes only 3% of the nuclear industry's total residues but represents 95% of residual radioactivity and remains potentially dangerous during thousands or even tens of thousands of years. Waste reprocessing is presently poorly justified due to abundant uranium reserves, synthesis of new potentially dangerous isotopes and high financial cost. Storage therefore is the option currently sustained by most countries. In-depth research is conducted for characterizing in a rational way the most suitable disposal systems: design of multi-barrier devices, tests of reversible storage at intermediate ground depth, implementation of underground experimental laboratories, investigation of rock behaviour in the presence of radioactive waste, and modelling of environmental changes due to potential modification in geological conditions. Pending a final decision, all high-level radioactive wastes are stored in surface or subsurface conditions and undergo an initial cooling phase.
- The current period is crucial. More than 60% of nuclear power plants will come before 2020 to the end of their optimal safe functioning. They will need to be decommissioned and dismantled. Their possible replacement should be largely anticipated and simultaneously induce the refinement of both technical competence and safety culture. The anti-nuclear protests that developed in the late 20th century have led several countries to give up the

nuclear option (Austria, Germany, Italy, Sweden) or to stop the extension programmes. On the other hand some countries facing massive energy needs have started to replace or to increase their stock of nuclear power plants (Armenia, Bulgaria, China, Japan, Russia, Ukraine). Due to the potential energy crisis, some highly industrialized countries envisage reactivating their civil nuclear strategy (USA). The galloping population growth requires production of more and more energy, which presently rules out the possibility of managing without nuclear power. The nuclear resource has become reliable, safe and clean. Its availability is immense relative to substitute resources, which still can play only a supporting role: hydroelectricity and tidal or aeolian energy, geothermal and sun energy, biomass or biogas power, etc. Four objective facts could help us consider with more hindsight nuclear-power advantages and drawbacks: (1) rise of Earth's global warming risk due to greenhouse-effect amplification by excessive non-nuclear fuel burning (coal and wood, oil and gas); (2) urgent energy demand by various developing countries, mainly in Asia; (3) safety-driven need for international organizations to solve the crucial question of replacing numerous reactors and nuclear power plants in the former Soviet Union; (4) thanks to the Nuclear Non-Proliferation Treaty, dissociation between the nuclear industry and defence, the evolution of which has for long been parallel and caused some confusion.

– High-level waste disposal constitutes the most worrying risk brought about by radioactivity exploitation. The distress caused by increasing amounts of industry-derived radioactive residues, which are difficult to control and will be passed on to our descendants, has overcome public opinion and also the decision makers. This is mainly the reason why a "reversible stocking" approach is currently favoured for "handling the future with care". Geologists, who know ground formation characteristics and are daily forced to assimilate the concept of very long term evolution (i.e. the geological duration) and assessing its consequences, should certainly get more involved in this society debate. Some geological formations deeply buried for several tens to hundreds of millions of years in tectonically stable regions have obviously the characteristics required for hosting radioactive materials and facilitating their harmless disintegration, as do radioisotope-containing natural rocks. The risk for such rocks to be naturally exhumed is objectively insignificant until radioactivity has diminished by itself. By contrast risks for nature-induced disruption increase at shallower burial depths, which are more exposed to short-term and aleatory changes induced by external geodynamical or human factors. Geologists should bear the triple responsibility of obtaining knowledge, diffusing information, and encouraging rationalization. This would certainly help better assessment and anticipation of the inter-reactions likely to occur between radioactive waste and the Earth's environment.

Further Reading

Bourgeois J., Tanguy P., Cogné F. & Petit J., 1996. La sûreté nucléaire en France et dans le monde. Polytechnica, Paris, 298 p.
Eisenbud M. & Gesell T. F., 1997. Environmental radioactivity: from natural, industrial and military sources. Academic Press, 4th ed., 656 p.
Foos J., Rimbert J.-N. & Bonfand E., 1993–1995. Manuel de radioactivité à l'usage des utilisateurs. Formascience, Orsay: vol. 1, 197 p.; vol. 2, 310 p.; vol. 3 (& G. Lemaire), 350 p.
van Loon A. J., 2000. Reversed mining and reversed-reversed mining: the irrational context of geological disposal of nuclear waste. Earth Science Reviews 50: 269–276
MacKenzie A. B., 2000. Environmental radioactivity: experience from the 20th century – trends and issues for the 21st century. The Science of the Total Environment 249: 313–329
Milnes A. G., 1985. Geology and radwaste. Academic Press, 328 p.
Nuclear Energy, 1995. Who's afraid of atomic power? Understanding Global Issues, European Schoolbooks Public., 18 p.

Some Websites

http://www/iaea.org/worldatom/: Site established under the auspices of the United Nations (International Atomic Energy Agency) and devoted to all questions related to nuclear energy, including sustainable development conditions in the vicinity of nuclear power plants
http://www.worldenergy.org/wec-geis/: Description, assessments, and perspectives of different energy types including nuclear energy. Site controlled by the World-Energy Council
http://www.ipsn.fr/: Essential information provided by the French Institute for Nuclear Safety and Protection about nuclear energy practical management: reactors, waste, safety and protection questions, priority files
http://www/andra/fr/: Site of the French Agency for Radio-Active Waste management (ANDRA: Agence Nationale pour la gestion des Déchets Radio-Actifs): waste types, disposal options, current research, underground laboratories, international collaborations
http://www/ecn.nl/: Netherlands Energy Research Foundation site, which is devoted to research on renewable energy sources, nuclear energy, and fossil fuels
http://www/nrc/gov/NRC/radwaste.html: Official American site dealing with nuclear energy waste

Part III

Earth Facing Man's Activities

Chapter 8

Soils

8.1. Haiti, Everglades: Loss, Conservation of Surface Formations

Hispaniola Island is located between Cuba and Puerto Rico in the Greater Antilles. Its western part constitutes the Republic of *Haiti*, a 27,500-km^2-large territory famous for its tropical luxury. Haiti comprises mountainous areas separated by fertile valleys, where bananas and coffee, sugar cane and cotton plants are grown. The country has been characterized for centuries by a balanced distribution of forests, meadows and cultivated areas, with fairly well preserved soils and dispersed human settlements.

The situation has progressively changed since the mid-20th century, because of an exceptionally high population growth. The annual growth rate has been of 3% during the last decades, each woman having six children on average. In 2000 the Haitian population attained 7.3 million inhabitants, among which 2 million live in the Port-au-Prince conurbation. This population increase has determined an exponentially increasing demand for food and fuel, which has led to accelerating *deforestation*. Most forests, woods, groves and scrub have been systematically cut down and replaced by vegetable and fruit plantations. This has favoured active surface run-off and soil erosion. During the 1980s the loss of wooded areas reached 30%, among the highest rate in the world. The Haitian green landscapes progressively turn into reddish and dry terrain formed by laterite soils devoid of their fertile top layer.

As a general result, the productivity of cultivated land has strongly decreased during the last decades, the population has become poorer and poorer, and under-nourishment has increased. In the late 1990s, Haitian inhabitants were receiving on average only 90% of necessary calories, and soil erosion was reaching alarming values. The excessive human pressure has deeply modified surface geo- and bio-systems, which have become strongly unbalanced. At the same time the economic status of the country has dramatically dropped. At the dawn of the third millennium the gross national product in Haiti is 57 times lower than in France. Consequently, the protection of the Haitian economy and society fundamentally depends on soil restoration and therefore on an active reforestation strategy.

On the other side of the Tropic of Cancer, about 1,000 km northwest of Haiti, the southernmost part of *Florida* in North America has also been subject to considerable anthropogenic changes during the 20th century. Very flat and swampy areas that were initially covering a surface of 200×150 km and periodically flooded during wet seasons were actively drained. A 2,000-km-long network of artificial canals was progressively implemented, which allowed cleaning up the region, irrigating the cultivated land gained over the marshes, and supplying with water the conurbation growing in the region of Miami. An immense, sub-tropical pseudo-savannah and mangrove ecosystem progressively turned into a densely urbanized and industrialized region.

This exponentially growing anthropization has been contained, starting in 1947, when the *Everglades* National Park (see Fig. 138) was set up, thanks to the action of a few people who wanted to preserve a unique natural heritage and limit urbanization and speculation frenzy. The Everglades Park has progressively been enlarged and now occupies 607,000 hectares of original soils, flora and fauna. The genuine character of the Everglades, which constitute the third biggest U.S.National Park after Death Valley and Yellowstone, has led to its inclusion in the UNESCO world heritage category and to consider it as a natural biosphere reserve. Chronic difficulties remain due to both unbalanced hydrographical balance and large farmland and urban centres causing pollutant discharge, fertilizer and pesticide contamination, and stresses due to intrusive species. South Florida nevertheless presents today an exemplary case of *large-scale preservation of original surface soils and plant cover.*

8.2. Deforestation

8.2.1. Human Demand and Forest Cover

Demographic Constraints

The exponential growth of world population is fairly recent: 1.6 billion people in 1900, 2.5 in 1950, and 6.3 in 2000 (Introduction, 1). For a few decades, this has meant an extraordinary increase in food needs, particularly farm products (cereals, soya bean, sugar, cattle and sheep), and the situation is critical in developing countries. Spectacular augmentation of farm productivity has been obtained during the 20th century thanks to massive fertilizer use, irrigation development, and also a huge increase in farmland surface. Extending farmland areas on a large scale necessitates exploiting loose woodland soils and therefore cutting forest trees extensively. *The extension of arable land is principally made to the detriment of forested zones.* This is particularly the case in tropical

Table 28: Correlation between 1961 and 1984 of world population, arable land surface, fertilizer amounts, and irrigation importance, in both developed and developing countries (after FAO; see Pimentel, 1992).

Criteria	1961		1984		Increase
Population (million inhabitants)					
developed countries	978		1,202		25%
developing countries		2,158		3,651	70%
Arable land (million hectares)					
developed countries	654		676		3%
developing countries		698		800	15%
Fertilizers (N, P, K; million tons)					
developed countries	27.3		82.3		200%
developing countries		3.8		48.4	1,100%
Irrigation (million hectares)					
developed countries	37		62		67%
developing countries		101		158	56%

regions, where some of the most impoverished and fastest growing populations live (Tables 28 and 29). The increase in farmland during the last quarter of the 20th century was 5 times larger in developing than in industrialized countries, the population growth being only 3 times higher.

Historical Evolution

The man-induced decrease in forested surfaces during the last few thousand years has been demonstrated by several studies (Introduction, 2). For instance the almost complete disappearance of woodland in Iceland and in some Pacific islands (e.g. Easter Island) has been associated with human invasion or colonization periods.

Nevertheless the respective importance of natural (e.g. climate change) and anthropogenic causes is still poorly quantified in most regions. Recent investigations point to dominant human influence during some historical deforestation phases. For instance the Montagnola Senese area in Tuscany (Italy) has been subject in some periods of the 12th, 16–17th, and 19th centuries, to intense forest soil exploitation, which is demonstrated by biological, sedimentological and archaeological convergent arguments; extensive tree

Table 29: World percentage increase between 1961 and 1984 of population, arable land, farm surface and productivity, fertilizer use, and irrigated surface (after FAO; see Pimentel, 1993).

Criteria	**Increase %**
Population	55
Arable land	9
Exploited land surface	
wheat	14
rice	26
corn	24
soya bean	120
sugar cane	78
palm oil	44
Production	
wheat	131
rice	117
corn	118
soya bean	236
sugar cane	106
palm oil	215
Fertilizer use	320
Irrigated surface	60

felling during these periods has caused strong soil erosion that is independent of climate variations (Hunt et al., 1992). In a similar way, intensive mineral and trace element analyses of laminated deposits from Holzmaar volcanic crater lake in the Eiffel region (Germany) reveal that during Holocene times specific sedimentary input phases resulted from deforestation and subsequent soil erosion; the anthropogenic augmentation of sedimentary fluxes due to forest clearance occurred as early as the Iron Age (4th century BC) and became important from the Middle Ages; the man-induced terrigenous input has been temporarily associated with natural sediment discharge due to forest fire or volcanic eruptions (Lottermoser et al., 1997).

On a large scale the compilation of data available for northwestern Europe shows that major deforestation stages started 3,000 years before the present, related to animal domestication, and increased about 1,000 years ago in accordance with farmland extension (Fig. 124).

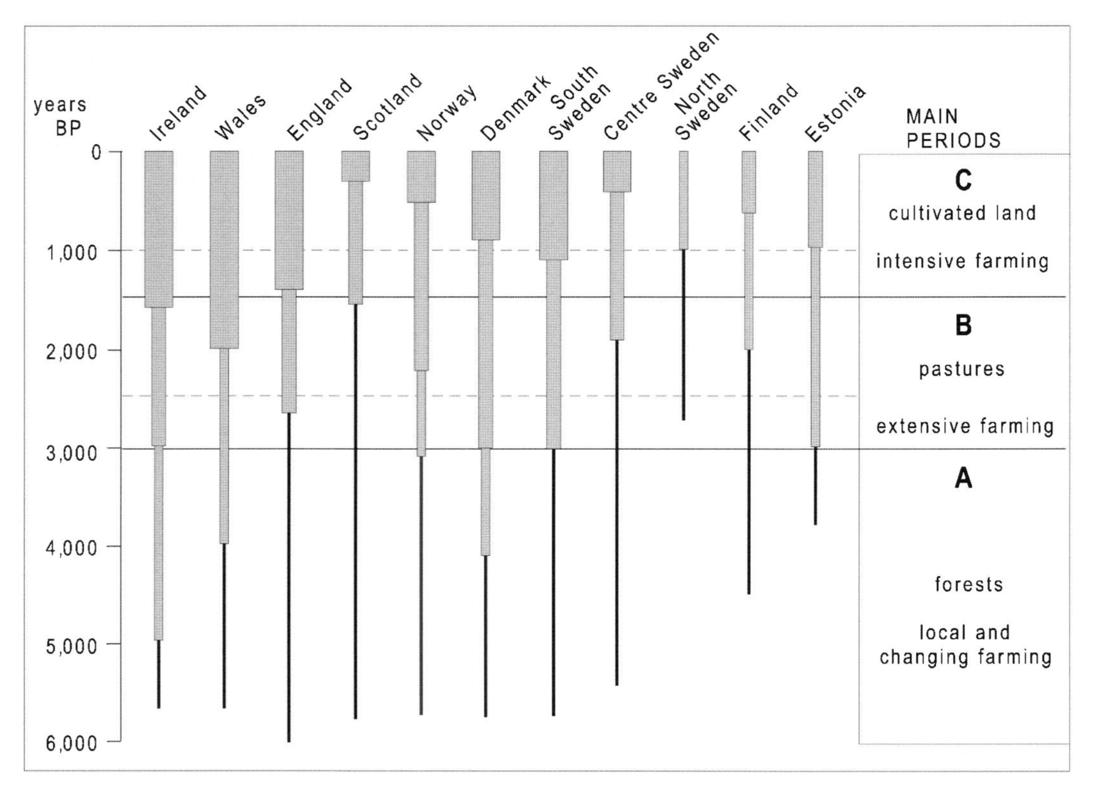

Figure 124: Main deforestation phases and human influence for 6,000 years on northwestern Europe soils (after Berglund et al.; see PAGES, 2000).

Natural deforestation periods have been mainly caused by *forest fire* that resulted from lightning, volcanic lava flows or excessive dryness. The importance of such short-term events has been illustrated recently by the 1988 forest fire that destroyed about one-third of Yellowstone National Park (Wyoming, USA), the size of which is close to that of Corsica. Forest fire induces the sudden availability of large soil surfaces devoid of plant cover and rich in mineral nutrients, predisposing to active erosion, but also to re-colonization and revitalization.

Geographical Variations

The importance of forest clearance mainly results from the need for new land devoted to livestock farming (more than 50%) *and to cultivation* (almost 10%). The worldwide demand for fuel results in less than 2% of forest clearance, and the rest concerns need for mines, building and extension of towns, artificial dams, communications, etc.

Extensive forest clearance and the reasons for it differ greatly according to the region of the world:

- Generally, deforestation is currently of a moderate extent in Europe, North America and Japan, where most clearance activities were performed during previous centuries and especially during major industrial expansion phases (i.e. 19th century). Assessments established for the last decades of the 20th century even point to reforestation trends in some of these countries (see Chapter 8.2.3). Recent forest clearance is mainly due to urban, industrial and commercial development, as well as to communications expansion.
- By contrast most tropical regions of South America, Africa and Southeastern Asia have suffered considerable and continuously increasing deforestation during the last century. The annual wood clearance commonly reaches 0.5–1.0% of forested areas. Deforestation in these regions is mainly determined by livestock farming and industrial cultivation needs, which are added to traditional slash-and-burn cultivation and wood cutting for domestic fuel. At the end of 20th century, the 10-year deforestation rate reached 20% of woodland in some countries (Ecuador, Nicaragua, Gambia, Paraguay, Vietnam), and even 30% in Haiti or Salvador. Some countries have lost a quarter, a third, a half, or even almost all their forests (Table 30). For instance the Sao Paulo state in Brazil, the forested area of which was 81.8% in 1500 and 79.7% in 1845, displayed only 58% of wooded areas in 1907, 18.2% in 1952, and 8.3% in 1973; about 10 years ago, estimates for 2000 were of 3% (Turner et al., 1992).

Table 30: Estimates in 2000 of tropical forest percentage loss due to 20th century man activities (after Park; see Rogers & Feiss, 1998).

Indonesia	–10%
Malaysia	–25%
Brazil, Colombia, Guatemala, Madagascar, Mexico	–33%
Ecuador, Honduras, Nicaragua	–50%
Thailand	–67%
Costa Rica	–80%
Ivory Coast, Nigeria	about –100%

8.2.2. Deforestation Impact

Soil Erosion

The presence of forest cover aids rainwater seepage in the soil, except in forwarding and fire zones. On the other hand, *forest removal induces accentuated rainwater flow at the soil surface, gully formation and loose-soil erosion*. Measurements performed on U.S. soils reveal that the average erosion rate on cultivated land is tenfold relative to woodland (Williams et al., 1993). Values become higher still if seasonal cultivation periods are short and separated by periods where bare soils are submitted to water and wind erosion (Chapter 4.2.2). Soil-erosion risks are amplified in regions characterized by strong seasonal rainfall; for instance, this is the case in Nepal and Northern India, where very large, hilly regions deprived of their original forest cover are submitted to intense flow and erosion during the summer monsoon.

The correlation observed between forest removal and soil erosion has also been established for past times. For instance the territories of present Germany during first centuries of the Christian era were characterized by a forest coverage exceeding 90% of the country's area; this resulted from numerous epidemics, famines and people migrations, which prevented people from widely exploiting soil resources. More prosperous conditions during Middle Ages induced the progressive replacement of forests by cultivated land and livestock farming meadows, which caused increasing soil erosion until the 13th century (Fig. 125). Disastrous soil loss occurred during the 14th and 18th centuries due to exceptionally heavy and long rainy periods. In the last century, the development of extensive cultivation practices determined increasing surface streaming, and mechanization-induced soil compaction prevented water seepage; this resulted in a further increase in the erosion rate.

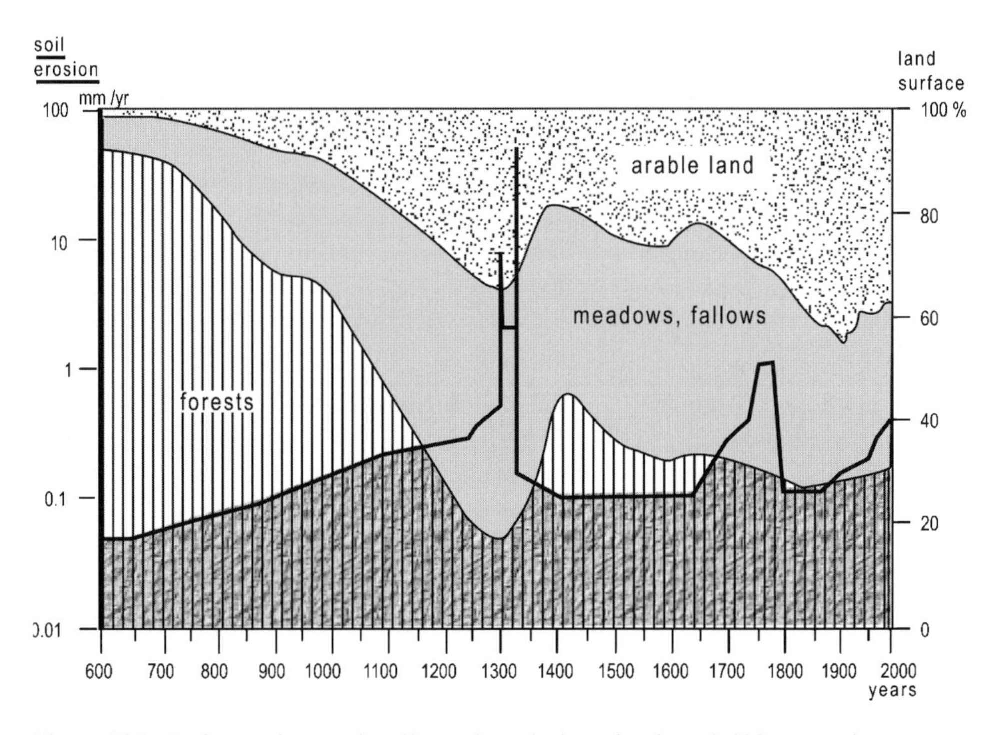

Figure 125: Deforestation and soil erosion during the last 1,400 years in present Germany territory, with the exception of alpine sectors (after Bork et al.; see PAGES, 2000).

Other Effects

An indirect consequence of deforestation and soil erosion is the *reworking and transportation of minerals and organic particles*. Huge sedimentary accumulations may form downstream from deforested regions, causing disruption of river, lake, estuarine or coastal ecosystems, and modifying the particle and dissolved-carbon balance (see Fig. 132).

On a regional scale, the plant cover removal induces an *environmental drying out* due to increasing evaporation from the topsoil layer (i.e. about the upper 30 cm). Disappearance of trees and their root network impedes deeper soil water to be brought to the ground surface and evaporation–transpiration processes to work. This tends to progressively induce the depletion of atmospheric humidity, the decrease in water recycling processes through evaporation–precipitation, and the corresponding increase in soil and mean air temperature. Resulting climatic changes may be serious. Simulation of environmental changes induced by the progressive destruction of the Amazon forests shows that average rainfall reduction could locally attain 800 mm/year, and that average surface temperature could increase by 0.5–2.5°C (Fig. 126). In the long term, deforestation therefore favours aridification, which in turn predisposes to desertification (see Chapter 8.4).

From a more global point of view deforestation leads to *disruption of the carbon and carbon dioxide (CO_2) balance*. Carbon "sinks" formed by forest trees and other plants tend to diminish due to less fixation of atmospheric CO_2. The wood combustion or deterioration increases the CO_2 emissions in the atmosphere and leads to amplification of the greenhouse effect (Chapter 11).

Indirect effects of forest removal *threaten the diversity of terrestrial living species*. Biological diversity is large in forested regions, particularly in rainy tropical and equatorial regions. Inter-tropical forests occupy only 6% of continental surface area but comprise about half of all plant and animal species on our planet. Potential consequences of man-induced biodiversity decrease are very important, mainly from ecological, genetic, agricultural and medical points of view.

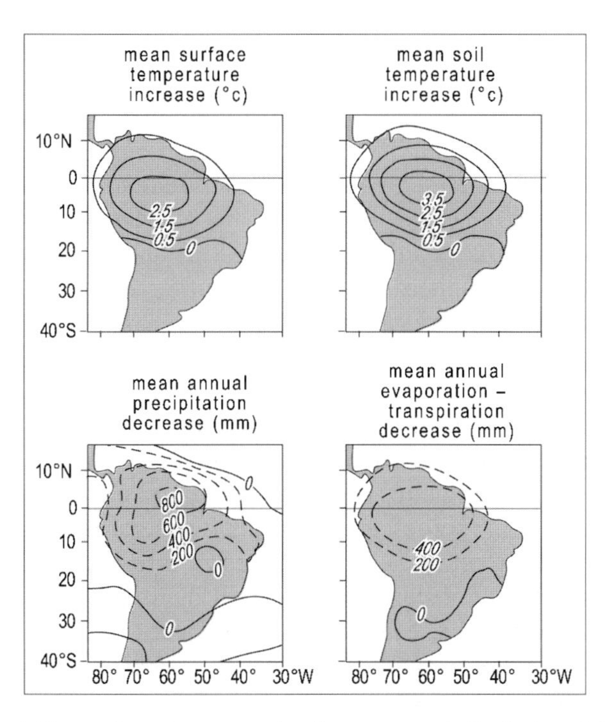

Figure 126: Modelling of environmental effects potentially induced by the Amazon forest removal (after Shukla et al.; see Allen, 1997).

8.2.3. Perspectives, Remedial Measures

Evolutionary Trends

The situations considerably differ according to forest types. *Temperate forests*, which are principally located in developed countries of the Northern Hemisphere, were essentially exploited during the 19th and early 20th centuries for meeting agriculture, urbanization and industrialization demands. Significant *reforestation* measures have been conducted during the 20th century in Europe and the former Soviet Union. For instance Iceland was almost deprived of forest at the end of the 19th century. The tree growth in this country is slow due to sub-arctic climate conditions. An extensive reforestation programme was implemented during the 20th century, which should lead to recovery, in the course of the 21st century, a wooded surface area amounting to one-third of the country. This proportion is similar to that which characterized the 17–18th centuries.

Reforestation has also been developed in Japan and North America during the last decades. The area of forest recovery in Canada has attained 5.8% during the eighties. In the USA the eastern states are currently subject to progressive reafforestation whereas north-western regions suffer active tree down-cutting; the average situation is balanced in this country. In southern hemisphere reforestation measures particularly concern New Zealand where forest surface has increased by 4.3% during the 1980s; Australia displays an almost balanced situation.

Sub-tropical to equatorial regions are continuously subject to active deforestation (Table 30), which gives cause for concern particularly in Latin America (–9% during the eighties), Africa (–8%) and southern Asia (–12%). Temperate countries of Southern Europe (south of Spain, Greece) also continue losing forests, and some arid tropical countries (Sahel region, etc.) often approach a status of almost total tree absence.

Despite their high plant regeneration potential permitted by appropriate climatic conditions, *humid tropical forests are especially threatened* by deforestation practices. This is largely due to the fact that nutritive conditions are restricted to a few decimetres thickness of topsoil close to ground surface; the thickest part of soil profiles is deprived of mineral and organic nutrients because of continuously acting leaching processes (see Chapter 8.3.2). Lateritic soils of warm, humid regions are characterized by a very thin organic surface layer that is barely cohesive, easily eroded, and very difficult and slow to be renewed. Due to the excess of deforestation practices, some low-latitude countries are presently almost devoid of real forest. This is the case in Bangladesh, Granada, Pakistan and the Philippines. Other countries such as Jamaica and Haiti (Chapter 8.1) are submitted to very high deforestation rates.

These humid, tropical regions are subject to dramatic human pressure, which necessitates particularly urgent measures for maintaining and restoring the vegetation cover.

Perspectives of Tropical Forest Restoration

Reafforestation measures are very difficult to implement in developing countries due to high demographic pressure and crucial need for food and fuel. Only a *voluntary policy* based on in-depth information and long-term anticipation may reverse the present trend. International assistance from developed countries may help to reach such objectives, mainly by means of national debt reduction. For instance 1.5 billion hectares of Amazon forest in Bolivia have recently been declared protected against forest clearance in exchange for a 650,000 U.S.dollars debt remission.

Other measures consist of developing *agro-forestry methods*: replanting of tree species characterized by high international market value, combination of farming and wooded areas, etc.

Development of economic *profit sharing and sense of responsibility* for threatened regional populations constitute other important potential tracks for restoring the forest heritage. Such an aim is on its way to becoming a success in *Niger*, one of the Sahel countries where survival mainly depends on the availability of wood for cooking. Tiger bush (*Brousse tigrée*) in Niger constitutes a natural ecosystem of small tree groves that are scattered on gently sloped surfaces. The trees collect rainwater coming both from local precipitation and from downslope flow, which allows them to grow despite the very arid climate. Bush develops by spreading slowly in the upstream direction, whereas trees progressively wither in downstream sectors, where they may be used as fire wood. Population involvement is being developed for maintaining and increasing this natural plant ecosystem, which is perfectly suited to local climate and topographic conditions. The large-scale restoration of 'brousse tigrée' allows envisaging a return to traditional livelihood conditions in a much more efficient way than by developing newly introduced plant species such as eucalyptus trees. The return to the 'brousse tigrée' cover should allow Niger to be self-sufficient in 2010 for 85% of the country's wood requirements. Similar measures soon could be extended to Mali, Chad, Burkina Faso, etc.

8.3. Soil Exploitation

8.3.1. Human Pressure Throughout History

For geologists and pedologists soils represent all types of surface formations that result from rock weathering and bear terrestrial plant life. Soils have been the object of human interest and appropriation since the time when primitive societies began to get organized (Introduction, 2). Hunting and gathering followed by land and livestock farming have determined increasingly soil appropriation at the expense of forest. As early as the first century before Christianity, Plato was bemoaning the fact that the Attica mountains and valleys in Greece had been laid bare of their forest cover, which caused soil stripping, loss of rainwater seepage, and surface water flow towards the sea; some Attica sectors were considered as being only fit for feeding the bees.

Soil exploitation and early stages of deterioration started in numerous regions as early as Mesolithic time. This is illustrated by the reconstruction of land erosion and reworking processes that occurred during the last 6,000 years in southern Sweden (Fig. 127). Changes in soil erosion and sedimentation yield have resulted from various causes that were mainly due to man activities: intensity of plant clearance and tree felling, types of cultivation practices (e.g. scattered versus dense planting), population growth periods, human migrations, epidemics, armed conflicts, etc.

Changes in soil exploitation practices and environmental impact have become particularly important in developed countries since the end of the 17th century. These changes are closely related to population growth and progressive industrialization (Barnier, 1992):

- last stages, in the 17th century, of a feudal regime that was characterized by extensive farming, fallow land decline and abundant farm labour;
- transitional regime, during the 18th century marked by active use of all available farmland, and farm labour becoming progressively in excess;
- 19th century industrial revolution associated with rationalization and intensification measures on cultivated land, and initiation of rural depopulation;
- second leap in industrial development, at the 19th–20th century transition: intensive farming, mechanization of soil exploitation, increase in rural depopulation, and beginning of over-production;
- rapid rise during the 20th century in the loss of balance between different interconnected parameters: intensive farming, large-scale agricultural mechanization and fertilizer use, systematic irrigation, intense rural depopulation and frantic urbanization, accumulation of gigantic agricultural stocks difficult to use up, increasingly unbalanced situation between developing and developed countries, in-depth disruption of world agricultural economy.

Soil erosion and deterioration spread as this evolution went on, and evolved parallel to the increase in forest clearance, exploited surfaces and agricultural production.

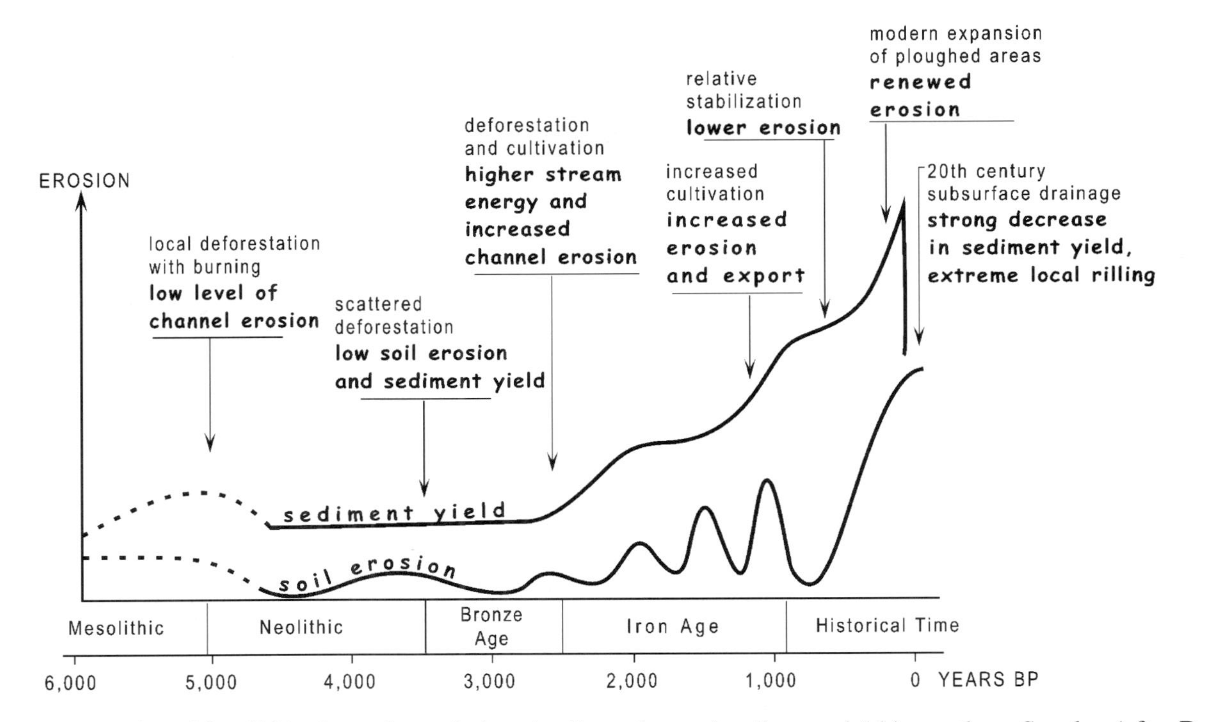

Figure 127: Changes since Mesolithic time of man-induced soil erosion and sediment yield in southern Sweden (after Dearing; see Pickering & Owen, 1997).

8.3.2. Main Soil Characteristics

Nature and Origin

Soils constitute a surface weathering blanket developing at the expense of outcropping geological formations. They result from long-lasting interactions between hard or soft surface rocks (i.e. lithosphere) and external geodynamical agents: water (i.e. hydrosphere), air (i.e. atmosphere), and living or dead organisms (i.e. biosphere). *Weathering processes lead to soil formation* by following two different and complementary ways:

- *Physical weathering* consists of the progressive, mechanical disintegration of outcropping rocks, which diminishes particle size without appreciable chemical change. Physical weathering is principally exerted in sloped regions marked by strong temperature and/or humidity changes, and comprises different modalities:
  - rock surface exfoliation due to periodic differential contraction through cooling during night relative to day, or during winter relative to summer;
  - rock rupture and dissociation due to water freezing and ice expansion within cracks and fissures;
  - void expansion due to salt crystallization during desiccation stages;
  - water and wind erosion;
  - gravity-driven rock fall, ground disruption by animal activity (grazing, dwelling, burying, etc.).
- *Chemical weathering* results in the progressive destruction of rock minerals. It acts in most surface environments and is of major importance in soil-forming processes. Chemical weathering often leads to generating new mineral species, especially *clay minerals*. It determines the export of various chemical elements that are dissolved in water and may combine to form new soils downstream from initial weathering zones. Chemical weathering principally develops under *leaching* processes controlled by water action. Water action is responsible for both mineral solution and dissolved matter transportation. Mineral solution by water (i.e. *hydrolysis*) is enhanced by several other weathering factors that differ according to local conditions: salt, carbonic acid (HCO_3), plant-derived organic acids (humus, roots), oxidizing products (e.g. iron-bearing compounds), and bacterial strains leading to chemical attack and dissociation.

Chemical weathering develops primarily at the expense of the more soluble minerals (e.g. salts, carbonates), but is responsible for potentially dissolving all mineral groups, including silicates, silica, alumina, etc. Fissures and cracks, small grain-size, and high porosity facilitate rock weathering. Optimum climatic conditions comprise heavy and long-lasting rainfall, high temperature and good drainage conditions, which aid continuous renewal of dissolving and transporting water.

According to the nature and intensity of weathering factors, resulting soils range from millimetre-thick surface *coating* or flaking to little-evolved *weathering complexes* (e.g. granite sand or arenite, the thickness of which ranges from a few decimetres to metres), and finally to *well-structured soils.*

Typical soils comprise several superimposed zones called *horizons* that have progressively developed at the expense of parent rocks (R): weathering horizon (C), marked by fissured and fragmented rock pieces; accumulation horizon (B), where clay or carbonate, iron oxides, silica, etc. are concentrated; leaching horizon (A), rich or poor in organic matter; and topsoil organic horizon (0) formed by plant debris, macro- to micro-fauna, bacteria, etc. (Fig. 128).

The developmental stage attained by soil cover depends mainly on climate parameters: rainfall, temperature and their seasonal distribution. Additional factors include mineral nature, diversity and abundance in parent rocks, drainage conditions, rate of relief tectonic rejuvenation, etc. Soil-forming processes work over *several hundred years* to produce well-organized pedological profiles that are balanced with regard to climate conditions. Some soils result from weathering processes that have acted over thousands and even millions of years. The thickness above parent rock of well-developed soils varies from a few

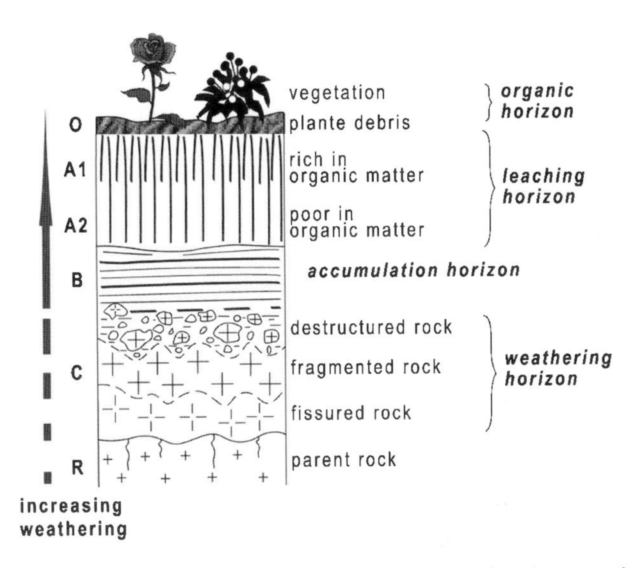

Figure 128: Horizon distribution in a typical soil submitted to active weathering processes (after Chamley, 2000).

decimetres to more than 100 m, depending on climatic and morphodynamic conditions. Best-differentiated and thickest soil blankets occur in tectonically stable regions that are subject to warm and humid climate conditions, and mainly characterize equatorial and tropical humid latitudes.

Soil Nomenclature

Several classification systems of soil types are used. They are both complex and organized into a hierarchy that is based on the main forming processes, physicochemical factors, and weathering. Soil classifications are structured in divisions, classes, subclasses, groups and subgroups, or in orders, suborders, families and series. Here we present three examples of such nomenclatures, two of which are classically used by pedologists. The third one is often employed by non-specialists.

The *French classification* comprises 12 classes that are based on the soil profile evolution and differentiation stages, the weathering and clay mineral formation types, and the basic physicochemical processes responsible for presence or absence of organic–mineral complexes (Duchaufour, 1983):

- poorly evolved soils displaying thin coating at parent-rock surface, i.e. lithosols;
- poorly differentiated, humus-bearing and desaturated soils containing abundant organic–metal compounds and detrital clay minerals: rankers developing on silica-rich rocks, and andosols forming on volcanic substrate;
- calcium- and magnesium-rich soils comprising much humus and resulting from moderate weathering: rendzines developing at the expense of limestone, and humus-bearing soils;
- iso-humic soils comprising organic matter that has stabilized through long-lasting climatic maturation, and containing a small weathered detrital clay fraction: brownish soils, chernozems;
- greyish, expansible soils rich in swelling clay (smectite minerals) and stable organic–mineral compounds, forming under warm climate marked by strongly alternating dry and wet seasons: vertisols;
- soils forming under temperate-humid conditions that have induced significant acid weathering, production of hydrated iron and barely soluble organic matter, and formation of moderately deteriorated clay minerals (slightly disorganized illite and chlorite, random mixed-layer minerals, poorly crystalline smectite, vermiculite): clay-rich, significantly leached brown soils;

- greyish soils rich in mobile organic matter and forming through active chemical weathering in acid silica-rich environments, ranging from fairly high to low latitudes: podzolic soils;
- iron-, silica- and aluminium-rich reddish soils, the fine-grained fraction of which still contains many clay minerals, and which form under temperate-warm conditions: fersiallitic soils;
- soils rich in crystalline iron oxides and significantly weathered detrital clay minerals, which form under warm-temperate conditions: ferruginous soils;
- iron- and aluminium-rich soils resulting from intense weathering under warm and humid climates and marked by a silica-depleted clay fraction (kaolinite): ferrallitic soils;
- hydromorphic soils developing in the presence of a permanent or long-lasting water layer at ground level, determining local oxidation–reduction processes and little clay-mineral deterioration: gleys, pseudogleys;
- salty and sodium-bearing soils that depend on the presence of sodium minerals and solutions: saline and alkaline soils.

The *American classification* comprises 11 orders, which are based on soil-forming processes and on the specific characters of leaching (A) and accumulation (B) horizons (Brady, 1990):
- entisols: thin, very poorly developed soils devoid of differentiated zones and with an ochreous surface. Cold climate;
- inceptisols: poorly differentiated soils showing a thin A horizon and the preliminary formation of an accumulation horizon (B). Ochreous to brownish surface, and the presence of significant amounts of organic matter. Humid climate in forest environment, under various latitudes;
- mollisols: dark-coloured soils saturated with alkaline chemical elements (Ca, Mg), rich in organic matter, sometimes with a clay- or salt-containing horizon. Meadows developing under semi-arid to sub-humid climate;
- alfisols: moderate base saturation, fairly high iron and aluminium content, rather thin A horizon and thick B, clay-rich horizon. Mid-latitude soils forming in forests under temperate-humid conditions;
- ultisols: red-yellow to reddish-brown soils characterized by strong weathering of parent rock, low base saturation (less than 35% cations), and thick, clayey B horizon. Warm and humid subtropical to tropical climate;
- oxisols: highly-oxidized A and B horizons (Fe, Al), thick B horizon characterized by silica-poor clay (kaolinite) and iron oxides often concentrated to form a crust. Warm and humid inter-tropical climate;
- vertisols: grey, organic matter-bearing soils rich in swelling clay (more than 35% smectite) that expands and contracts depending on humidity. Warm climate with short but strong humid season, and pronounced dry season;

– aridisols: ochreous surface, thin A and B horizons, occasional presence of carbonate crust and/or evaporation-forming salt. Arid to desert climate.
– spodosols: acid, humus- and silica-rich, light-grey soils with local accumulation of poorly crystalline iron and aluminium oxides. Forest environment in cool-temperate and wet regions;
– histosols: soils rich in plant debris (more than 30% of particulate organic matter) and forming on water-saturated, peaty substrate. Humid climate, poor drainage conditions;
– andisols: dark, clayey soils developing at the expense of volcaniclastic parent-rocks, and characterized by sub-amorphous clay minerals (allophane) or other chemically active minerals (e.g. aluminium compounds) that easily adsorb organic carbon or phosphorus. Rather humid climate, various temperatures.

Applying such classifications directly in the field allows detailed mapping of the soil types, transitional forms, and weathering profiles, but necessitates thorough soil knowledge. Scientists who are not pedologists often use *simpler and less precise nomenclature*, as for instance the following one that also considers the characteristic clay-mineral composition:

– lithosols: thin coating constituting the preliminary stage of rock weathering under high-latitude or high-altitude climate, as well as in desert conditions. Detrital clay minerals;
– brown soils of temperate-wet regions. Moderately altered detrital clay with transitional species (random mixed-layers, vermiculite, degradation smectite);
– laterites of warm and very humid regions. Neoformed kaolinite and hydrated aluminium oxides (gibbsite);
– podzols of humid, sandy, organic matter- and/or silica-rich environments in a cool to warm climate. Highly degraded clay, free silica;
– vertisols of warm, sub-arid regions marked by a pronounced alternation of wet and dry seasons. Aluminous-ferriferous authigenic smectite (swelling clay);
– calcretes developing on hill slopes in warm arid regions. Authigenic magnesium-rich smectite and/or fibrous clay minerals (palygorskite, sepiolite);
– gleys and pseudogleys of poorly drained hydromorphic regions under various climates. Detrital clay species;
– andosols developed at the expense of volcanic rocks that are highly sensitive to weathering processes. Authigenic clays of smectite, halloysite, or allophane types;
– uncompleted soils that are chronically eroded and reworked in tectonically active regions, under any climatic regime. Partly weathered clay minerals.

Agricultural Use

The various types of soils at the Earth's surface display a *very complex distribution that is mainly controlled by climatic factors.* Fertile soils characterized by balanced inorganic and organic composition dominantly occur in temperate-humid regions where most people of high living standards are located. Thin and organic matter-depleted soils characterize principally high-latitude, high-altitude and arid climatic regions where population density is among the lowest. Soils of tropical-humid and equatorial regions are generally characterized by a very thin organic surface horizon overlying thick iron oxide and/or clay-accumulation horizons (i.e. laterite soils): these nutrient-depleted soils occur in world regions often marked by soaring populations and a low standard of living.

The main climatic and other environmental factors controlling the soils agricultural value and exploitation combine in different ways in the principal regions of the world. Water excess or depletion, soil thickness and salt content, and presence or absence of a permanent frozen horizon (permafrost), largely depend on soil geographic distribution (Fig. 129). Soils particularly appropriate to agricultural use occur for instance in Europe, in North and Central America, and in mid-latitude Russia. The lesser productive soils mainly characterize many regions of north and central Asia, and of Oceania.

8.3.3. Soil Deterioration

Natural Causes and Anthropogenic Amplification: Consequences

Aeolian erosion Ground surfaces devoid of plant cover display noticeable roughness and are vulnerable to wind effects. Wind speed on such surfaces is potentially maximum and induces active bare soil erosion, which greatly differs from ground surfaces covered by a blanket of vegetation (Fig. 130). The presence of a forest cover determines the mechanical protection of soil surface, and therefore both an augmentation of the speed threshold necessary for particle displacement and a slowing down of wind velocity; its absence induces *accelerated erosion of loose ground* and especially of soft soils.

Erosion by wind potentially affects 550 Mha of soils at the Earth's surface. Most threatened regions are situated in subtropical Africa and on eastern Asian plateaus (Table 31). For instance, 1.3 Mkm^2 of soils and soft sediments in China are presently subject to significant aeolian erosion (Pimentel, 1993).

Water erosion Soil reworking by water mainly depends on rain intensity, slope incline and length, surface material erodibility relative to streaming strength, and the nature and density of plant cover. *Soil erosion is amplified by*

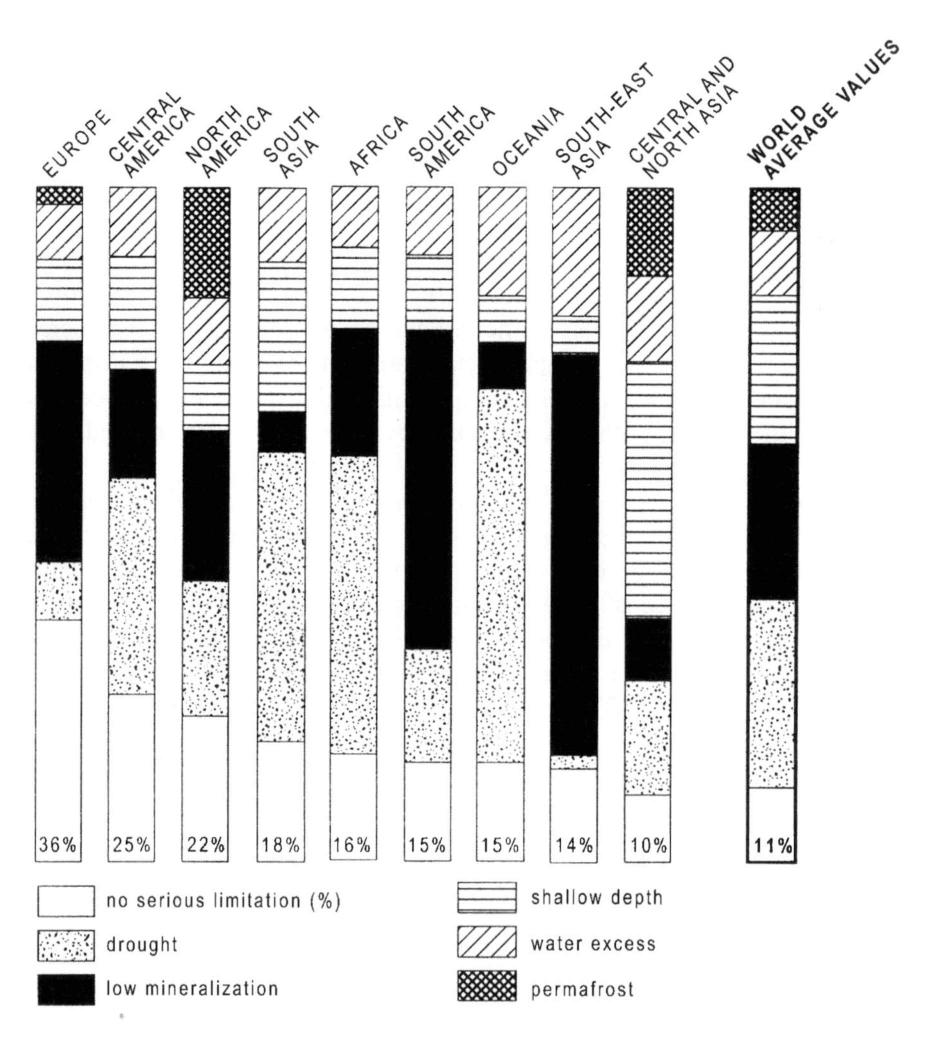

Figure 129: Distribution of the main factors limiting agricultural use in surface soils of major world regions (after Allen; see Pimentel, 1993).

deforestation and by some cultivation practices such as seasonal fallow. Diminishing the natural roughness of the ground's surface also accentuates the speed of streaming water; this is the case when stones are systematically removed from ploughed fields. Various estimates report that the reworking rate of world soils annually amounts to 0.7%. Some sloped regions subject to both tropical humid climate and active deforestation and cultivation show annual erosion rates exceeding 100 or even 1000 t/ha per year (Pimentel, 1993). This is, for instance, the case in numerous regions in Guatemala, Nigeria and southeastern Asia.

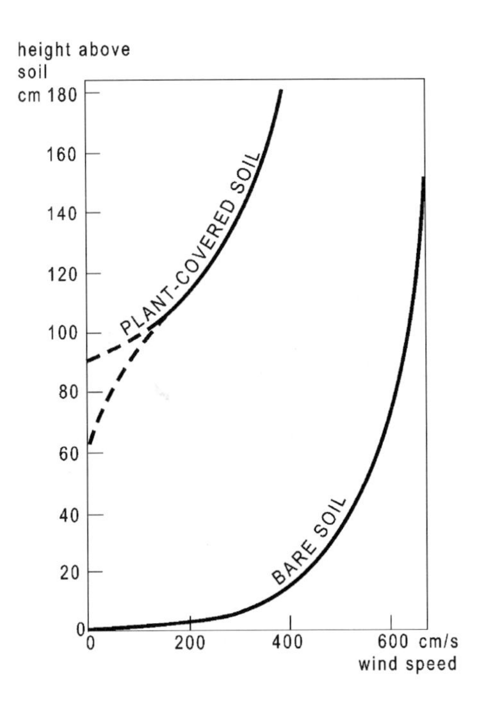

Figure 130: Comparative profiles of wind velocity and height above bare and vegetated surfaces (after Troeh et al.; see Morgan, 1986).

Table 31: Continental areas affected by accelerated soil erosion (after World Resources Institute; see Lal et al., 1995). Values in million hectares (Mha).

Continent	**Water erosion**	**Wind erosion**
Africa	227.3	187.8
Asia	433.2	224.1
Europe	113.9	41.6
North and Central America	106.6	38.8
South America	83.4	16.4
Oceania, Pacific Islands	124.1	41.4
Total	1,0885	550.1

Ground surface areas subject to intense erosion processes through water action are 2 times larger than those by aeolian erosion (Table 31). Inter-tropical Africa and southern Asian regions are particularly affected due to heavy rainfall and active deforestation. The average amounts of eroded soils are estimated to reach 40–100 t/ha per year in Africa and Asia, and only 13 t/ha per year in Europe (e.g. Germany) and 18.1 t/ha per year in North America (Pimentel, 1993).

In formerly industrialized and urbanized temperate regions of northwestern Europe (e.g. France, Germany, Great Britain), soil erosion mainly results from water action, and secondarily from aeolian processes. In these regions erosion is generally scattered and of local importance only, as shown for instance by the present situation in England and Wales (Fig. 131). Vulnerability to gully erosion and surface reworking is fairly low due to various natural and anthropogenic features: gently undulating relief and moderate slopes, traditional land division in small, hedge-bordered farm or meadow areas, frequent hedged farmland, a dense country road and path network. The agricultural soil quality in these countries tends to dwindle due to human pressure rather than to natural factors.

Salt capillary rise Groundwater rises actively in sub-arid to sub-humid regions, marked by an excess of evaporation relative to precipitation. Soil-water capillary rise aids salt concentration at ground level, which may greatly *affect agricultural potential.* Salt-bearing soils impede germination processes; salt crusts forming on the surface tend to diminish soil permeability and prevent normal water seepage. Salt minerals that commonly precipitate in dry-climate soil-surface horizons mainly comprise sodium chlorides, sulphates and carbonates ($NaCl$, Na_2SO_4, Na_2CO_3), and secondarily potassium or magnesium sulphates and potassium chlorides (K_2SO_4, $MgSO_4$, KCl). Natural salination mechanisms are often amplified by groundwater extraction and large-scale irrigation (Chapter 6.4.3).

Salt capillary-rise affects almost 7% of the world's terrestrial areas. It mainly takes effect in dry tropical and subtropical regions, but may also develop in temperate regions due to aquifer over-pumping. For instance soil salination periodically threatens large areas of U.S. High Plains and southern Canadian prairies, and develops on almost 5,500 km^2 of soils in Australia.

Environmental consequences Natural weathering and erosion mechanisms are strongly amplified by human activity. These convergent actions lead to *reworking and exporting very large amounts of minerals and organic material.* Massive soil export determines the increase in water suspended matter and turbidity, the disruption of biological productivity mainly in coastal areas, and especially considerable *sedimentary accumulation in downstream rivers, lakes and seas.*

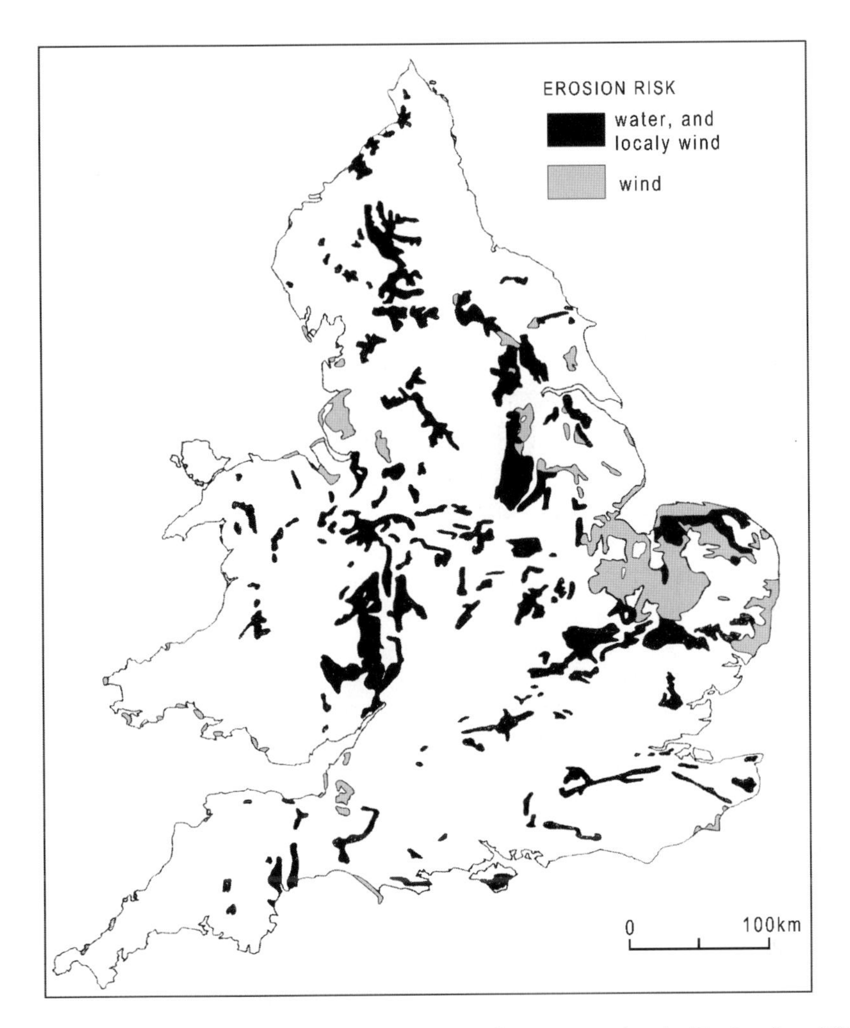

Figure 131: Areas potentially subject to wind and water erosion in England and Wales (after Morgan; see Morgan, 1986).

Exported materials comprise both particulate and dissolved compounds, the global balance of which significantly changes in soil and sediment reservoirs at the Earth's surface. This is especially the case for organic carbon, which is principally stored in terrestrial soils (ca. 1,500 billion t). Massive soil reworking leads to disruption of the dynamics of carbon and of derived products (carbon dioxide, carbonate minerals, etc.). *Potential disequilibria develop between organic carbon sources and sinks*, namely in terrestrial domains where soil erosion processes are very active: arid regions bordering the Mediterranean and in Central America, humid regions of southern Asia, deforested areas of Africa and South America, etc. (Fig. 132).

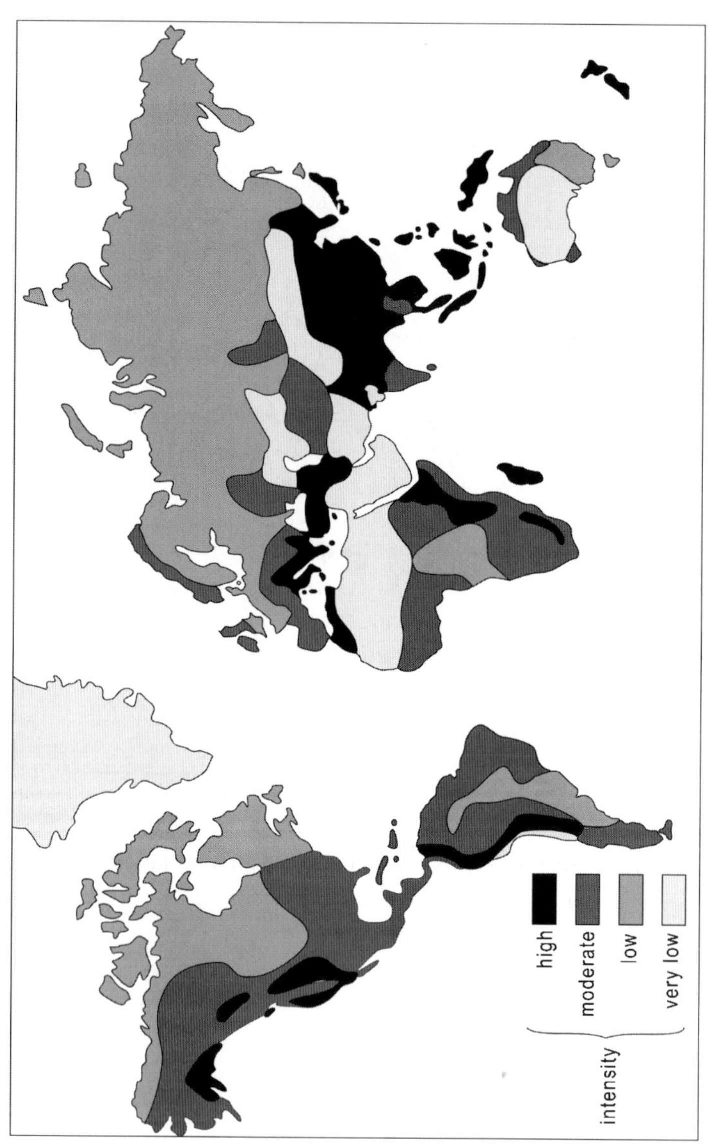

Figure 132: Correlation on a world scale between several inter-related land parameters: soil erosion, water streaming, surface sedimentary fluxes, and particle carbon export (after Lal; see Lal et al., 1995).

Strictly Anthropogenic Causes

Irrigation River exploitation and aquifer rock pumping have increased in an exponential way since the late 19th century, first in developed countries and then in developing ones. *Systematic irrigation induces major changes in the surface's natural hydrologic cycle*, mainly by increasing the amount and modifying the seasonal distribution of water supply to the soils. This causes *evaporation* to dramatically increase (see Fig. 98). Poorly controlled irrigation may determine significant deterioration of soil structure and properties: *desiccation* outside watering periods, progressive aridification, and greater sensitivity to *wind erosion*.

Various other surface environmental changes result from excessive soil irrigation: *freshwater shortage* or even exhaustion, water and soil *salination* (e.g. Aral Sea; Chapter 6.4.3), *disruption of drainage conditions* on a regional scale, etc. For instance extension work in the irrigation system of Santa Cruz valley, Arizona, has given way during the last decades to various hydrographical changes: channelisation, dam building for water storage, modification of flowing network, water course obstruction by civil engineering constructions, etc. This has substantially altered the natural drainage system, caused an increased instability of channel route, and especially caused a growing sensitivity to flooding events and subsequent soil erosion. In 1983 floods in the valley began to occur outside the normal periods calculated on a centenary scale (Rhoads, 1991). Recent investigations show that lake Chad in Central Africa has undergone a surface reduction of 94% during the last 40 years. The lake area has fallen from 25,000 to 1,350 km^2, due to a combination of excessive irrigation and rainfall decline.

Notice that over-irrigation induces increasing evaporation processes principally in surface soil horizons, which tend to become more and more salt-rich. Deeper soil horizons instead display a salt dilution due to freshwater seepage followed by capillary rise and root pumping.

Ploughing *Repeated soil working to meet cultivation needs leads to deterioration and even destruction of the vertical organization* of the superimposed weathering, accumulation, and leaching horizons (Chapter 8.3.2). Ploughing is particularly responsible for disrupting 0, A, and B upper soil horizons, and for modifying chemical weathering conditions and soil–subsoil exchanges. The lower layer of ploughed field soils get mechanically compacted and smoothed, and subsequently tends to loose permeability properties because of both rainfall and evaporation processes (Fig. 133). Resulting compaction impedes plant germination and root growth. Similar effects proceed from soil compacting through large cattle or sheep herd displacements.

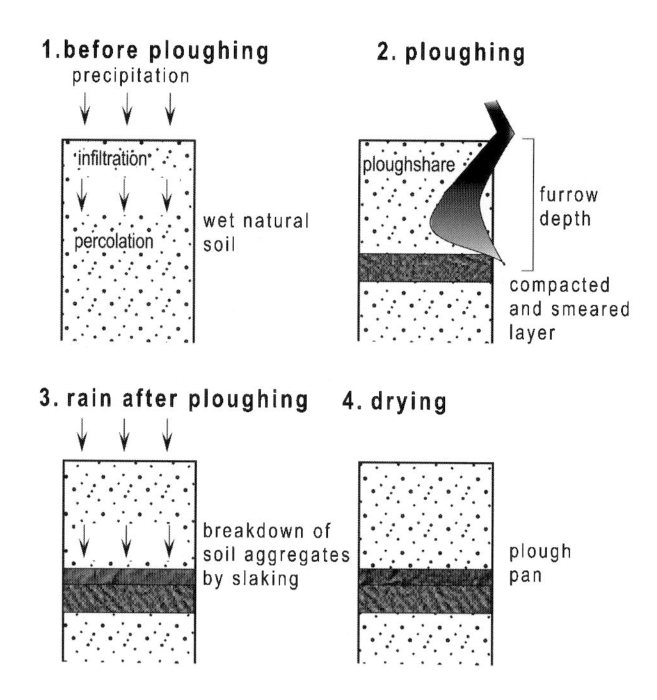

Figure 133: Physical deterioration process of cultivated soil due to ploughing work (after White et al.; see Pickering & Owen, 1997).

Plough- and other tool-induced soil deterioration has been dramatically amplified in numerous world regions during the 20th century, due to both agricultural mechanization development and ploughing depth. Such physical deterioration is especially dangerous in arid climate countries where thin, poorly cohesive and nutrient-depleted soils occur very commonly and become easily disorganized and eroded.

Slash-and-burn cultivation, overgrazing Extending the traditional practices of forest burning for getting new cultivable land leads to increasing the soil content in mineral compounds and nutrients. But at the same time soils tend to loose their original structure and organic matter compounds, and therefore become directly exposed to water and wind erosion. Resulting risks develop on a vast scale in Central and South America, in tropical Africa, and in southern Asia.

The *loss of soil cohesion*, their *fertility decrease and propensity to gravity movements* are also amplified by excessive pasturing, which causes plant-cover deprivation and root network removal. Overgrazing effects are observed in many regions, and induce especially worrying risks in sub-arid prairie and steppe regions. Large parts of the world's terrestrial areas are currently exposed to such excessive land and livestock farming practices (Fig. 134).

Fertilizers and pesticides Systematic use of fertilizers and treatment of plant diseases by insecticides and other pesticides has allowed, together with thorough plant variety selection and widespread irrigation, a *considerable increase in the productivity of cultivated land* during the 20th century. This has also permitted more regular vegetable and fruit production and related income. For instance the wheat production in France was of about 20 quintals per hectare (qu/ha) in the 1950s and reached threefold values in the 1990s. Wheat yields of 140 qu/ha have been obtained in some cereal fields from the Great Plains (e.g. Nebraska). Selection and planting of very high yield rice have allowed the eradication of serious food shortages in India. Similar results have been obtained in other Asian countries as well as in Central America.

Such progress nevertheless is often associated with over-production and useless or even wasteful competition, especially in developed countries. From an environmental point of view, *excessive farmland exploitation determines the long-term serious deterioration of the soil quality.* Fertilizers frequently contain trace amounts of arsenic and heavy metals (cadmium, lead, etc); their repeated use may cause toxicity of soils. Over-exploited soils tend to become nutrient-depleted, which may induce yield limitation or even reduction (Introduction, 4; see Fig. 6). Regular soil spraying with pesticides both diminishes the natural resistance of cultivated plants and increases resistance to parasites; more abundant chemicals are therefore used and subsequently stored in ground minerals and water. The combined and excessive use of fertilizers and pesticides may cause some soils to become practically barren and unfit for further cultivation.

Fertilizers are often difficult to adsorb by soil components; this is especially the case in nitrogenous compounds that are negatively charged, as are the clay minerals abundantly present in surface soil horizons. Nitrogen fertilizers are therefore easily leached and extracted from soil by rain or irrigation water, before they finally become fixed by plants; they may then either be converted into *nitrates* responsible for water pollution, or induce uncontrolled biological blooms in downstream river and coastal zones (i.e. *eutrophication*; Chapter 10).

Surface soil contamination by inappropriate or excessive agricultural practices sometimes extends to underground formations and may affect aquiferous rocks. This happens particularly when soils are thin and established on permeable geological formations. In western Ireland intensive farming practices sustained by the European Union have led to increasing fertilizer and

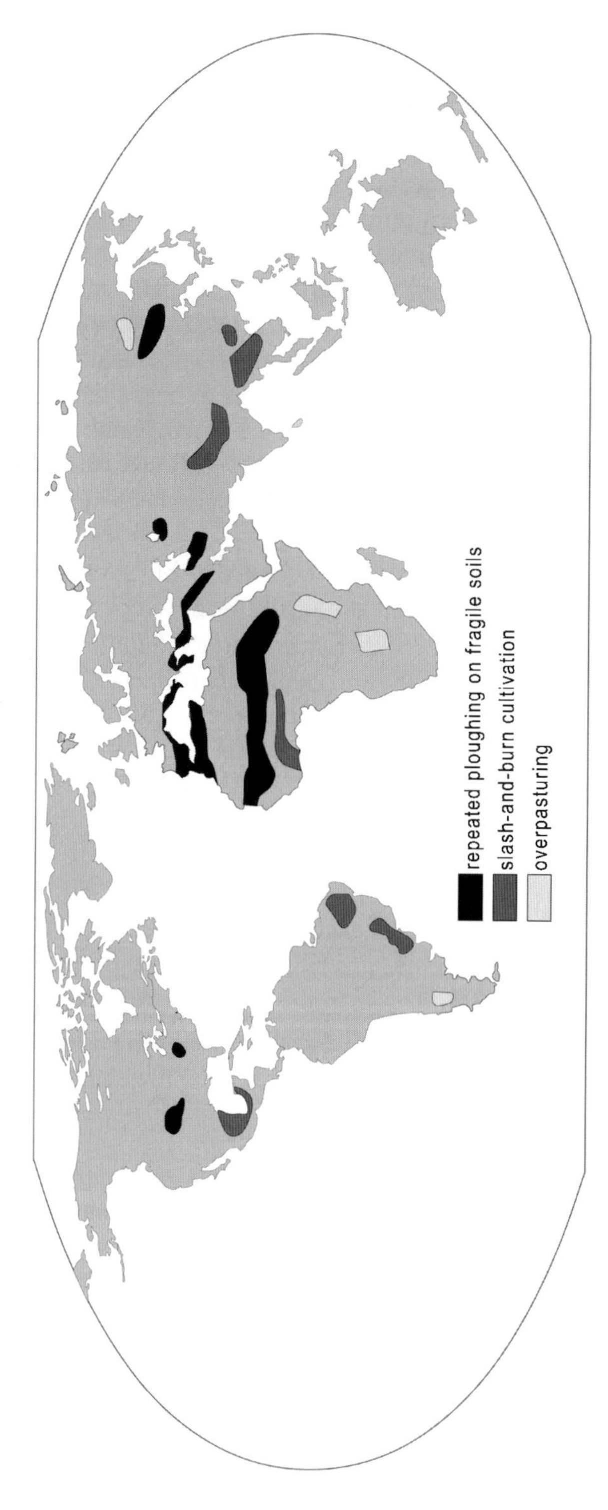

Figure 134: Main world regions affected by soil anthropogenic deterioration due to excess ploughing, slash-and-burn cultivation, and overpasturing (after Barnier, 1992).

organic pollutant leaching; these anthropogenic compounds are locally seeping from thin soil blankets down into vulnerable karstified limestone (Drew, 1995).

Acidification Soil pH lowering may be facilitated by some human actions of local to global impact:

- excess spreading of *nitrogenous fertilizers* that turn into nitrates or even nitrites;
- *acid rain* forming through the atmospheric discharge of industrial smoke rich in sulphur dioxide (Chapter 11.3).

Soil acidification develops easily in regions where bedrock formations are siliceous (e.g. granite, metamorphic rocks, silica-cemented sandstones) and free of buffering limestone. Such regions are widespread in various countries, as for instance northeastern Canada and Scandinavia. Acid soils facilitate toxic aluminium and manganese to be mobilized and incorporated in interstitial water, which may inhibit plant growth, diminish the vegetation cover, and cause increasing soil erosion.

Conclusions On a worldwide scale, *intensive soil exploitation for land and livestock farming causes altogether a surface and volume reduction through erosion, and agricultural quality to diminish through excessive use of irrigation, fertilizers and pesticides.*

The geographic distribution of soil deterioration is highly variable. Arid regions comprise thin and unproductive soils that are exposed to periodical droughts and are highly threatened by excessive exploitation: subtropical Africa, southernmost America, southwest USA, Australia, central Asian plateaus, and the Middle East (Fig. 135).

Tropical humid regions are characterized by thick but infertile soils, which are highly vulnerable due to active forest clearance and over-exploitation: Central and South American countries, sub-equatorial Africa and Asia. Soils from developed countries are generally saved from excessive anthropogenic pressure in present times, but have suffered intensive farming exploitation for several centuries. In addition they display important environmental disequilibria due to intense rural depopulation and excess agricultural fallow land.

Man-made impacts during the last quarter of a century have resulted in an *increasing reduction of forest soils for the benefit of agricultural soils*. These changes increasingly affect surface formations that are characterized by moderate, low or even very low productivity (Table 32). Soil-forming processes take hundreds and sometimes thousands of years to reach an optimum and well-balanced status. Soils therefore constitute *non-renewable resources for humans* if they become seriously damaged.

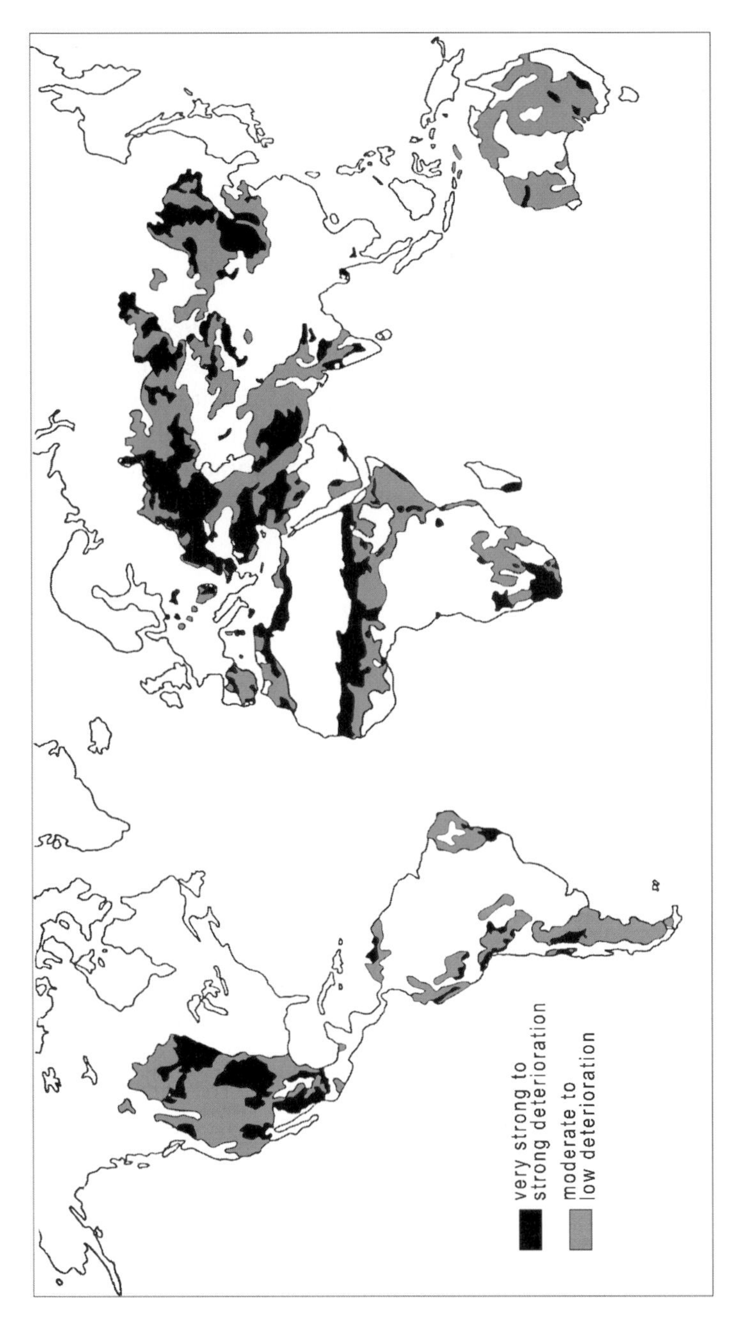

Figure 135: Worldwide distributions of arid climate soils, the deterioration of which is amplified by man's activities (after Thomas & Middleton, 1995).

Table 32: Estimates of areal changes having characterized between 1975 and 2000 the world soils exploited for land farming, livestock farming, and forestation (after Buringh & Dudal; see Pimentel, 1993). Several productivity levels are considered. Values in million hectares (Mha).

	Soil productivity				
Use	High	Moderate	Low	Very Low	Total
Cultivation					
1975	400	500	600	0	1,500
2000	345	745	710	0	1,800
Pastures					
1975	200	300	500	2,000	3,000
2000	170	320	510	2,000	3,000
Forests					
1975	100	300	400	3,300	4,100
2000	30	100	230	3,140	3,500
Other					
1975	0	0	0	4,800	4,800
2000	0	0	0	5,100	5,100
Total					
1975	700	1,100	1,500	10,100	13,400
2000	545	1,165	1,450	10,240	13,400

8.3.4. Protective Measures, Soil Restoration

Soils constitute the major nutritional basis of our planet. They should be preserved on a long-term scale for ensuring normal functioning of the biosphere. This aim necessitates progressive mitigation of soil destruction that have been exacerbated during the past century. The main objectives are diminishing erosion of soft surface formations, reducing nutrient leaching and loss, controlling surface drainage, preventing capillary salination and limiting the use of artificial fertilizers and pesticides. Various *remediation methods* exist which are readily able to induce positive results.

Agronomical measures Return to *agricultural rotation* leads to better protection of soils against water and wind erosion, allowing surface horizons to recover both organic matter and mineral nutrients, and aiding physical restructuring of the soil. A suitable method consists in alternating high-yield cultivation (e.g. cereal or tuber planting) with fallow land or leguminous planting at least every 5 years. The use of *natural manures* associated with reduced amounts of artificial fertilizers (nitrogen and phosphorus compounds) helps soils keep their structure and facilitates water seepage; such soft treatments permit both maintenance of high productivity values and limitation of acidification and particle reworking. Diverse alternative cultivation techniques lessen the erosive effects of rain, streaming and wind (Table 33): seasonal seedbed coverage, laying of plant remains and compost on the bare soil surface after harvesting (i.e. *mulching*), alternation of vegetable- and grass-ploughed furrows, or field mulching with straw before plant germination.

Erosion control Slope planting that closely follows the mountain or hill contour lines is called *contouring*. This technique induces to breaking up the streaming of flow dynamics, increasing the relief roughness, and limiting water loss and wasting. *Terrace cultivation* is a traditional procedure that consists of reshaping sloped terrains into successive horizontal planted planes and sub-vertical stone low walls; it keeps soil, water and fertilizers on cultivated surfaces, resisting streaming, and aiding rainwater seepage. Tree *hedges*, stone walls and windbreak networks facing the prevailing wind also help to prevent aeolian erosion (Table 33).

Terrain restoration following active gully erosion Arresting the deterioration of hillsides full of ravines necessitates setting up water gutters and drainage pipes, and replanting the sloped surfaces. Such actions may help to prevent further erosion on a permanent basis. Alternate solutions consist of diverting the water route and collecting rainwater within artificial pipe networks whether they are buried or not.

Table 33: Effect of soil treatment and preservation measures on particle detachment (D), and transportation (T) according to different erosion factors (after Voetberg, Morgan; see Morgan, 1986). Effect intensity: + high, = moderate, – low to nil.

Practice	**Control over**					
	Rainsplash		**Runoff**		**Wind**	
	D	**T**	**D**	**T**	**D**	**T**
Measures						
Agronomic						
covering soil surface	+	+	+	+	+	+
increasing surface roughness	–	–	+	+	+	+
increasing surface depression storage	=	=	+	+	–	–
Soil management						
fertilizers, manures	=	=	=	+	=	+
subsoiling, drainage	–	–	=	+	–	–
Mechanical measures						
contouring, ridging	–	=	=	+	=	+
terraces	–	=	=	+	–	–
shelterbeds	–	–	–	–	+	+
waterways	–	–	–	+	–	–

Table 34: Effects of low-cost soil conservation techniques on erosion decrease and crop yield increase (after Doolette & Smyle; see Middleton, 1999).

Technique	**Decrease in erosion %**	**Increase in yield %**
mulching	73–98	7–188
contour cultivation	50–86	6–66
grass contour hedges	40–70	38–73

Conclusions *Protective measures often considerably reduce soil erosion rates*, as well as maintaining or even increasing agricultural yields (Table 34). The techniques employed are often of moderate cost and correspond to more or less traditional practices. They basically imply diminishing intensive farming methods.

Encouraging results have been observed in several countries and are clearly expressed by the reduction of sediment derived from soil erosion and subsequently discharged in the river and estuary courses (Fig. 136). Important remedial effects mainly are recorded in some developed countries. For instance soil conservation efforts recently deployed in the USA reduced the mean erosion rate by 42% between 1982 and 1997. Sediment reworking from cultivated land has decreased in 15 years from 8 to 5.5 t per acre (1 acre = 40.47 ares; Uri & Lewis, 1998). Nevertheless the cost of soil loss due to excess erosion in the USA still amounts to 36.7 billion dollars per year (Uri, 2001). Consolidating such positive results and applying restoration techniques to densely populated regions characterized by thin and/or fragile soil formations requires huge additional research, assistance and education programmes.

8.4. Desertification

8.4.1. Sahel, a Potential Desert

The Sahel climatic zone comprises the regions situated between the Sahara desert and the north tropical humid zone. Sahel extends over 6,000 km from the Atlantic to the Indian Ocean, and includes various countries: Senegal, Mauritania, Gambia, Mali, Burkina Faso, Niger, Nigeria, Chad, Sudan, Ethiopia, Somalia (Fig. 137). Annual rainfall is low and restricted to a very short humid season. Precipitation decreases by about 1 mm/km from south (600 mm/year) to north (100 mm/year). Sahel soils are thin and scattered on the ground surface; aquifers are of limited extent and capacity. More than 20 million people live in the Sahel, where subsistence conditions closely depend on resources controlled by water, soil and plant availability: drinking and cooking water, land and livestock farming, wood for fuel. *The whole ecosystem is very fragile and highly dependent on climatic factors.*

Droughts tend to disrupt the environmental balance of the Sahel zone. *A long dry period of natural origin affected the whole region between 1968 and 1973*, causing wells to dry up, plants and atmosphere to dry out, and soils to be eroded actively. These changes have induced disastrous effects: thirst and famine, dysentery, innumerable deaths (more than 100,000 people), slaughter of cattle (loss of 40%, i.e. more than 3.5 million animals), disorderly southward human migrations, etc. Ethiopia and Sudan were particularly struck, the southern Sahara border advancing locally to the south by eight to several tens of kilometres in a few years. The late 1970s became wetter, and temporary rainfall return induced soil gullying and erosion. Drought periods resumed during the 1980s.

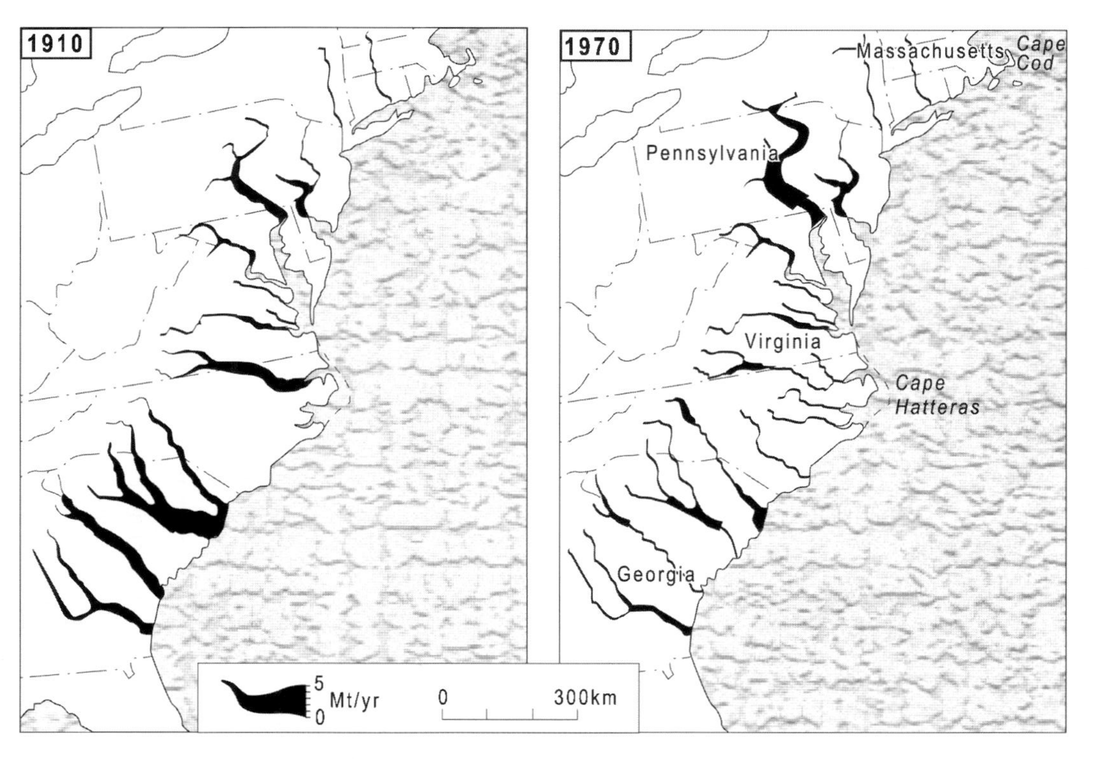

Figure 136: Reduction of sediment discharge towards USA Atlantic coasts due to soil conservation measures and cultivation practice changes (after Meade & Trimble; see Pickering & Owen, 1997). Values in million tons per year (Mt/yr).

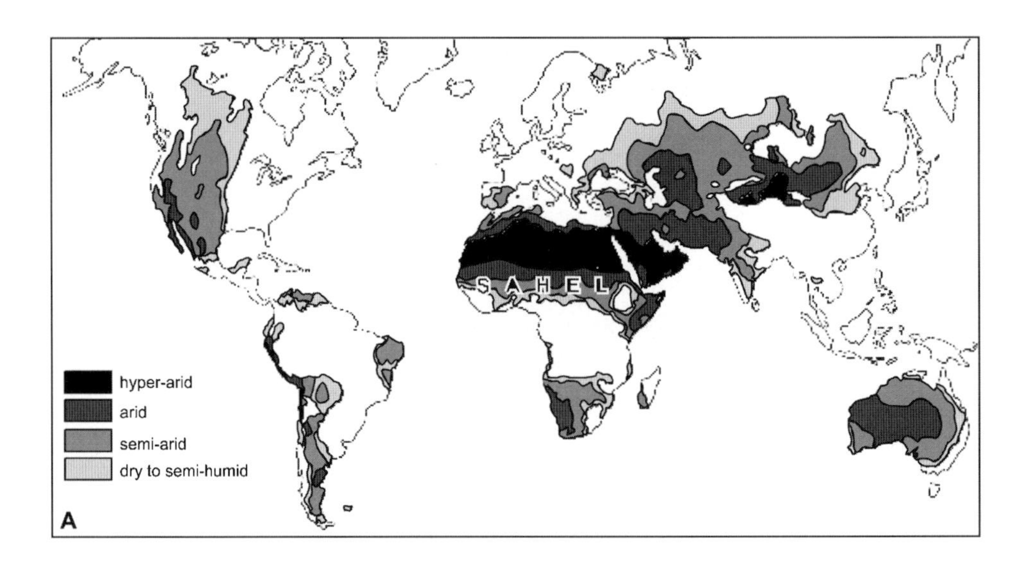

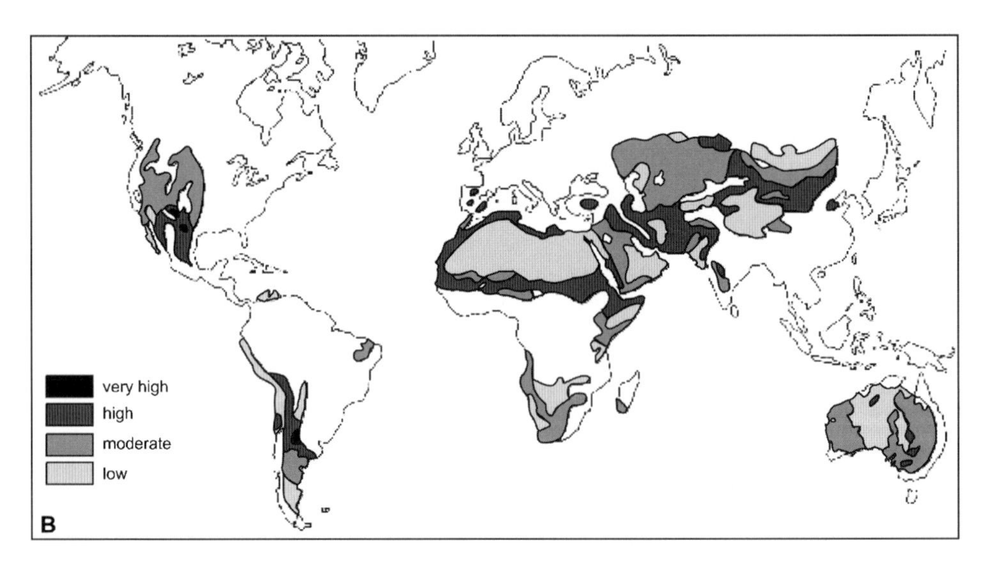

Figure 137: World distribution at the end of 20th century of, (A) desert and arid zones, and (B) regions threatened by desertification (after UNEP, and Dregne; see Thomas & Middleton, 1995).

The Sahel currently represents one of the most threatened regions in the world and suffers chronic food shortage because of *desert spreading*. By the end of 2000, Ethiopia had been deprived of rain for almost 3 years, more than 800,000 children in this country being subject to acute malnutrition. The year 2001 was a bit wetter, which allowed partial vegetation resumption.

Droughts similar to that prevailing since the end of the 1960s in the Sahel regions had not been known in past centuries, including those encompassing historical times. But a much more serious drought had struck the Sahel around 5,500 years ago, during the mid-Holocene; this nature-induced, dry climatic stage probably resulted from minor changes in Earth's orbital parameters (see IGBP, 2000).

Natural effects of Sahel drought on human subsistence and environmental balance have been considerably amplified during the last decades by strictly anthropogenic causes, which have combined in a tragic way:

- *population growth* has reached proportions that largely exceed the soil nutritional capacity. The number of Sahel inhabitants presently exceeds 20 million and increased more than twofold between the 1950s and the 1980s;
- *widespread extension of cultivation* on poorly productive and rapidly exhausted soils (see Table 32) has led to increasing artificial fertilization. Such action has favoured durable sterilization of surface formations. Permanent concern for cultivating new terrains and for increasing soil productivity by shortening the fallow periods has induced environmental situations that are incompatible with regional climatic constraints. During the last four decades, cultivated land in Niger has been extended northwards in the direction of the Sahara from latitude 15°to 16°20 north; this corresponds to a rainfall decrease from 400 to less than 250 mm/year, and to a shift from normal to very sparse bush zones;
- excessive *increase in cattle* numbers and livestock grouping in large herds have caused varied deterioration: over-grazing and plant disappearance, over-standing about and soil compacting, well and underground water exhaustion and contamination;
- *over-exploitation of wood* for fuel has exacerbated the series of deteriorating effects caused by deforestation:
  - loss of water seepage, infiltration and retention in the ground,
  - increase in soil erodibility,
  - intensification of reflected solar radiation,
  - atmospheric drying and heating,
  - rainfall decrease.

Some Sahel regions have become practically deprived of trees, mainly around the largest cities such as Addis-Ababa in Ethiopia and Khartoum in Sudan;

- *poorly controlled irrigation* has helped to exhaust aquifer reserves and concentrate groundwater salt. Subsequent salt-bearing water evaporation has led to the acceleration of soil salination, plant destruction and loss of suitability for drinking. Soils submitted to excessive irrigation have

temporarily become soaked in water, which causes nutrient dissolution and export;
- *settling* development has led to permanently exploiting both surface ground and watering places. This has impeded the soil, water and plant regeneration that was formerly encouraged by traditional nomadic habits;
- *amplification of* ethnic, religious, political and social *points of contention* has dramatically complicated the search for long-lasting solutions, and handicapped international assistance measures concerning food, water, medicine, seeds, techniques, etc.

8.4.2. Desert Expansion

From an environmental point of view desertification mainly results in deteriorating and destroying biological activity through excessive dryness. Desertification is basically a *natural and reversible process* induced by temporal rainfall fluctuations in arid regions. This phenomenon has been identified and described for different periods of Earth's geological history.

Desertification tends to be amplified by human action and has been especially exacerbated during the last century due to excessive demographic pressure. As shown in the Sahel, the desert expansion is principally illustrated by aquifer exhaustion, water and soil salination, surface erosion increase and destruction of plant cover.

Desert proper includes hyper-arid to semi-arid regions (Table 35, 1–3) and is characterized by very low surface soil productivity. Organic matter production

Table 35: Extent of different types of dryness subject to world regions (after UNEP; see Thomas & Middleton, 1995).

Climatic zone	Continent surface concerned (MHa)	Proportion relative to total exposed landmasses (%)
1 – hyper-arid (desert)	978	7.5
2 – arid (semi-desert)	1,570	12.1
3 – semi-arid	2,305	17.7
4 – dry to sub-humid	1,295	9.9
Total surface subject to desertification (2 + 3 + 4)	5,170	39.7

in desert soils is frequently lower than 0.3 kg/m^3 per year. Desert regions extend over about 30% of Earth's continental surface, and dominantly consist of two wide latitudinal belts situated around 30° in the northern and southern hemispheres (see Fig. 137, A). These regions correspond to permanent anticyclone conditions, causing partial absence of cloud cover and therefore of precipitation. Anticyclone conditions are amplified when winds dominantly blow from land towards the ocean, which determines increasing pressure, temperature and aridity on landmasses. This is the case in the Sahara and Kalahari in West Africa, of Central Australia, and of low-latitude western American landmasses. These regions constitute *warm deserts*.

Other natural desert regions are due either to a location remote from marine evaporation zones, or to a location protected by high mountain chains from long-distance humid atmospheric fluxes. These regions are labelled as *cold to temperate deserts*.

Man-induced desertification threatens especially the present desert periphery, where climate tends to get more and more arid due to ground over-occupancy and soil over-exploitation. Such human pressures affect mainly the Sahel, the Maghreb and Machreq regions around the Sahara, the Middle East, as well as South America and southwestern U.S. steppes (see Fig. 137, B). Almost 10% of world landmasses are currently subject to desert expansion (Table 35, 4), which brings to about 40% the Earth's terrestrial surfaces exposed to heavy periods of drought. More than 60 countries are involved to varying degrees. About one billion world inhabitants are facing desertification-induced famine and health problems. Desert expansion in some African countries has reached several kilometres per year, due to the long-lasting anthropogenic pressure exerted during the three last excessively dry decades. Desertification every year newly propagates more than 6 Mha. About 3.3 billion hectares of farmland in moderately humid regions are exposed to significant aridification risks.

8.4.3. Causes, Mitigation

Desertification is usually a consequence of natural and human actions that converge to cause deforestation, plant removal, soil erosion and exploitation, water evaporation and contamination (Chapters 8.2, 8.3, and 8.4.1). The world's driest regions display population growth rates that are similar to other regions, but they suffer in an exacerbated way the effects of such human pressure. More than 13% of the world population is presently living in dry regions of the Earth where they cause a very excessive exploitation of surface soils. Most current desertification factors are therefore of an *anthropogenic* origin.

All world regions where evaporation significantly exceeds precipitation display a *chain of environmental modifications that lead to accentuating aridity.* This is for instance the case in excessive livestock farming, which sets up the following deterioration chain: over-pasturing — excess trampling — loose-soil destructuring — plant-cover removal — aeolian erosion — gully-water erosion — water-table lowering — capillary rise and salt concentration — soil salination — plant-restoration inability — atmospheric drying out and temperature increase — desert propagation.

Restoration designed to limit desertification in semi-arid regions mainly consists of *reafforestation and soil conservation* (Chapters 8.2.3, 8.3.4), as well as appropriate *water management.* Positive results are recorded in various countries: tiger bush (i.e. 'brousse tigrée') restoration in Niger, agro-forestry development in Tanzania, aeolian dune stabilization through planting on some Sahara and Kalahari desert borders, etc.

Numerous measures simply consist of *re-implementing and encouraging some traditional habits* that take into account the natural low productivity and high fragility of surface formations subject to severe dryness constraints:

- size limitation of cattle herds and flocks;
- synchronous cultivation of different plant species;
- extensive farming associated with plant species rotation and frequent fallow periods;
- policy implementation for long-term balance of aquifer exploitation;
- search for more deeply located aquiferous rocks;
- soil desalination;
- wind-break installation and fight against gully erosion.

Due to the complexity and interlinked character of intervening factors most restoration solutions are of a *local applicability.* Transferring them to other regions is often difficult, especially when drought problems occur in a situation of human famine and disease. It is almost impossible to envisage large-scale and long-term environmental restoration in desert-expanding countries as long as population growth continues. This concern applies particularly to world regions where many inhabitants suffer dramatic food shortage, epidemics and disease.

A Few Landmarks, Perspectives

- World population growth is associated with an exponential increase in land and livestock farming needs, which necessitates increasing agricultural areas at the expense of forest cover. Such a drift was already initiated several thousands years ago but has attained considerable proportions since 1750 in

Europe, 1850 in North America, and 1950 in more recently developing countries. Deforestation in the early 21st century predominantly affects rather poor countries that are facing high birth rates and rapidly growing agriculture and energy needs: tropical humid regions, the dense forest of which hosts fragile, erodible and nutrient-poor surface soils; desert peripheral regions, the soil and tree blanket of which are scattered and thin. Most industrialized countries that had been deforested formerly, presently tend to have a balanced vegetation cover or even experience some reafforestation.

- Ancient deforestation events have mainly resulted from nature-induced forest fire, volcanic eruptions or accentuated drought periods. By contrast recent and current deforestation predominantly proceeds from man-induced actions which lead to disrupting local to global environments: increase in soil erosion, diminishing of organic carbon storage and water seepage properties, increase in CO_2 discharge, surface evaporation and temperature, threat on biological diversity, intensification of greenhouse effect. Such trends may be reversed in a few decades through reafforestation, provided that surface soils have not suffered irretrievable deterioration.
- Soils constitute the Earth's fragile epidermis, where the continental biological chain starts developing. They display a large array of types, thickness, components and evolutionary stages, which depend on the nature and characteristics of parent rocks, climate, slopes, drainage and tectonic activity. Soils are generally thicker and more evolved when rock minerals' hydrolysis has been longer and more intense; this corresponds to humid, warm and well-drained conditions that have prevailed durably in a stable environment. Thick inter-tropical soils nevertheless are nowhere near the most fertile, since they dramatically lack bio-available mineral and organic nutrients. Evolved and balanced soil formation takes several hundreds or even thousands of years to be completed. Deeply deteriorated or reworked fertile soils therefore constitute non-renewable resources on a scale of a few human generations.
- Soil exploitation tends to amplify some nature-induced effects: wind and water erosion of cleared ground surfaces, increased downstream particle export, disruption of organic carbon dynamics, and amplification of arid soil salination. Some strictly man-induced physicochemical changes often occur that exacerbate soil deterioration: over-compacting and destructuring, dissociation through over-pasturing or inadequate irrigation, toxic elements' incorporation, acidification, soil exhaustion and sterilization. Soil restoration necessitates long-term actions and may be implemented only if previous qualitative and quantitative anthropogenic changes have been moderate. It mainly implies returning to extensive farming practices which means a fight against massive erosion, excessive yield and artificial fertilizers, and rotating cultivation with significant fallow phases, etc.

- More than 30% of terrestrial surfaces are naturally exposed to desert conditions; this proportion fluctuates considerably in the course of time according to drought intensity and duration. Arid surfaces tend to increase considerably due to human pressure. Over-population and food shortage in desert surrounding sub-arid regions instigates a chain of severe environmental disruptions. Scattered bush clearance, poor soil cultivation, aquifer exhaustion, and human and cattle settlement cause soil denudation, drying up and sterilization. This chain ends in diminishing the soil yield and accentuating human poverty.
- Recent desert expansion mainly results from human activities and corresponds to the coupling of excess deforestation and soil exploitation practices. The fight against the deterioration of vulnerable inter-tropical and high-latitude or high-altitude soils is often of a very local, provisional, or even aleatory impact. Restoration or conservation measures are technically rather easy to implement since they are based largely on traditional habits. They are nevertheless dramatically impeded by both excessive population growth and the chronic destitution of people.

Further Reading

Brady N. C. & Weill R. R., 1999. The nature and properties of soils. Prentice Hall, 12th ed., 881 p.

Duchaufour P., 1983. Pédologie. 1- Pédogenèse et classification. Masson, Paris, 2ème éd., 491 p.

Lal R., Kimble J. Levine E. & Stewart B. E. ed., 1995, Soils and global change, CRC, Lewis Publ., 440 p.

Mainguet M., 1991. Desertification. Natural background and human mismanagement. Springer, 306 p.

Morgan R. P. C., 1986. Soil erosion and conservation. Longman Scientific & Technical, 298 P.

Pimentel D., 1993. World soil erosion and conservation. Cambridge University Press, 349 p.

Thomas D. S. G. & Middleton N. J., 1995. Desertification: exploding the myth. Wiley, 194 p.

Some Websites

http://www.isric.nl/: various international information on soil research and technology documented by the International Soils Reference and Information Centre

http://ag.arizona.edu/OALS/IALC: research and systematic measurements concerning desertification mechanisms. This site is under the auspices of the International Arid Lands Consortium

http://www.fao.org/: site under the aegis of United Nations Organization (Food & Agriculture Organization) that largely consider desertification processes, especially in tropical regions

http://www.inra.fr/: site of the main French institution in charge of soil and plant investigations

Chapter 9

Cities, Industries, and Communications

9.1. Florida Keys: A Concreted Complex in the Heart of a Natural Park

The Keys archipelago extends over 350 km from southeastern Florida as far as Dry Tortuga islands in the Gulf of Mexico (Fig. 138). The Florida Keys comprises 800 islands, forming a thin NE-SW chain from Key Largo to Key West. Most islands are small, the largest being Key Largo, with a length of 42 km and a maximum width of 2 km. The Keys chain corresponds to the relics of barrier reefs and associated deposits that grew since the Middle Miocene age over several geological periods marked by a higher sea level than presently.

Until the last quarter of the 19th century, Florida Keys formed a barren and inhospitable territory exposed to hurricanes and mosquitoes. People living on the islands comprised a few Indian communities, some farmers, fishermen and wreck pillagers. In the late 19th century, cigar factories were built and progressively became very important. Keys' tobacco factories in 1890 were the first in the world, which allowed Key West to temporarily hold the U.S. record of the largest income per inhabitant. In the early 20th century a railway was constructed that connected continental Florida to Key West. The *Overseas Railroad* expanded the islands' trade over 23 years, before the railway was destroyed by a violent hurricane in 1935. A road link was then constructed to replace the rail tracks. The *Overseas Highway* was completed in 1938. It is a 188-km-long road comprising 43 bridges that cross the multiple straits separating the islands.

The *human hold* on the Keys rapidly increased during the second half of the 20th century. The narrow island chain housed 14,000 people in 1960, and 50,000 in 2000. Almost the total land area is nowadays occupied by residences and resorts, shops, businesses and small factories, harbours and marinas, roads and parkings, airports and military bases, etc. Beaches are fairly rare, mostly private, and equipped for tourism and boating. The plant-covered areas remaining outside private property consist only of a Virginia deer reserve on Big Pine Key Island, and barely accessible and unhealthy intertidal domains occupied by

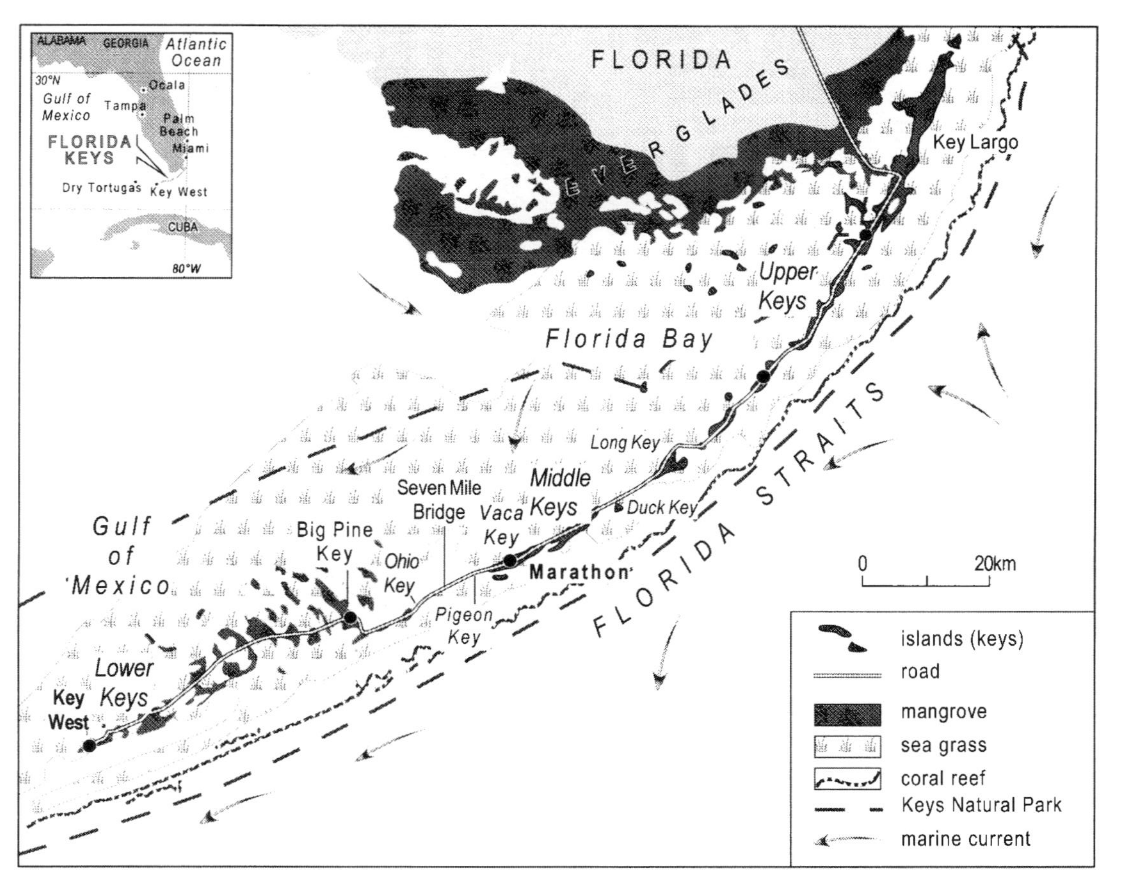

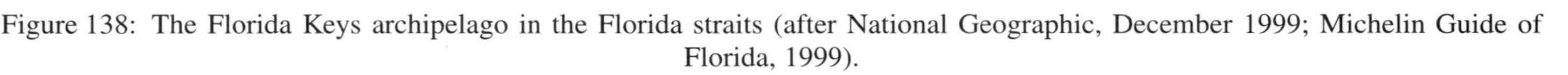
Figure 138: The Florida Keys archipelago in the Florida straits (after National Geographic, December 1999; Michelin Guide of Florida, 1999).

mangrove trees and associated vegetation. The Florida Keys have therefore changed in about one century from a status of deprived island and islet chain to that of a concreted and populated range causing much nuisance and pollution, and tending to extend towards the sea.

Beyond the coastline and below sea level, the Florida Keys are surrounded by a very large natural domain comprising *three interdependent ecosystems* (Fig. 138):

- *Mangrove* woods mainly grow on the island's northern border and at the southern edge of the Everglades National Park (Chapter 8.1). Mangrove trees are adapted to coastal hydrodynamics and the saline environment because of their dense branch tracery and aerial root system. They protect the shore against storms, aid the deposition of abundant organic matter, provide safe fauna nurseries, and host various animal groups.
- *Sea grass* is made of marine phanerogams that constitute large subtidal fields in Florida Bay and extend westward in the Gulf of Mexico. Sea-grass leaves and stems help to reduce wave energy, assist organic and inorganic sedimentation, and provide shelter for an abundant seafloor fauna: turtles, fish, crustaceans, molluscs, echinoderms, etc.
- A long *barrier reef* has developed on the Atlantic, southern side of the Keys. It borders the Florida Strait to a length approaching 320 km, at a depth extending from the lowest tide level to 25 m below sea level. It constitutes the largest coral reef domain in North America, and the third in the world after the Australian Great Barrier and Belize Barrier in the Antilles Sea. The Florida Keys barrier comprises numerous madrepore varieties that are associated with very diversified flora and fauna groups.

Maintaining the coexistence of a densely anthropogenic island complex and of particularly fragile and interreacting biosystems constitutes a difficult challenge. Various threats are posed to the Keys' marine ecosystem, which are exacerbated by both the importance of the private sector and huge financial interests: mangrove cutting to get more shore space; channel dredging in sea-grass fields; excessive fishing; seafloor damage by boats and anchors, divers and sea-penetrating devices; excess discharge of nutriments responsible for eutrophication; waste dumping responsible for threatening marine life, etc.

Until now the fragile balance existing in the Keys region between urban–domestic pressure and ecosystem preservation has been roughly maintained. This is due to the implementation of several converging measures: strict legislation and efficient control, diversified and attractive educational practices, restricted access to the most sensitive sectors, and the rather well-accepted incentive of a civic sense of responsibility.

9.2. An Increasing Demand for Artificial Ground Surfaces

The exponential growth in the human population since 1900 (see Fig. 2) has induced a dramatically increasing need for ground-surface occupancy by buildings, transport links, parking and storing zones, etc. (Introduction, 2). A growing area of Earth's solid envelopes is therefore isolated from external fluid envelopes by a thin layer made of concrete, cement, asphalt, stone and diverse other materials. This artificial blanket determines *profound modifications in exchange modalities* between underground lithosphere and the hydrosphere, atmosphere, and biosphere. The ground often becomes much more impervious, which causes *strong diminishing and/or concentrating of the vertical transfer of both water and particulate and dissolved material.*

The artificial character of terrestrial surfaces has been particularly augmented in *urban and suburban areas.* During the last decades, the rate of urban ground waterproofing has increased especially rapidly in developing countries and/or nearby coastal zones. Let us remember that the ten biggest megalopolises of our planet, each of which hosted about 10 million people in 1990, are expected to comprise 20 million inhabitants in 2015 (see Fig. 3). People living in cities represented 34.2% of the world's inhabitants in 1960, 45.2% in 1990, and almost 50% in 2000. Of the coastal population, 75% live in urbanized regions. The urban hold on ground is particularly characterized by unbridled asphalting of surface terrains: streets, ring-roads, car parks, pavements, pedestrianized districts and building roofs. It is generally associated with the reduction of natural forest and soil surface (Chapter 8), the increasing exploitation of aquifers (Chapter 6), and the discharge of growing amounts of various effluents and domestic waste.

Industrial growth during the 19th and 20th centuries has accompanied the urban growth and strongly participated in enlarging artificial surfaces. The major phases of industrial expansion have occurred at different periods and rhythms according to the countries considered (Table 36). Western European countries strongly developed their industrial capacity as early as the mid-19th century, England being ahead and displaying a particularly regular growth. North America was strongly industrialized, starting from the late 19th century, Russia from 1930, and Japan from 1960. Various developing countries started encouraging their industrial potential through subsidies during the second half of the 20th century, mostly during the few last decades.

Some industrial complexes that have been implemented during the last decades, namely in North America, Western Europe and the former Soviet Union, presently occupy huge surfaces and cause nuisance and pollution on a level that is similar to that of big cities.

Table 36: Development of industrial capacity in different countries since the 19th century (after Bairoch; see Turner II et al., 1993). Annual relative values based on a number of 100 for England in 1900. First phases of strong industrial expansion in each country indicated in bold characters.

	England	Germany	USA	France	Italy	ex-Soviet Union	Japan	China	India
1800	6	5	1	6	4	8	5	49	29
1830	18	7	5	10	4	10	5	55	33
1860	**45**	11	16	18	6	16	6	44	19
1880	**73**	**27**	**47**	25	8	25	8	40	9
1900	**100**	**71**	**128**	**37**	14	48	13	34	9
1913	**127**	**138**	**298**	**57**	**23**	77	25	33	13
1928	135	158	**533**	**82**	**37**	72	45	46	26
1938	181	214	528	74	46	**152**	88	52	40
1953	258	186	1,373	98	71	**328**	**88**	71	52
1963	330	330	1,804	194	150	**760**	**264**	**178**	**91**
1973	462	550	3,089	328	258	1,345	819	**369**	**194**
1980	441	590	3,475	362	319	1,630	1,001	**553**	**254**

The extension of transport links, which is necessary for developing and exploiting new territories, handling goods and merchandise, and serving urban and industrial districts, also attained considerable proportions during the 20th century. Roads, railway tracks, airport runways, inland waterways and sea routes have been dramatically multiplied and currently constitute an extensive surface network, the mesh of which is heterogeneous and punctuated by numerous operation centres and car parks. Innumerable examples of artificial landscapes due to communications development are recorded, which disrupt natural equilibrium and induce diverse nuisances: complex highway and motorway tracery in densely populated regions, traffic problems in big cities and surrounding districts, air pollution in poorly ventilated megalopolises (e.g. Bangkok, Bombay, Los Angeles, Mexico City), air-traffic congestion and delays, excessive tree clearance for facilitating the access to wet tropical forests, etc.

The development of river *waterways and canals* is responsible for moderate disruption in the chemical exchanges occurring between Earth's surface envelopes, but causes serious modifications in the hydrological balance (Chapter 6.4.1, see Table 23). In addition it strongly increases the anthropogenic effects on the continental surfaces. By the end of the 19th century, numerous canals were already constructed in Europe and North America; but during the 20th century the number and length of these waterways was dramatically augmented in almost all countries (Fig. 139). Strong modifications in hydrodynamic conditions, coastal sedimentary transfer, and ecosystem functioning have also been caused by digging artificial channels and by their maintenance, and by harbour construction and development.

In numerous urbanized and industrialized regions, the artificial character of terrestrial surfaces extends fairly deep underground. A very complex and diffuse occupancy of sub-soil terrains has often progressively developed down to 100 m depth. In some urban areas, three-dimensional artificial developments concern huge underground volumes that extend beneath hundreds of square kilometres. Multiple disruptions therefore occur that result from a dense network of cables, pipes and pipelines, sewage and drainage systems, deep

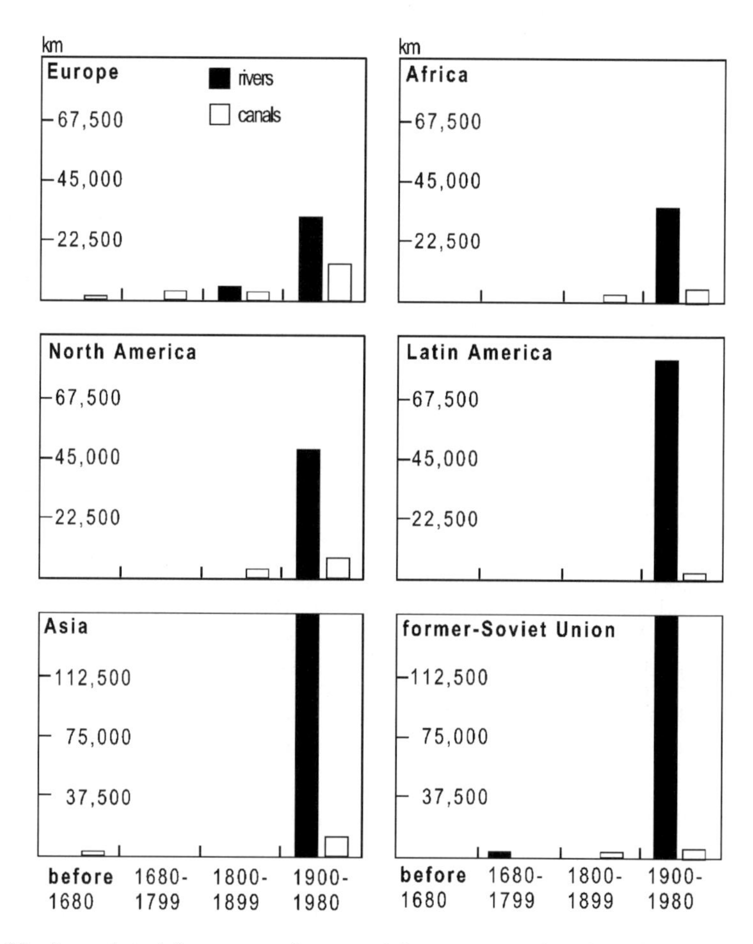

Figure 139: Length in kilometres of terrestrial waterways developed between 1860 and 1980 in major world regions (after Turner II et al., 1993).

building foundations, underground parking and commercial complexes, tunnels, etc.

9.3. Human Impact On Surface Environments

9.3.1. Covering of Natural Terrains

Urban, industrial and road influences cause surface soils and outcropping rocks to be extensively sealed by asphalt, various constructions and buildings, and other artificial materials. This tends to *diminish natural permeability* in a proportion that depends on the nature of material covering the ground surface: unpaved paths, gravel roads, asphalt, etc. (see Table 9). The occupancy is larger and the impervious surfaces are more widespread. Extensive covering of natural ground determines increasing risk of flooding and inundation (Chapter 4.1.4), as well as strong modification in hydrographical functioning and costs.

Ground surface waterproofing is mainly associated with increasing rainwater streaming and decreasing local infiltration, which result in diminishing soil humidity and local aquifer recharge. Estimates made on a worldwide scale show that cities in 1800 were hosting only 3% of the total population and therefore had almost no impact on surface water flow characters. By contrast more than one-third of Earth's inhabitants were living in megalopolises and other big cities by the end of the 20th century, causing widespread soil waterproofing and various subsequent disruptions: excessive surface streaming, rapid downwards flow of sheets of water ending in inadequately shaped river courses, and lack of underground infiltration (Fig. 140, A). The annual augmentation of surface water streaming is estimated to have averaged 163 km^3 between 1800 and 1980; the corresponding reduction of aquifer recharge has amounted to 26 km^3/year (Turner II et al., 1993).

This tendency has been amplified during the last decades, which have been marked by the concentration in urban areas of almost half of the world's population. For instance in the Moscow region (Russia), surface flow rates increase in an exponential way (up to 260%) when passing from countryside to densely urbanized sectors. In a symmetrical way, local infiltration tends to diminish down to rates that are 50% lower in downtown Moscow than in the rural environment (Fig. 140, B).

9.3.2. Aquifer Exploitation

Exhaustion of Water Reserves, Ground Compacting and Subsidence

Aquifers that get over-exploited because of excessive domestic, urban and/or industrial needs tend to run dry in densely populated regions where natural

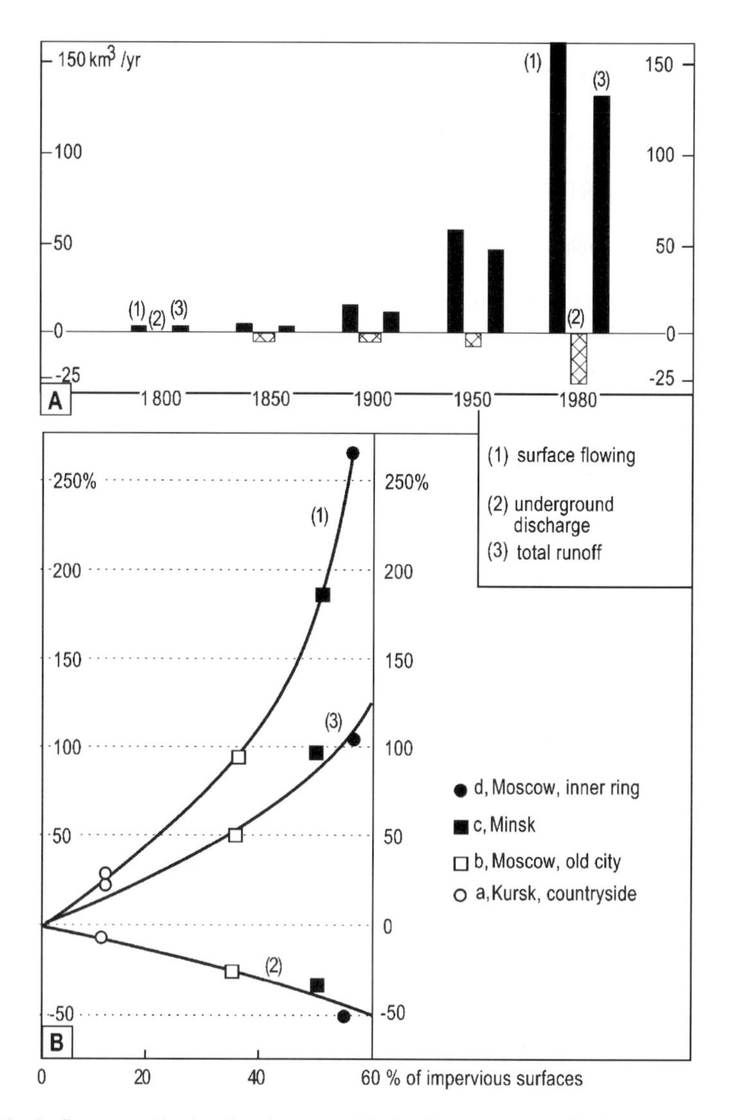

Figure 140: Influence of urbanization on: (1) the importance of rainwater streaming on surface ground, (2) the amount of water seepage in soil and underground, and (3) the amount of water flowing rapidly in the downstream direction (after Turner et al., 1993). (A) Assessment of worldwide changes having occurred between 1800 and 1980. Water amounts in km^3/yr. (B) Relative changes recorded in Russia for increasing rates of waterproofed ground, in: (a) a countryside area, and in (b) moderately, (c) strongly and (d) very densely urbanized sectors.

water recharge is insufficient. This is especially the case in rapidly growing towns submitted to arid climatic conditions such as Middle East cities (e.g.

Amman in Jordan, San'a in Yemen, Islamabad in Pakistan), or in North African conurbations, where deep fossil and/or more or less unrenewable groundwater is actively extracted. Excessive depletion of underground interstitial water causes aquiferous rocks to become over-compacted. This may induce superimposed terrains to suffer active subsidence; compacting may propagate up to the ground surface and cause morphological disruptions and building deterioration (Chapter 3.2).

Ground Over-flowing

Temporal variations in groundwater extraction rhythm or poor aquiferous rocks management may induce noticeable water table fluctuations, which predisposes to ground impregnation, pollutant diffusion and soil contamination. Such damage may result from various man-induced situations:

- pumping cessation in abandoned mines. This has happened widely during the last decades in northwestern European coal basins, in iron ore fields of Lorraine (NE France), etc.;
- seasonal diminution of irrigation water needs, because of sufficient rainfall amount. This frequently occurs in temperate-humid regions during late Spring;
- use of upstream, sometimes remotely originating water carried by pipelines, to the detriment of local aquifers that become excessively recharged. This situation periodically characterizes some urban areas of Persian Gulf borders and Egypt;
- important and long-lasting water escape from buried pipes that have deteriorated through corrosion, tectonic or seismic instability, etc.;
- wastewater discharge within subsurface aquiferous rocks;
- water accumulation in soil through pipe over-flowing or accidental sealing, or because of excessive irrigation.

Local Pollution

Urban and industrial water discharge tends to threaten the aquifer water quality especially if ground formations are porous and highly permeable due to rock fissuring, faulting or dissolution (i.e. karstification). Numerous examples of aquifer pollution in densely populated urban perimeters are regularly published in scientific journals devoted to soil and sub-soil environments (Chapter 10). In China for instance, the groundwater of more than 50 big urban regions contains excessive amounts of phenols, cyanides, mercury and chromium, or arsenides and fluorine. In India the aquifers located beneath numerous towns display excessive and even dangerous amounts of sulphates, nitrates and various heavy metals (see Maud & Eddleston, 1998). In the Nord-Pas de Calais region (Northern France), numerous sites that had been used for extracting groundwater from Cretaceous chalk or Tertiary sand aquifers were subsequently closed by the Regional Agency for Water Management (Agence de l'Eau Artois-Picardie) because of excessive chemical contamination.

Sometimes anthropogenic action leads to combining the effects of groundwater exhaustion, aquifer recharge and soil pollution, resulting in propagating chemical contamination at greater depth within the ground. This is the case in Santa Cruz, in Bolivia, a town that has experienced very rapid population growth and hosted about 700,000 inhabitants in the late 20th century. Santa Cruz is built in a region where semi-confined aquifers consist of sandy alluvial bodies intercalated with impervious clay lenses (Fig. 141). Pumping underlying

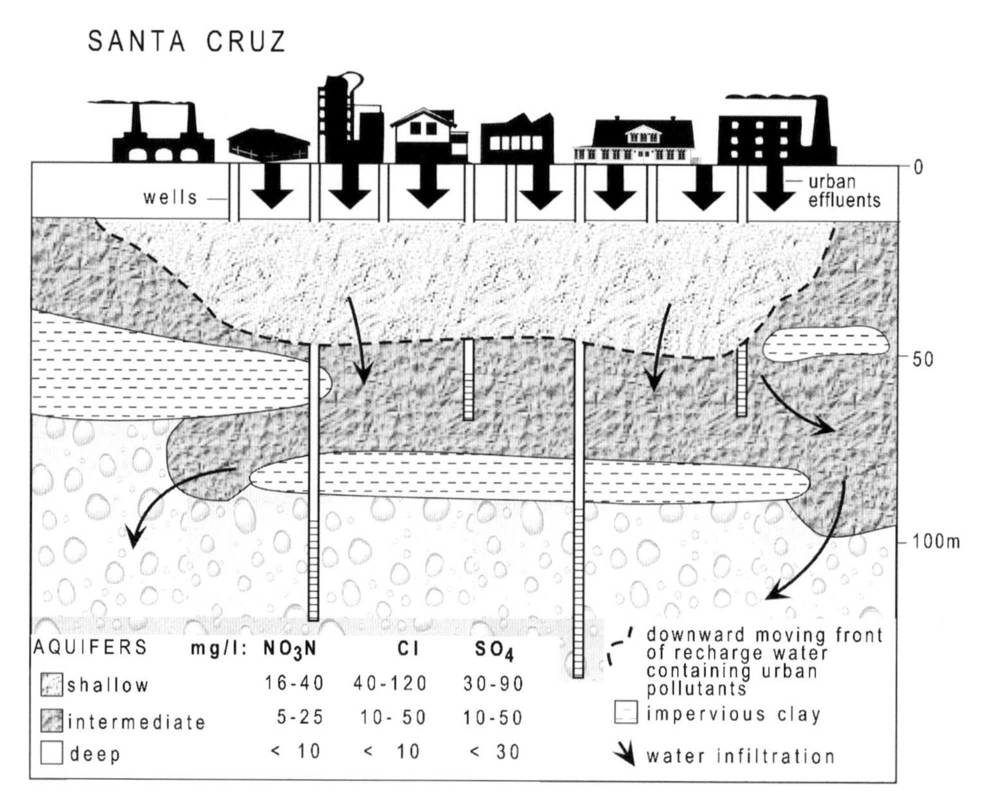

Figure 141: Pollution of Santa Cruz city's underground formations, Bolivia, due to subsurface aquifer recharge (less than 50 m depth) from deep aquifer pumping (90–500 m depth), and mixture with urban effluents seeping from ground surface (after Lawrence et al.; see Maud & Eddleston, 1993).

aquifers that are located deeper than 100 m artificially recharges subsurface groundwater. This water gets contaminated by infiltration of urban effluents and then diffuses again towards deeper ground. Anomalous levels of nitrates, chlorine and sulphates have been measured as deep as 90 m below ground level.

Gas Exchanges

Over-exploitation of porous, hard-rock aquifers augments water-free interstitial voids that may potentially be filled by gaseous products and induce unexpected secondary effects. In the Tokyo conurbation, excessive groundwater pumping during the 1960s caused the water table to drop as deep as 60 m below the soil surface. This considerable lowering caused porous ground rocks to be impregnated by air that was artificially blown by pneumatic machines for supplying oxygen to people in underground building sites. The air circulation in galleries and cavities caused active oxidation of ferrous rock components, which in turn was responsible for oxygen depletion. Some workers became asphyxiated and a significant number of deaths occurred (Hayashi & Ishii, 1989). The progressive recharge of aquifers suppressed this danger. However, excessive rock pressure resulting from re-injected water locally led to expulsion of gaseous methane that formed through rock–air exchanges, which induced unexpected underground explosions and fire during the 1970s and 1980s.

9.3.3. Local Chemical Discharge

Industrial and Urban Wasteland

Mortagne-du-Nord Local chemical contamination induced by urban and industrial activities affects various environments (i.e. air, water, soil, underground) and various mineral and organic waste products. Resulting pollution may present extremely diverse modalities, some of them constituting case studies as far as risks, environmental impact and restoration processes are concerned.

The site of Mortagne-du-Nord in northern France constitutes such an example. This rural sector hosted three factories during the 20th century. One of them was producing sulphuric acid and the two others, zinc and lead. The three plants, which are now closed, have been responsible over sixty years for the atmospheric discharge of cadmium (Cd), copper (Cu), lead (Pb) and zinc (Zn) over 3,500 ha of surrounding countryside.

Solid industrial waste removed from Mortagne-du-Nord factories was dumped on a 25-ha surface containing the same heavy metals (Cd, Cu, Pb, Zn) and a mixture of various other compounds. The industrial dump has been isolated from two local river courses (Scarpe and Détours rivers) by impervious dykes. The waste consists of 350,000 m^3 of industrial slag, sterile gangue, and various other materials, which have been covered by a protective blanket made of clay and limestone (Fig. 142). Mineralogical and geochemical investigations have revealed the presence in the waste of more than 30 different metal sulphides and oxidation products, the alteration of which has caused local ground acidification (Thiry et al.; see Schmitt et al., 2001). Most abundant sulphides include sphalerite (ZnS), galena (PbS), greenockite and hawleyite (CdS). The amount of heavy metal oxides stored in the wasteland is estimated at 6,000 t of ZnO, 2,000 t of PbO, and 400 t of CdO, the total dump volume being 200,000 t.

The industrial dump at Mortagne-du-Nord is partly soaked in subsurface groundwater that may contain up to 1,500 mg/l of zinc, 600 mg/l of cadmium, and 670 mg/l of lead, the pH being between 2.5 and 7. Fortunately this groundwater is confined, raised and isolated from deeper and unconfined aquifers (i.e. the Ostricourt Palaeocene sand) by barely permeable, clay-rich alluvia formerly deposited by the Scarpe River. Nevertheless the water table fluctuates up and down according to rainfall conditions; its level may rise during spring or autumn (see Fig. 142) to over-flowing the protection dykes. Several restorative measures have recently been proposed. The most effective probably consist of implementing "chemical barriers" that induce soil buffering and in situ precipitation of metal compounds.

Countryside surface formations outcropping around Mortagne-du-Nord include forest, meadow and farming soils which had been contaminated in a

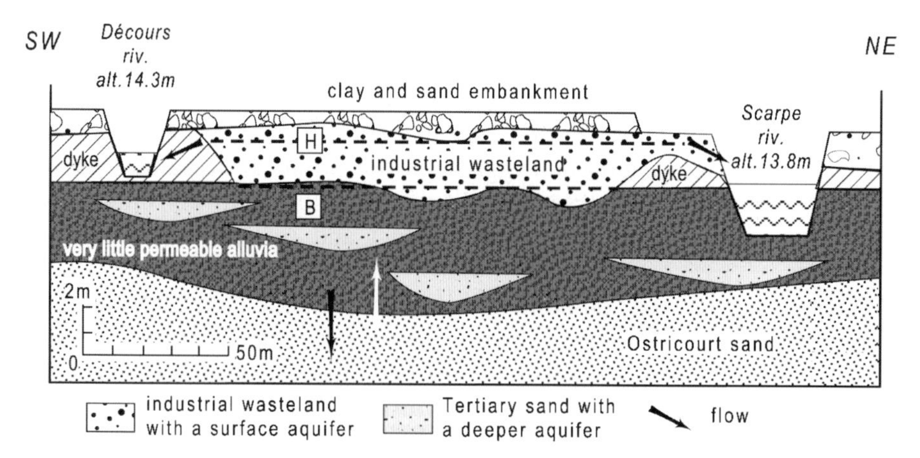

Figure 142: Section of industrial wasteland at Mortagne-du-Nord (northern France), and local hydrological functioning (after Thiry et al.; see Schmitt et al., 2001). High (H) and low (B) water table levels of subsurface groundwater are indicated.

homogeneous way by heavy metals emitted during periods of industrial activity. Each of these soil types nevertheless nowadays presents a very specific distribution of toxic metals, according to dominant farming practices that were conducted during the past decades (Fig. 143, A–C). Regularly cultivated, fertilized and calcium-enriched soils are alkaline; lead and zinc are therefore almost immobile, they do not migrate at depth and basically remain concentrated in the upper few decimetres (A). Prairie soils are actively bio-turbated by an abundant endogenous fauna (earthworms, insects, etc.) which has induced heavy metal homogenisation down to 1 m deep (B). In forested sectors, the acidic character of soils caused metal contaminants to be dissolved and leached; this was mainly the case for zinc, the amount of which has strongly decreased in surface horizons relative to lead (C); most lead compounds have been immobilized in the few upper centimetres by organic matter, with which they form complex substances.

Restoration Techniques which decontaminate industrial and urban wasteland and surrounding soils are currently subject to active research, which takes into account various parameters:

- lithology, texture, structure, mineral and geochemical characters of polluted terrains;
- migration of pollutants within the soils, to be stored or fixed by physical or chemical agents, or to be chemically transformed and subsequently dissolved or stabilized;
- influence on waste dispersal of climate, hydrological balance, biological and micro-biological activity.

Decontamination techniques are very diverse and are becoming more and more efficient (Lecomte, 1998):

- mechanical removal of polluted material through excavation pumping or washing;
- contaminant blocking by physical isolation, capsule sealing, etc.;
- chemical or electrochemical extraction;
- incineration, pyrolysis or vitrification;
- biological fixation or transformation by using natural agents or man-made treatments.

Some polluted soil-processing techniques are based on the ability of certain plant species to react with inorganic compounds. These techniques are labelled as *phytoremediation*. They are currently very actively investigated (e.g. Vangronsveld; see Schmitt et al., 2001) and are especially helpful for fighting moderate pollution or finishing decontamination processes. Phytoremediation techniques include three types of mechanisms:

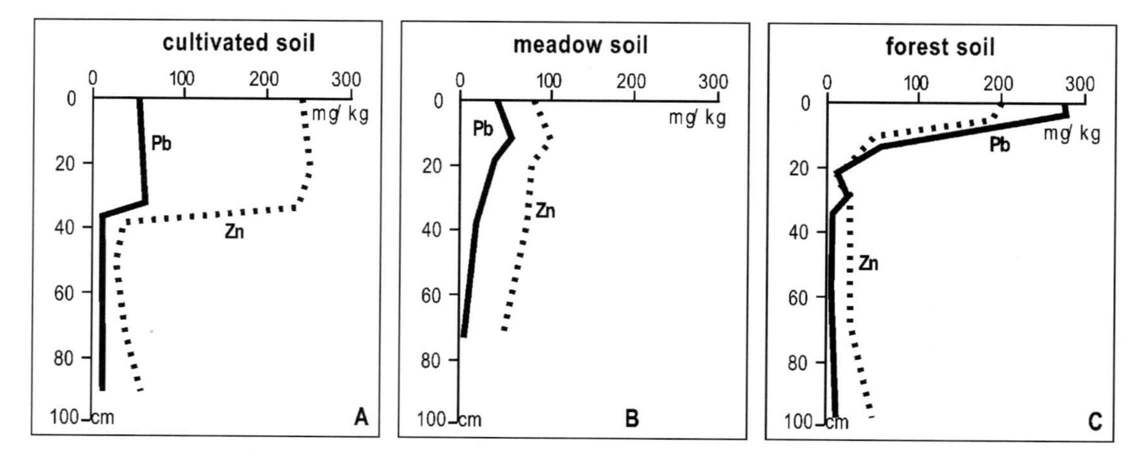

Figure 143: Lead and zinc vertical distribution in countryside soils surrounding the formerly industrial sector of Mortagne-du-Nord, northern France (after Van Oort et al.; see Schmitt et al., 2001). (A) Cultivated soil; (B) meadow soil; (C) forest soil. Values in mg/kg.

- *phytostabilization* consists of fixing pollutants within the plant root network. It is mainly employed for treating heavily contaminated soils;
- *phytoextraction* results in removing the soil pollutants by the root system and transferring them biologically towards plant aerial organs (stems, leaves). This technique is particularly applied to moderately contaminated soils;
- *phytodegradation* is based on the property of some plant species to destroy or transform some contaminants through metabolic processes.

Various tests and experiences are currently conducted for identifying plant species that may be used for adequately treating the different soil and pollutant types, determining tolerance thresholds, and defining suitable methods for definitively recovering and neutralizing the pollutants (plant weeding, cutting or incineration).

Notice that processes other than phytoremediation tend to lead to similar results. They consist for instance in using the adsorbing properties of some minerals such as *zeolites* for fixing the soil pollutants, or in employing the aptitude of some *bacterial strains* for feeding, stabilizing or neutralizing the contaminants (e.g. nitrogen, sulphur, heavy metals).

Soil and Sediment Run-off from Major Trunk Roads

Cultivated and forested soils situated nearby major road links have, during the last decades, been subject to numerous measures and experiments for quantifying the nature and intensity of traffic-induced pollution. For instance the soils located close to a road used by 3,200 vehicles per day in a forested sector of the Dortmund region, western Germany, are strongly enriched in specific pollutants emitted by car exhausts: lead and cadmium are fivefold more abundant than in soils remote from the road; chromium, nickel, vanadium and zinc display an enrichment by a factor 2–4; and polycyclic aromatic hydrocarbons (PAH) are 10–100 times more abundant (Münch, 1993). South of Paris, in France, an important enrichment of lead content is registered in pond sediments located on both sides of the north–south-oriented motorway crossing Fontainebleau forest; the heavy metal augmentation extends over 4 km to the east and over only 1 km to the west, due to pollutant transportation by prevailing easterly winds (Fig. 144).

Between Lille and Paris in northern France, the concentration of heavy metals has been measured over several years in soils, fungi, lichens and air from 12 parking and rest areas distributed along the A1 motorway, which is the most densely used in the country. All rest areas display an increase in cadmium, lead and zinc (Cuny et al., 2001). The heavy metal contamination results from intense truck and car traffic, and also from gas emission by diverse factories established in the motorway vicinity. Lead levels are progressively diminishing

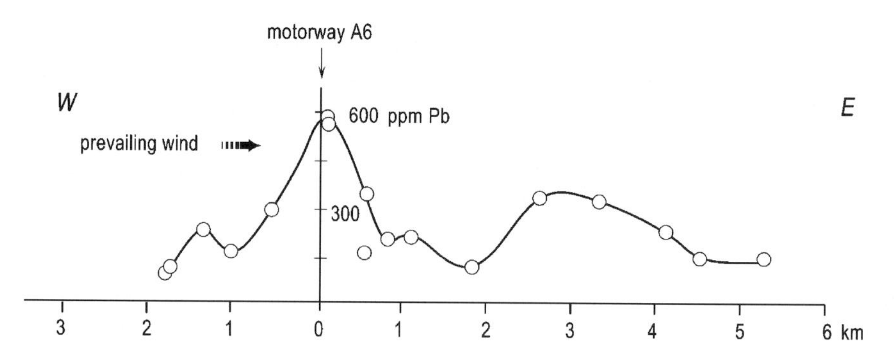

Figure 144: Distribution of lead in pond sediments of Fontainebleau forest, South of Paris, on either side of A6 motorway (after Liron & Thiry; see Schmitt et al., 2001).

due to increasing use of unleaded fuel. The absorption of heavy metals by fungi and epiphytic lichen allows the degree of pollution and its variations on a long-term scale to be estimated accurately.

Dredging Sediments

Sediment removal by dredging ensures continuous boat circulation and anchoring in rivers, canals, harbours and access channels, but results in the reworking of considerable amounts of more or less polluted materials. Sediment dredging is actively conducted in navigable river courses marked by abundant suspended or bottom-transported matter, as well as in maritime harbours located at or close to river mouths. In France the volume of sediments annually dredged is about 5 Mm^3 for the Seine estuary (Rouen city), 7 Mm^3 for the Gironde (Garonne estuary, Bordeaux), and 10 Mm^3 for the Loire estuary (Nantes, Saint-Nazaire).

Dredged sediments are stored on land, or are dumped in aquatic environments where they do not hamper boat navigation and are expected to be dispersed by currents. Sediment dumping at sea is often made at fairly shallow depths close to the coastal zone, which reduces the dredgers' rotation time and therefore the dredging cost. For instance 41% of the sediments dredged in 1980 in world maritime harbours have been dumped on continental shelves at less than 5.5 km from shorelines (Testa, 1993).

Harbour-extracted waste may contain various contaminants that are responsible for local pollution (Table 37):

- arsenic and heavy metals: cadmium, chromium, copper, mercury, nickel, lead, zinc, etc.;
- tributyltin (TBT) that constitutes a component of antifouling paints;

Table 37: Minimum and maximum concentrations of various contaminants in harbour sediments dredged between 1986 and 1993 along the three French maritime façades (after Géode; see Alzieu, 1999). Values in milligrams per kilogram (mg/kg).

Contaminant	The Channel, North Sea	Atlantic Ocean	Mediterranean Sea
arsenic	3.9–13.8	4.4–28.7	10.4–11.2
cadmium	0.5–0.95	0.27–0.64	1–1.25
chromium	38–65	37–75	56–74
copper	18–35	10–53	107–745
mercury	0.15–1.45	0.05–0.19	1.16–2.51
nickel	12–17	6–39	25
lead	36–59	41–75	93–357
zinc	105–175	60–180	274–506
PCB	0.01–0.14	0.005–0.1	0.1–0.81

- polychlorobiphenyls (PCB) that are employed as additives or primers;
- polycyclic aromatic hydrocarbons (PAH) that represent residues from hydrocarbon, coal and wood combustion;
- pharmaceutical and faecal contaminants;
- bacterial and viral strains; etc.

The risks posed by contaminant-bearing dredged sediments waste may be serious. On land they are similar to those linked to the presence of industrial wasteland. Dumping at sea may be especially dangerous if discharge is carried out at a too short a distance from shore, if marine currents tend to transport waste products back to the coast, or if the food chain concentrates toxic compounds in molluscs, crustaceans, fish, etc. Dumping of waste at sea may determine various environmental modifications (Monbet; see Gérard, 1999):

- accumulation of organic mud enriched with sulphur (H_2S, HS^-), iron and/or manganese compounds, inducing oxygen-depleted conditions on the marine floor and hampering biological activity;
- reworking of waste deposits by marine currents or swell, causing increasing turbidity, deterioration of water quality, and dispersal of toxic substances;
- excessive supply of nutrients stored in the waste (e.g. agricultural fertilizer residues or re-mineralised organic waste), determining eutrophication and anoxia;
- contaminant dispersal by marine currents, responsible for pollution far away from dumping sites.

Diverse *prevention measures* fight the problems caused by dredged-sediment storage or discharge:

- Toxicity tests are especially conducted to assess the capacity of waste dumped at sea to induce toxin diffusion and concentration in the food chain where aquaculture resources are involved (Alzieu, 1999). Tests consist for instance of evaluating the toxicity of mollusc eggs or larvae (e.g. oysters, mussels) through the study of biological developmental abnormalities: embryo growth blocking, shell anomalous shape or size, or mantle hypertrophy. The measures recommended by the ad hoc committee of Oslo and Paris (OsPar) include implementing the following experiments: (1) growth-inhibition tests on a marine algae (*Skeletonema*), (2) toxicity tests on crustaceans of Copepod (*Acartia*) and Amphipod (*Corophium*) families, and (3) toxicity tests on the juvenile population of bottom-living fish (turbot, *Cyprinodon*).
- Monitoring and modelling of dispersal conditions for dredging waste dumped at sea allow us to check the local character of the impact and to prevent potential large-scale contamination.
- Establishment of severe regulation and control rules are mainly designed to prevent dumping close to the coast or within particularly fragile or threatened ecosystems.
- Land-stored waste treatments are similar to those applied to industrial or urban wasteland: incineration, confinement, controlled dumping, chemical neutralization, etc. The management of dredging residues on land is particularly difficult to achieve adequately for various reasons that often combine: considerable waste amounts are scattered along river courses and therefore uneasy to manage; waste inventory and handling in the past have frequently been poorly conducted; subsequent leaching of barely biodegradable pollutants may occur and re-contaminate the rivers (e.g. heavy metals, hydrocarbons); transfer processes of contaminants are still incompletely known, and regulations are sometimes poorly adapted. The decision to perform new sediment dredging operations in river courses is therefore often postponed, which increases the future risk and potential damage.

9.4. Urban and Industrial Waste

9.4.1. Past and Future of Fresh Kills Urban Dump, New York City

The Fresh Kills dump in New York City was established in 1948 on Staten Island, at about 15 km southwest of Manhattan's southern edge and of the Statue of Liberty. The dump is one of the largest in the world and occupies an area of 7,500 ha. It has been the *principal waste disposal site of New York's conurbation*, where about 20 million people live. During the 1980s, more than 21,000 t of solid waste were dumped and compacted each day on the Fresh Kills site. Due to progressive implementation of both household-waste sorting and recycling techniques, the amount of waste discharged dropped to about 13,000 t/day in the 1990s. Waste-storage capacity nevertheless tended to become dramatically reduced during the last decade. In 2000 less than one-quarter of the site was still receiving additional dump material. The Fresh Kills urban waste site was therefore planned for closure in 2001. The site was temporarily reactivated at the end of that year for storing the rubble of the World Trade Center resulting from the September 11th terrorist act.

A considerable *restoration* programme has been implemented, which is supposedly allowing this part of Staten Island to recover its original aspect of wood, meadow and saline soils–mixed coastal landscape. The huge waste pile is systematically compacted and blanketed by a thick clay layer and should give way to the setting up in New York suburb of a new 100-m-high hill that will be planted with local vegetation. Urban refuse is progressively isolated from Island underground by an impervious layer of concrete, and a dense network of pipes and wells drains the dump. A treatment centre has been implemented that should decontaminate 3,800 m^3 of liquids each day resulting from dump leaching, as well as purifying and using for heating buildings 283,000 m^3 of methane produced by fermentation of waste organic matter. Restoration tests prove to be encouraging; local contamination is contained or controlled, and the re-implanted vegetation develops without apparent anomaly. The real capacity for Fresh Kills anthropogenic site to pass from the status of a huge urban dump mass to that of a nature and leisure place will nevertheless take many years before it can be realised.

9.4.2. Importance and Risks

Waste quantities Urban and industrial refuse represent, respectively, 5% and 3% of the world's total solid waste due to human activity. Most anthropogenic waste results from livestock farming (39%), land farming (14%) and mining activities (38%). Despite its moderate proportion, urban and industrial waste amounts to considerable quantities, and in addition presents two specific characters that amplify the risks. First, the refuse is concentrated over fairly limited areas, which increases local nuisance. Second, they often contain some noxious substances, the toxicity, contamination and diffusion power of which may be considerable.

In the USA about 640,000 t of urban and industrial solid waste are produced each day, which correspond to a non-compacted mass extending over a surface of 1.6 km^2 and a thickness of 5 m. Twenty-five percent of former Soviet Union inhabitants live in regions where the average concentration of toxic chemical substances is 10 times higher than recommended values.

Urban waste dumped annually on the Earth's surface represents about 900 Mt, 196 Mt of which are produced in North America, 118 in China, 114 in southern Asia, 108 in Europe, and 102 in Latin America (Barnier, 1992). Urban waste largely consists of paper, packaging, and organic matter (Fig. 145, A); various other materials are associated, the nature and concentration of which differ according to the countries, the standards of life, etc. For instance Walloon area of Belgium at the end of the 20th century was discharging about 1 Mt of household refuse each year, which represented 40% more waste than a quarter century before (i.e. the 1970s). The waste distribution was as follows: 45% of organic matter, 30% of paper and cardboard, 8% of plastics, 8% of glass, 4% of metals, and 5% of miscellaneous refuse (Berger, 1992).

The average amount of urban waste produced per inhabitant and per year considerably differs depending on the country considered: 750 km for an American, 450 for an Australian, 335 for a European, and 62 for people of Central Africa. These amounts tend to systematically increase each year in most countries, parallel to the augmentation of urban surfaces and of demand for goods by people.

Industrial waste principally comprises chemicals, paper and packaging materials, as well as food processing refuse. Other compounds are less abundant and of diverse types and risks (Fig. 145, B). For instance the factories of Ontario province in Canada are responsible on average for the discharge of 8% of cardboard and packaging, 14% of paper, 21% of wood, 11% of other plant refuses, 5% of glass, 3% of plastics, 11% of metal compounds, 2% of tires, and almost 25% of partly toxic chemicals (Murck et al., 1997).

Potentially dangerous waste discharged from worldwide industrial activity annually amounts to 500 Mt, among which 335 Mt issue from developed countries, 278 Mt from the sole North America, 30 Mt from former Soviet Union, and 21.8 Mt from European Union. The mitigation of such huge quantities of diversely noxious and toxic compounds necessitates special treatments and specific storage processes.

Underground contamination *The risk for soil and sub-soil to be contaminated* by urban, industrial, domestic or sanitary waste *mainly results from noxious substances leaching by rainwater*, and from subsequent drainage and seepage. Numerous regions in the world comprise countless waste accumulations that are poorly controlled or uncontrolled, and which may be located on fissured, fractured, karstified or simply permeable ground areas. Rainwater tends to pervade the waste, and dissolved products are transferred to aquiferous rocks directly through infiltration or indirectly via surface streaming (Fig. 146, A). If waste has been dumped on impervious (e.g. clay) or watertight indented

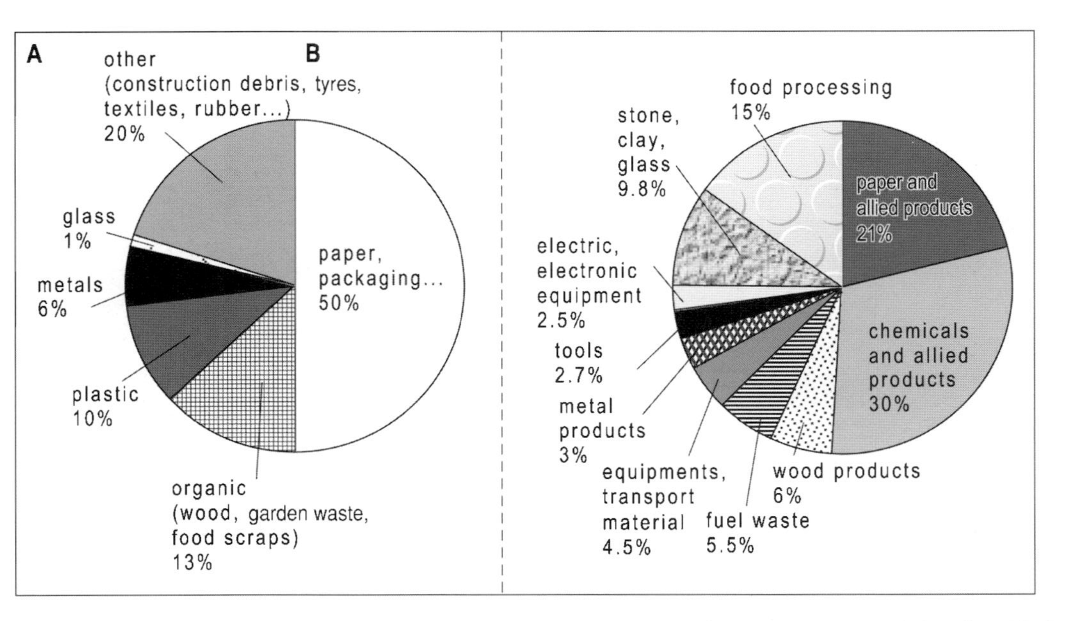

Figure 145: Proportion of main urban (A) and industrial (B) waste in the world (after Environmental Quality, National Geographic; see Montgomery, 2000).

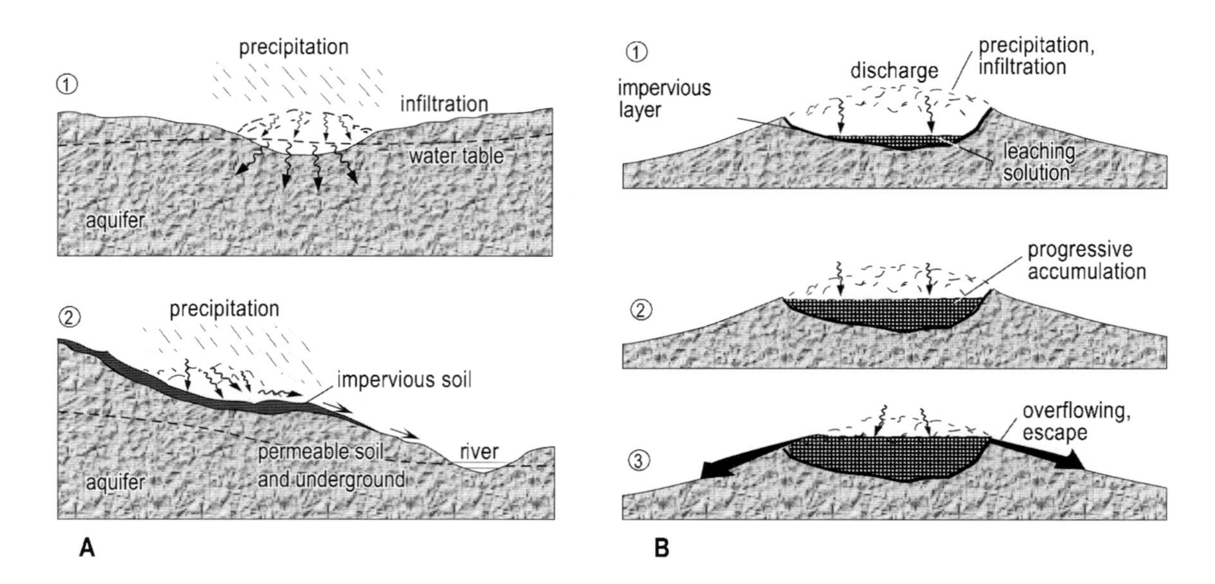

Figure 146: Examples of leachate dispersal in poorly controlled waste landfills submitted to rainwater accumulation (after Montgomery 2000). (A) Rainwater infiltration in underground aquifer (1), and in downstream soil and river (2). (B) "Bathtub effect" due to water seepage (1) and accumulation (2) above an impermeable liner, and subsequent leachate overflow (3).

terrains (e.g. plastic-film layer), rainwater tends to accumulate and finally overflow, which determines the dispersal of pollutants and their subsequent seepage in downstream areas (Fig. 146, B).

These diverse nuisances add to those arising locally from the waste mass itself: refuse fermentation and decay, vermin and bacterial development, disease spread by birds and other animals, unpleasant smells, spontaneous combustion, etc. Underground pollution due to waste leaching may induce widespread aquifer and downstream ground contamination (Chapter 10). Note that aquifer cleanup is particularly difficult to conduct, is sometimes even almost impossible, and always necessitates very long-lasting actions.

9.4.3. Waste Control and Processing

Waste Disposal

The best solution for handling solid residues produced by domestic, urban and industrial activities is to reuse them as much and as many times as possible. Appropriate efforts are presently deployed in various countries for improving waste-disposal reduction: paper, glass, metal and plastic recycling; organic-matter composting for soil fertilization; modification of manufacturing techniques for ending the production of "ultimate residues". Despite this positive tendency, *waste storage techniques* are by far still the most widely used. Waste disposal is currently employed for handling almost *two-thirds* of all residues produced as a result of human activity in the world.

Implementing a well-controlled landfill site for receiving ordinary waste is much less restrictive than for storing radioactive waste (Chapter 7), but nevertheless necessitates some specific measures and precautions, the most important of which are:

- Landfill location above an impervious geological substrate: compact clay, non-fissured and non-weathered crystalline rock, etc. If the ground is permeable, an impermeable layer made of tear-proof and rot-proof artificial liner must first be implemented. Most secure systems comprise both a natural and an artificial impervious barrier preventing the pollutants from infiltrating the ground (Fig. 147).
- Periodic waste-covering by clay or neutral material as dumping continues. Such a blanketing operation is especially important when landfill closure and final sealing time become closer. The use of clay has several advantages: it is very slightly permeable, retains heavy metals and other toxic elements through ionic adsorption, is ease to handle, and is inexpensive.

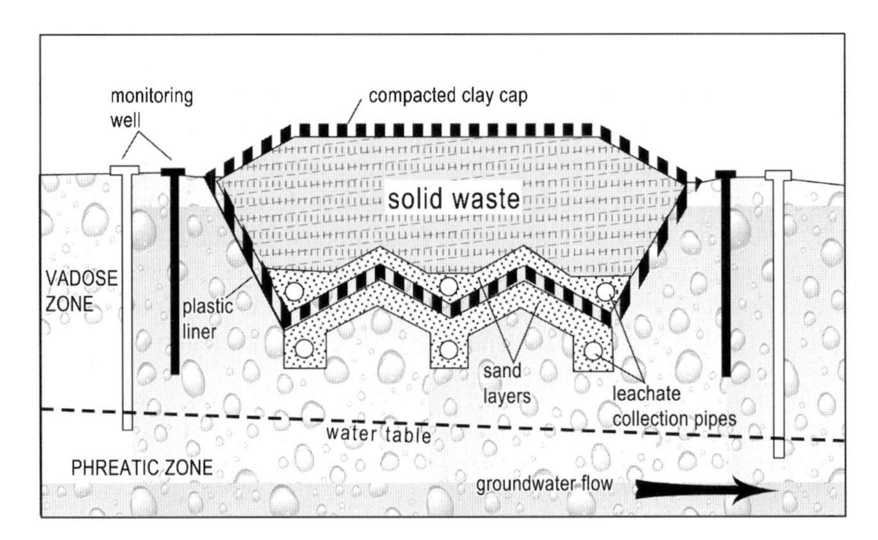

Figure 147: Cross section of an exemplary landfill site marked by a double impervious liner, a leachate collection system, and monitoring wells (after Keller 2000).

- Construction of an underground system for the recovery of seeping water. Infiltration water is then decontaminated and purified before being re-injected in the regional hydrological network.
- Implementation of a catchment system designed to recover methane and other gasses produced by organic fermentation. Some organic-rich landfills are the source of fairly abundant combustible gas. Methane is subsequently purified and concentrated, and may be used for feeding domestic or urban heating systems.
- Continuous monitoring of water table level and water quality by means of control wells established upstream and downstream to the waste reception centre.

Regions that are theoretically the most suitable for hosting extensive waste-disposal sites are characterized by both a dry climate, allowing limited rainwater infiltration, and impermeable soil and rock formations, preventing groundwater contamination. The less-favourable conditions characterize very wet regions, where ground formations are fractured, fissured or highly porous, and where landfills are close to alluvial plains, marshes, lakes or seashores.

Other Techniques

Incineration of solid waste during the last decades has tended to become increasingly used for eliminating domestic, hospital, urban and even industrial

residues. The use of incineration techniques helps to limit the amount of refuse produced in large conurbations, mainly in the developed countries. It theoretically allows reduction by 75–95% of the volume of combustible residues, which get burned within large furnaces, where the temperature is about 900–1,000°C. Residual ash and non-combustible waste are regularly removed from incineration plants and stored in strictly controlled landfills.
Incineration nevertheless constitutes a potentially polluting process because of the discharge of various gases into the atmosphere, volatile substances and flying ash (Table 38). Some of these compounds are stable and possibly toxic (e.g. dioxins), and/or may be fixed and accumulated by living organisms. Incineration techniques therefore necessitate being associated with physical filtration and chemical treatments preventing the emission of noxious volatile compounds. Notice that gaseous products discharged from incineration plants contribute to generating acid rain (sulphur compounds) and increasing the greenhouse effect (carbon and nitrogen compounds); such risks currently tend to limit the extensive implementation of incineration plants (Chapter 11).

Liquid waste is usually stored in waterproofed basins and subsequently reprocessed by *physicochemical treatments*: decantation, filtration, extraction of contaminants and purification. Deep *underground storage* is sometimes carried out, mainly by injecting toxic liquids in previously exploited hydrocarbon or groundwater fields. Liquid waste may then prevent further extension of ground compaction and subsidence (Chapter 3.2). Precautionary measures should be taken to avoid widespread ground contamination or artificial earthquakes; they include preliminary investigations and modelling experiments, or experimenting with the possibility of reversible storage.

9.5. Stone and Building Decay

9.5.1. The "Disease" of Acropolis Caryatids

The Rock of Acropolis at Athens was crowned in the 5th century BC by a Greek citadel comprising various limestone and marble buildings and walls surrounding the Parthenon. The Erechtheum temple, which was consecrated to Athena and Poseidon, constitutes a jewel of this famous architectural complex. The Erechtheum's southeastern portico is sustained by marble pillars that represent female statues called the caryatids.

In 1816 one caryatid was removed from its pedestal and transferred to the British Museum at London. In the last quarter of the 20th century the state of preservation of this statue was compared with that of the caryatids still present on the Acropolis site (Gauri, 1990). The statue stored in the British Museum was recognized as perfectly preserved, whereas those remaining in the atmosphere

Table 38: Examples of contaminants produced by urban waste incineration (after Murck et al., 1997). Furans are carbon hydrates contained in plant tars. PAH, polycyclic aromatic hydrocarbons; ppm, part per million; mg, milligrams; μg, micrograms; ng, nanograms.

Contaminants	Air without treatment	Air after treatment	Flying ash	Non-flying ash
chlorides	430 ppm	50 ppm (75 mg/m^3)		
CO_2	150 ppm	50 ppm (57 mg/m^3)		
SO_2	260 mg/m^3	260 mg/m^3		
nitrogen oxides	400 mg/m^3	400 mg/m^3		
dioxins and furans	250 ng/m^3	0.5 ng/m^3	1–1,400 ng/g	0.16 ng/g
PAH	70 μg/m^3	0.5 μg/m^3	18–5,640 ng/g	0.23–968 ng/g
arsenic	130 μg/m^3	1 μg/m^3		
cadmium	1,500 μg/m^3	100 μg/m^3	23–1,080 μg/g	18 μg/g
chromium	2,000 μg/m^3	10 μg/m^3	86–1,070 μg/g	984–3,170 μg/g
mercury	320 μg/m^3	200 μg/m^3	8–54 μg/g	2.1–3 4 μg/g
lead	34,000 μg/m^3	50 μg/m^3	1,400–26,000 μg/g	1,000–9,000 μg/g
zinc			47,000–70,000 μg/g	1,300–5,210 μg/g

of Athens had lost details in their chiselling and delicate drapery. Microscopic and chemical analyses revealed that the caryatid's marble on the Acropolis had undergone serious decay through man-induced sulphation. The calcium carbonate minerals ($CaCO_3$) on the statue surface had been partly dissolved and replaced by gypsum encrustations ($CaSO_4$, $2H_2O$), the crystals of which had propagated in rock micro-cavities to a thickness of 0.5 cm. The Antiquities Department of Athens therefore decided to remove all remaining caryatids from the Acropolis to prevent irreversible deterioration. In 1980 the statues were placed in a town museum under non-polluted air conditions.

Erechtheum caryatids in the open-air in Athens have therefore suffered significant stone decay in a few decades, whereas they had not undergone any appreciable deterioration during the 25 preceding centuries. Investigations showed that the accelerated marble decay was directly related to a dramatically increasing discharge of noxious gas in the atmosphere produced by industrial, urban and domestic activity, and was therefore indirectly caused by the exponential population growth in the Athens conurbation. The dominant noxious gas proved to be sulphur dioxide (SO_2), which was mainly released by coal and hydrocarbon combustion. SO_2 reacted with rainwater to dissolve surface limestone on the statues, which caused the precipitation of gypsum.

The deterioration of the Acropolis caryatids corresponds to one of the "stone diseases" or *stone-decaying processes* which have been identified and scientifically investigated during the few last decades, and which are partly due to dramatically increasing human pressure. The Great Sphinx at Giza in Egypt, the Taj Mahal in North India, the marble palaces at Venice in Italy, numerous European cathedrals, and many other monuments in the world give evidence for such stone decay. Stone diseases affect countless buildings of very diverse petrographical nature, geographical location and age.

9.5.2. Weathering Sensitivity of Building Stone

Stone and Mortars

Building stone is used in rough blocks or after specific shaping into dressed stone. Rock formations utilised for constructions are chosen according to various parameters: regional geology, access to quarries and mines, extraction constraints, mechanical properties and sensitivity to weathering, aesthetic concerns, value and exploitation cost, etc. The most frequently employed rock types are *limestone* and *sandstones*, the varieties of which are numerous and present diverse petrophysical properties; their resistance to weathering and decay increases if diagenetic cementation is strong and residual porosity is low.

Eruptive and metamorphic rocks are less frequently used, since they do not outcrop as commonly as sedimentary limestone and sandstones, and are often more expensive; they nevertheless often present adequate physical properties, especially if they are homogenous, compact and non-fissured, e.g. dolerite, quartzite, basalt, granite and microgranite. *Bricks* and *tiles* result from the firing of clay-rich materials; they present various degrees of resistance against weathering, which depend on mineral composition and manufacturing treatments.

Stone blocks are generally assembled with a *mortar* that may participate to further physicochemical evolution of the building. Mortar consists of a mixture of sand and binder. Binder may be of diverse types: slaked lime ($Ca(OH)_2$) made by hydrated quicklime (CaO); hydraulic lime resulting from the setting, in the presence of water, of a mixture of calcium silicates, aluminates and oxides that were fired beforehand at 1,000–1,200°C; cements made of a combination of limestone, silica, alumina, iron compounds, gypsum, etc.

Vulnerability to Weathering

Stone blocks that have been extracted from their natural underground environment and put in open-air conditions for building purposes become exposed to weathering by rain, wind, cold, heat, etc. Weathering results from natural physicochemical exchanges occurring between the lithosphere and the Earth's external fluid envelopes (Chapter 8.3.2). Building stone therefore undergoes *surface changes* that are quite normal and basically concern all rural, urban, industrial and coastal environments (e.g. Fig. 148). These changes are generally of moderate importance, but tend to be amplified if exposure to bad weather is strong, direct, and very frequent. This is the case for wall fronts facing dominant rain and/or wind, for delicate outdoor columns or statues, especially in coastal regions exposed to saline spray. The effect of weathering on building surfaces may be predicted by knowing physical properties of the stone: rock porosity and micro-porosity, coefficient of saturation by infiltrating water, degree of capillarity, etc.

Types of Stone Decay

Building stone subject to open-air conditions is potentially weathered in two different ways:

- Acquisition of a surface *patina*, the thickness of which does not exceed a few hundredths to tenths of millimetres. Patina results from the interaction of rock blocks at ground level in current climate conditions. It corresponds to a

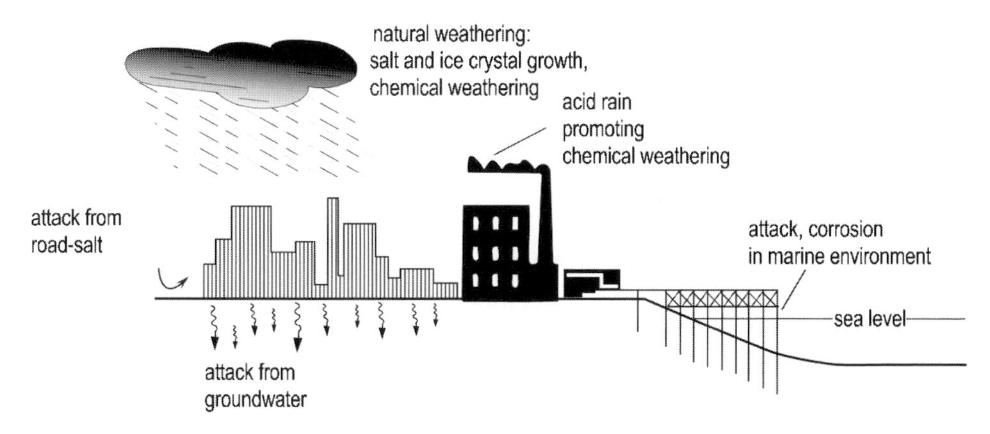

Figure 148: Stone exposure conditions to different surface environments (after Bennett & Doyle, 1997).

slowly evolving surface deterioration and helps to give stone its definitive aesthetic cachet. Patina generally consists of salt compounds that were dissolved and evaporated from natural rock cement, and which subsequently precipitated on the stone's surface. It may comprise various minerals: microcrystals of calcite precipitating on the limestone surface; gypsum forming at the expense of carbonate-cemented sandstone; mixture of calcite and clay film typical of silica- and carbonate-cemented sandstone; metal oxides (iron, manganese) responsible for the final stone shade.

Recent investigations performed on various monuments from the Mediterranean range (Spain, Southern Italy, Sicily) show that patinas often comprise several centripetal films made of organic matter, calcite and gypsum, the nature and formation of which depend on microflora activity at stone surface (Garcia-Vallès et al., 1998).

- Development of deeper *stone lesions* that are responsible for rock decay at a millimetre to centimetre scale. Lesions propagate gradually in a continuous way and may be of two types, which sometimes coexist:
  - *alveolisation* is the more frequent phenomenon encountered. It consists of the selective dissolution of most fragile rock parts: micro-cavities and micro-fractures where water circulates easily, unstable minerals of bioclastic limestone, carbonate cement of sandstones, heterogeneous and porous stone blocks;
  - rock surface *desquamation* is less widespread. It leads to removal of a thin, decayed film called "plaque". Desquamation progressively propagates into

the inner, non-weathered stone. It typically affects some sparite-cemented limestone, eruptive rocks and marble (e.g. Erechtheum caryatids; Chapter 9.5.1).

9.5.3. Natural and Anthropogenic Causes

Salt Crystallization

The migration of dissolved salt and its crystallization in rock pores and joints is responsible for most lesions affecting building stone (Table 39). Salt precipitation determines significant volume augmentation, which causes considerable increase in the internal pore pressure. This physical change helps to crush rock surface layers into micro-fragments. Salt crystallization goes on as long as some liquid remains inside the rock micro-cavities and fissures.

Table 39: Main processes of building stone decay, and resulting effects (after Honeybourne; see Bennett & Doyle 1997).

Process	Stone types	Effects
salt crystallization	limestone, sandstone	efflorescence, pore pressure increase, microfracturing
frost damage	limestone, sandstone	surface desquamation, ice growth within pore voids, expansion pressure
heating and cooling	marble	differential expansion of calcite crystals, slab bending
	granite	differential expansion of micas, quartz and feldspar causes microcracking
	limestone, calcite-cemented sandstone	differential expansion of gypsum crust leading to surface loss
wetting and drying	porous rocks	salt crystallization, expansion and contraction causes fatigue failure
acid decay	limestone, marble	calcite dissolution, precipitation of gypsum skin
	dolomite	similar to limestone but with Mg sulphate causing blistering
	sandstone	loss of calcite cement, growth of subsurface gypsum causing differential thermal expansion

Salt intervening in stone decay may result from *natural processes* such as chemical weathering of some unstable rock components (carbonates, oxides), capillary rise from soil and underground, marine spray, or bacterial reactions. Salt may also issue from *artificial sources*: concrete and cement leachates, road-salt products, pollutants derived from air or water transportation (Table 40).

Salt minerals responsible for stone deterioration mainly belong to sulphate and nitrate groups. Most frequently encountered minerals include *gypsum*, a hydrated calcium sulphate, and *saltpetre* which is a mixture of sodium and potassium nitrates. Gypsum is particularly widespread and forms either through natural oxidation of rock-bearing sulphides (pyrite, marcassite) or from reactions developing between rock carbonates and airborne anthropogenic sulphuric acid derived from fossil fuel combustion (Chapter 11).

Salt migration and crystallization depend on the *presence of water and its periodic evaporation*. Stone decay by salt therefore tends to be accentuated under sub-arid climatic conditions marked by important seasonal variations in rainfall regime and/or by strong day-and-night alternations. This is especially the case if rock porosity favours capillary rise of salted water, as described for

Table 40: Salt types potentially harmful to building stone, and main origins (after Honeybourne; see Bennett & Doyle, 1997).

Salt	**Common source**
sodium sulphate	washing powder, solid fuels, some soils and bricks
sodium carbonate	washing powder, domestic cleaning aids, fresh concrete, cement-based mortar
magnesium carbonate	some bricks, acid rain wash from dolomite
potassium carbonate	fresh concrete, cement-based mortar, fuel ashes
sodium and potassium chloride	seawater, road-salt, salt in general, soil
calcium sulphate (gypsum)	bricks, limestone and dolomite affected by acidic atmosphere, gypsum-based wall plaster
sodium and potassium nitrate (saltpetre)	soils, fertilizers

Eocene limestone, from which the Great Sphinx of Giza is carved (see Gauri, 1990).

In addition the *grain size and lithology heterogeneity of stone blocks* condition salt migration and crystallisation rate, as does the block orientation relative to water infiltration and rain and wind direction. For instance heterometric stone blocks placed in an outdoor wall allow salt-bearing water to easily penetrate the rock when sedimentary beds are arranged parallel to the water seepage direction; by contrast the water infiltration gets difficult when stone beds are placed perpendicular to the wall.

Action of Frost and Thermal Variations

Frost setting corresponds to an augmentation by 9% of liquid water volume, which induces pore pressure to increase and facilitates stone microfracturing. At low temperature the pore pressure may reach very high values (e.g. 2,200 kg/cm^2 at –22°C). *Stone disintegration* is kept going by frost and thaw alternating phases, and tends to be amplified through continuous growth of ice crystals from large to small inner rock pores. In the same way as for salt crystallization, ice formation is facilitated by the presence of coarse and porous sedimentary beds and joints; the orientation of stone blocks in building walls and ornamentations therefore also participates in conditioning weathering intensity and speed.

High and low temperature alternation means stone minerals undergo expansion and contraction cycles, which induce volume variations of rock outer parts relative to the inner parts. Such a process contributes to stone desquamation and disintegration, and particularly affects most heterogeneous rock types: granite, clayey to carbonated sandstone, etc. Temperature-driven mechanisms give way to ball or plate weathering facies, and develop especially in warm desert and arid regions that experience strong day-and-night thermal alternation.

Wind Action

Contrary to classical belief, wind does not significantly induce stone alveolisation and weathering. Most decaying processes are of a chemical nature and induced by wind-transported water and salt (marine spray, aerosols, etc). Repeated aeolian sandblasting essentially participates in mechanically forming the frosted and *matt surface* typical of wind-exposed rocks. In addition wind action helps to clear away disintegration and desquamation products.

Role of Organisms

Biological processes responsible for stone disintegration mainly result from the development of *vegetation* on building surfaces:

- maintenance of wall humidity by higher plants, green and brown algae, moss or mushrooms. The permanence of wetness predisposes to mineral weathering and dissociation;
- rootlets penetration in rock cracks and fissures, which favours mechanical disjunction and pore-water propagation;
- limestone encrustation, corrosion and dissolution through lichen and blue algae growth;
- drilling of micro-cavities by endolith microflora, and subsequent rock fissuring or patina desquamation and erosion (Garcia-Vallès et al., 1998);
- salt formation through bacterial activity, which contributes to deterioration of indoor building and monument stones. This is, for instance, the case for some *Thiobacillus* species, which induce sulphide oxidation and gypsum crystallization, and of nitrifying germs using atmospheric nitrogen and rock sodium and potassium for causing saltpetre precipitation.

Amplification in a Polluted Environment

Soot, tar and some other urban and industrial pollutants contribute to the *stone patina blackening* on buildings, the front of which must periodically undergo cleaning operations. Stone *lesions* tend to expand through the action of residues from combustion (SO_4, NO_x) that are discharged by factories, heating systems, car and truck engines, etc.

The proportion of stone decay specifically produced in a densely urbanized environment may be estimated by measuring the *stone mass loss* in a given time interval. Portland limestone blocks exposed over 10 years to open-air conditions in southeastern England have shown a weight loss of 1–1.5% in the English countryside, and of more than 3% at Whitehall, in the heart of London (Fig. 149). The bigger weight losses registered in the urban atmosphere were attributed to acidic attack of stone carbonate cement by sulphur dioxides (SO_2), which were abundantly emitted by fossil fuel combustion. Notice that SO_2 levels measured in the air were more than 5 times higher in central London than in the countryside, whereas the stone weight loss was only twofold. This difference reflects the complexity of decay mechanisms, which depend on factors other than the total amount of sulphur gas; the rainfall, the duration of stone impregnation by water, and specific interactions of SO_2 with other gasses (NO_2) also participate in controlling the stone's deterioration.

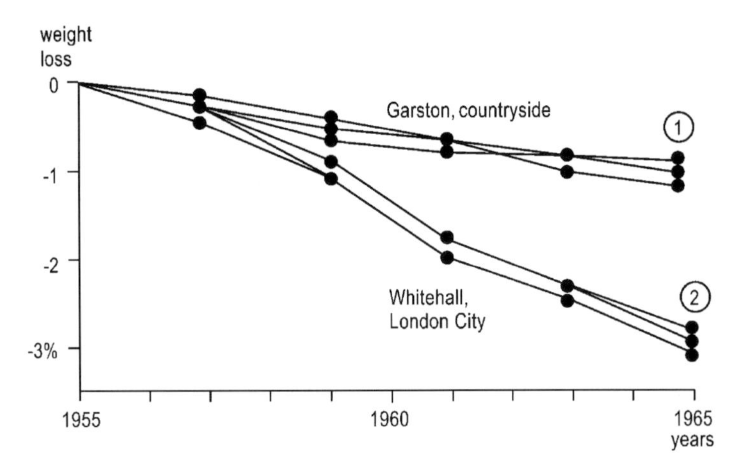

Figure 149: Weight loss rate of Portland limestone samples exposed during 10 years to open air conditions in (1) countryside and (2) urban conditions (see Cooke & Doornkamp, 1990). Each curve corresponds to a given sample.

The density of urban traffic largely conditions the extent of dirt on buildings, of patina deterioration and of stone-lesion propagation. Micro-morphological, microscopic and geochemical investigations have been performed on dressed stone from 390 buildings in Stockholm city districts subject to different traffic conditions. Measurements have clearly shown the dominant influence of car circulation density on deterioration rate of the building façades (Fig. 150). The

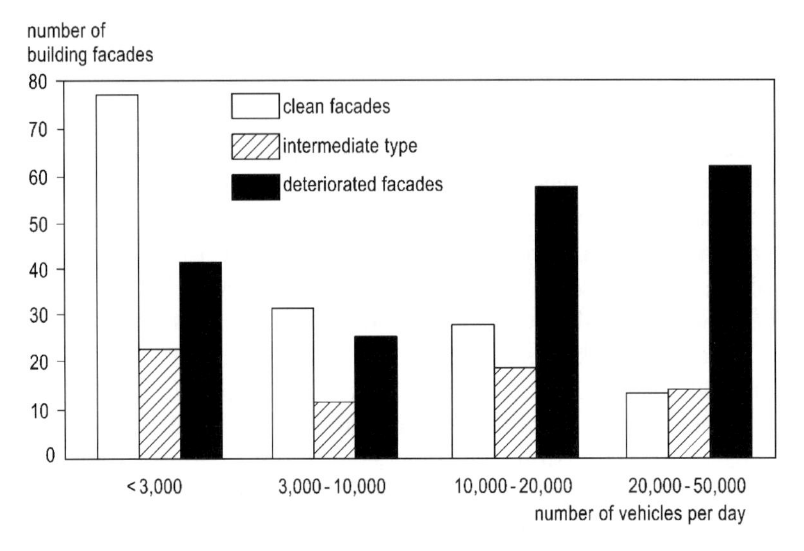

Figure 150: Relationships between front deterioration degree and car traffic intensity in Stockholm City, Sweden (after Nord & Holeyni, 1999).

penetration of sulphur compounds in stone blocks varies from 0.2 to 4 mm depth depending on daily traffic intensity, rock type (i.e. maximum vulnerability of limestone rocks) and rainfall exposure.

9.5.4. Restoration

Fighting stone decay is essentially justified in dense urban and industrial areas where natural rock deterioration suffers significant anthropogenic acceleration. The simpler and more efficient way for reducing stone decay consists in *limiting the emission into the atmosphere of polluting gas, aerosols, and salt,* which induce rock acidification and corrosion. These measures have been successfully implemented during the last decades in various conurbations of developed countries. The basic purpose was to both diminish the sanitary risks induced by air pollution, and preserve the architectural heritage. The methods employed necessitate using costly chemical treatments and filtration processes, which can rarely be afforded by most developing countries.

Other techniques exist, which are implemented in various environmental contexts:

- *Stone-decay prevention* during building construction or renovation work. The techniques employed aim to minimize the risk of decay when conceiving the building shape and size, locating façades and ornamentation, and choosing, adjusting and orienting the stone blocks. Work consists of controlling at different locations the building exposure to bad weather and pollutants, of selecting rock types of low micro-porosity and saturation ratio, and of applying successive insulating liners as construction goes on. These techniques constitute part of *building design,* which involves several other environmental aspects of urban and industrial architecture: building location relative to prevailing wind direction; prevention of sand and dust accumulation in arid regions; foundations size and extent in unstable, very sloped or saturated ground; etc. All these concerns belong to the expanding scientific field of *urban geology.*
- *Façade renovating* by scouring and stripping. The treatments employed aim to restore the original stone aspect and colour. Techniques are diverse and often used in a complementary way: stone washing by water jet, abrasion by sand-blasting, brushing, impregnation by diluted alkaline or acidic solutions, etc. Two drawbacks are associated with renovating treatments: first they contribute to eroding the rock blocks on building façades; second they remove the stone patina and therefore cause fresh surface decay.
- *Elimination of water* infiltration and capillary rise. Diverse methods may be used, such as: installation of gutters, drainpipes or dripstones; insertion of air cavities or watertight liners; electro-osmosis techniques. These actions rely on

constructions being subject to regular and strict control and maintenance work: replacement of porous and decayed stone blocks, checking of mortar, binder or lime, etc.
- *Surface treatment and stone strengthening*. Some of the techniques employed aim to increase the rock or mortar cohesion and resistance by using mineral or organic compounds (e.g. synthetic resins), watertight films, or filters. Other techniques consist of using water-repellent substances to waterproof rock blocks: beeswax, casein, latex or synthetic wax. Notice that certain insulating products are not really watertight, do not penetrate deep enough into rock pores, or have harmful micro-heterogeneity properties. Detailed investigations show that insulating substances should penetrate the rock pores at a depth exceeding 25 mm below the stone surface, constitute coatings rather than sealing the grains, and display low viscosity (Bell, 1993). Appropriate products comprise some monomer vinyl, epoxy resins and alkoxysilanes.

9.5.5. Weathering of Prehistoric Rock-art Caves

Some Facts

Underground natural cavities frequently contain carbonate concretions that result from karsts processes: stalactites, stalagmites, delicate curtains, etc. Sometimes the cavities have been visited or inhabited during prehistoric times by people who have left *drawings, paintings or engravings* on cave walls and roofs. These various natural and man-made cave and rock ornamentations are potentially subject to *weathering* and decay by water and air. This is for instance the case of the paintings ornamenting the Triassic rock walls and caves of Albarracin cultural park in northeastern Spain; ornaments were painted during Levantine prehistoric times (8,000–3,000 years before present), and have subsequently deteriorated by salt crystallization and wetting–drying climatic alternations (Benito et al., 1993). Decay proves to be less where iron oxide coatings, the impervious character of which prevents salt penetration, cover the painted surface of the cave.

Natural deterioration of prehistoric paintings tends to be amplified by temperature, humidity and luminosity changes that are associated with cave incursions by speleologists, tourists and other visitors. Rock and cave decorations are particularly threatened by anomalous patina development, fungus and algae growth, and local micro-climatic change caused by an increasing human presence. Some ornate caves have been closed and prohibited to visitors to avoid accelerated deterioration of an inestimable social and artistic heritage. This is for instance the case of Lascaux cave in the Black Périgord region of southwestern France, which is decorated with 1,500 drawings and paintings of the Magdalenian Age (17,000–15,000 years BC). Animal and everyday figures ornamenting the cave walls and roof had been kept intact for more than 150 centuries, but were succumbing to serious deterioration in the mid-20th century, i.e. only 15 years after the site opened to the public. Lascaux prehistoric paintings have been progressively altered by algae and moss (i.e. "green disease") and covered by a thin coating of calcium carbonate ("white disease"). Lascaux cave, which had been visited by more than one million people in less than two decades, was closed again in 1963; its most richly decorated part has been artificially reproduced in a nearby disused quarry.

Underground cavities are naturally characterized by buffered and particularly stable temperature and humidity conditions. These parameters tend to be disrupted under the pressure of visiting tourists, who are responsible for noticeable daily, weekly, seasonal and annual cyclic changes. Such modifications of physicochemical parameters have been evidenced in the main underground room of Grotta Grande del Vento in Ancona province, central-eastern Italy. The number of daily visitors to this cave may exceed 8,000 people. The tourist pressure causes significant increase in mean temperature and carbon dioxide concentration, and the humidity rate tends to reach saturation values (Fig. 151). Calculation shows that the average heat annually released by the 500,000 visitors who stay about 1.5 h in Grotta Grande del Vento reaches 4.59×10^{11} J/sec (i.e. 128 MWh). This causes surface wall dissolution and weathering processes to significantly increase.

Prevention and Control Measures

Maintaining decoration in prehistoric caves necessarily implies that the most precious and vulnerable sites are kept permanently closed and forbidden to visitors, with the exception, under control, for scientific research or special grounds. Several measures allow limiting of deterioration in caves that remain open to tourist visits. These measures are particularly important in underground rooms and galleries of small volume, where temperature values and dissolved gas levels tend to increase rapidly:

- Determination of a maximum number of daily visitors, which is based on long-term measurements of several parameters including temperature, humidity, CO_2 content, dissolved carbonate content. This supposes implementing a continuously working monitoring system in the cave.
- Setting of airlock-equipped entrances that prevent thermal and hydric shocks and limit the spore and bacteria penetration.
- Wall-drainage improvements and joint sealing for fighting salt crystallization and impregnation of decorations by water.
- Implementation of air-conditioning and of CO_2^- and dust-filtration systems.

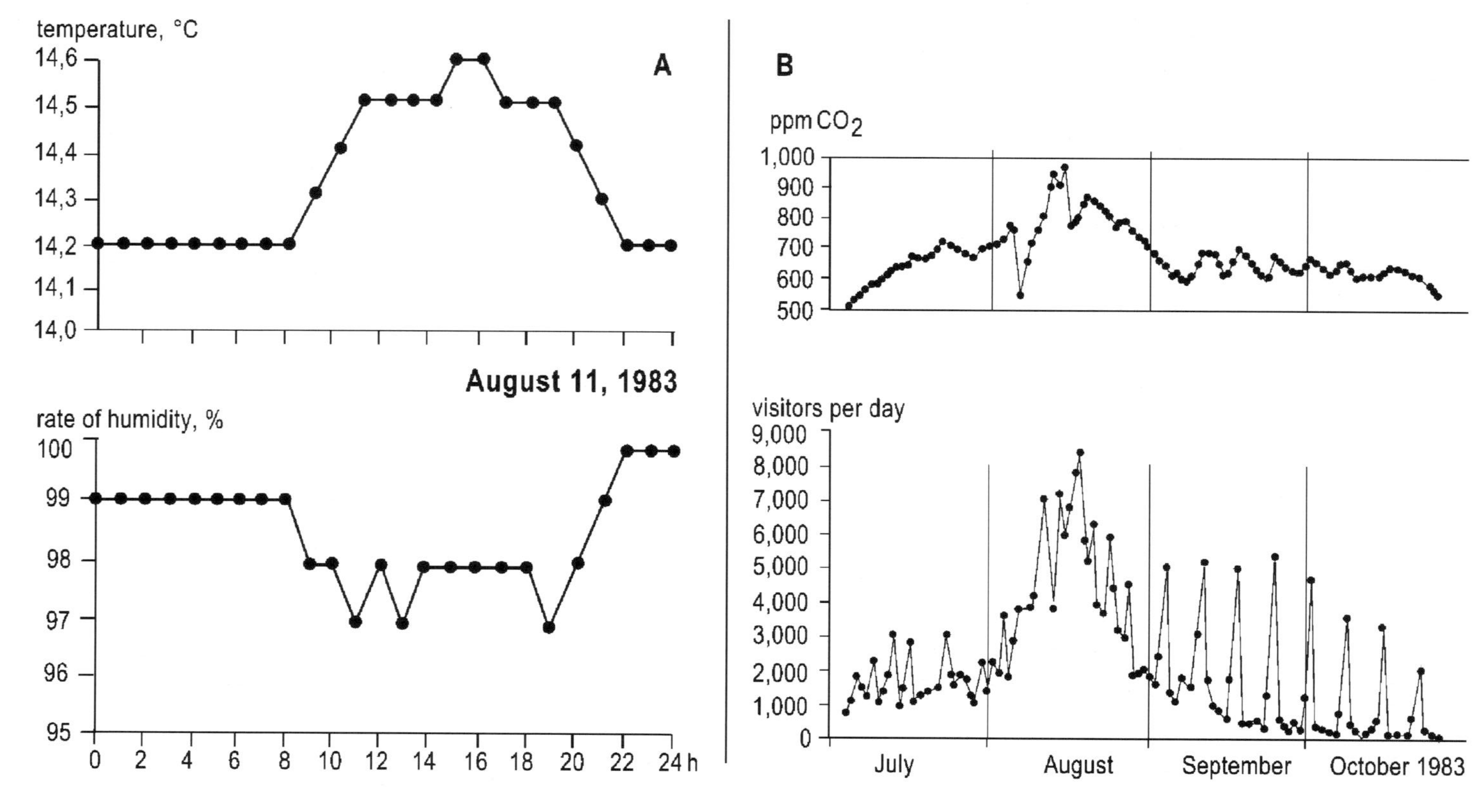

Figure 151: Correlation between tourist pressure and air physicochemical characteristics in Ancona room, Grotta Grande del Vento cave, Ancona province, Italy (after Cigna, 1993). (A) Temperature and humidity rate during a 24-hr cycle (7,011 visitors). (B) Carbon dioxide (CO_2) levels in summer and fall periods; maximum values correspond to the weekends.

- Use of dim-light systems and bulbs with cold-filament, narrow-emission spectrum, and photosynthesis-preventing wavelength.

A Few Landmarks, Perspectives

- The extraordinary population growth characteristic of the last century has combined with a very dense grouping of people in continuously expanding conurbations to cause an unprecedented soil artificialization on Earth's surface. In addition to the massive removal of natural plant cover for farmland use, man has appropriated during the last decades gigantic urban areas marked by large-scale soil asphalting and concreting. The extent of the artificial soil has been amplified by three urbanization-related developments caused by increasing demand for energy, goods and working efficiency: (1) industrialization that expanded from local to widespread factory implementations; (2) huge increase in road, rail, water and air communications that always induced denser networks of streets, trails, canals, railroads, cables, pipelines, etc.; (3) growing occupancy of vertical space for satisfying needs for more building space, additional underground networks, wiring connections, etc.
- Urban and industrial soil artificialization is responsible for extensive modifications in physicochemical exchanges occurring between the lithosphere and Earth's external envelopes (hydrosphere, atmosphere, biosphere). The widespread implementation at soil surface of a sort of watertight and airtight screen determines various environmental disorders: change of rainwater and subsurface groundwater route and transfer; blocking of water infiltration and capillary rise; worsening of surface streaming, downstream overflowing, and flooding risks; potential drying up and subsidence of aquifer terrains. Water management is not easy to conduct in a countryside environment but becomes terribly complex in densely artificial regions. Urban geology developments include in an increasing way thorough expertise in urban hydrogeology.
- Urban, industrial and domestic wastewater discharge is sometimes associated with the emission of gas, dust, flying ash or scoria. All these materials pose multiple risks for local or even regional contamination of groundwater, soils and surface-to-subsurface geological formations by heavy metals, hydrocarbons, bacteria and viruses, etc. The most spectacular expression of local pollution is represented by urban and industrial wasteland, but most dangerous effects result from chemical contamination of groundwater. Groundwater cleaning processes are complicated, long and expensive, especially if aquifers are located deep in the ground, suffer extensive

karstification, or experience heterogeneous and irregular water flow. Industrial-wasteland restoration and management are all the more efficient when dumpsites are drainage-free, and when transfer and cleanup mechanisms are adequately controlled. Promising research paths lie in the use of living plants for extracting and stabilizing soil chemical pollutants (i.e. phytoremediation). Positive results are also expected from using some highly reactive minerals such as zeolites.

- Increasing emission in surface lithosphere of mineral and organic waste partly results from the huge extension of transport links and means. Pollution due to terrestrial transportation is usually restricted to soils and sediments situated close to major trunk roads, but may induce significant contamination in densely populated regions and suburban zones. Local pollution due to river, lake or sea transportation mainly results from channel and harbour dredging. Dredged waste is destined to be either stored on land or dumped at sea, which potentially determines various contaminants to be subsequently dispersed. Dredging residues released in marine (or lacustrine) environment are susceptible to modify local ecosystems, to be concentrated by local fauna and flora, or to be exported by tide or swell currents. Dredging waste management necessitates implementing ecotoxicity tests, dispersal monitoring experiments and modelling, appropriate chemical treatments and strict regulations.
- Urban and industrial waste constitutes widespread, locally gigantic accumulations that induce considerable proximity nuisance and may contaminate both soils and underground terrains. Groundwater pollution risks are particularly due to leaching by rainwater and get exacerbated in humid regions marked by very porous, fissured or karstified rocks. People's realization is growing of long-term issues linked to solid and liquid waste disposal, especially in developed countries where dumps are both more abundant and better controlled. Recycling, incineration, sorting and deep burial techniques are associated in an increasing and more efficient way with methods involving dump monitoring, treatment, exploitation (e.g. secondary production of methane) and even rehabilitation.
- Stone decay results from physical and chemical weathering of underground materials extracted for construction or rock sculpture and ornamentation. Stone deterioration develops naturally by forming patinas and surface lesions induced by salt crystallization and by the action of water, micro-organisms, and micro-climatic changes. Rock decay occurs in all surface environments exposed to open-air and humidity variations, including the caves containing natural carbonate concretions and/or man-made prehistoric paintings and drawings. Building-stone "disease" gets noticeably amplified in urban and industrial environments where human activities are responsible for the discharge of abundant acid fallout, combustion gas, exhaust fumes and other

atmospheric pollutants. Various stone-decay prevention techniques are deployed that range from scouring to artificial waterproofing. They are rarely of long-standing effect because of the permanence of weathering actions.

Further Reading

Alzieu C. coord., 1999. Dragages et environnement marin. Ifremer Public., Paris, 223 p.

Eyles N., 1997. Environmental geology of urban areas. Geol. Assoc. Canada, 590 p.

Lecomte P., Les sites pollués. Traitement des sols & des eaux souterraines. TEC & DOC, Paris, 2nd ed., 204 p.

Maund J. G. & Eddleston M., 1998. Geohazards in Engineering Geology. Geological Society of Engineering Geologists, sp. pub. 15, 448 p.

Said-Jimenez C. ed., 1995. The deterioration of monuments. The Science of the Total Environment, 167, Sp. Iss., 400 p.

Turner B. L. II, Clark W. C., Kates R. W., Richards J. F., Mathews J. T. & Meyer W. B., 1993. The Earth as transformed by human action. Cambridge University press, 713 p.

Williams P. T., 2000. Waste treatment and disposal. Wiley, 417 p.

Some Websites

http://www.urbanecology.org/: group of associations based to San Francisco and concerned by worldwide urban environmental questions. Very large cluster of Websites

http://www.epa.gov/epahome/Citizen.html: official site of the U.S. Environmental Protection Agency, which covers almost all types of energy and pollution problems (water, air, and soil) in urban and rural environment

http://www.indigodev.com/Industry.html: large review of ecology problems and solutions linked to urban and industrial environments, collected by the International's Industrial Ecology Research Centre, which is affiliated to RPP International (Research–Policy–Practice)

http://www.worldbank.org/urban/solid_wm/swm.htm: various aspects of urban waste problems and management aspects, especially in developing countries. The site is held under the auspices of the World Bank

Chapter 10

Chemical Contamination of Earth's Surface Formations

10.1. Fresh Water, Soil and Underground Terrain

10.1.1. Contaminant Recording in Continental Water and Sediments

The Thames River Facing Human Pressure in Greater London

The lower course of the Thames River in southeastern England collects most of the water draining the London megalopolis. Measuring the chemical characteristics of Thames water downstream of London therefore records the environmental changes caused by human influence. This has been done since the beginning of the 19th century for dissolved oxygen content, and constitutes one of the longer-term freshwater records so far (Fig. 152).

The quality of Thames water had already deteriorated by the early 19th century and displayed strong oxygen depletion a long time before the British industrial revolution started. The river ecosystems remained deeply disrupted until 1880, which corresponded to a period marked by a population growth from 2.75 million in 1850 to 4.75 million in 1880, as well as by successive cholera epidemics between 1830 and 1871. London was one of the first large cities to be equipped with an extensive sewer network (1860) and subsequently by water-treatment systems (1885). This progressively restored the river water quality, in spite of the continuously growing population (more than 6 million inhabitants in 1900). The water oxygen content increased regularly during the second half of the 19th century and reached a saturation rate of 50% around 1900. This allowed the return of freshwater fauna in the river, including several fish species. The continuation of demographic growth (8 million people in 1940) caused new shortfalls in cleaning and resumption of water oxygen depletion in the 20th century (less than 2 mg/l O_2 during the 1940s–1960s). Further extension of the sewer network (1940) and of water treatment systems (1965) has allowed, since the 1970s, the fluvial water quality to be improved again. The biological properties of Thames river water have continued to become safer during the last decades.

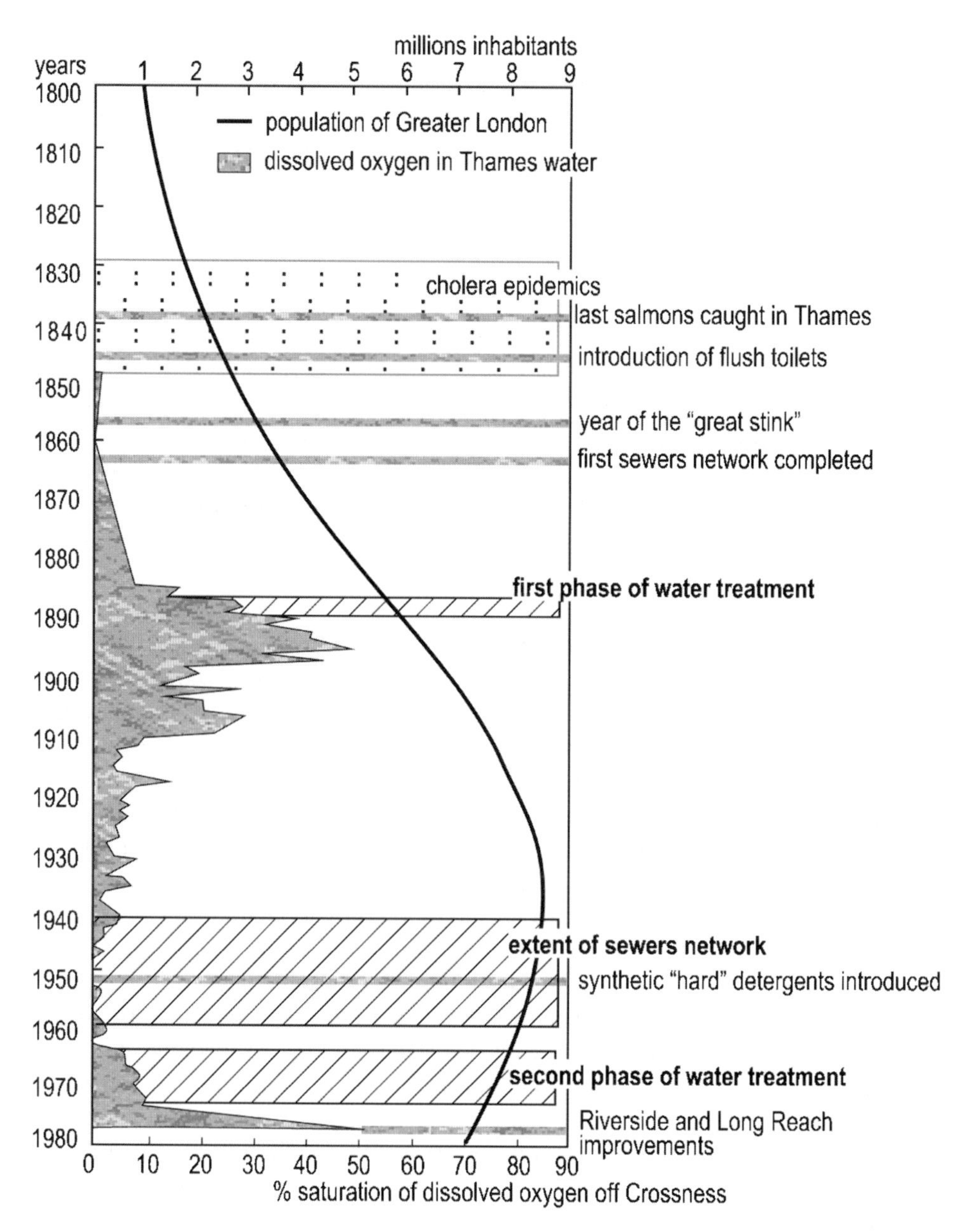

Figure 152: Variations since 1800 of the dissolved oxygen content in Lower Thames river water. Comparison with the Greater London demographic changes, and major sewage and cleaning developments (after Schwartz et al.; see Turner II et al., 1993).

This example shows that anthropogenic pollution of freshwater river courses is sometimes very ancient and not systematically related to population growth. River pollution was sometimes much worse formerly than recently, and may be effectively fought by appropriate cleaning and treatment measures.

Recording, using Malter Reservoir Sediments, of Erzgebirge Agricultural and Industrial History

The Malter artificial lake was built in 1913 to the south of Dresden in Eastern Germany to prevent flooding and produce hydroelectricity. During the 20th century the reservoir accumulated particulate and dissolved materials carried by the Weisseritz River, which drains a 104.6 km^2-wide mining and farming region. The sediments trapped in the Malter reservoir have been cored, which helps in reconstituting the anthropogenic modifications successively undergone by the river basin due to river and lake management, to mining industry development and closure, and to initial environmental improvement measures (Müller et al., 2000).

Mining activities in the Weisseritz River basin started as early as the 13th century and were devoted to the extraction of tin, copper, molybdenum, bismuth and silver. Before being stopped in 1992, the metal-ore extraction had brought about important industrial activities, which developed during most of the 20th century simultaneously with urban and agricultural expansion. This is clearly expressed by the fluctuating, globally increasing discharge in the Malter reservoir of various pollutants: tin (Sn), copper (Cu), tungsten (W), molybdenum (Mo), bismuth (Bi), silver (Ag), zinc (Zn), lead (Pb), cobalt (Co), cadmium (Cd), chromium (Cr), antimony (Sb), organic carbon, fertilizers, acid compounds, etc.

Several *local events* are reflected by the heavy metal and fertilizer content of lake sediments, as well as by sedimentation-rate fluctuations: major phases of mining development and recession, waste discharge linked to major flooding, increasing agricultural influence, etc. Some *global events* are also recorded, such as the stopping of military nuclear tests in 1963, the Chernobyl nuclear plant explosion in 1986, and the increase in acid rain fallouts in the late 20th century. The heavy metal content in the Malter reservoir sediments has temporarily reached values 290 times higher than continental-shale average values for Cd, 140 times for Ag, 90 times for Bi, 25 times for Sb, and 21 times for Pb. Very high values have been locally recorded for various heavy metals: 27 mg/kg for Ag, 37 mg/kg for Bi, 91 mg/kg for Cd, 410 mg/kg for Cr, 740 mg/kg for Pb, and 1,900 mg/kg for Zn. The peak values have been attributed to specific human activities: discharge of mining waste, slagheap leaching exacerbated by heavy rainfall, and industrial waste drainage.

A considerable increase in sediments of phosphorus (up to 14,000 mg/kg) and organic carbon, as well as of siliceous algae tests (diatoms), has corresponded during the 1940s and 1950s to the transition from extensive to intensive farming practices, responsible for massive nutrient leaching and lake water eutrophication. Finally a decrease in Cu, Zn, Cd and Cr levels is recorded in Malter reservoir sediments deposited since the late 1980s; it correlates with both the closure of some polluting plants and the implementation of new water-treatment systems, which followed the reunification of Western and Eastern Germany.

The contaminants stored in the Malter artificial lake sediments are relatively stable and slightly mobile, thanks to rather high pH values preventing chemical solubilization, to moderate permeability of sediments, and to the rareness of sulphides that could potentially induce the microenvironment to be reduced. The risk nevertheless remains for these pollutants to be reworked and exported downstream in the Weisseritz River, mainly because of possible lake sediment erosion by flooding or by lateral channel migration.

10.1.2. Types and Modes of Contamination

Water Quality, Sanitary Risks

Surface and underground water destined for drinking and domestic use must contain dissolved and particulate elements the nature and abundance of which do not affect health. Strict *drinking water standards* have been defined, which essentially concern the contaminants released in water through human activities: heavy metals and some other inorganic substances (arsenic, alkaline cyanides, fluorine), insecticides and pesticides, hydrocarbons and derived products, *E. Coli* bacteria and viruses (Table 41). Man-released pollutants in water may either be of natural origin such as polycyclic aromatic hydrocarbons (PAH), or result from artificial synthesis as is the case of xenobiotic compounds such as polychlorinated biphenyl (PCB) or dichloro-diphenyl-trichlorethane (DDT).

The presence of contaminants generally does not affect the water aspect, colour, smell or taste. *Olfactory and taste characteristics* mainly depend on the presence in water of other non-toxic chemical substances, which may be of natural or human origin. These substances must not exceed certain amounts for guaranteeing the water drinking quality. This is the case of chlorides that give a salty taste to water (Cl <250 mg/l), of iron responsible for reddish colouration and ferruginous taste (Fe <0.3 mg/l), of manganese that gets easily oxidized and forms a blackish coating (Mn <0.05 mg/l), or of hydrogen sulphide, the smell of which is putrid ($H_2S < 0.05$ mg/l). In addition the drinking water pH must be between 5 and 9.

Three groups of freshwater contaminants may be defined by taking into account the *potential toxicity level* relative to *concentration*:

Table 41: Example of water drinking standards and potential sanitary risks (after Chapman, 1992; Merritts et al., 1997; Bell, 1999; Keller, 1999). Mg/l, milligrams per litre. Mfl, million fibres larger than 10 micrometers per litre. n/l, number per litre.

Contaminants	**Maximum value**		**Risks**
	European standards	**Other standards**	
inorganic chemicals			
asbestos (Mfl)	–	7	minor cancer risk
arsenic (mg/l)	0.05	0.05–0.005	dermal and nervous systems toxicity effects
barium (mg/l)	1.0	1.0	circulatory system effects
cadmium (mg/l)	0.01	0.01	kidney deficiency
chromium (mg/l)	0.05	0.05	liver and kidney effects
cyanides (mg/l)	0 05	0.05	poisoning
fluorine (mg/l)	–	4.0	skeletal damage
mercury (mg/l)	0.01	0.01–0.02	nervous system and kidney effects
lead (mg/l)	0.1	0.015–0.1	nervous system and kidney effects
selenium (mg/l)	0.01	0.01	digestive system effects
dissolved organics			
pesticides (mg/l) (endrin, lindane, methoxychlor)	–	0.0002–0.1	nervous system, liver and kidney effects
herbicides (mg/l) (2,4D, silvex)	–	0.05–0.07	nervous system, liver and kidney effects
volatile organics (mg/l)			
benzene, carbon tetrachloride, trichloroethene, vinyl chloride	0.001	0.005	cancer risk
	–	0.002	cancer risk
micro-organics (c/l)			
faecal contaminants (*Escherichia*, *Streptococcus*, *Clostridium*)	100	10	indicators of digestive system infections
virus (n/l)	1	–	various risks

- Existence of some toxicity at *any level of concentration* inducing a risk increase parallel to the contaminant content. The chemicals concerned lack a minimum toxicity threshold. They include various substances susceptible to inducing cancers, genetic mutations, of growth disorders: arsenic, pesticides, and chloride organic compounds.
- Existence of a *minimum level of concentration* above which a potential risk appears and increases parallel to the contaminant concentration. This is the case for nitrate ionic compounds (NO_3^-) that induce potential toxicity for values exceeding 50 mg/l. Nitrates may for instance cause unbalanced blood composition marked by haemoglobin insufficiency.

  Most toxic elements in water belong to one of these two groups.
- Existence of both a *minimum and a maximum concentration level.* The elements concerned may be dangerous for human health if their amount is either too low or too high. This is the case for fluorine, iodine and selenium.

Methods of Contamination

Man-induced deterioration of surface water quality develops either directly by waste discharge through domestic, farming, urban, hospital and/or industrial activity, or indirectly through dump leaching by rainwater and streaming. In the case of scattered settlement, contamination *sources* are multiple, diffuse, of little and limited impact and difficult to put right; noticeable pollution risks occur in the long term only. By contrast dense soil occupancy by farming and other human activities induces pollution that is well defined, often serious, and potentially of large and rapid impact.

The propagation underground of polluted water (Fig. 153) is facilitated both by subsurface and deep aquifer natural connections (e.g. faults, fissures, karsts, coarse-grained rocks; Chapter 6), and by some human devices employed for water extraction or injection (e.g. wells, pumps).

Natural and Anthropogenic Contaminant Fluxes

The geochemical composition of groundwater first depends on the nature and chemistry of geological formations constituting aquiferous rocks. For instance eruptive and metamorphic rocks may include various heavy metals such as lead, chromium, nickel, cobalt, copper, and zinc, which are partly dissolved and incorporated in groundwater. By contrast these metals tend to be rarely present in water draining clay or limestone sedimentary formations.

Natural mineral fluxes are modified by the artificial supply in water of chemical elements resulting from mining activities, energy production, industrial processes, and farming or urban pressure (Table 42). Some man-provided

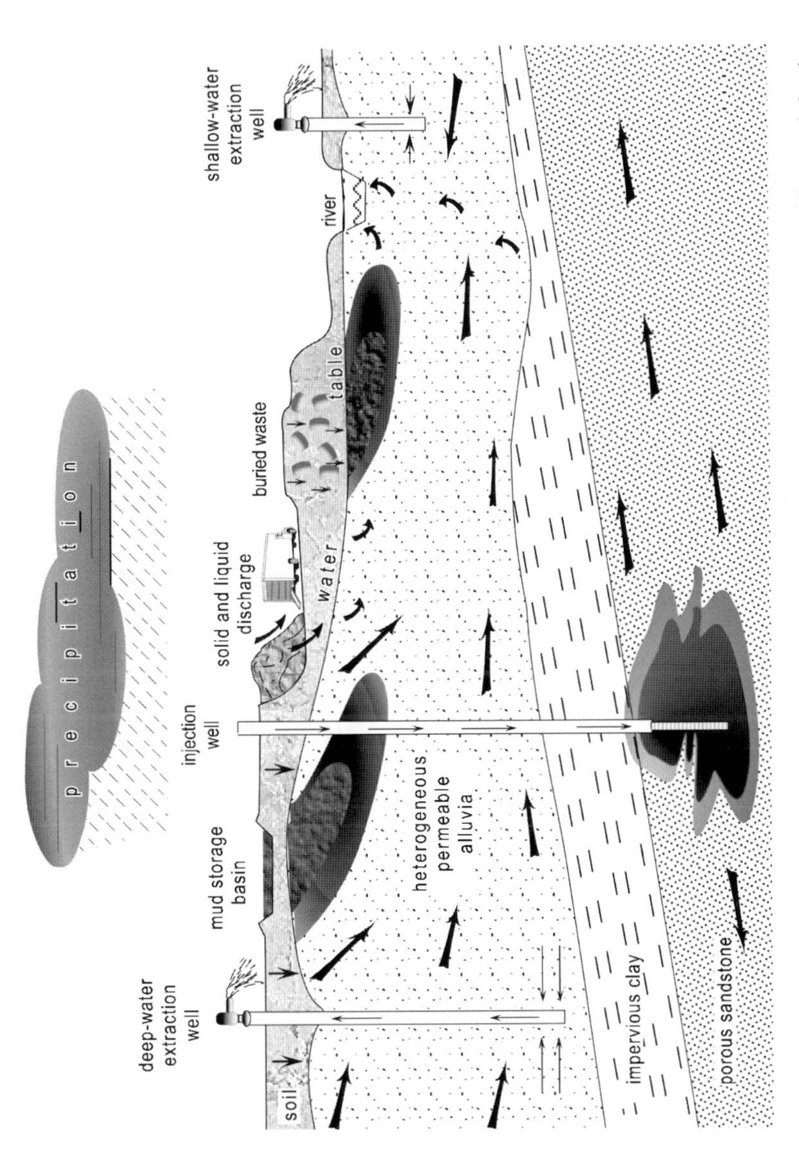

Figure 153: Routes of surface and ground water contamination from waste storage sites and aquifer exploitation systems (after National Research Council, 1993; Botkin & Keller, 1995). Arrows indicate the underground course of polluted water due to infiltration, percolation, lateral migration, pumping or injection. Dark spots indicate the pollutants diffusion zones in the underground.

Table 42: Comparison of chemical element fluxes resulting from human activities (A) and from natural discharge in non-anthropized freshwater (B). C = A/B ratio (after Mineral Commodities Summaries, Holland & Petersen, 1995). Values in grams per year (g/yr).

Element	**A World production 1990**	**B Flux in non-anthropogenic rivers**	**C A : B ratio**
iron Fe	$500 \cdot 10^{12}$	$1{,}000 \cdot 10^{12}$	0.50
aluminium Al	$18 \cdot 10^{12}$	$1{,}600 \cdot 10^{12}$	0.01
copper Cu	$9 \cdot 10^{12}$	$1.1 \cdot 10^{12}$	8.2
zinc Zn	$7.3 \cdot 10^{12}$	$1.4 \cdot 10^{12}$	5.2
lead Pb	$3.3 \cdot 10^{12}$	$0.26 \cdot 10^{12}$	13
chromium Cr	$11.7 \cdot 10^{12}$	$2.0 \cdot 10^{12}$	5.8
nickel Ni	$1.0 \cdot 10^{12}$	$6.5 \cdot 10^{12}$	0.15
tin Sn	$220 \cdot 10^{9}$	$60 \cdot 10^{9}$	3.7
magnesium Mg	$380 \cdot 10^{9}$	$400{,}000 \cdot 10^{9}$	0.0001
molybdenum Mb	$114 \cdot 10^{9}$	$30 \cdot 10^{9}$	3.8
uranium U	$20 \cdot 10^{9}$	$36 \cdot 10^{9}$	0.55
cadmium Cd	$21 \cdot 10^{9}$	$4 \cdot 10^{9}$	5.2
mercury Hg	$6.0 \cdot 10^{9}$	$1.6 \cdot 10^{9}$	3.7
gold Au	$2.0 \cdot 10^{9}$	$0.08 \cdot 10^{9}$	25
platinum group Pt	$285 \cdot 10^{6}$	$200 \cdot 10^{6}$	1.4
carbon C	$5{,}000 \cdot 10^{12}$	$650 \cdot 10^{12}$	7.7
sulphur S	$58 \cdot 10^{12}$	$100 \cdot 10^{12}$	0.58
nitrogen N	$110 \cdot 10^{12}$	$10 \cdot 10^{12}$	11
phosphorus P	$22 \cdot 10^{12}$	$2 \cdot 10^{12}$	11

elements are discharged within rivers in concentrations up to tens of times larger than natural fluxes; this is the case for copper, lead, chromium, cadmium, mercury, gold, carbon, nitrogen and phosphorus. Some other elements such as iron, sulphur, aluminium, nickel, magnesium and uranium are released within rivers greatly affected by Man in total amounts that are much lower than the amounts carried by rivers not so affected by human activity.

Monitoring in a rigorous manner the nature, amount and impact of chemicals released by man in fresh water represents a difficult task, for several reasons. Some elements may be either dissolved and therefore easily exportable, or be under a particular or fixed form and consequently be rather stable; changes in this status may happen during groundwater migration due to physical or chemical causes, and induce further modification of the potential contamination risk. The amount of anthropogenic contaminants discharged in surface water may vary greatly on a temporal scale, and the location of pollution sources may also change; water contamination is therefore potentially of variable intensity and environmental impact, which necessitates implementing continuous, sensitive and costly measurements. In fact long-term monitoring of river and lake water and sediments is still rarely conducted (e.g. Thames River; Chapter 10.1.1), and concerns only a very limited number of organic and/or inorganic parameters.

The way by which contaminants may affect the human health is often only partly understood. The toxicity thresholds characterizing the different types of a given chemical element depend on complex criteria and on varying receptivity conditions, and are therefore difficult to assess. For instance *mercury*, which potentially constitutes a dangerous element together with arsenic and lead, is usually insoluble and therefore very slightly noxious. This heavy metal may nevertheless be activated and become toxic in fresh water, mainly because of bacterial reactions inducing the formation of mercury methyl or di-methyl (CH_3Hg^+, (CH_3) 2Hg); these compounds are soluble in fresh water, may be concentrated in the food chain, and determine mercury poisoning, especially through fish ingestion.

By contrast, *gold*, a heavy metal that is 25 times more abundant in anthropogenically affected than in natural fresh water (see Table 42), is very difficult to extract and concentrate in organic tissues; despite its relative abundance this element does not therefore cause any dangerous contamination. But some chemical elements are employed for extracting and purifying gold, are easily released in fresh water and may be a serious toxic risk; this is especially the case for mercury (Chapter 5).

Contamination by Metals

Water, soil and underground metal contamination potentially concerns all regions where dense mining and industrial activities are conducted. The *Rhine river* basin constitutes a typical example. This western European river extends over an area of 185,000 km^2 and a length of 1,320 km in various industrial, mining and port regions of Austria, Switzerland, Germany, France and the Netherlands. The heavy-metal load in Rhine water, suspended particles and sediments has considerably increased since the late 19th century, both from upstream to downstream regions and over the course of time. The total heavy metal concentration reached its highest values in the early 1970s (Fig. 154). The difference measured between Lake Constance in Switzerland and Biesbosch in the Netherlands has risen from 0.4 to 18 ppm for mercury, from 30 to 850 ppm for lead, from 119 to 760 ppm for chromium, and from 185 to 3,900 ppm for nickel (after Förstner & Müller; see Betz, 1975). In 1970 the anthropogenic pollution due to Rhine River discharge in the southern North Sea represented 35% of total cobalt, 60% of nickel, 80% of chromium, 95% of copper and almost 100% of cadmium.

Realization of this strong and increasing pollution, which also concerned non-metal substances such as arsenic, mine-derived salt compounds and fertilizers, led in the late 1970s to drastically reducing the waste discharge in the Rhine, modifying some industrial processes and implementing severe legislation rules. As a result the amounts of heavy metal and other mineral toxic elements strongly decreased in Rhine water and sediments between 1975 and 1985. The contents measured downstream from the German part of Rhine River have dropped from 4.5 to 1.8 μg/l for arsenic, from 2.3 to 0.1 μg/l for cadmium, from 35 to 8 μg/l for chromium, from 0.4 to 0.07 μg/l for mercury, from 22 to 4 μg/l for lead, from 10 to 5 μg/l for nickel, and from 135 to 50 μg/l for zinc (after

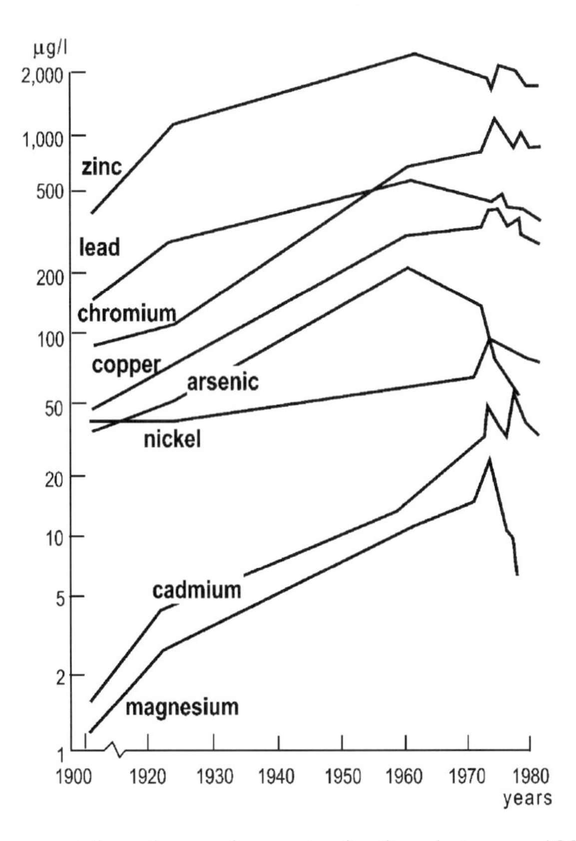

Figure 154: Heavy metal and arsenic contamination between 1900 and 1980 in suspended matter of lower Rhine River (after World Resources Institute; see Pickering & Owen, 1997). Values in micrograms per litre (μg/l).

UNEP; see Holland & Petersen, 1995). During the 1990s the contaminants in the Rhine River dropped to values lower than those measured in 1940; for a few elements nevertheless the contents remained significantly higher than those typical of the natural geochemical background.

Numerous other examples of metal contamination of rivers and lakes have been published during the last decade in specialized scientific journals such as *Environmental Geology* (Springer), *Environmental Geosciences* (Blackwell), *The Science of the Total Environment* (Elsevier), and *Water, Air & Soil Pollution* (Kluwer; see also Chapter 11.3.3). Freshwater environments subjected to mining activity or industrialization are particularly threatened due to the acceleration of demographic pressure and subsequent needs for more resources and goods. This is especially the case in recently industrialized developing countries.

Significant metal pollution also characterizes some developed countries where remedial measures have not been extensively implemented so far. In France the Seine River displays an upstream to downstream increase in heavy metal and other contaminant levels (Fig. 155). The rate of contamination tends to augment regularly from uphill-forested sub-basins and historical deposits

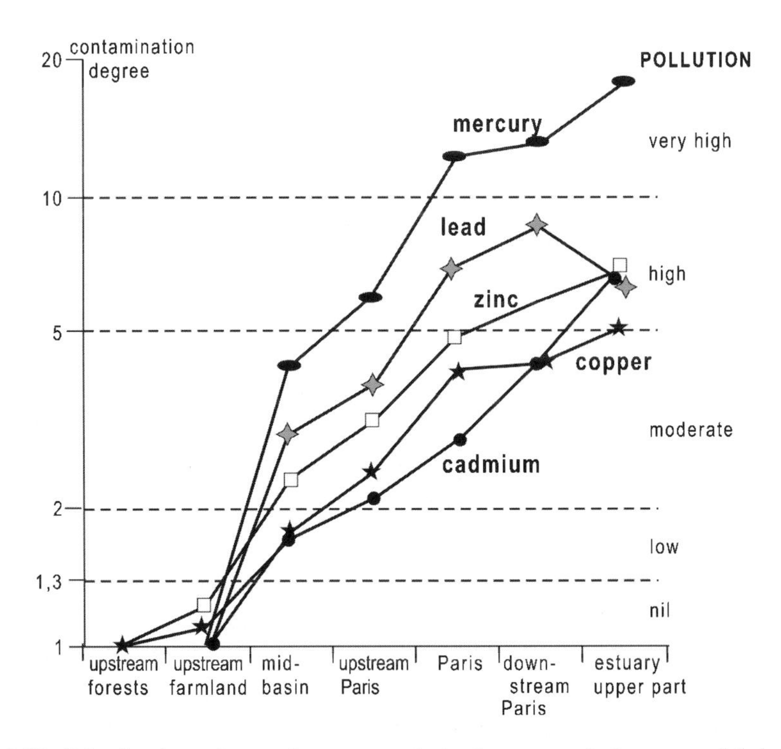

Figure 155: Distribution of some heavy metals in the suspended matter of Seine River, France (after Meybeck, 2001).

down to estuary suspended matter and sediments near Rouen city. Investigations performed downstream from Paris reveal a large diversity of situations depending on pollutant nature and type (Meybeck, 2001). Mercury has the highest concentration and strontium the lowest, according to the following succession:

$$\mathrm{Hg} = \mathrm{Cd} > \mathrm{Pb} > \mathrm{Zn} = \mathrm{P} > \mathrm{Cu} = \mathrm{Sb} > \mathrm{C}_{\mathrm{org.partic.}} = \mathrm{Cr} = \mathrm{Se} > \mathrm{Ni}$$
$$= \mathrm{As} > \mathrm{Li} = \mathrm{Be} = \mathrm{La} = \mathrm{Co} = \mathrm{Ba} = \mathrm{V} = \mathrm{Sr}$$

In Bordeaux harbour upstream of the Gironde estuary (SW France), sediments deposited since the 1930s have been cored and dated by radioisotope methods (^{137}Cs, radioactive elements released from 1950–1960 military nuclear tests and from the 1986 Chernobyl nuclear accident). The port deposits show a pollution by lead, cadmium, zinc and tin; the contaminant amounts tended to decrease until 1980 due to mine and foundry closure along the Lot river, which is an uphill tributary of the main Garonne river. Some chromium and vanadium are still discharged from upstream tannery and electrolytic treatment plants (Tarn and Garonne river valleys), but no appreciable heavy-metal contamination is identified that could result from Bordeaux harbour activity (Grousset et al., 1999).

Pollution by Nitrogen and Phosphorus

Nitrogen and phosphorus released in surface water tend to increase the primary productivity of algae groups (phytoplankton, macrophyts), which result both in developing complex eutrophication processes and causing strong oxygen depletion within freshwater and coastal environments. The lack of oxygen often induces massive flora and fauna destruction, water and bottom reduction, and deterioration of aquatic ecosystems.

On the other hand, the ingestion by man of nitrate-containing water poses various sanitary risks: diminution of blood capacity to carry dissolved oxygen, especially where very young children are concerned (i.e. blue baby syndrome); formation in the digestive tract of potentially carcinogenic nitro-amines that result from NO_3^- and amine combination via action of bacterial flora.

Most continental water and sediment contamination by nitrogen and phosphorus results from the release of fertilizers that are used in a dramatically increasing way for meeting agricultural needs. The world consumption of fertilizers rose between 1940 and 1999 from 14 to 140 million tons, more than one-half of which were made up of nitrogen compounds, and more than one-tenth of phosphates. Non-agricultural sources of anthropogenic nitrogen and phosphorus comprise livestock farming waste and excrement, domestic orthophosphates employed for clothes washing, etc. Nitrogen compounds released in surface environments mainly consist of nitrates, which in downstream parts of large river basins display an abundance that is proportional to population density and agricultural needs (Fig. 156). Other fertilizers besides nitrogen and phosphorus are used, namely potassium compounds such as potash.

Many river basins are characterized by high nitrate contents. This is especially the case of some Western European rivers. The water of Meuse River (France, Belgium, the Netherlands) carries more than 4 mg of nitrogen/l, that of Rhine (Austria to the Netherlands) and Guadalquivir (Spain) more than 3 mg/l, that of Loire (France) and Po (Italy) more than 2 mg/l, and that of Danube (Germany to Romania) and Rhone (Austria to France) more than 1.5 mg/l. Nitrate levels are lower in river water of most other countries including North America (e.g. Delaware and Mississippi Rivers, about 1 mg of nitrogen/l; Saint Lawrence, 0.25 mg/l).

Anthropogenic discharge of fertilizers in freshwater systems is often abundant and poorly controlled. This is the result both of the agricultural use in excessive amounts of nitrates and phosphates, and of artificial drainage practices that are conducted at the expense of natural soil infiltration. In addition *nitrates* are easily leached by rainwater, since they are negatively charged, as are the soil clay-mineral particles; they therefore easily get released from clayey soils as

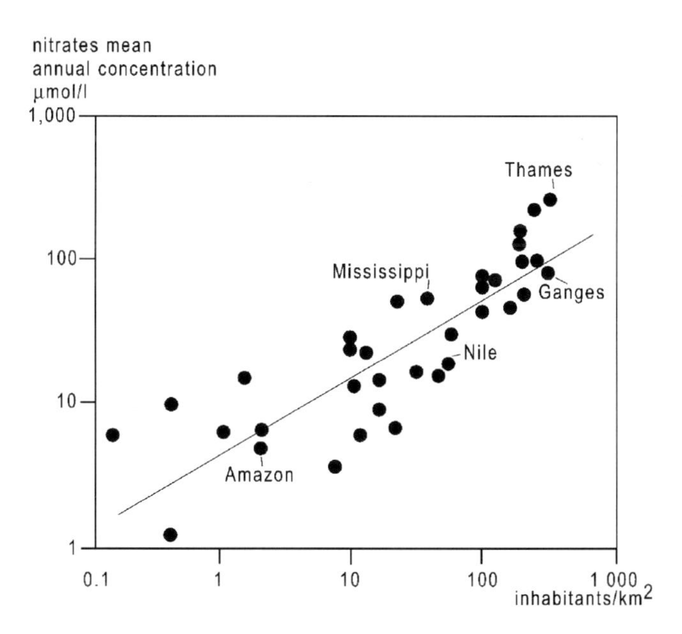

Figure 156: Correlation between nitrate concentration in freshwater and population density of some large river basins (after Peierls et al.; see Mackenzie & Mackenzie, 1995).

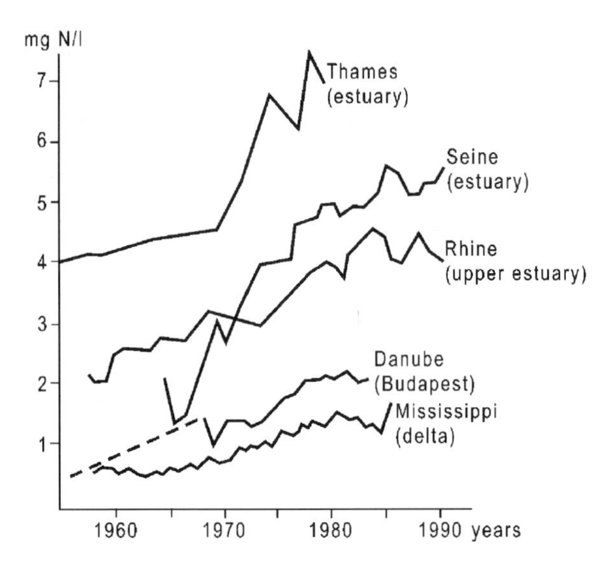

Figure 157: Variations since 1960 of nitrate concentration in water of several Europe and North America rivers (diverse sources: Meybeck; see von Bodungen & Turner, 2001).

long as the plants themselves do not fix them. Many river and groundwater systems are currently characterized by high nitrate concentrations (Fig. 157). For instance coastal rivers of Brittany, western France, have suffered an average nitrogen increase amounting annually to 4% during the last two decades. At best, the nitrogen content has been stabilized.

Phosphate discharge in freshwater systems is sometimes better controlled and managed. For instance the concentration of dissolved phosphate in Lake Geneva, which had first increased from 10 μg/l in 1960 to 90 μg/l in the late 1970s, started decreasing from the 1980s to values that became lower than 50 μg/l in the 1990s (see Holland & Petersen, 1995). This evolution resulted from 1985 legislation measures prohibiting the use of phosphate-containing detergents.

Notice that underground diffusion of fertilizers, especially of nitrates, is very slow and usually progresses by less than 1m/year. This is responsible for a delay of one decade or more in groundwater contamination, after massive fertilizer spreading on the soil.

Other Contamination Sources and Evolution Trends

Various compounds other than heavy metals and fertilizers have been or are still currently released in surface or underground freshwater systems. This is particularly the case for hydrocarbons, pesticides, herbicides and pathogenic organisms.

Whatever the pollutants considered, *freshwater contamination in strongly industrialized countries was generally maximum during 1950–1975*. The level of pollutants started decreasing in the late 1970s or early 1980s, due to the implementation in an increasing way of limitation and restoration (Chapter 10.1.3). Specialized journals and scientific handbooks provide numerous examples of this evolution. For instance the amount of oil, grease, heavy metals and arsenic in the Hudson-Raritan basin (NE USA), which had continuously increased during a large part of 20th century, started decreasing during the 1970s (Fig. 158). In Lake Geneva the 1960–1975 period was characterized by maximum amounts of hydrocarbons of poly- and hexa-chlorobenzene types (PCB, HCB), as well as of DDT insecticides (dichlorodiphenyltrichloroethane) and their metabolic by-products; similar trends are registered by sediments of

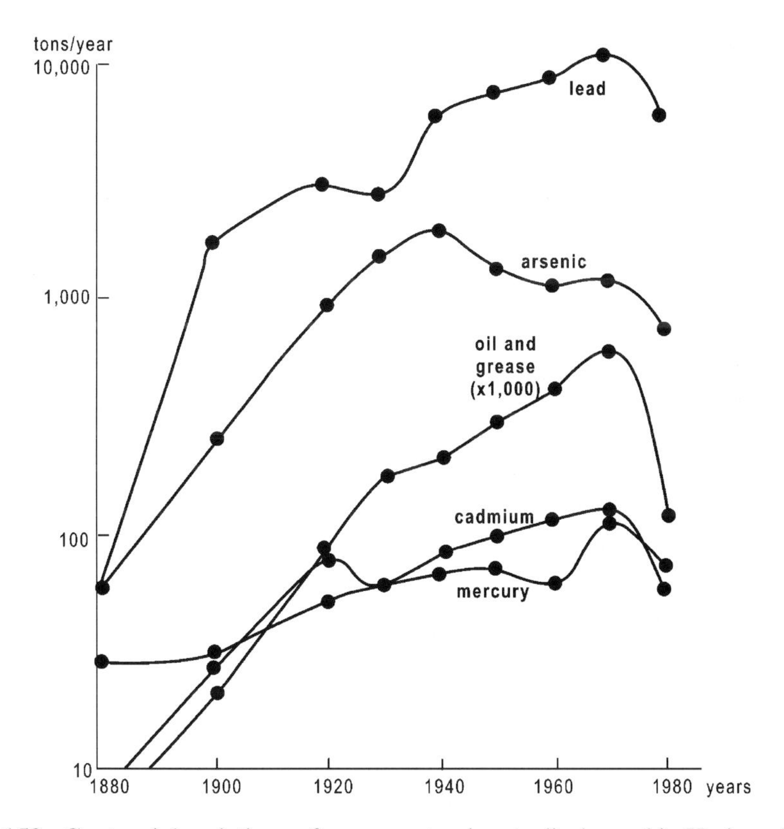

Figure 158: Centennial variations of some contaminants discharged in Hudson–Raritan river basin, northeastern USA (after Tarr & Ayres; see Holland & Petersen, 1995). Values in tons per year (t/yr).

the American Great Lakes, where water contamination started decreasing in the late 1970s (Thomas et al., 1984). In 1975 the petroleum industry-derived geolipids constituted up to 90% of aliphatic hydrocarbons stored in Saginaw Bay, Lake Huron (Meyers & Takeuchi, 1981); the values started decreasing during the late 20th century (see Holland & Petersen, 1995). In the eastern part of Lake Ontario, maximum hydrocarbon levels characterize the sediments deposited during 1970–1975; after that values started diminishing (Bourbonniere & Meyers, 1996).

Many developing countries presently suffer increasing freshwater, soil and underground contamination, which contrasts with the situation encountered in various developed countries. This worsening trend in poor countries results from a dramatic population growth that necessitates increasing the productivity but precludes financing a fight against pollution. Pathogens are particularly widespread in freshwater systems of Third World countries, where high mortality rate and serious illnesses result from "hydric diseases": cholera, hepatitis, malaria, bilharzias, filariasis, etc. The poor bacteriological quality of continental water is responsible for 30% of deaths and for 80% of all diseases encountered in inter-tropical developing countries. Malaria, which is still largely spread in warm swampy regions and is transmitted to man by mosquitoes, may be effectively contained by using DDT; but the massive use of this pesticide has proven to be both polluting and potentially toxic, which sets up an unexpected environmental problem.

10.1.3. Mitigation and Difficulties

Limitation of Waste Discharge

Most rational ways for fighting contamination of freshwater and ground formations consist of *reducing the emission and discharge of pollutants* linked to farming, domestic, industrial and other human activities, especially in densely populated regions. Positive results from such restrictions are observed mainly if they are associated with severe legislation; standards for the maximum amounts of waste allowed to be discharged, application of the principle "polluter = payer", requirement for a 10-year or even longer guarantee when potentially polluting activities are implemented, etc.

Promising techniques include *wastewater recycling*, which allows reduction of the contaminant contents. Remarkable results have already been obtained for limiting the organic or inorganic contaminants discharge in river systems issuing either in lakes (e.g. Lake Geneva, American Great Lakes) or in coastal bays (e.g. northeastern USA; Fig. 158).

Acquisition of durable progress from pollutant discharge limitation necessitates that both environmental monitoring and observance of restrictions are conducted long term under the responsibility of independent organisms. Restrictions are unfortunately of limited impact on the still-widespread practices of uncontrolled waste discharge. Actual diminution of scattered disposal and dumping implies implementing very long-term information measures that progressively lead to increasing individual responsibility and modifying people's mentality regarding environmental damage.

General Mitigation Potential

Continental environments where anthropogenic contaminants are released and migrate are characterized by *natural purification properties* of various types. *Soil and ground* formations present the most diverse and efficient possibilities for mitigation: physical retention of pollutants during contaminated water percolation; adsorption by clay minerals and organic matter; long-term fixation by chemical precipitation or transformation; pathogenic action of bacteria or fungi. *Surface and karst freshwater* is usually marked by limited mitigation potential, which essentially consists of diluting and dispersing the pollutants at increasing distances from discharge points. *River and lake sediments* have little power for reducing contamination, especially when they are permanently impregnated by almost static water.

All types of mitigation processes recognized may act strictly *naturally* or be amplified by *human intervention.*

Interactions of Fresh Water, Sediments and Pollutants

Running water in river and karst systems helps to decontaminate essentially by dilution and dispersion. Water may also be an oxidizing agent that contributes to fixing, destroying or transforming some organic contaminants. The effectiveness of such chemical processes is nevertheless often unpredictable, since it depends on water composition and oxidation-reduction status, which may significantly vary over the course of time. Water contained in *underground aquifers* generally moves slowly and under reducing conditions, which prevents diminishing the contamination rate. An exception occurs when sub-surface groundwater migrates fairly rapidly in suboxic conditions.

River, lake and artificial reservoir sedimentary deposits tend to retain the pollutants adsorbed on particle surfaces or trapped in inter-particle voids. *Contaminants' storage* by freshwater sediments therefore constitutes a natural mitigation process, the consequences of which lead to three main comments:

(1) *The risk of contamination in sediments is* not eliminated but *only postponed.* Regularly conducted cleaning operations of river courses and of man-made

lakes are inevitable (e.g. dredging, flushing), which subsequently necessitate processing of the polluted sediments removed from the bottom (e.g. specific storage, partial incineration; Chapter 9.3.3). Sedimentary deposits in numerous rivers and canals contain huge amounts of noxious or even toxic compounds, which have sometimes accumulated over more than one century, and the removal of which has been chronically delayed. Continuous sedimentation in river and canal courses increases both the amounts of temporarily stored pollutants and the water-contamination risk induced by water from the bottom of watercourses mixing through propeller and boat movements.

(2) *Bottom erosion during flooding* determines a chronic risk for river and lake sediments to be reworked and for pollutants to be remobilised and discharged in fresh-water. Fluvial sediments of Carson basin on the eastern flank of the Sierra Nevada in North America have stored large amounts of mercury since the 19th century, which was used for gold-ore processing. During mining periods, mercury-containing waste was systematically dumped in the valley and progressively incorporated in the sediments deposited along the 150 km of the river course. The toxic metal is now stored in a complex network of active braided banks and channels, as well as in abandoned meanders. Mercury concentration in the Carson River alluvia ranges between 0.018 and 887,000 μg/l, depending on the geographic, topographic, sedimentological and stratigraphic situation. The average mercury content in historical deposits is close to 100 mg/l (99,966 μg/l). Sediment erosion by flooding water is locally responsible for a serious risk of mercury dispersal in downstream regions. Mitigation measures have been recommended that comprise two steps: (1) precise location of major mercury accumulation sectors by means of integrated sedimentological investigations; (2) artificial modification of the Carson river course for preventing the erosion of these critical sectors (Miller et al., 1998).

(3) *The chemical reactivity of freshwater sedimentary environments varies greatly* depending on pH conditions as well as on deposition, burial and diagenesis characteristics. The mobility of particulate metal elements in recent sediments from Loch Dee in southwestern Scotland is largely controlled by local geochemical fluctuations. Below 15 cm depth, most metals are of a natural origin and in various forms: iron, magnesium, copper, cadmium, cobalt, lead and nickel are incorporated in silicates or organic–inorganic complexes, whereas manganese, calcium and zinc are adsorbed on particle surfaces as oxides that tend being oxygen-depleted. By contrast the upper 15 cm of lake sediments are characterized by fairly high amounts of copper, lead and zinc that result from industrial contamination.

Lead in surface deposits displays important fractioning and tends to accumulate in unstable oxides and organic complexes. The difficulty for lead and zinc to constitute stable early diagenetic carbonates appears to result from thermodynamic instability at the sediment–acid freshwater interface. Most recent sediments are enriched in cobalt and manganese, which probably proceeds from secondary precipitation due to vertical migration of the oxidation-reduction boundary. A relatively high concentration of dissolved zinc and cadmium is identified at the sediment–water interface, which points to some discharge of heavy metals and may set up a problem of belated pollution (Williams, 1993). This example illustrates the *complexity* of some freshwater sedimentary environments, inducing much difficulty for understanding the detailed kinetics and chronology of geochemical exchange processes, and leading to still poorly constrained methods for controlling the contaminant mobility.

Pollutants' Retention in Terrestrial Environment

The soil's ability for retaining the pollutants strongly depends on the vegetation-cover types. The possibility for certain plant species to stabilize and assimilate the contaminants has given way to phyto-remediation techniques, which are employed in an increasing way for treating urban and industrial wasteland (Chapter 9.3.3). As shown by recent investigations at Mortagne-du-Nord in northwestern France, forest soils and, to a lesser degree, prairie soils retain the pollutants better than cultivated soils (see Fig. 142).

In the Alsace plain, northeastern France, intensive cereal cultivation associated with increasing use of fertilizers has led to pollution of the Rhine Valley groundwater through nitrate infiltration. Groundwater pollution beneath cultivated soils is much stronger than beneath forest and prairie soils, where pollutants are retained by hydromorphic and peaty formations, or display natural *denitrification* of NO_3^- into N_2O (Bernhard et al., 1992). *Restoration of alluvial and riparian forests* therefore constitutes an efficient means for limiting the contaminant export. The definitive elimination of pollutants may be obtained by implementing successive operations in a cyclic way: reforestation, plant growing, tree cutting, incineration, and "ultimate waste" storage.

Retention of pollutants by underground formations is controlled by rock permeability, which determines the possibility for contaminated water to percolate or not. In Brittany for instance granite rocks are densely fractured and weathered, which allows contaminated water to infiltrate the rocks. By contrast metamorphic schists are cohesive and little fissured, which prevents seepage of contaminated water. As a result the nitrates and phosphates derived from land and livestock farming in Brittany tend to be trapped in granitic ground formations, whereas they are exported at soil surface in schistose areas and favour downstream pollution as well as estuarine and coastal eutrophication.

As underground ability to retain water-transported contaminants fundamentally depends on local *geological characteristics*, specific maps are progressively established that summarize the self-purification power of ground formations. These maps, which take into account field and laboratory data, are particularly useful for assessing the capacity of sensitive, potentially threatened regions to mitigate the pollution impact. For instance Réunion, which constitutes a sloped territory subject to growing agricultural, urban and tourist pressure, produced in 2000 such a map, designed by the French Geological Survey (BRGM).

Chemical Adsorption, Oxidation and Reduction Processes

The chemical reactions occurring within soil and subsoil formations partly contribute to neutralizing the effects of pollutants. This is mainly the case in *tropical and equatorial soils* formed in warm and humid climates. Organic activity concentrates in surface horizons of these acid and reducing soils, which favours degradation of alkaline elements and adsorption of chemical compounds. Such natural mitigation processes work in Amajà State, northern Brazil (Amazon Basin), where surface horizons of aluminiferous–ferruginous soils retain noticeable amounts of *mercury*. This heavy metal, the average amount of which in drainage water ranges between 100 and 300 μg/kg, and locally reaches 1,000 μg/kg, is dominantly of an anthropogenic origin (95%) and issued from atmospheric fallout. In the soil, mercury is stabilized through iron oxy-hydroxide adsorption, and its content in seeping water decreases down to 1.2–6.1 μg/l (Oliveira et al., 2001).

The property of tropical forest soils to filtrate and regulate the metal pollutant fluxes tends to be impaired by deforestation and modern cultivation practices. Soil denudation and resulting erosion cause mercury destabilization and release towards river tributaries, and its potential concentration through food chain processes. Freshwater fish in French Guyana rivers may contain noticeable amounts of mercury originating from gold mining activities and concentrated by successive predators. Carnivorous fish species contain an average of 48 μg/l Hg, which induces a possible health risk for people by reference to European standards (see Table 41), especially as far as large-sized individuals are concerned (Richard et al., 2000).

Deeper horizons in tropical soils are usually much too old to have been contaminated by human activities linked to industrial expansion. These ferruginous, partly indurate horizons are able to retain some contaminants carried by drainage water via *natural oxidation* processes. This property has led to predicting implemention of specific mitigation methods in the Ganges deltaic plain (India), where groundwater locally contains large amounts of natural and anthropogenic arsenic. Arsenic oxidation and retention by ferrallitic soils could help to diminish the risks of dermatosis, hyperkeratosis, and melanosis in this region where more than thirty million people live (Bhattacharyya et al.; see Schmitt et al., 2001).

Other methods are envisaged for stabilizing contaminants in the soils, namely by using exogenous mineral compounds. Experiments on mercury-contaminated soils from Greece have shown that adding to soils 1–5% of *zeolites* extracted from natural ores from Thrace helps retain 58–86% of the toxic metal accumulated in esparto-grass and rye-grass roots and shoots (Haidouti, 1997).

Action of Bacteria

Soil decontamination properties depend to some extent *on bacterial, fungal and viral activity.* The metabolic, ecological and pathogenic action of soil microorganisms has been insufficiently investigated and is therefore imperfectly understood. The bacterial mitigation efficiency is deduced from results showing the decrease in toxicity of pore water that has percolated through natural soils characterized by a different organic load.

Scientific investigations are currently performed increasingly to better-characterize mitigation by bacteria and other microorganisms. The Bemidji aquifer in Minnesota, USA, has been contaminated by mono-aromatic hydrocarbons (MAH) of benzene and alkyl benzene types, from the oil industry. These rather soluble compounds have been transformed into organic acids as groundwater was migrating within geological formations. The MAH transformation and resulting mitigation effects have been attributed to an in situ degradation caused by anaerobic bacterial activity (Cozzarelli et al., 1990).

Possible applications of bacterial action for chemical decontamination of soil give rise to multiple experiments, some of which tend to become operational: digestion by microorganisms of agriculture-derived nitrates (e.g. livestock excretions, fertilizers) that have been stored by infiltrating water in groundrock fissures and fractures; bacterial bio-lixiviation permitting extraction of gold from its gangue without using toxic metals like mercury; etc. Recent researches suggest that the "sleeping micro-biosphere" within underground rocks constitutes an important biomass and could be used for cleaning contaminated ground and soil formations. Some microphyte specialists place great hope on the potential for underground bacterial strains to be reactivated and used for mitigating the diffuse and almost inaccessible pollution affecting deep aquifers and other groundrocks.

10.2. Coasts and Seas

10.2.1. Coastal Towns and Harbours Mirror Anthropogenic Contamination

1975, Gulf of Saronikos, Greece

Located between the Peloponnese and Attica in southern Greece, the Gulf of Saronikos is limited to the northeast by conurbation of the coast of Athens and the harbour of Piraeus (Fig. 159). Keratsini and Elefsis bays form the inner part of the gulf, and constitute a semi-closed and shallow area (water depth <30 m) submitted to moderate hydrodynamics (tidal range <60 cm). A dense port, industrial and urban complex responsible for active effluent discharge borders the two bays.

The evolution of the seafloor environment in the northern Saronikos Gulf was followed from 1972 to 1975 by means of various observations and analyses (Griggs et al., 1978). Spectacular changes of surface *sediment aspect and colour* were recorded, mainly in Keratsini bay, where original greyish-to-brownish clayey mud was progressively replaced by blackish-to-red sediments. In less than 3 years, blackish mud rich in hydrogen sulphide (H_2S) spread over 5 km to

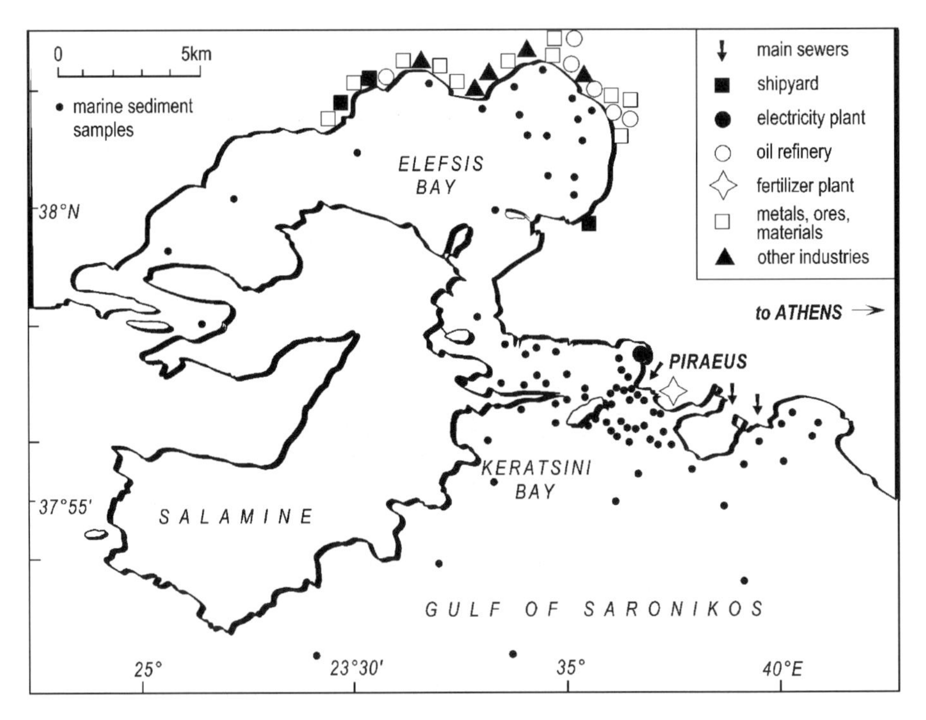

Figure 159: Industrial and urban pressure in the northern part of Saronikos Gulf, Greece (after Griggs et al., 1978).

the west, south and southeast of the bay, starting from a major sewer outlet that drained Athens's wastewater. Reddish mud formed an area larger than 1 km^2 that developed to the south of a chemical fertilizer plant (Fig. 160). Nearby the main sewer-heated water was artificially discharged at a rate of 19 m/sec from the cooling system of an electric power plant; this was responsible for a rise by 10°C of seawater temperature and caused serious disruptions in the local ecosystem. In 1975 the whole area subject to various anthropogenic waste discharge was devoid of benthos life.

The *chemical composition* of surface sediments in Elefsis and Keratsini bays was analysed during a 3-year investigation. Important changes were recorded

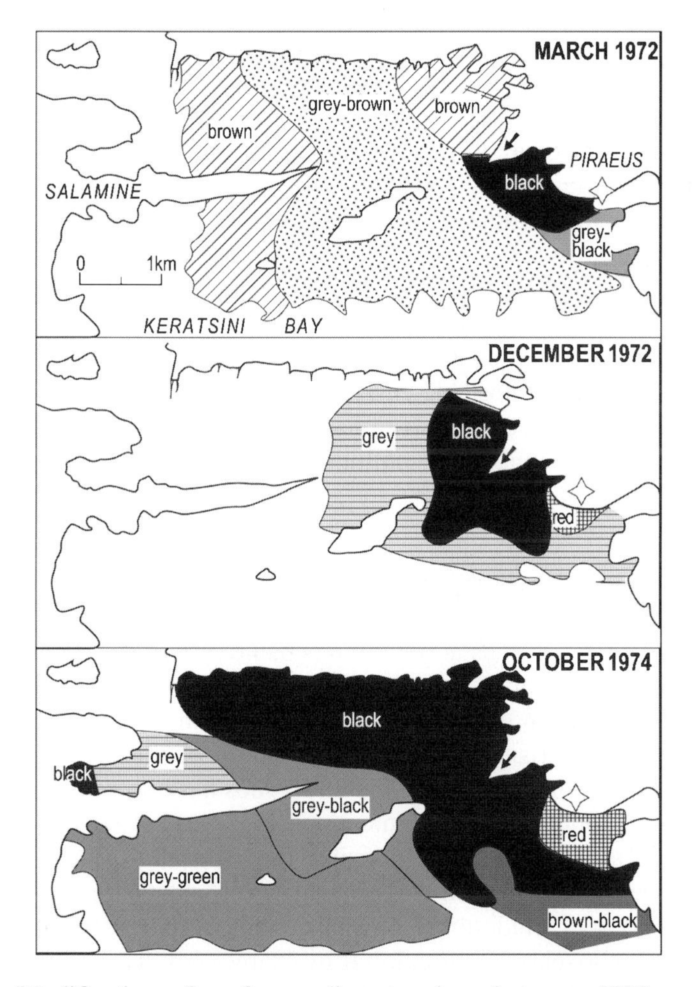

Figure 160: Modification of surface sediment colour between 1972 and 1975 off Piraeus harbour (after Griggs et al., 1978). Arrows, main sewers. Star, fertilizer plant.

that point to the importance of coastal pollution. By comparison with non-contaminated zones, the proportion of organic carbon in polluted areas has increased from 0.4 to 6.0%, mercury (Hg) from 0.3 to 10 ppm, arsenic (As) from 5 to 1,500 ppm, chromium (Cr) from 50 to 1,200 ppm, zinc (Zn) from 40 to 2,992 ppm, antimony (Sb) from 0.3 to 75 ppm, and cobalt (Co) from 4.9 to 426 ppm. The growth rate of pollutant levels attained and sometimes exceeded 25 for Hg, 215 for As, 17 for Cr, 37 for Zn, 108 for Sb, and 47 for Co. Concentrations of Hg, As, Zn, Sb and Co have particularly augmented in red mud linked to the fertilizer plant discharge, and those of Cr were maximum off the main sewer outlet. Local contamination has been recorded in many other sectors, mainly in the vicinity of the various plants and factories established along the bay shore: oil refineries, shipyards, metal-working and food-processing plants, chemical and electrochemical plants, manufacturing complexes for producing asphalt, cement, aluminium, glass, paper, weapons, etc.

The paper published in 1978 by G. B. Griggs and colleagues illustrates the scientific approach developed since the 1970s for identifying and characterizing anthropogenic contamination of a coastal zone. In the conclusion the authors insisted that it was in the interest of the scientific community to ensure the monitoring of contaminated zones for better identification of heavy-metal sources, determination of the pollutant dispersal modes and causes, and understanding the modalities and kinetics of ecosystem modifications. Almost no mention was made of possible environmental consequences of man-induced disruptions having affected the Gulf of Saronikos. This points to the fairly recent realization by most scientists of pollution risks caused by excessive human pressure on coastal environments.

1985, Major Contaminant Discharge in Mediterranean Coastal Zones

The impact of Athens on coastal pollution is by no means an isolated case in the Mediterranean area. This semi-closed sea is facing considerable human pressure, which has been dramatically amplified during the last decades due to huge expansion of tourism activities. Compilation of the main industrial and urban pollution zones recognized in the mid-1980s has allowed identification of about 30 major sources of contamination (Fig. 161). Coastal currents and longshore drifts are chronically responsible for *important offshore dispersal of contaminants*, especially in northwestern and easternmost parts of the basin. The most affected coasts include those of Spain, France, Italy, Lebanon and Israel. Nitrogen and/or phosphorus agricultural discharge may participate in seawater contamination, which periodically determines coastal *eutrophication*. This is especially the case along Adriatic Sea coasts.

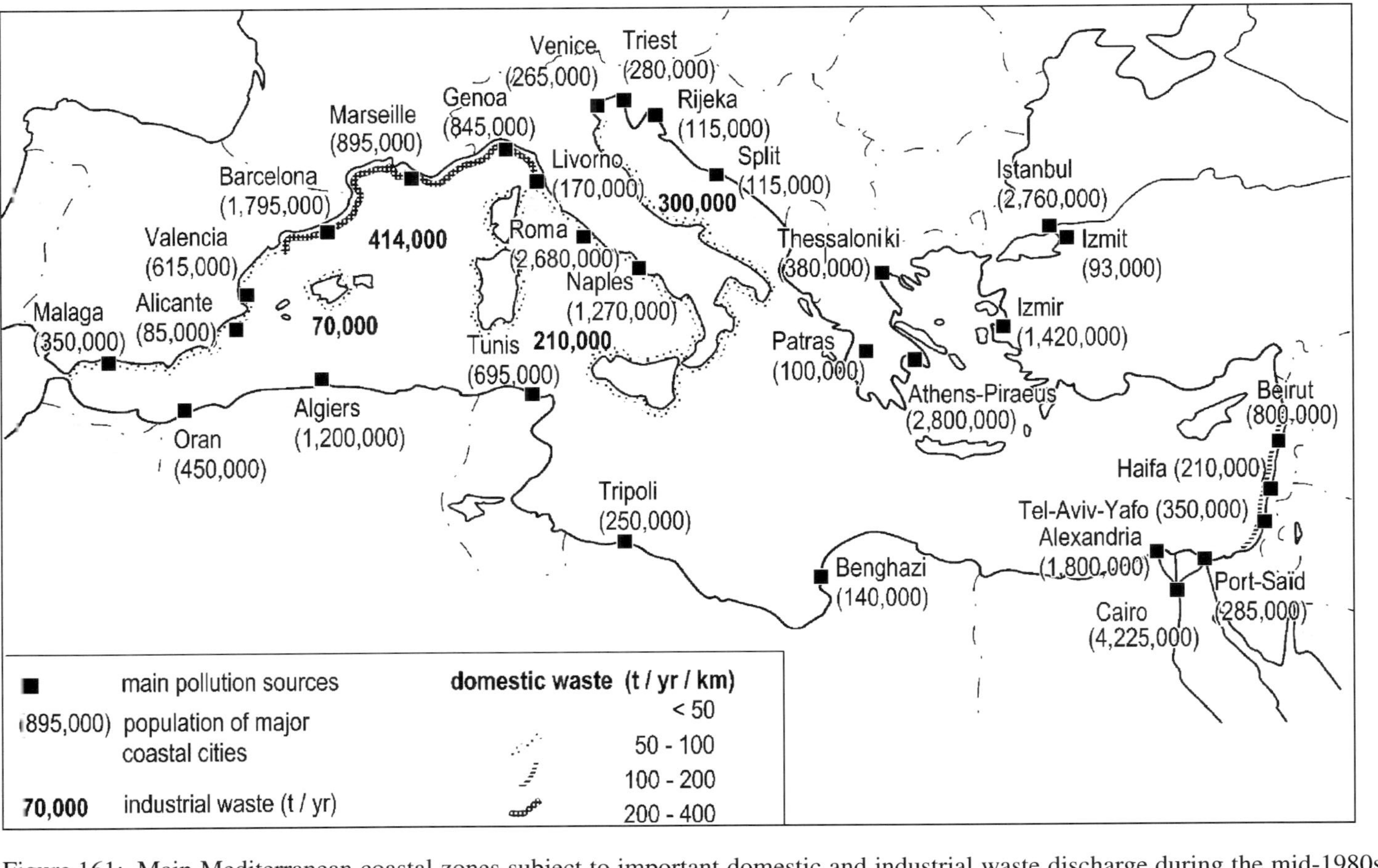

Figure 161: Main Mediterranean coastal zones subject to important domestic and industrial waste discharge during the mid-1980s (after Clark; see Viles & Spencer, 1995).

Note that man-induced *coastline retreat* frequently develops in Mediterranean regions and adds to anthropogenic nuisance. This is mainly the case in the vicinity of some major river mouths such as Ebro, Rhone, Po, Vardar and Nile. These rivers are characterized by a strong diminution of suspended and bottom particle load, due to active sediment trapping in the numerous artificial reservoirs implemented along hydrographical basins (Chapter 4.3.4; Fig. 77).

As a consequence costal zones in the Mediterranean range belong to the most threatened regions in the world (see Fig. 77). Realization for a few decades of both high contamination rate and environmental fragility in many Mediterranean sectors has given way to active research and mitigation experiments. It is hoped that such measures, combined with the necessity to maintain a high potential of attractiveness, will help most Mediterranean coasts recover adequate environmental quality.

1995, Contamination of Sydney Harbour, Australia

Port Jackson estuary in southeastern Australia is located in the heart of Sydney's urban and industrial complex. Sydney's conurbation hosts four million inhabitants, which represents a quarter of the country's population. Exhaustive investigations on recent sediment contamination have been performed in the perspective of the 2000 Olympic Games, an important part of which was planned to be held along the estuary. The levels of eight heavy metals (Cd, Co, Cu, Fe, Mn, Ni, Pb and Zn) were measured in 1,700 surface sediments sampled in diverse parts of the 30 km-long estuary, as well as in river tributaries, harbour annexes, and adjacent canals (Birch & Taylor, 1999). Results showed a *strong pollution by copper, lead and zinc*, particularly in upstream sectors marked by intensive industrial and commercial activity (Fig. 162). Noticeable reworking of contaminated deposits from tributary canals and subsequent transfer pollutants occurred frequently, due to thundery showers and flooding. Aquatic flora and fauna were strongly affected by this chemical pollution. The heavy metal content in sediment diminished rapidly from the upper to the lower part of the estuary. But the total amount of contaminants stored in estuarine sediments was considerable: 1,900 t of copper, 3,500 t of lead, and 7,300 t of zinc.

The strong metal pollution of Sydney harbour bottom resulted from uncontrolled industry waste discharge during numerous decades in the Port Jackson estuary. A large-scale programme was established in the late 1990s for reducing waste discharge and treating the contaminated deposits. The large volumes of contaminated sediments nevertheless make remediation very difficult, even on a long-term scale.

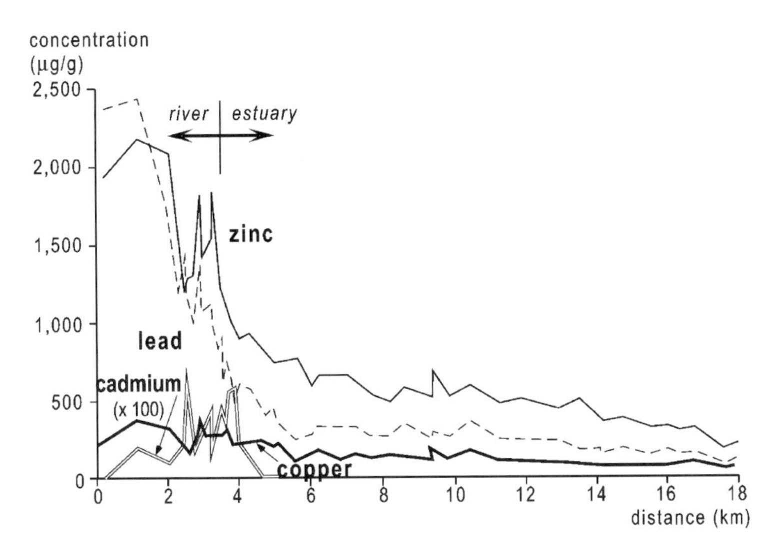

Figure 162: Distribution of heavy metals in Port Jackson estuarine sediments, Sydney harbour, Australia (after Birch & Taylor, 1999).

10.2.2. Nature and Distribution of Contaminants

Ocean Pollution

Sources

River discharge Most dissolved and particulate contaminants carried from land to sea result from river discharge associated with coastal streaming. Whatever their location and size, numerous river valleys display dense human population generating abundant farm, urban and industrial waste that are subject to leaching and drainage. Rivers therefore play a predominant role in collecting the contaminants and transferring them to the coastal zone (Fig. 163). This transfer is nevertheless artificially reduced by particle trapping in the lakes that have been implemented by Man along many river courses (i.e. "neocastorization"; Chapter 6.4.1). The retention of sediments by artificial reservoir dams is estimated to amount to 15% for all world river basins, 30% for all regulated basins, and from 50% to almost 100% for densely equipped river basins (e.g. Colorado, Indus and Nile rivers; Vörösmarty et al., 1997). River valleys comprising most artificial reservoirs responsible for massive sedimentary particle and pollutant trapping are primarily located in Europe, and secondarily in North America, Africa and Australia.

Atmospheric fallout The second way for contaminants to reach the ocean consists of wind transportation and subsequent fallout due to density effects or

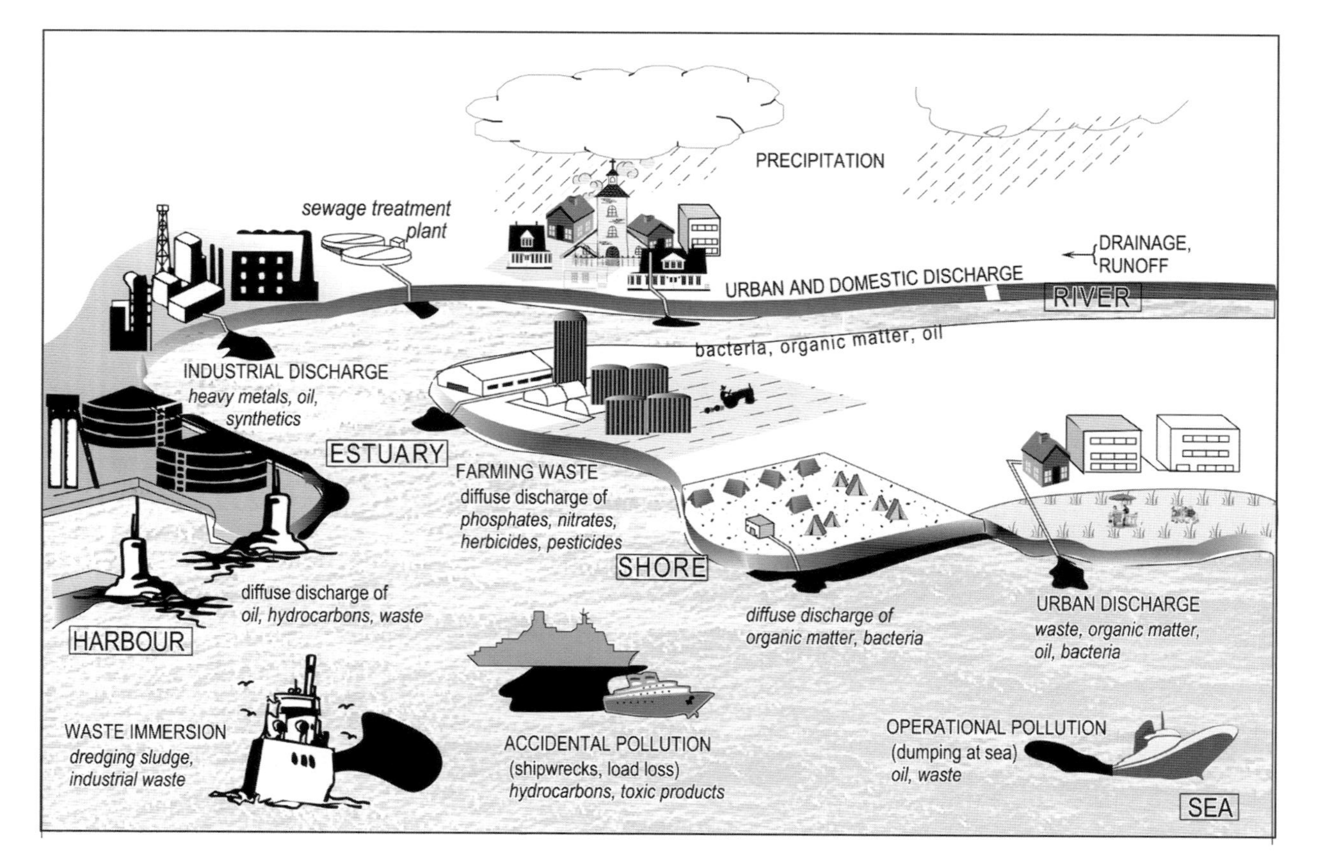

Figure 163: Main contamination sources in coastal environments. Importance of river input relative to other discharge modes (after Massoud & Piboubès, 1994). Dark patches indicate sectors of abundant local discharge.

rain trapping. Atmospheric transportation of particles and aerosols depends on the wind regime and is generally diffuse. Its quantification necessitates combining long-term measurements and heavy numerical simulations, and is therefore difficult to achieve. Some experts estimate that nitrates and mercury in marine water and sediments result in comparable quantities from aeolian and river fluxes.

Waste dumping Direct immersion of solid or liquid residues constitutes the third way for man-produced contaminants to be released in the marine environment. The nature and packaging of contaminants as well as their abundance are very diverse, changing and often still poorly known and controlled. Dumping at sea was without official regulation until the 1970s. Most dumped materials consist of: (1) building construction and demolition products; (2) industrial, farm and domestic waste; (3) chemical and radioactive products; and (4) various devices, weapons and explosives. Most authorized dumping areas in the ocean are located in fairly deep seas (water depth > 1,000 m), except for some dredging residues (Chapter 9.3.3). Major dumping zones are located in the Atlantic Ocean around England and east of USA and Canada.

Distribution

Anthropogenic contamination of marine environment depends on several major parameters:

- distance of exposed landmasses where most waste are produced and discharged;
- location of river mouths where terrestrial products are released from most densely populated basins;
- location of major coastal conurbations and industrial complexes from which important pollution may derive;
- location of sea-bordering countries where population growth is exponential and where waste mitigation and processing techniques are insufficiently developed (e.g. Brazil, China, India, Indonesia, Nigeria);
- possibility for contaminants to be dispersed and diluted in epicontinental and semi-closed seas where marine currents tend being sluggish or of local effect only.

Most heavily contaminated marine zones are situated on marine continental shelves and slopes, and especially on platforms enclosed by landmasses or bordering densely urbanized coastal regions (Fig. 164).

World regions displaying high marine contamination comprise the Mediterranean Sea (namely the French and western Italian continental margins), the

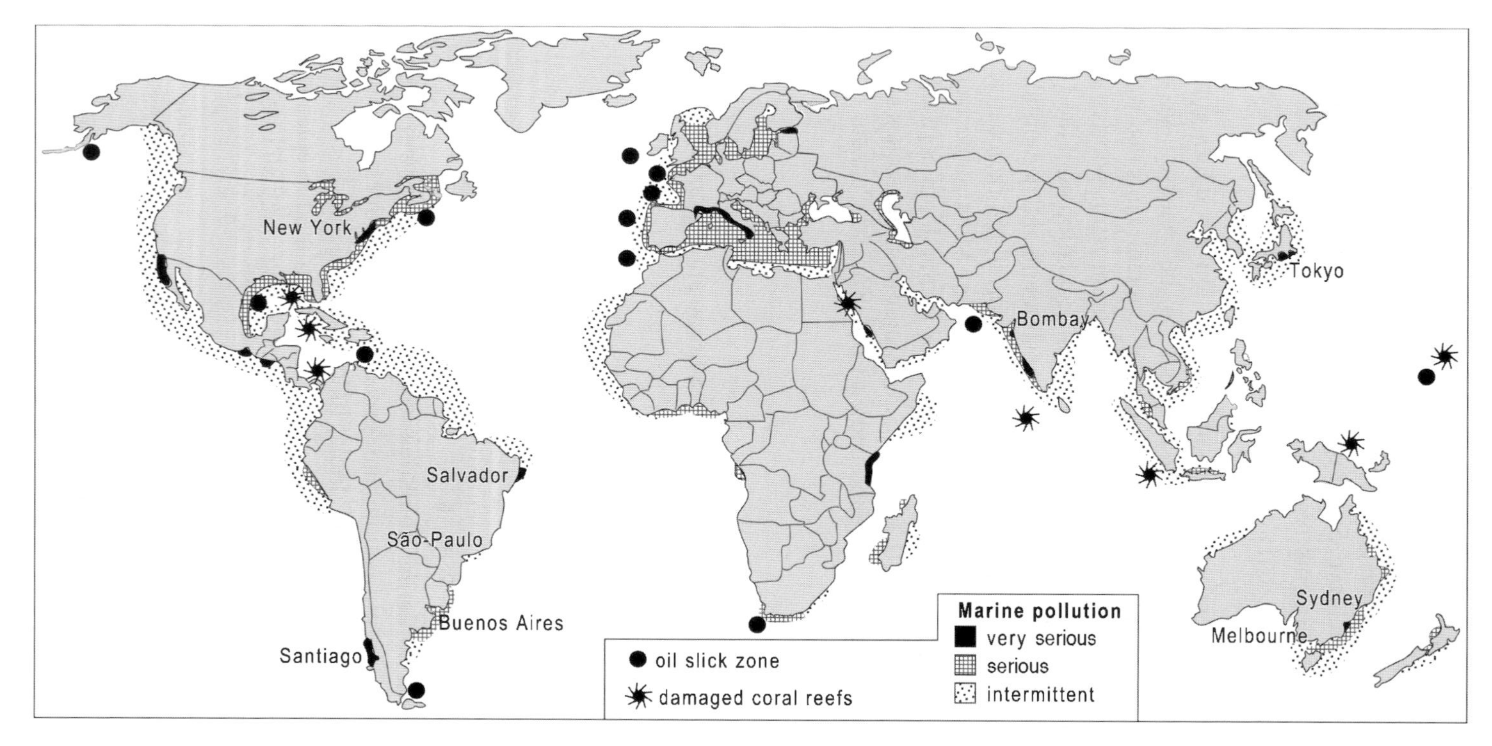

Figure 164: Main polluted areas in the ocean. Location of main zones subject to oil spills and to coral reef deterioration (after Barnier, 1992; Keller, 2000).

North and Baltic Seas, the Gulf of Mexico, the Caribbean Sea, the Red Sea, and the Sea of Japan. Other heavily contaminated zones are located on continental shelves bordering some megalopolises: Los Angeles–San Diego and New York City in North America; Buenos Aires, Salvador and Santiago in South America; Bombay and Tokyo in Asia; Melbourne and Sydney in Australia. The *Mediterranean* (see Fig. 161) *and Baltic Seas* are particularly subject to strong pollution for they are surrounded by densely anthropized countries and fed by diversely contaminated rivers. Contaminants in both seas hardly get diluted since seawater is difficult to be renewed due to straits narrowness and shallowness, low tide and current regime, and temperature and salinity stratification limiting vertical exchanges.

The seas bordering developed countries that became industrialized since about one century ago tend to display noticeable contamination, which spreads largely towards offshore zones: North America, Western Europe, Japan, Eastern Australia, etc. The same tendency affects some more recently industrialized countries where strong economic development and extremely rapid population growth combine; such a situation characterizes particularly some South American, African and Southeastern Asian countries.

Impact

Potential nuisance induced by contaminant discharge at sea concerns primarily the *marine biota and ecosystems*: habitat disruption, excessive bio-stimulation, fixation of metals or organic compounds impeding growth and reproduction functions, etc. Various examples of such biological disruptions are regularly published in scientific journals and often tend to be amplified by the media: wild birds' sterilization due to insecticides (DDT) ingestion, increased viral dependency of seals having accumulated polychlorobiphenyls (PCB), contamination of shellfish beds (oysters, mussels), asphyxiation of coastal plant and animal communities suffering eutrophication due to excessive fertilizer discharge, etc.

Secondary effects may potentially affect *human health and practices*: food poisoning by contaminants that have been concentrated through the food chain; pollution and decimation of fish- or shellfish-farming populations; temporary loss of littoral use due to excessive algal blooms or beach pollution; etc.

Metal Discharge

Heavy-metal fluxes in the North Sea The North Sea is exposed to important contamination sources and effects since its basin is fairly shallow, surrounded by several industrial countries, poorly connected to the Atlantic Ocean, subject to noticeable river input (Elbe, Rhine, Scheldt, Thames) and crossed by countless merchant ships. The discharge of contaminants into the North Sea has been considerable during 1950–1980: 5,000,000 t of nitrogen, 100,000 t of phosphorus, 11,000 t of lead, 335 t of cadmium, 4,500 t of copper, and 75 t of mercury have been annually released in this basin.

A general *pollution inventory* was established during the late 1980s. It allowed assessing the diversity, abundance and distribution of discharged elements, identifying the contaminant sources, defining transfer and storage modes, and starting implementation of mitigation measures (Salomons et al., 1988). Heavy metals have been considered in detail, since they were abundantly released from the countless factories established along river courses and Western European coasts. Lead, cadmium and mercury contents have especially given cause for concern. Due to dilution the heavy-metal concentration tends to decrease from coastal to offshore environments, both in the air and in water. The heavy-metal distribution is proven to be little dependent on salinity and suspended matter load, which is attributed to the diversity of sources and transportation.

Few correlations also arise from comparing heavy metals in seawater and sediments, as exemplified by the distribution of *lead* (Fig. 165, A and B). Lead fluxes in seawater are especially important in southern regions, whereas metal accumulation in bottom deposits predominates in the central part of the basin and correlates with the amount of fine-grained sedimentary fraction. Fairly high contents of lead occur in some isolated areas, mainly eastwards of England, which could result from local bioaccumulation processes. During the 1990s more severe regulations have been implemented, namely under the auspices of the European Union. Recent investigations show that heavy-metal fluxes currently tend to diminish in North Sea water and surface sediments.

Heavy-metal exchanges *Chemical exchanges and interactions* between heavy metals and both seawater and suspended or deposited sedimentary particles are crucial to know for understanding and potentially controlling the contaminants' concentration, migration or solubilization. In the *Hudson estuary* close to New York City, cadmium, lead, copper and zinc released from upstream industrial discharge display maximum concentration both in most turbid water where river and marine currents get mixing, and in fine-grained mud deposited in Haverstraw Bay. The toxic metals get firmly adsorbed on clay-mineral suspended particles, which settle in the lesser hydrodynamic parts of Hudson estuary. Contaminants trapped in clay-rich mud are nevertheless susceptible to be reworked and dispersed downstream, mainly through harbour dredging or bottom erosion by maritime traffic (Menon et al., 1998).

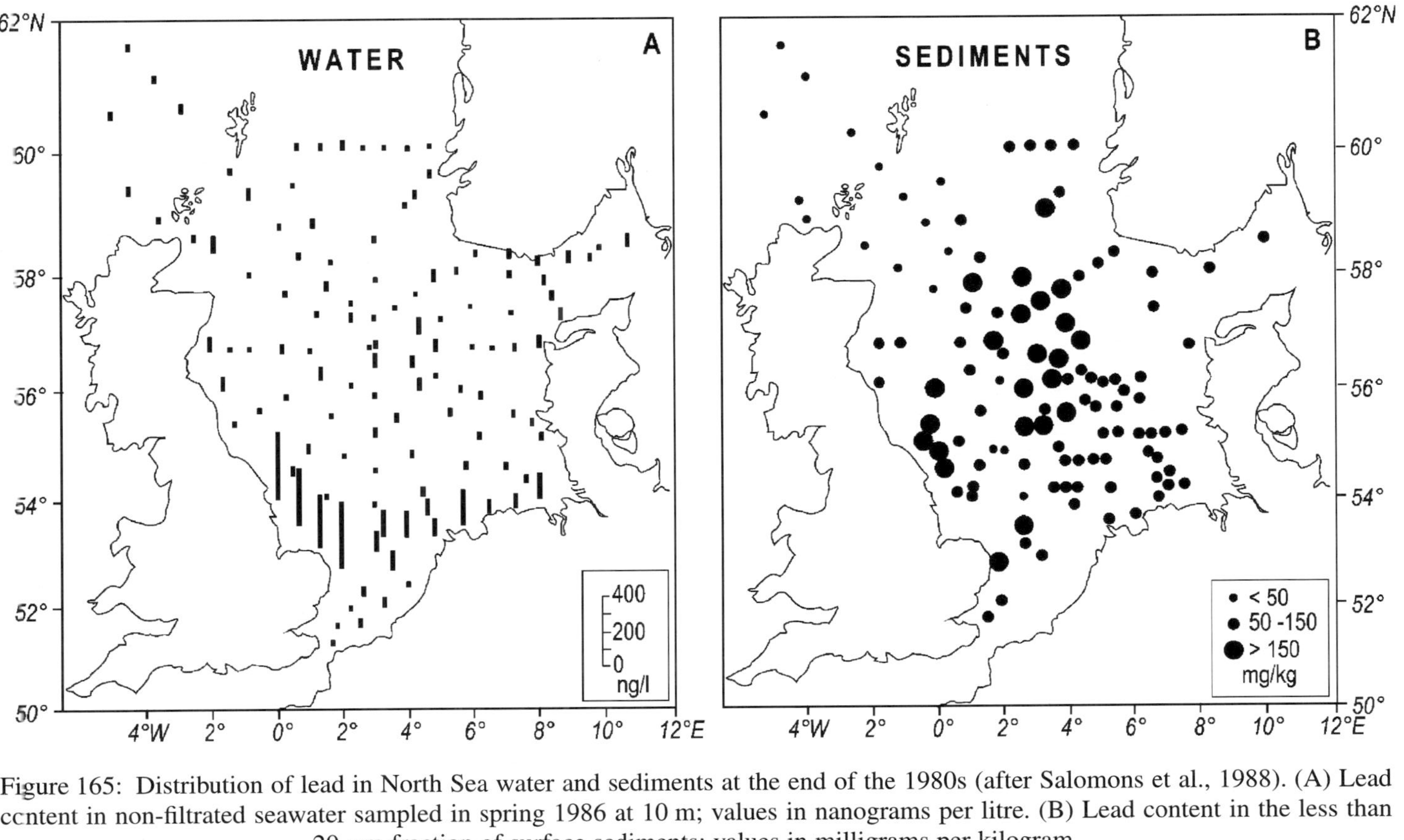

Figure 165: Distribution of lead in North Sea water and sediments at the end of the 1980s (after Salomons et al., 1988). (A) Lead content in non-filtrated seawater sampled in spring 1986 at 10 m; values in nanograms per litre. (B) Lead content in the less than 20 μm fraction of surface sediments; values in milligrams per kilogram.

In the southern part of the *Kara Sea*, between northern Russia coasts and the Arctic Ocean, surface sediments of the St. Anna trough contain significant amounts of mercury (up to 3,915 ppm) and arsenic (up to 710 ppm). These toxic elements are essentially of an industrial origin and partly result from continental drainage of the huge and complex Ob and Yenisey river systems. Another fraction of these heavy metals could have been released in seawater from military and industrial metal waste dumped between 1950 and 1991 offshore from the Novaya Zemlya archipelago. Due to bottom oxidizing conditions, early diagenetic processes in the St. Anna trough allow arsenic to remain stable in the form of arsenate sorbed onto iron oxy-hydroxides. On the other hand mercury, which was initially fixed by organic carbon and clay minerals (smectite, kaolinite), tends to be oxidized by bottom currents and released in seawater (Siegel et al., 2001). Toxic elements therefore undergo various, somewhat opposite chemical exchanges. Such reactions may induce ecosystem pollution and be partly responsible for the accumulation of toxic mercury by seabirds, seals and polar bears of Arctic European zones.

Evolution during recent decades Various analyses published during recent years report that *metal contaminant levels have stabilized or even diminished* in marine sediments deposited during the last quarter of the 20th century. This is for instance the case for silver, cadmium, mercury, and lead stored in Puget Sound deposits off Seattle, Oregon (Turner II et al., 1993), and of heavy-metal concentrations in Mersey Estuary sediments, Great Britain (Chapter 10.2.4, see Fig. 170).

On the *Biscay Bay* platform, the amounts of mercury increased during most of the 20th century and were maximum in the late 1970s with values locally reaching tenfold the pre-industrial European standards. Since the 1980s the mercury content stored in sediments has tended to diminish rapidly (Fig. 166).

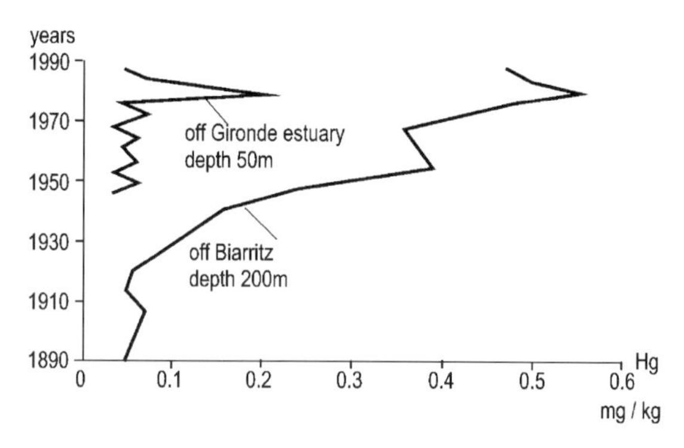

Figure 166: Mercury concentration profiles in sediments cored on the Biscay Bay shelf (after Alzieu; see OSPAR, 2001).

The toxic metal levels are lower in mud accumulating off the Gironde estuary than in the outer shelf deposits cored off Biarritz–Capbreton in the southern part of the bay. This points to the preponderance of local sources such as Basque Country factories relative to the drainage of large hydrographical networks such as Garonne River basin. The diminution of mercury in cored sediments mainly results from voluntary reduction of air-and-water contaminant discharge, with possible additional effects of early diagenetic natural changes (e.g. increasing bottom currents inducing oxidation at the sediment–water interface).

It is still unclear whether or not the results published so far in scientific journals reflect the average trend of the metal contamination status in the marine environment. Most published data arise from offshore zones situated along developed countries, where mitigation measures are effective. Many coastal regions subject to heavy metal discharge have not led so far to appropriate investigations, especially as far as some recently industrialized countries are concerned. Many large-scale, synthetic studies involving the temporal evolution (i.e. analyses of cored sediments) and being associated with modelling data are needed before getting reasonable appraisal of heavy metal contamination status and trends in the world's oceans.

Discharge of Organic Contaminants

Many environmental disruptions may result from the discharge at sea of organic compounds such as fertilizers, hydrocarbons and insecticides. Most common and widespread damage consists of uncontrolled biological developments and subsequent *eutrophication* induced by the anthropogenic release of nutrients. Eutrophication is particularly triggered by the supply of nitrates and/or phosphates that have been excessively, untimely or inappropriately spread on cultivated land, have been leached by rainwater and streaming, and finally have been discharged by rivers. Nitrates and phosphates tend to fulfill their fertilizing function downstream from farmland areas, either in river courses and lakes, or in estuarine, deltaic and coastal environment. These artificial nutrients help to stimulate the primary productivity and cause uncontrolled algal blooms. A chain of ecological disruptions results from such excessive plant development: strong oxygen consumption and subsequent depletion, carbon accumulation, chemical reduction of water and sediment, asphyxia and massive *mortality* of biota.

Some experts are wondering if eutrophication effects are of local importance only or if they may potentially disrupt more regionally the oxygen and carbon balance at continent-to-ocean and sediment-to-seawater interfaces. On a global scale, the anthropogenic input of inorganic nitrogen to the coastal zone has been multiplied by 2.5 in the last decades, and that of phosphorus by 2. Maximum values characterize the coastal regions, and especially those situated off Western European countries.

The *Baltic Sea* is a noteworthy example of fertilizer-contaminated marine environment. This large, semi-closed basin is surrounded by densely cultivated farmland drained by about 250 rivers from which large amounts of nitrates and phosphates are seasonally released. About one million tons of land-derived nitrogen is annually discharged in the Baltic Sea. Nitrate and phosphate levels have increased continuously since the 1960s in both seawater and sediments. These nutrients occur usually more abundantly in bottom water than close to the sea surface, where phytoplankton is responsible for primary nitrogen and phosphorus consumption. The Baltic Sea periodically undergoes eutrophication phases that induce important ecosystem disruptions: uncontrolled biological development for the benefit of very few species, organic matter accumulation, water and surface sediment reduction, and massive plankton and benthos mortality.

The *lagoon of Papeete* in Tahiti, French Polynesia, constitutes a tropical coastal environment bordered by both anthropized land and fragile coral-reef ecosystems. The distribution of phosphorus (P) has been measured in sediments deposited since 1860 in the harbour sector of this lagoon. Precise dating by ^{210}Pb excess activity has allowed identifying successive steps of the nutrient distribution in its different chemical forms: exchangeable P, iron oxide sorbed P, bio carbonate-contained P, marine organic P, and terrigenous P. A strong increase in phosphorus accumulation rate occurred in 1957 (Fig. 167) and was associated with a doubling of sedimentation rate, a diminution of carbonate P, and an augmentation of other P forms except the exchangeable one. This major change reflects a sudden increase in terrestrial phosphorus supply, which correlates with strong increase in soil erosion and domestic waste discharge caused by human pressure growth. Land-derived phosphorus constituted 54% of all the element phases contained in marine sediments deposited in 1995, whereas it formed only 30% of total P in the early 20th century (Harris et al., 2001). This example illustrates the *close dependence* linking man's activities on land and functioning of the coastal ecosystem in adjacent seas.

10.2.3. Accidental Dumping at Sea

Amoco Cadiz, Erika: Two Examples of Oil Spills off French Coasts

On March 17, 1978, the oil super tanker *Amoco Cadiz* scraped its keel on the rocky bottom of the Brittany coast, 2 km from Portsall on the English Channel southern border. The ship ran aground and a large hole opened in the hull; crude oil started leaking. Bad weather conditions prevented refloating, laying of sea-surface barriers against hydrocarbon dispersal or tank emptying. The onslaught

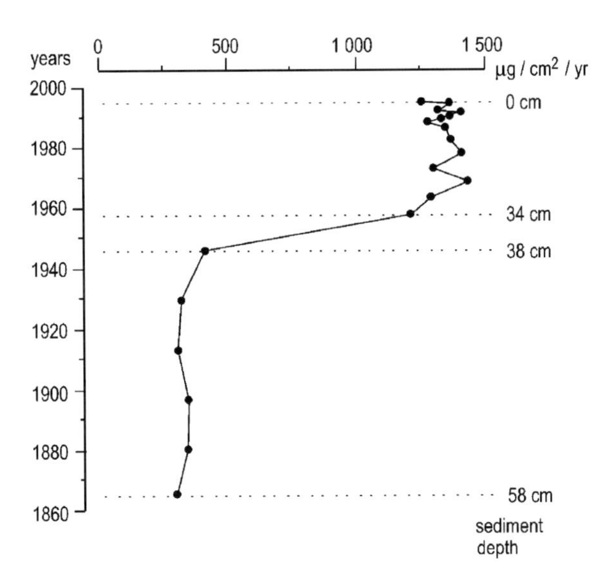

Figure 167: Phosphorus accumulation rate between 1865 and 1995 in Papeete lagoon, Tahiti, French Polynesia (after Harris et al., 2001).

of the storm finally caused the ship to break up and the release at sea of 227,000 t of light crude oil. Wind and currents pushed hydrocarbons towards the coast. A huge, viscous oil-slick invaded and polluted the water column, shallow-water bottom rocks and sediments, as well as sandy beaches and cliffs.

Of the North Brittany shoreline, 198 km were affected by the oil spill, and the seafloor was polluted as far as 60 km from the coast. Marine fauna and flora were decimated. North Brittany shores remained heavily damaged for more than 5 years, inducing considerable ecological and economic losses.

On December 12, 1999, the oil tanker *Erika*, which was chartered by Total Oil Company from an Italian ship owner, was facing a violent storm in the North Biscay Bay, fairly close to the Southern Brittany coast. The ship broke in two and sank. Twenty thousand tons of dense crude-oil escaped from the tanks and were pushed by waves towards southeastern Brittany and Vendée coasts, which became severely polluted.

An extensive plan of intervention was rapidly implemented, based on an agreement between French authorities and the petroleum group, and associated with numerous voluntary groups. This allowed largely restoring the coastal environment before the 2000 summer season. The 11,245 t of heavy fuel remaining in Erica tanks were pumped and recycled, 15,000 birds were saved and cleaned, sandy and rocky beaches were actively cleaned up, and 200,000 t of shore and marine sediments were removed and stored in safe conditions.

Cleaning, restoring and preventing operations continued in 2001 and 2002 by different methods:

- continuation of shore and seafloor cleaning up;
- construction and starting of a treatment plant destined to extract and recycle the oil-slick waste stored near Donges in the Loire-Atlantique department;
- implementation of public and private research programmes for better understanding the pollutant–environment interactions, preventing future oil slicks, and improving the management measures to be taken in case of similar disasters;
- public information for restoring the genuine tourist picture of disrupted coastal zones.

Importance of Oil Spills Among Other Marine Pollution Events

Hydrocarbons and other sea-transported pollutants The Amoco Cadiz and Erika oil-spill disasters belong to a series of major accidents that began in 1967 with the Torrey Canyon oil tanker sinking off Cornwall. These accidents progressively increased the public's awareness of the *environmental risks induced by maritime transportation* of toxic or potentially noxious substances. *English Channel and northeastern Atlantic coasts* are particularly threatened, because of both the high frequency of seaward storm winds and the density of maritime merchant traffic. The large number of tankers sinking, oil spills and maritime load-losses that have happened since the late 1960s demonstrates the strong risk of accidental dumping in these regions (Fig. 168). Other coastal regions in the world have suffered serious pollution due to oil-tanker shipwrecks (see Fig. 164): Gulf of Mexico, Alaska and Nova Scotia in North America; The Magellan Straits at the edge of South America; Arabian Sea, etc.

Environmental risks induced by maritime transportation and amplified by bad weather and operational conditions are certainly not restricted to oil spills. Until the mid-20th century ecological disruptions were essentially caused by fortuitous ship transportation and intercontinental spread of allochthonous *invasive plant or animal species*: coastal Graminacae such as *Spartina*, Gastropods such as *Crepidula*, micro-organisms (e.g. bacteria, viruses), etc. In recent decades, the exponential increase in maritime merchant transportation has caused new environmental risks. Numerous *chemical substances* transported by sea are likely to cause contamination: acids, ammonia, heavy-metal compounds, fertilizers, pesticides, and corrosive or pharmaceutical products, etc. The environmental impact of accidental dumping tends to be accentuated at close proximity to the shore and when marine currents are heading to the coast. Numerous in-depth investigations are still necessary for better understanding the

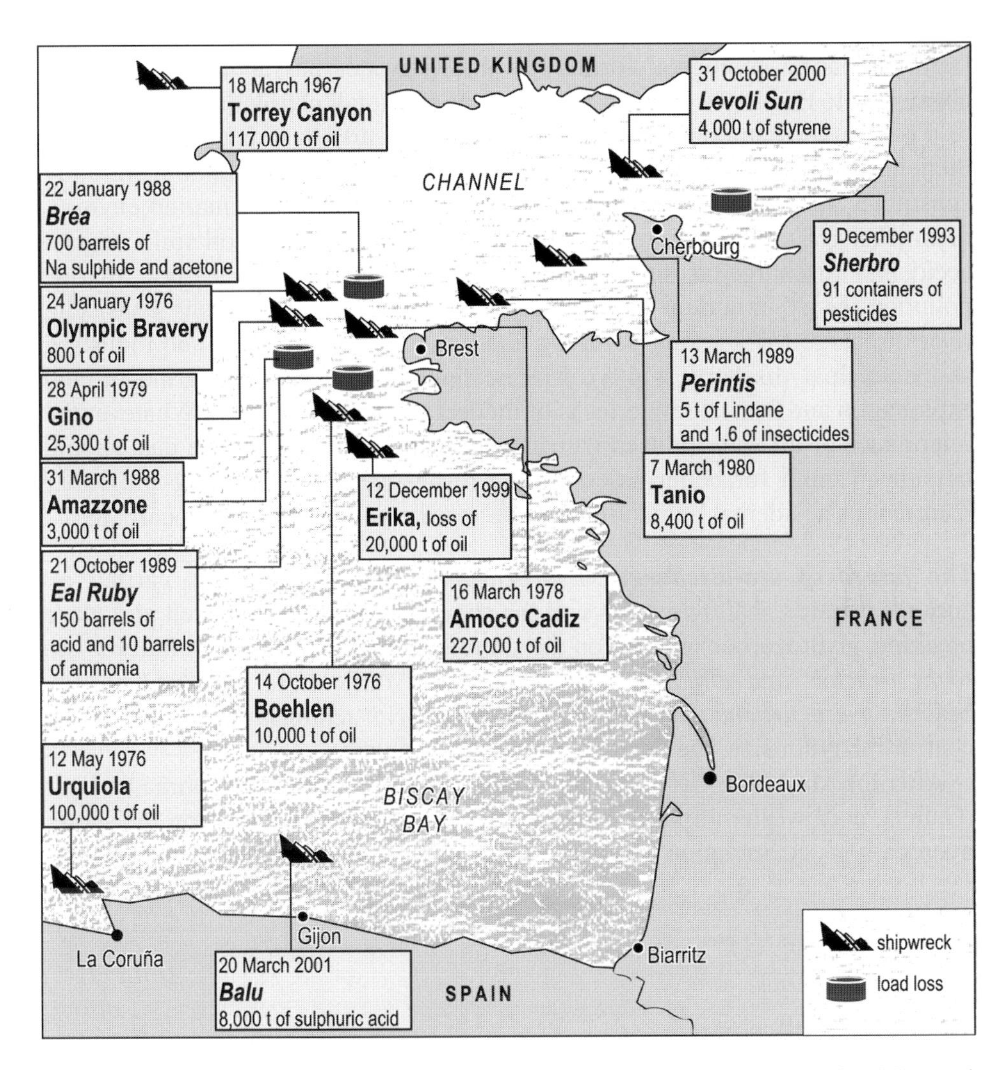

Figure 168: Major oil slicks and load losses off French Atlantic coasts (after Massoud & Piboubès, 1994; La Voix du Nord, 2001).

dispersal and settling modes of mass-discharged chemical products, the conditions and duration of neutralization, and the long-term environmental consequences.

Diversity of hydrocarbon pollution sources The local abundance of crude-oil discharge by oil spills and the dramatic character of public and private damage resulting from accidental oil spills tend to overshadow the actual *importance of other modes of hydrocarbon release in the sea*. Natural upward diffusion of oil from submarine geological formations into seawater has always existed and always will. Estimates suggest that half of the total petroleum in the ocean results from the natural upward diffusion of hydrocarbons trapped in submarine oil fields (Owen et al., 1998). Crude oil seepage from the ocean bottom currently constitutes about 15% of liquid hydrocarbons released in the marine environment (Fig. 169), which represents 3 times the amounts due to oil spills; in fact accidental discharge at sea is responsible for 5% of hydrocarbon marine pollution. Most important anthropogenic sources of hydrocarbons consist in river discharge (41%), tank dumping and washing at sea (15%), and industrial and municipal discharge (11%). Marine pollution due to submarine oilfield exploitation and coastal refineries is of moderate importance (6%). Whatever the importance of oil spill-induced damages and inconveniences, these data should be kept in mind for better assessing the respective part of natural and either recurrent or accidental human impacts on marine pollution by hydrocarbons.

The future of sea-discharged hydrocarbons Crude oil layers carried by currents towards shallow waters and the shore are subjected to swell and wave breaking. Hydrocarbons spilled on the shore tend to *impregnate and blacken the rocky beaches and cliffs*, which become very difficult to clean. On sandy beaches *hydrocarbons tend to mix* more or less *with sedimentary particles*. The coarser the grain size, the deeper the pollution penetrates.

After the May 1976 Urquiola shipwreck at the entrance of La Coruña harbour (Galicia, northwestern Spain), 30,000 t of crude oil were released on the shore; hydrocarbon accumulation was maximum where swell were converging in high-

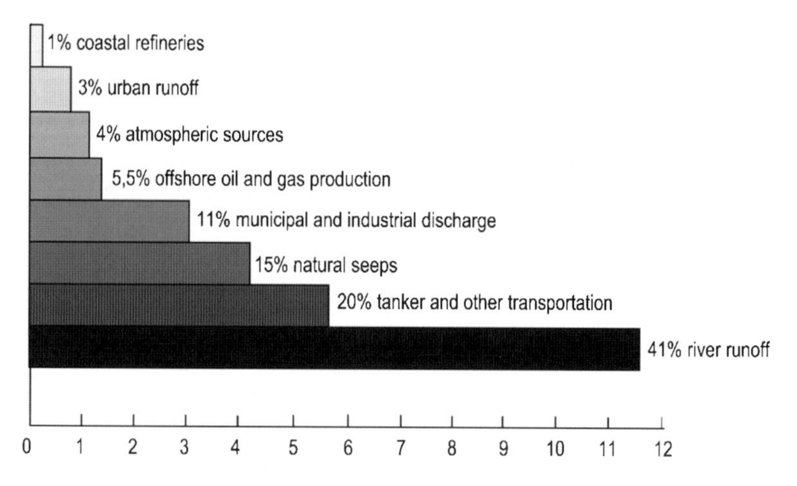

Figure 169: Proportion of the different sources of marine pollution by hydrocarbons (after Owen et al., 1998). Discharge due to marine transportation (20%) principally results from oil dumping (15%), secondarily from oil spills (5%).

tide breaking-wave sectors. Oil accumulated mainly in the form of irregular cake layers intercalated between the sediments, at a depth reaching 30 cm in fine sand and up to 100 cm in coarse sand. Intense mixtures of oil and sediments locally developed and spread down to 10 cm depth in fine sand and to 65 cm in coarser deposits (Gundlach et al., 1978).

On a longer-term scale, the hydrocarbons released at sea tend to be naturally degraded by bacteria and assimilated in the biosphere. They rarely cause any durable deterioration of marine ecosystems, and indirectly help feed the marine food chain.

10.2.4. Mitigation and Related Problems

Limitation of Pollutant Discharge

At sea, as on land, the best way for reducing contamination consists of *containing the pollutants at source*. This may be performed by adding to industrial processes specific pollutant-treatment and -storage systems, by spreading farm fertilizers in phase with plant sorption periods and in accordance with climate conditions, etc.

For instance the industrial treatment practices implemented since the 1970s in the Liverpool-Runcorn district, western England, have allowed reducing progressively the amount of heavy metals released in the Mersey River estuary. The levels of cobalt, chromium, copper, mercury, nickel, lead and zinc in sediments deposited during recent decades display either very low or constantly decreasing rates (Fig. 170). Some recurrent pollution risks nevertheless remain due to the possible erosion during flooding periods of estuary levees and salt meadows, where toxic metals have accumulated for more than one century. Such an erosion phase was responsible for the temporary resumption of mercury increase that was recorded between 1989 and 1992 in lower Mersey estuary deposits (Harland et al., 2000).

Implementing *strict regulation measures and periodical controls* is as important at sea as on land, but all the more difficult to conduct at increasing distances from the coast. In a general way the sanitary quality of beaches and shallow marine zones leads to systematic measurements and monitoring, and has improved in many countries since the last quarter of the 20th century. The 1975 London Convention regulates waste dumping in international waters. In addition each country or federation of countries has enacted rules and procedures that are specific to its territorial maritime domain. In the USA the Environmental Protection Agency is responsible for controlling various products forbidden to be dumped at sea: radioactive waste, toxic chemicals, bio-noxious substances, floating waste, mercury-, cadmium- and fluorine-containing compounds, hydrocarbons and their by-products. In Europe the Community

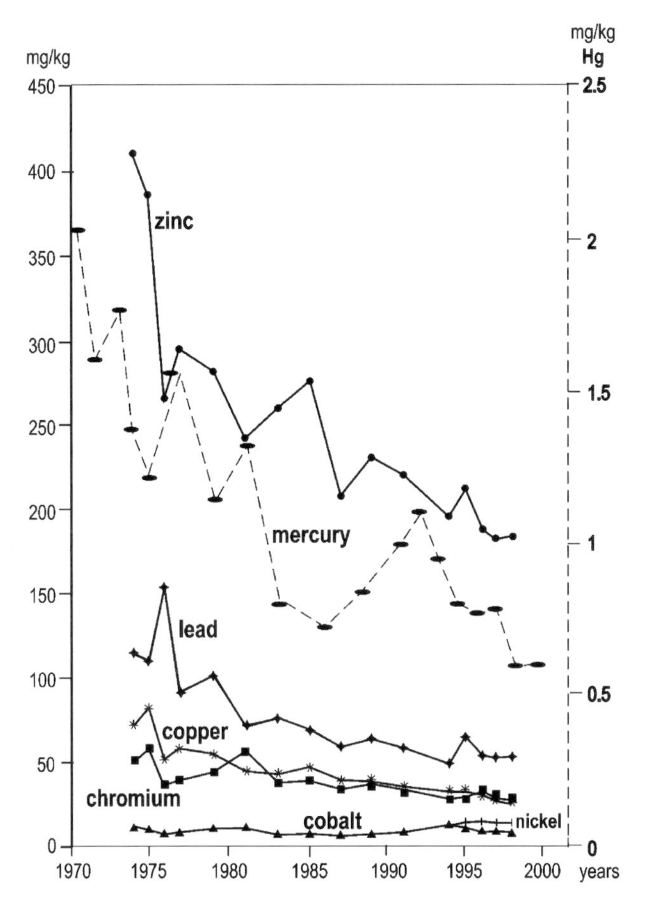

Figure 170: Variations between 1974 and 1998 of the average concentration of several heavy metals in Mersey estuary, Liverpool-Runcorn district, England (after Harland et al., 2000). Values are normalized to 40% silt-clay content. Values in milligrams per dry weight kilogram (mg/kg).

directives lead to homogenizing pollution standards and regulations for all member countries.

In France the surveillance of coastal domains is under the responsibility of national institutions such as IFREMER (French Institute for Marine Research) and implies several measurement networks devoted to specific watch: RNO for water and sediment chemical composition, REPHY for noxious and toxic algal blooms, REMI for the microbiological quality of coastal water. Improvements of the methods ensuring the quality of coastal geo- and bio-systems rely on current fundamental and applied investigations (e.g. PNEC, LITEAU programmes), which allow better understanding of the coastal ecosystem's functioning, assessing the significance of environmental quality and modification indicators, and finally better managing coastal and adjacent marine domains.

Mitigation Through Dispersal and Natural Treatment

The ocean has a huge *natural potential of contaminant dilution and assimilation*, a property that has been intuitively exploited for centuries by humans, who have always discharged waste at sea. Purification properties of marine water and sediments tend to become impoverished in some marine regions such as the Baltic and Mediterranean Seas, which are characterized by restricted connections with the open ocean and undergo excessive anthropogenic pressure.

Ocean purification potential is still imperfectly understood and quantified. It appears to largely depend on *bacterial activity*. For instance the amount of toxic insecticides of the DDT group that are released off the California coast display decreasing amounts in sediments deposited since 1970; this is attributed to bacterial biodegradation associated with dilution phenomena (Zeng & Venkatesan, 1999). DDT mobilized in seawater may undergo further particle trapping and diffusion outside initial deposition zones. Various researches are currently conducted for better constraining the ocean decontamination potential and establishing pollutant transfer balances; these researches should help to implement long-term management measures.

Incineration at Sea

Deliberate incineration of combustible waste from ships stationed in open sea has for instance been practised in the North Sea for destroying organo-chlorinated products (vinyl chlorides, PCB; Campan; see Salomons et al., 1988). This method has been violently criticized by some associations because of the potential dispersal in the atmosphere of noxious smoke and particles. In fact environmental concerns are similar to those set up by terrestrial incineration procedures (Chapter 9.4.3): discharge of dust, flying ash, or scoria; release of acid compounds and of gas increasing the greenhouse effect; need for neutralizing the potentially toxic combustion residues (e.g. chlorine, fluorine, arsenic, heavy metals). Additional risks at sea may arise from transportation means used for carrying the toxic substances to incineration sites, the location of which is crucial in relation to prevailing winds.

A Few Landmarks, Perspectives

- The countless contaminants discharged in water, soil, sediment and underground systems mainly result from mining and industrial activities as far as metal and other mineral substances are concerned, and from farming, domestic and sanitary activities for organic materials. Pollutant sources are either punctual and well located, or diffuse and scattered. Noxious or toxic properties of contaminants may be proportional to their abundance, start acting above a certain threshold, or have an effect both below a minimum and above a maximum amount. Water is the major means by which pollutants are transported on land towards either terrestrial underground or lake and marine basins. Rivers play an essential role in the continent-to-ocean continuum. River discharge of particulate and dissolved contaminants in the ocean is much larger than fallout from air-transported substances. River input to the sea was nevertheless significantly reduced during the 20th century due to the implementation along hydrographical basins of numerous, sometimes huge artificial lakes and reservoirs that are responsible for active sediment trapping.
- Terrestrial contamination reaches critical values and huge expansion in numerous densely populated and heavily industrialized regions. Contaminant infiltration and storage in groundwater and ground rocks causes growing environmental problems, the control of which remains poor in all situations. Efficient mitigation methods consist of recycling or reducing the pollutant discharge at source, and much progress is being obtained in high-technology countries. Other solutions potentially exist that need to be further investigated for better understanding of effective mechanisms and efficient decontamination: retention by plants or soil components, chemical exchanges and stabilization, activation of the subterranean micro-biosphere.
- Sea and ocean contamination results from rapid lateral dispersion of pollutants by marine waves and currents, followed by slow vertical storage through sedimentation. The dominant impact of river discharge and coastal streaming is supplemented by more diffuse but regionally significant atmospheric fallout, and also by accidental or deliberate dumping of hydrocarbons and various chemical substances. Oil slicks caused by petroleum-tanker shipwrecks induce disastrous local pollution on coastal and beach zones. Oil spills nevertheless contribute only 5% to total hydrocarbons discharged at sea, are destined to be assimilated by the organic matter cycle, and sometimes prove to be less noxious than some chemicals dumped at sea.
- The coastal zone, which extends from estuaries and deltas to outer continental shelves, is particularly threatened by anthropogenic pollution. This is mainly the case in semi-closed, shallow and current-depleted epicontinental basins

that are tributaries of rivers draining densely populated, urbanized and industrialized regions. Contamination rates in such marine environments tend to induce major disruptions, the mid-term reversibility of which may be seriously impaired. Mitigation of marine pollution is mainly obtained by limiting waste discharge at sea, which often allows restoring significantly the environmental quality. Self-purification properties of ocean water and sediments are also traditionally exploited, but are still insufficiently understood and tend to locally and even regionally be offset by excessive waste and contaminant dumping.

- Amplification of contaminant-induced ecological and sanitary risks may arise from various effects that either combine or follow each other: pollutant infiltration underground and re-emergence in previously non-contaminated sectors; polluted-sediment reworking by erosion, flushing or dredging, responsible for downstream river contamination; development of coastal algal blooms induced by agricultural fertilizer discharge and leading to eutrophication, under-oxygenation, and massive mortality; mineral and organic pollutant bio-accumulation through food chains causing poisoning of fish and seafood species. Such "boomerang-effects" are characteristic of environmental disruptions linking hydrosphere, biosphere and pollution.
- The importance and distribution of anthropogenic contamination of surface terrestrial envelopes greatly vary with time. Heavy metal, organic compounds or micro-organic pollution intensity does not evolve parallel to population growth but reflects successive man-induced deterioration and restoration stages. Some recent, historic or even prehistoric periods have undergone more serious contamination than the present time. An increasing number of efficiently cleaned up continental and coastal sectors have been reported in the last quarter of the 20th century especially in developed countries, which tend to better control the waste discharge and disposal processes. Nevertheless the information provided by scientific journals still has frequently a local and partial character. Limited consideration is provided to the complex, intermingling and linking contamination, economic development, technological control, demographic pressure, and actual realization by human societies of environmental risks.
- Further alleviation of rock, soil and sediment contamination necessitates facing particularly critical challenges implying new and ample scientific and technological investments. This differs from most other environmental concerns that already benefit greatly from previous research results. New fundamental knowledge should be acquired involving multidisciplinary and inter-connected investigations, as well as simultaneous efforts for developing public information and sense of responsibility. Scientific progress should be acquired in several crucial fields: physicochemical conditions of contaminant

transfer, retention and transformation in water, soil, subsoil and groundrock networks; importance and reactivity of the deeply buried micro-biosphere; residence and self-purification time of organic and inorganic chemical compounds stored in different surface reservoirs. Long-term monitoring systems involving inter-connected measurement systems should be implemented in most threatened regions of either developed or developing countries, for feeding the databases necessary for predictive modelling operations. Such monitoring should be applied to different case-study problems: contaminant fluxes in river basins subjected to specific or multiple anthropogenic impacts; respective effect on the environment of recurrent and factual or exceptional pollution; quantification of the role played on contaminant migration and transformation by free- and pore-water dynamics.

Further Reading

Andrews J. E., Brimblecombe P., Jickells T. D.& Liss P. S., 1995. Introduction to environmental geochemistry. Blackwell, 232 p.

Bodungen B. von, Turner R. K. ed, 2001. Science and integrated coastal management, Wiley, Dahlem Series 85, 378 p.

Bradshaw A. D., Southwood R. (Sir) & Warner F. (Sir), 1992. The treatment and handling of wastes. Chapman & Hall, The Royal Society, London, 302 p.

Chapman D. V., 1992. Water quality assessment: a guide to the use of biota, sediments and water in environmental monitoring. UNESCO Public., E & F. N. Spon, London, 2nd ed., 626 p.

Clark R. B., 2001. Marine pollution. Oxford University Press, 5th ed., 237 p.

Price M., 1996. Introducing groundwater. Chapman & Hall, 2nd ed., 278 p.

Salomons W., Bayne B. L., Duursma E. K. & Förstner U., 1988. Pollution of the North Sea: an assessment. Springer, 687 p.

Testa S. M., 1993. Geological aspects of hazardous waste management. CRC Press, Lewis Publishers, 537 p.

Turner B. L. II, Clark W. C., Kates R. W., Richards J. F., Mathews J. T. & Meyer W. B., 1993. The Earth as transformed by human action. Cambridge University press, 713 p.

Some Websites

http://www.epa.gov/epahome/Citizen.html: official site of the U.S. Environmental Protection Agency, which covers almost all types of energy and pollution problems (water, air, soil, and ground) in urban and rural environment

http://biotech.icmb.utexas.edu/pages/wildlife.html: examination of main agricultural, industrial and urban contaminants: characteristics, risks, control; natural, bacterial or man-induced mitigation and remediation means. Web site of the Veterinary School at the University of Texas

http://www.100toppollutionsites.com/: grouping under the auspices of the World Environmental Organization, of 100 Websites devoted to synthetic data regarding pollution: soils, rivers and wet zones, oceans and continental water planes, air, noise, etc.

http://oceanlink.island.net/ask/pollution.html: scientific answers to the various questions set up by marine pollution. Cross-reference to more specialized Web sites

Chapter 11

Regional to Global Change of Earth's Fluid Envelopes, and Impact on the Solid Earth

11.1. Modification of Ocean–Atmosphere Interactions

11.1.1. 1997–1998, Major Climatic Disturbances in the Pacific Ocean

The easternmost part of the Pacific Ocean bordering the coast of Ecuador and Northern Peru coast is famous for its high fish productivity, which results from hydrological ascent of nutrient-rich water favouring active development of the marine food chain. Called upwelling, this water ascent tends to diminish each year around Christmas time; this is due to the temporary influence of a warm water current that migrates from the north and stops fairly cold and deep nutritive water from reaching the ocean surface. The fishing season therefore stops, and Christian fishermen traditionally give thanks to the Christ Child (*El Niño* in Spanish).

In some years, however, the fishing season proves to be mediocre. The waters of Ecuador and Peru bordering the Pacific are not well stocked with fish, since upwelling is barely active and prevents nitrogen- and phosphorus-containing cold water to reach the surface productivity zone. Scientists have called such episodic disruptions of the annual oceanic cycle off South American coasts *El Niño*.

In 1997–1998 the expression "El Niño" became familiar worldwide, since the climatic disturbance responsible for dwindling of the East Pacific upwelling was of an especially strong intensity. A huge belt of warm surface water extended in the ocean from South American coasts over most of the equatorial zone (Fig. 171) and prevented on a large scale the nutrient-bearing deep water from ascent. This induced a major fish-economy crisis in Latin America. The westward latitudinal extension of surface warm water was associated with very dry climatic conditions in western Pacific Ocean regions, which caused devastating forest fire to spread in Indonesia and northeastern Australia. In contrast the

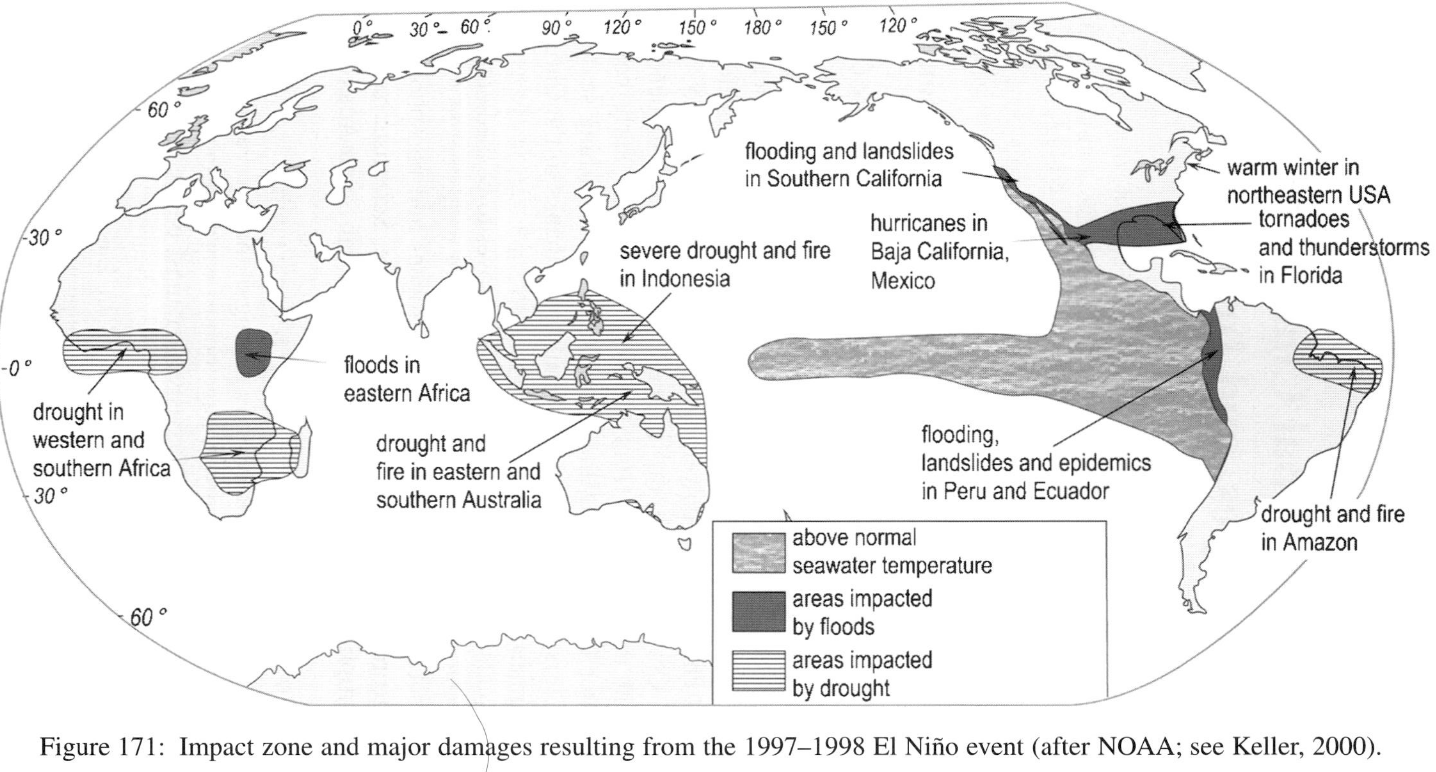

Figure 171: Impact zone and major damages resulting from the 1997–1998 El Niño event (after NOAA; see Keller, 2000).

eastern Pacific regions experienced an abnormally wet climate, which induced disastrous flooding in Central and South America; this resulted in giant landslides, debris avalanches, plagues of vermin and epidemics. Finally the 1997–1998 El Niño event was considered largely responsible for either intense drought or heavy rainfall in several other inter-tropical regions of America, Africa, Asia and Oceania (Fig. 171). This major climatic disturbance is thought to have caused 2,000 human deaths. Damage to property has been estimated at 30 million U.S. dollars.

Such warm El Niño events last about 12–18 months and display highly variable intensity and diverse modalities. They recur non-periodically every 2–10 years, the average recurrence time being close to 4 years. During the 20th century the Peruvian upwelling underwent particularly intense blocking and associated disruptions in 1900, 1911, 1918 and 1925, then in 1939–1941, 1982–1983, 1990–1995, and finally in 1997–1998.

11.1.2. The El Niño Phenomenon

Southern Oscillation

The inter-tropical zone of the ocean is subject to intense evaporation of warm water masses and periodically violent winds (cyclones, hurricanes and typhoons; Chapter 4.3). The dynamics of both liquid and gaseous Earth envelopes in this zone is controlled by the functioning of two groups of atmospheric cells that move in either east-west or north-south direction (Fig. 172).

- *Latitude-oriented cells*, called Walker cells, are driven by easterly trade winds, which blow on the ocean surface and push marine water westwards parallel to the equator. The return air current moves eastwards at several kilometres altitude and is connected with sea-surface wind by both a western ascending branch responsible for low pressure and an eastern descending branch causing high pressure. This Walker cell extends over the whole equatorial Pacific, where trade winds push surface water from high-pressure American coasts to low-pressure Indonesian regions, and lead to a high-altitude return current with two vertical branches. Distinct Walker cells operate over the Atlantic and Indian oceans, and move in a similar way but less actively than in the Pacific. The whole east-west cell system constitutes *Walker circulation*.
- *Meridian-oriented atmospheric cells*, called Hadley cells, develop on each side of the equator. Warm and humid air fed by seawater evaporation progressively rises at altitude and moves via the Coriolis effect to the north in the northern hemisphere, and to the south in the southern hemisphere. Due to

heavy rainfall in low-pressure equatorial and humid tropical zones, the moving air masses get progressively dryer and tend to subside when reaching the high-pressure arid tropical zones situated at about 30°north and south. Combination of both northern and southern Hadley cells constitutes *Hadley circulation*. Convergence in the equatorial zone of each Hadley cell-ascending branch constitutes the *inter-tropical convergence zone* (ITCZ).

During winter, the ITCZ tends to migrate southwards by a few degrees of latitude. This air-mass displacement induces a southwards migration of ocean-surface warm water, which therefore around Christmas occupies the coastal zones of Ecuador and northern Peru. This is the reason why in ordinary years upwelling of cold, nutrient-bearing deep water is impeded, which causes the fishing season to end (Chapter 11.1.1). ITCZ southward migration itself results from a seasonal decrease in atmospheric pressure in the easternmost region of the equatorial Pacific (i.e. off Southern America) relative to western ocean borders (i.e. off Indonesian and Australian coasts). These seasonal fluctuations of the atmospheric pressure field in eastern and western parts of the Pacific

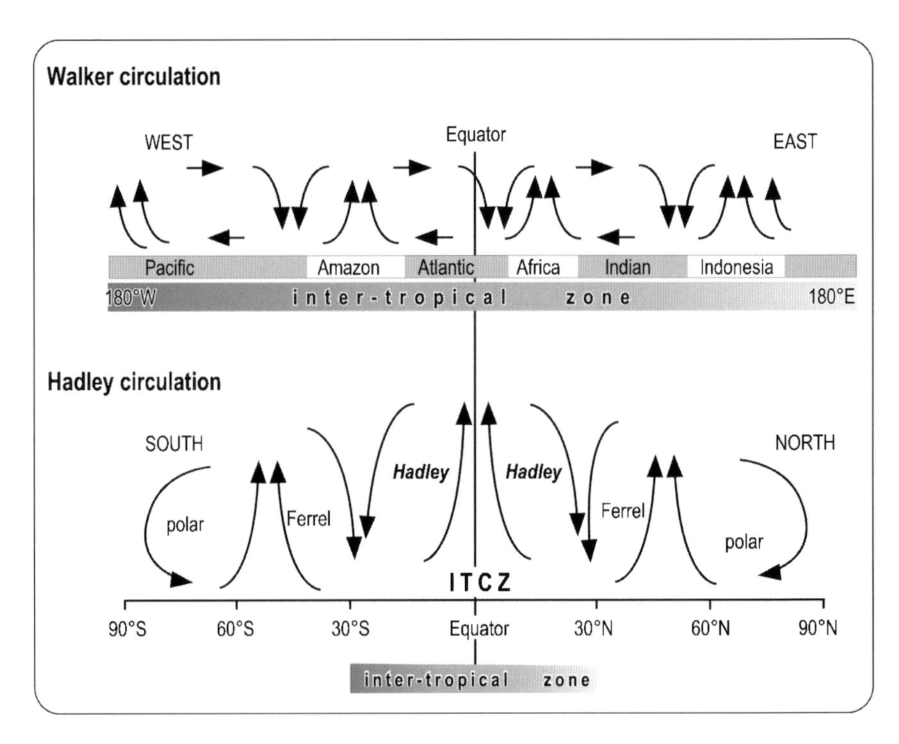

Figure 172: East–west Walker and north–south Hadley atmospheric circulation patterns in the inter-tropical zone (after Chapel et al., 1996). Meridian-oriented Ferrel and polar atmospheric cells develop beyond northern and southern tropical zones.

range are located immediately south of the equator between 0° and about 10°S and are therefore referred to as the *Southern Oscillation*.

El Niño–Southern Oscillation — ENSO

Ordinary years The high pressures usually established on the western side of South America's inter-tropical zone are responsible for both dry climate and powerful trade winds, which blow from continent to Pacific Ocean. Winds moving above the ocean surface push seawater to the west and progressively get wetter. The thermocline, which constitutes the boundary between the warm surface and a colder subsurface water, tends to compensate the eastern density depletion by rising. This allows deeper, colder and nutrient-rich water reaching the ocean surface off South American coasts, giving way to the upwelling that aids increasing productivity and fruitful fish catches. By contrast the ocean surface pushed by easterlies in the Western Pacific rises by a few tenths of centimetres, while the thermocline moves deeper in the water column; atmospheric pressure is low and heavy precipitation falls on Indonesian regions (Fig. 173, A). This situation goes on during most of the year; it lessens only around Christmas, due to ITCZ southward migration.

El Niño years *The year preceding* an El Niño episode is marked by increased trade winds and upwelling in the eastern part of the equatorial Pacific, a westward accentuation of ocean-surface rise, and a considerable accumulation of warm water in Indonesian regions (i.e. several thousands of cubic kilometres).

Then a *sudden weakening of trade-winds intensity* occurs on the American side of the equatorial Pacific and combines with both an atmospheric pressure drop to the east and a symmetric rise to the west. This allows warm water accumulated around Indonesia to flow rapidly eastwards. The relative sea level therefore drops to the west and rises to the east, whereas the thermocline tends to become horizontal; this induces eastern upwelling to slow down and even to stop. A progressive increase in ocean surface temperature in the central and then in the eastern Pacific causes a low-pressure zone to move eastwards together with the ascending branch of the Walker cell and heavy rainfall: the climate dries out in Indonesia and other West Pacific equatorial regions, whereas precipitation dramatically increases in Polynesian regions (Fig. 173, B). All these ocean–atmosphere inter-related modifications are referred to as *El Niño –Southern Oscillation (ENSO)*.

Towards the end of an El Niño episode, the abnormally warm surface water in the eastern equatorial Pacific extends both in latitude and longitude, and tends to occupy in a few months a large part of the inter-tropical zone (see Fig. 171).

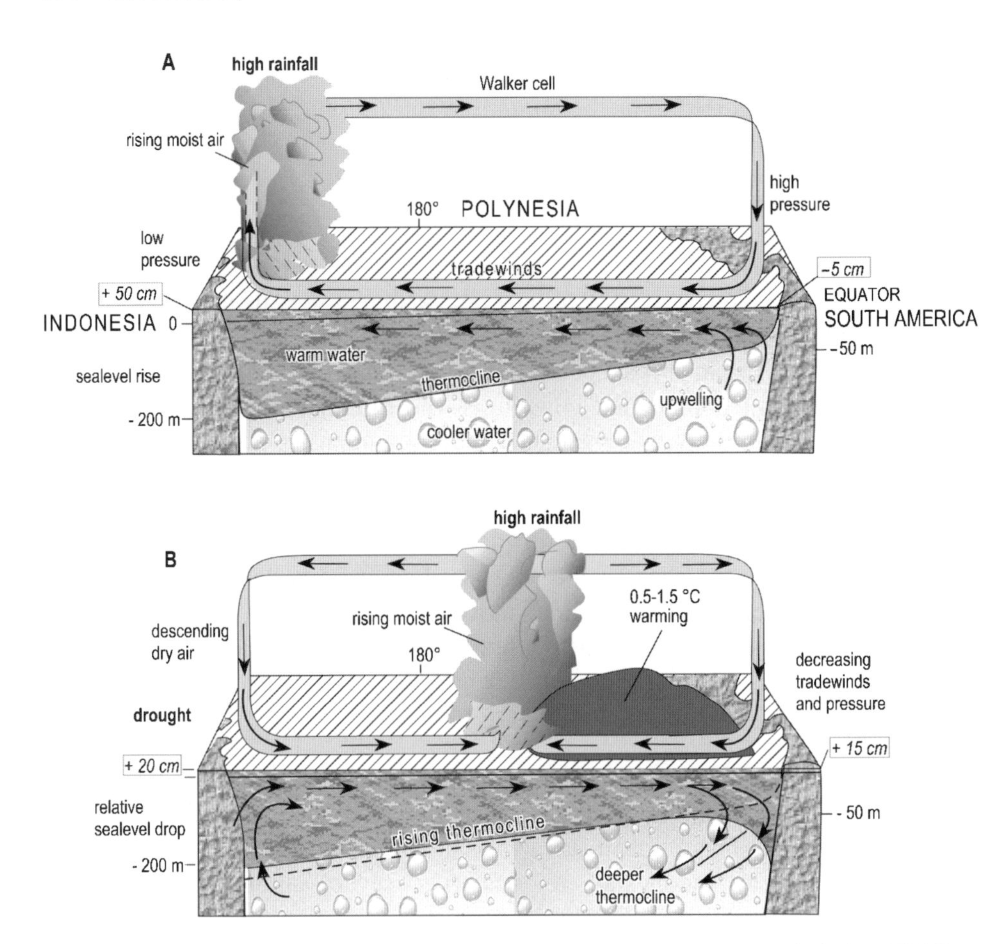

Figure 173: Ocean-atmosphere interactions in ordinary years (A) and during El Niño episodes (B) (after Chapel et al., 1996, Murck et al., 1996).

Return to ordinary conditions starts in the eastern Pacific, where atmospheric pressure increases and upwelling resumes. Low pressure and heavy rainfall resume somewhat later in the western ocean.

La Niña years Sometimes the *dominant phase of the Southern Oscillation intensifies* over a few months or years, which determines particularly high pressure in the eastern Pacific, stronger trade winds, reinforced upwelling, and increased marine bio-productivity. Such excessive conditions are symmetrical to those characterizing El Niño episodes and are therefore called La Niña ("little girl"). Very active upwelling off Southern American coasts may locally induce dissolved oxygen depletion in ocean water and subsequent eutrophication. The Ecuadorean and Peruvian climate becomes particularly dry, which favours forest fire. The atmospheric pressure gradient diminishes strongly towards western Pacific regions, where torrential rainfall tends to cause flooding in Indonesia and adjacent regions.

Environmental Impact

Strong El Nino episodes like the one that raged in 1997–1998 *cause, in the Pacific equatorial range, a series of disruptions*, the most spectacular of which are the following:

- tragic drop of anchovy catches off the coasts of Ecuador and Peru;
- strong precipitation over Central and South America leading to floods, slope movements, mud flows, debris avalanches, and epidemics;
- torrential rain and cyclones over Polynesia, responsible for deterioration of landscape and property;
- long-lasting drought in Indonesia and northeastern Australia causing huge forest fire, dense smoke and dust clouds (see Fig. 171), and subsequent deforestation, wind intensification and soil erosion.

Latitudinal zones adjacent to the Pacific equatorial zone also become affected, since the Walker cell submitted to Southern oscillation disruption inter-reacts with nearby tropical Hadley cells (see Fig. 172). This results in various *climatic disturbances and geomorphologic deteriorations*, extending as far as latitudes 30° north and south. The most common expression of such climatic change consists of heavy rain and floods in tropical American coastal regions, and of drought and dense dust clouds over eastern Oceania and northern Indonesia.

Impact of ENSO disturbances may possibly spread over *Pacific temperate zones*, where Hadley cells react in turn with Ferrel cells (see Fig. 172), which induce mid-latitude, low-pressure conditions. El Niño's indirect effects are nevertheless still poorly demonstrated in these regions. It seems that either positive or negative atmospheric pressure anomalies affect some oceanic zones, causing distortion or depressions relative to ordinary conditions. In the Central Pacific the depressions' movement tends to be deflected southwards and to induce rainfall decrease in the southern ocean; in the eastern Pacific, depressions tend to be deflected to the north, which favours climatic warming in Canada and Alaska (see Chapel et al., 1996).

Causes

El Niño episodes correspond to complex interactions of ocean–atmosphere coupling, the initial causes of which are still *poorly known*. The nature, intensity, duration and geographical extent of water-air exchanges vary greatly, which dramatically complicates understanding of the functioning of El Niño's mechanisms. We are not even quite sure whether corresponding disturbances start working in the atmosphere or in the ocean.

Some scientists have suggested that temporary *ocean-water warming induced by submarine volcanism and hydrothermalism* could significantly participate in triggering El Niño episodes. Volcanic and hydrothermal phenomena are active and of variable intensity in the inter-tropical zone of the eastern Pacific Ocean (i.e. East Pacific, Galapagos and Cocos ridges) where El Niño disturbances typically form. It is recognized that submarine seismic resumption usually corresponds to renewed volcanic activity, and that starting of El Niño episodes tends to correlate with increased seismic activity in Eastern Pacific regions. In addition the aperiodic development of submarine volcanism in oceanic crust ridges fits well the non-cyclic character of El Niño events. Such an explanation would imply strong linkage between lithosphere, hydrosphere and atmosphere functioning. The biosphere, which is the fourth major component of Earth, would be indirectly controlled by this coupling through important disruption in ocean productivity.

11.1.3. Geographic Extension, Other Ocean–Atmosphere Coupling Disturbances, and Impact

Equatorial Extension of El Niño

The succession of three major Walker cells along the inter-tropical belt (see Fig. 172, top) determines some connections between both the El Niño–La Niña episodes characteristic of the Pacific range and the climatic disturbances affecting other low-latitude regions of the world. The Indian Ocean appears moderately affected by such global changes, since it is strongly susceptible to active regional cycles of alternating winter terrestrial and summer marine monsoons.

The Atlantic Ocean is much more involved for it comprises a coupled high- and low-pressure atmospheric system that strongly resembles that of the Pacific Ocean; high pressure and upwelling predominate on African Atlantic margins, whereas low pressure and rainfall are more important on the American side of the ocean. The contrast between eastern and western oceanic regions is nevertheless attenuate since the Atlantic Ocean is much narrower than the Pacific (6,500 km wide relative to 20,000 km) and marked by important freshwater discharge to the west (e.g. Niger and Congo rivers). During the 1983 El Niño episode, characterized by abnormally warm water in the eastern tropical Pacific, the surface ocean temperature in the eastern tropical Atlantic was low and typical of a La Niña episode. The situation was reversed in 1984, when eastern Pacific water was colder (e.g. La Niña episode) while eastern Atlantic water was warmer and of an El Niño type (see Chapel et al., 1996). El Niño disturbances in the Pacific range therefore induce some symmetrical but attenuate disruptions in the Atlantic. This indicates that major disruptions in Southern Oscillation functioning are responsible for climatic impact and environmental disorders (e.g. flooding, drought, storms) in a large part of the world's inter-tropical belt.

Northern and Southern Extension. Relation with the North Atlantic Oscillation — NAO

A coupling system similar to the low-latitude Walker circulation works with somewhat less energy in a north-south direction, due to the "drive belt" relay provided by Hadley, Ferrel and successive polar cells (see Fig. 172, bottom). Strong El Niño episodes may therefore propagate with attenuate energy and effect toward northern and southern hemispheres, especially in the Pacific range.

North Atlantic regions are subject to aperiodic fluctuations of atmosphere–ocean coupling, which are referred to as the *North Atlantic Oscillation (NAO).* NAO results from atmospheric-pressure variations developing between the Azores anticyclone and Icelandic depression, and may induce large-scale climatic and environmental consequences. When the pressure gradient significantly exceeds average values (positive NAO phase), the climate gets dryer as far as the Middle East; the Euphrates River flow rate may decrease by 50% and threaten Turkey's freshwater resources. When the pressure gradient is abnormally low (negative NAO phase), climate conditions become rainier in mid-latitude regions and colder at high latitudes; the human-induced energy consumption may then increase by 30% in Norway. The exceptionally violent storms that hit France in October 1987 and December 1999 (Chapter 4.2.1) have been associated with unusual fluctuations of NAO.

NAO fluctuations vary at a 10-year to several tens of years' aperiodic rhythm; depending on the authors, the NAO maximum frequency is estimated to average either 2–3 years, 7–8 years or 5–10 years. The NAO function has been less investigated so far than ENSO but indisputably undergoes different variability patterns. Relationships appear to be rather loose between the two oscillatory systems and mainly controlled by El Nino episodes. Years marked by a *combination of intense ENSO and positive NAO events* seem to undergo particularly strong temperature, rainfall–drought and wind gradients.

Global Impact, Perspectives

El Niño atmosphere and ocean disturbances determine serious climatic disruptions, the environmental and economic consequences of which are multiple. *Solid surface formations* are often involved in a less intense but more durable way than fluid envelopes. For instance the 1982–1983 El Niño event has been responsible for the following major damages:

- six hurricanes hit Polynesia, which were the first for 75 years. They caused considerable damage to plantations, soils and buildings of numerous islands from the Tahitian to the Hawaiian archipelagoes;
- major flooding occurred in Bolivia, Ecuador, Northern Peru, Cuba, the USA, etc. and induced countless landslides and other slope movements;
- intense droughts caused dramatic famine in Zimbabwe due to deterioration of agriculture (e.g. destruction of corn fields). Huge fires were responsible for massive forest destruction and soil erosion in numerous regions: South Africa, Southern India, Sri Lanka, the Philippines, Indonesia, Australia, southern Peru and Central America.

Notice that some beneficial effects were also recorded, as for instance spectacular increase in cereal production in North America and Russia.

Current research is based on enlarged observation and measurement networks, modelling developments, and growing interdisciplinary scientific collaborations. Most useful information for forecasting ENSO and NAO events and predicting major related risks necessitates long-term data series collected from various devices: satellites, aeroplanes and ships equipped with appropriate sensors, radars, buoys and off-line measurement modules. Particularly useful data arise from measurements conducted continuously over several years at a 10- to 100-km scale. This was performed during the French–American Topex–Poseidon satellite programme, which from 1992 to 2000 followed and transmitted sea-level fluctuations with a 1-cm precision. Data processing has allowed measuring of ENSO and NOA effects during the whole study period and supplying numerical models for predicting future major ocean–atmosphere coupled disturbances.

11.2. Greenhouse Gases

11.2.1. Carbon Dioxide Increase at Mauna Loa Summit, Hawaii

Big Island in the Hawaiian archipelago is located in the central part of the northern tropical Pacific Ocean, far away from large urban and industrial centres. Mauna Loa constitutes the highest point of this volcanic island, which is 4,170 m above sea level. The atmosphere around this extinct volcanic cone is remarkably protected from pollution sources and other man-induced nuisance, which has led to establishing on the summit the largest international complex devoted to astronomical and astrophysical observations. Mauna Loa air characteristics are considered as representative of the *average composition of Earth's atmosphere*.

Measurements of carbon dioxide concentration (% CO_2) in Mauna Loa's atmosphere have been performed since 1745 (Fig. 174, A). They revealed a *continuous CO_2 increase during the 19th century, and an exponential augmentation during the 20th century.* CO_2 concentration was lower than 280 parts per million (ppm) before 1800, reached 290 ppm in 1880, exceeded 350 ppm in 1980, and continued to increase during recent decades of the 20th century. CO_2 levels have therefore augmented by about 30% in less than 200 years, which corresponds to twice the variation interval registered during the last 160,000 years by either polar ice or marine sediments. This points to acceleration by two orders of magnitude of CO_2 incorporation in the atmosphere during the last glacial–interglacial cycle.

The hypothesis that *growing human activities progressively could trigger such a sharp CO_2 increase* arose since the 1980s because of several convergent observations:

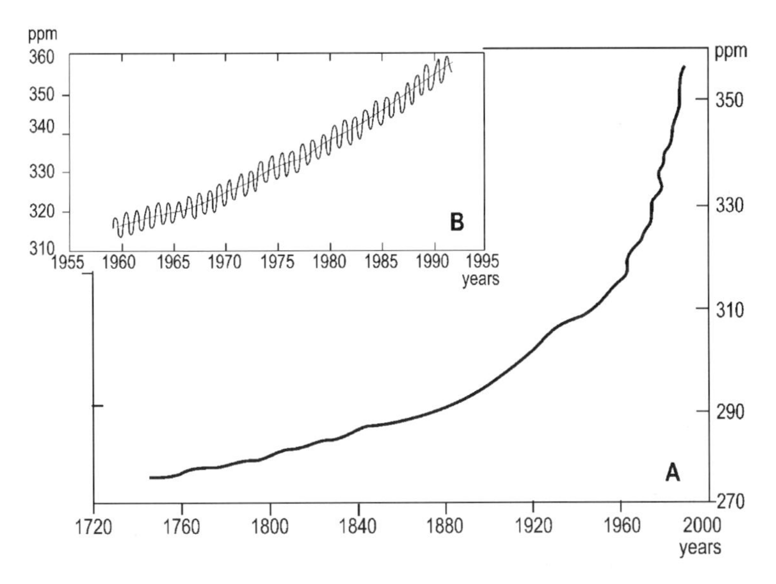

Figure 174: Variations of CO_2 atmospheric concentration at Mauna Loa observatory, Big Island, Hawaii. (A) Average distribution since 1745. (B) Seasonal and annual variations from 1958 to 1993 (after Mackenzie & Mackenzie, 1995). Density values in parts per million (ppm).

- fluctuations of atmospheric CO_2 concentration correlate directly with air temperature variations;
- air and water average temperatures have significantly increased during the 20th century;
- CO_2 concentration and temperature values have increased synchronously with the industrial expansion that started in the 19th century.

More detailed measurements of CO_2 concentration in the Mauna Loa atmosphere were performed during the second half of the 20th century, which gave two major results (Fig. 174, B). First, the *CO_2 atmospheric fluctuations closely reflect temperature seasonal variations*: the gas amounts diminish in the air during spring and summer due to absorption for feeding vegetation growth, and increase again during autumn and winter due to release through organic matter decaying. Second, *the average CO_2 concentration increased* between 1960 and 1993 from 315 to about 350 ppm, which correlated with an increase in average air temperature. CO_2 concentration values in recent decades exceeded by about 100 ppm those of prehistoric and historic times, during which human impact was very low on Earth's atmosphere and terrestrial ecosystems.

The hypothesis of a world climatic warming triggered by human activities therefore became more carefully considered. Continuously growing accumulation in the atmosphere of CO_2 during the past century was tentatively attributed to both excessive release and insufficient stocking of CO_2, due to various and increasing human actions characterized by an increasing impact: fossil fuel consumption (coal, petroleum, natural gas), wood burning, deforestation, lime manufacturing from limestone reserves, etc.

11.2.2. Global Warming

Evidence for Earth's warming has already been mentioned and illustrated in the Introduction (Section 4; Fig. 7, E). *An average augmentation by 0.5 to 0.6°C of terrestrial air temperature has been registered over the past century*, with several fluctuations: strong and irregular increase from 1920 to 1940, stabilization until the 1970s, severe augmentation starting again in 1975. The three last decades were characterized by the strongest temperature increase in the whole of the 20th century. The 1986–1998 period constitutes the warmest of the last 140 years.

Warming intensity has varied during 20th century according to various regional parameters: altitude, vicinity of ocean and semi-closed seawater masses, location of main atmospheric and marine currents, etc. For instance the numerous long-term series data obtained between 1901 and 2000 for metropolitan France reveal an average warming of 1.2°C in western regions located close to the Atlantic Ocean and British Channel, and only of 0.6–0.8°C to the east in the Alps and Vosges mountainous regions. Along a meridian transect the centennial warming has increased from 0.2°C in northern France to more than 1.0°C in southern regions (Météo-France data, 2001).

The global character of surface warming is testified by various observations and measurements gathered by the Intergovernmental Panel on Climate Change (IPCC), which works under the auspices of the United Nations. Two groups of examples are:

- The world's snow and ice cover have diminished by 10% since the 1960s. The average duration of sea ice in the Arctic Ocean has dropped since 1966 from 2.1 to about 1.6 months per year (see Demek, 1994). Most mountain glaciers have suffered significant retreat during recent decades, as shown by various measurements performed in Italy, Switzerland and northwestern Rocky Mountains (Table 43). Sonde-balloon records show that the freezing point in inter-tropical mountain regions tends to rise by 4.5 m/year; ice cover at 5,895 m altitude on Mount Kilimanjaro's summit (Tanzania) diminished by 85% between 1912 and 2000.
- The mean sea level has risen by more than 10 cm in recent decades, and the mean temperature of surface ocean water has increased by 0.5°C. During the same period rain amounts have increased by 0.5 to 1% per decade in northern hemisphere mid- and high-latitude regions, and by 0.2 to 0.3% in inter-tropical zones.

Table 43: Movement of some European and North American glaciers at the end of the 20th century (diverse sources; see Keller, 2000). Figures indicate the number of glaciers considered.

	Years	**Retreat**	**Immobiiity**	**Advance**
Italy	1981	10	10	25
	1988	92	13	26
	1993	127	8	6
Switzerland	1967	55	14	31
	1986	13	9	42
	1993	73	0	6
North Cascades, USA	1967	7	8	7
	1985	32	10	5
	1995	47	0	0

11.2.3. The Greenhouse Effect and its Augmentation

Solar Radiation and Energy Balance

Incident solar radiation (340 W/m^2) is partly reflected by cloud upper surface and ground surface (100 W/m^2) or adsorbed by stratospheric ozone (80 W/m^2), and partly transmitted to ocean and continent surfaces (160 W/m^2). Solar radiation directly reaching Earth's surface (160 W/m^2) is absorbed by water, soils and surface rocks, as is infrared radiation indirectly transmitted by the cloud cover (330 W/m^2). Both these incident radiations are restored to the atmosphere in the form of latent and overt heat flux (100 W/m^2) and of massive infrared radiation (390 W/m^2). The balance of incident and reflected heat flux is balanced at each working level of this general energy exchange process (Fig. 175): in the upper atmosphere (i.e. the mesosphere), values are of $340 = 100 + 80 + 160$ W/m^2; at Earth's surface they are of $160 + 330 = 100 + 390$, i.e. 490 W/m^2. Incident radiation is especially reflected where the Earth's surface is lighter coloured and less energy absorbent; this is especially the case of ice, snow cover, and ocean or lake surfaces. By contrast dark surfaces such as forest cover absorb solar energy more abundantly.

Infrared radiation reflected by Earth's surface (390 W/m^2) tends to be retained in the atmosphere's lower layers (i.e. the troposphere) *by various gases*, which allows air warming. This energy-retention mechanism constitutes the *greenhouse effect*, which is *natural and essential for life's development*. Without the greenhouse effect, the average temperature at our planet surface

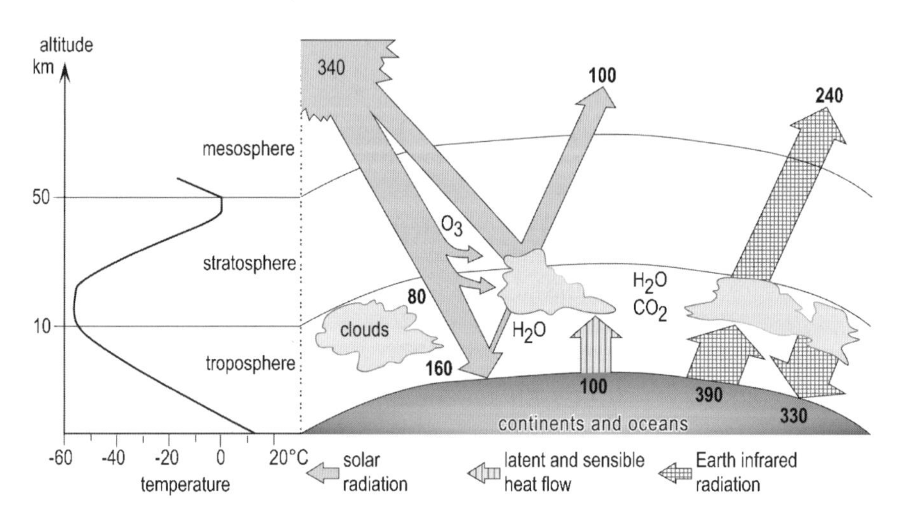

Figure 175: Radiation and energy balance of earth–atmosphere coupling system (after Dictionnaire du climat, Larousse ed.). Radiation balance expressed in watts per square metre (W/m^2).

would be of −18°C instead of +13°C, and therefore hardly compatible with current life.

Greenhouse Gases

Atmospheric gases absorbing infrared radiation are characterized by strong *radiation power.* They mainly comprise *water vapour* that is considered to induce about 80% of the natural greenhouse effect. *Other gases* are also involved, which occur in trace amounts in the atmosphere and the origins of which are often both natural and anthropogenic. In spite of their scarcity they significantly contribute to the greenhouse effect and its quantitative variations (Table 44).

Carbon dioxide — CO_2 Carbon dioxide constitutes only 0.01–0.1% of atmospheric gases but contributes about 60% to the anthropogenic greenhouse effect. CO_2 concentration in the atmosphere has increased parallel to industrial expansion (Chapter 11.2.1; Fig. 174), and currently displays an annual growth rate of 0.5%/year. Continuation of such an increase should induce in 2050 a concentration exceeding 450 ppm (i.e. 450,000 ppb), a value that is 1.5 times higher than before the industrial revolution.

Man's responsibility for the recent CO_2 increase mainly results from *massive combustion of fossil fuels*, where greenhouse gas has been stored for several hundreds of millions of years in the form of carbon. The largest amounts of CO_2 occur in *coal* deposits; the combustion of petroleum is responsible for 82% of the greenhouse effect induced by coal, and combustion of natural gas for 57%. Increasing concentration in the lower atmosphere of the isotope carbon-12 relative to carbon-14 clearly demonstrates the influence of Man's activities during recent decades. ^{14}C constitutes an unstable, short-period radioactive isotope (period $t = 5{,}730$ yrs), which has for long been disintegrated in fossil fuel ores currently exploited and burned. Massive discharge of the old ^{14}C isotope in the air induces $^{12}C : {}^{14}C$ ratio to increase.

About one-third of CO_2 released in the atmosphere results from wood burning linked to *deforestation*, the extent of which is growing dramatically and is barely compensated by new tree planting. This is mainly the case in inter-tropical countries subject to active population growth and rapid industrialization (Chapter 8.1). Other anthropogenic causes for CO_2 atmospheric discharge comprise industrial processes such as *lime* manufacturing (CaO) at the expense of natural carbonates (e.g. limestone calcite, CO_3Ca).

Altogether each Earth inhabitant contributes to the release each day into the atmosphere of an average of 15–20 kg of carbon dioxide. Fortunately the anthropogenic CO_2 is not totally stored in the atmosphere, which partly limits

Table 44: Atmospheric trace gases participating to the greenhouse effect (Murck et al., 1997). Values in parts per billion (ppb).

	Carbon dioxide CO_2	**Methane CH_4**	**Nitrous oxide N_2O**	**Chlorofluoro-carbons CFC**	**Tropospheric ozone O_3**	**Water vapour H_2O**
principal natural sources	balance between sources and sinks	wetlands	soils of tropical forests	–	hydrocarbons	evaporation transpiration
principal anthropogenic sources	fossil fuels deforestation biomass	rice culture, cattle, fossil fuels, burning	fertilizers, land-use conversion	refrigerants, aerosols, industrial processes	hydrocarbons with NO_X biomass burning	land conversion, irrigation
atmospheric lifetime	50–200 years	10 years	150 years	60–100 years	weeks to months	a few days
preindustrial concentration (ppb)	280,000	790	288	0	10	?
present atmospheric concentration (ppb)	353,000	1,720	310	0.284–0.48	20–40	3,000 to 6,000 (stratosphere)
present annual rate of increase	0.5%	1.1%	0.3%	5%	0.5–2.0%	?
relative contribution to the anthropogenic greenhouse effect	60%	15%	5%	12%	8%	?

increasing the greenhouse effect. Part of the gas is retained by both biosphere and hydrosphere, namely within the ocean, in quantities and through mechanisms that are still imperfectly known.

Methane — CH_4 Methane is the major component of gaseous fossil fuels and the most abundant organic trace gas in the atmosphere. It is responsible for firedamp explosions in coalmines. It currently results from biological fermentation processes such as plant decomposition, ruminant digestion and marsh gas formation. CH_4 release in the atmosphere is increased by various human actions: coal mining, rice growing, livestock farming, organic waste storage in public dumps, bio-incineration, and gas escape from pipelines. CH_4 concentration in the atmosphere increases by about 1.1%/year. This has led since the 19th century to more than doubling its atmospheric content (Wuebbles & Hayhoe, 2002). Methane abundance is much lower than that of CO_2 but its radiation power is much higher. Methane contributes about 15% to the non-aqueous greenhouse effect, through chemical reactions that are not yet accurately documented.

Chlorofluorocarbons — CFCs CFCs constitute artificially synthesized gases that are used in producing a cold temperature (e.g. refrigerators, air-conditioners), or manufacturing aerosol propellants and other industrial systems. They comprise various compounds such as CCl_3F (CFC-11), CCl_2F_2 (CFC-12), and CCl_4. Chlorofluorocarbons started being released into the atmosphere during the late 1950s. They contribute to both increasing the greenhouse effect (12% of non-aqueous trace gases) and destroying the stratospheric ozone (O_3) that protects Earth's life against ultraviolet radiation. Significant increase during recent decades of the latter risk (i.e. high-latitude development of an "ozone hole") has conducted the implementation of international regulations through the Montreal protocol in 1987 and Rio (1996) and Kyoto (1999) conferences. These measures have allowed reducing significantly the use of CFCs and starting to reduce ozone depletion. The use of these artificial gases nevertheless remains strong in various developing countries. In addition substitute products still partly contribute to increasing the greenhouse effect.

Troposphere Ozone — O_3 O_3 concentration in the lower atmosphere depends largely on photochemical reactions associated with the functioning of internal combustion engines. The exponential augmentation during the last century of cars, trucks and other combustion engine devices has induced a threefold increase in troposphere ozone levels in Europe and North America. Troposphere ozone contributes to both increasing the greenhouse effect and deteriorating the air quality in the urban environment.

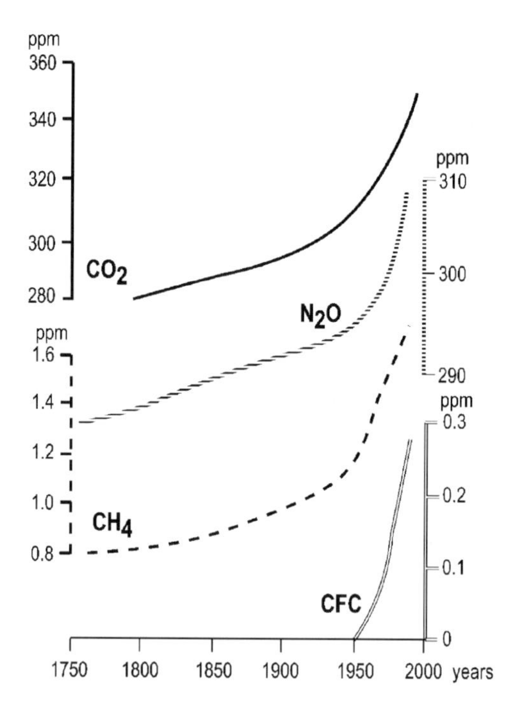

Figure 176: Variations since the mid-18th century of greenhouse trace gas concentration in the atmosphere (after Woodcock; see Bennett & Doyle, 1997).

Altogether *the concentration in the atmosphere of diverse non-aqueous greenhouse gases has augmented in a systematic and exponential way* during the last decades (Fig. 176). The increasingly curved slope tends to first be parallel to 19th century industrial development (CO_2, CH_4 and N_2O input), and then becomes more recently amplified by both new technological developments (CFCs) and urban hyper-concentrations (O_3). *Human influence in greenhouse effect escalation is therefore obvious.*

11.2.4. Causes of Global Warming, Perspectives

Natural Causes Versus Anthropogenic Causes

Realization of global warming and its recent acceleration in the 1980s determined the emergence among the scientific community of *two opposite interpretation groups*.

First, supporters of a *strictly natural control* of global warming largely based their line of argument on the geological variability of CO_2 concentration and temperature at Earth's surface. Most arguments relied on obvious historical facts:

- Atmospheric CO_2 has been stored or released many times and at various time scales during Earth's history, which has resulted in drastic warming or cooling events, strong sealevel rise or drop, and abundant organic matter stocking or destocking. The current increase in atmospheric CO_2 corresponds to values that are much lower than concentrations that characterized several geological periods (e.g. Early and Late Jurassic, or Middle to Late Cretaceous) and may therefore be little dependent on human activities.
- Earth's surface temperature is largely controlled by planet orbital parameters such as precession (period T = 23,000 years), obliquity (T = 41,000 years), and eccentricity (T = 400,000 years), which are referred to as Milankovich cycles. Such astronomically driven thermal changes are particularly important in Quaternary times and probably responsible for overwhelming the CO_2 variations, especially if they become combined with short-time events such as Heinrich or Dansgaard-Oeschger episodes.
- Some investigations reveal that variations of ocean surface temperature since the late 19th century correlate with sunspot density changes, which suggests significant control of solar activity on Earth warming and cooling stages. CO_2 variations might therefore not be necessary for inducing temperature changes at the planet's surface.

Further arguments have been indirectly provided by shortcomings arising from the rival interpretation: difficulty explaining the storage mode and location of all greenhouse gases; need to involve important feedback mechanisms responsible for temperature stabilization at values lower than those predicted; uncertainties about the quantitative impact on warming of water vapour relative to trace gases such as CO_2 and CH_4; inappropriateness for numerical models to provide reliable predictions, etc. Notice that some current hypotheses refer to the possible influence on global warming of solar activity, solar wind and cosmic radiation coupling, which could induce combined variations of water vapour condensation, cloud formation and atmospheric temperature.

Second, supporters of an *essentially anthropogenic control* of 19th–20th century global warming have developed explanations based on the correlations established between Earth's surface temperature, greenhouse-effect accentuation, population growth, and the increase in economic demands (Chapter 11.2.2, 11.2.3).

According to most recent observations and quantitative data, and notwithstanding the media escalation they have been hostage of, scientists promoting the *decisive influence of human activities on global warming* prove to be more correct than their opponents. Both refinements of climate evolution models and data validations published since the late 20th century under the aegis of the Intergovernmental Panel on Climate Change (IPCC) demonstrate the close relationship linking world temperature and economic development, and confirm the accentuated warming trend suspected in the early 1990s. The reliability of climate simulations has considerably improved, mainly due to the use of combined data on greenhouse gases and on air-released anthropogenic dust.

A weighty argument arises from noting that recent air-warming is less pronounced in the northern than in the southern hemisphere. Atmospheric dust tends to screen out direct solar radiation, which induces both the reduction of infrared radiation reflection and its subsequent retention by greenhouse gases. Air warming therefore tends to diminish above the northern hemisphere, polluting countries more than in the southern hemisphere. As atmospheric trajectories are mainly latitude-oriented, industrial dust limiting the greenhouse effect mainly falls on land and sea before being dispersed in the atmosphere of the southern latitude, which therefore without appreciable mitigation undergoes anthropogenic warming effects (see Duplessy, 1996).

Other convincing arguments arise from modelling data presented in 2001 under IPCC auspices. In particular the mean world temperature curves resulting from measurements performed during recent decades are almost superimposed to simulation curves obtained by adding up the model data controlled by both strictly natural and presumably strictly anthropogenic effects.

Major Projected Impacts

Temperature Environmental consequences of the enhanced greenhouse effect on Earth's solid and fluid envelopes basically depend on the intensity and geographic distribution of the expected warming. Recent IPCC simulations led to the prediction that *average world temperature in 2100 would be 2–5°C higher than the mean temperature that characterized the second millennium of the Christian era* (Fig. 177). The most recently forecasted atmospheric warming is stronger than those deduced from previous IPCC simulations, mainly because greenhouse gas concentrations have been re-evaluated on a greater value basis; in fact recent augmentation rates seem to exceed greatly the 20th century average trend. The rate of warming should be stronger in high- than in low-latitude regions, in the northern than in the southern hemisphere, and on landmasses than above the ocean. Warming is expected to become more important during the winter season at high latitudes, especially in Northern America and Asia. Earth's global warming should induce increasing flora and fauna migrations towards high-latitude regions.

Water balance Global warming should induce an increase in atmospheric content of water vapour responsible for *heavier precipitation*, particularly in northern hemisphere high-latitude regions, as well as on the Antarctic landmass. The general water balance on land is not necessarily increasing, some regions

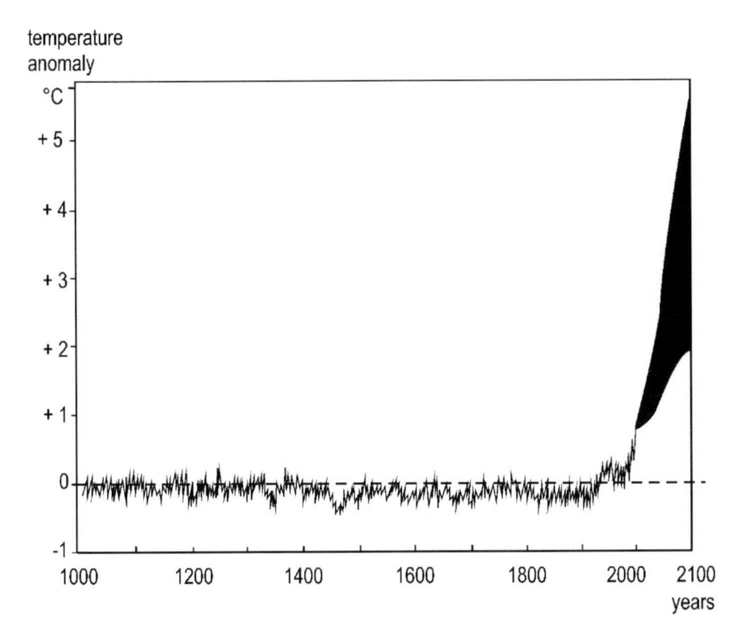

Figure 177: Estimate of the temperature anomaly range of the world's atmosphere during the 21st century, by comparison to average 2nd millennium values (after Mann et al., and Watson; see IGBP, 2000). Dashed line corresponds to the mean air temperature between 1000 and 2000 AD.

tend to become dryer; simulations for Western Europe suggest that water reserves in aquiferous rocks should rather diminish. On a worldwide scale, hydrological cycles are expected to intensify, humid regions become rainier and arid regions still dryer.

Such modifications in water balance should favour natural tendency for either active flooding and slope movements, or enhanced soil erosion and desertification (Chapters 3, 4, 8). On the other hand important productivity and economic gains should result from combined temperature and humidity increase in some regions, as suggested for instance by simulations performed about future productivity in Australia (Fig. 178).

Notice that no reliable argument presently arises from long-term observations or modelling data about the often-peddled idea that recent global warming is directly linked to storm intensity and number, or even to El Niño or NAO episodes (Chapter 11.2.3). For instance the maximum wind intensity and major storm frequency have been similar throughout the two last centuries, in spite of the global warming registered since the 19th century. Long-term data therefore contradict the impression that recent decades are marked by particularly intense and widespread weather deterioration.

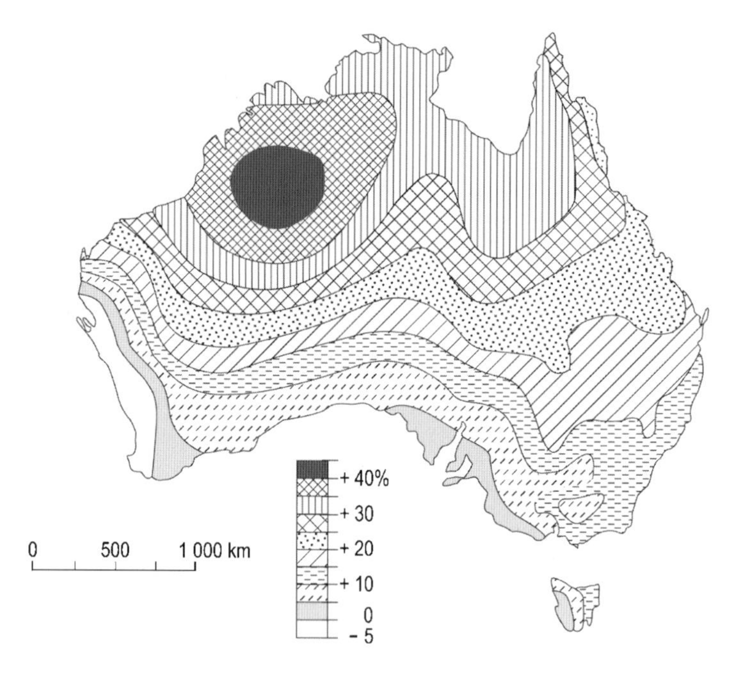

Figure 178: Predictive evolution of net primary productivity in Australia for an approximately double CO_2 concentration relative to the present (after Pittock & Nix; see Williams et al., 1993.

Sea level and ice cover Recent simulations indicate that average sea level during the 21st century should rise by at least 10 and at the most 90 cm, due to continuing glacier ice melting and ocean water rising. Given the inertia of glacier and ocean systems, this tendency is expected to go on for about 1,000 years, even if the greenhouse gas discharge in the atmosphere is strongly reduced. Note that the prediction given in 2001 about sea level change is lower than for previous estimates, since several scientific shortcomings have been identified concerning possible interactions and reactions between global warming, ocean water expansion, icecap melting, snowfall intensification, and water-stocking modes and location.

In any case, *solid water reserves should diminish in high-latitude and -altitude regions*. Continuing glacier melting, sea ice regression and snow-cover dwindling should express this trend. The drinking water balance would probably be affected and marked by either increasing or decreasing amounts, depending on geographical location. Similar modifications would probably concern the water exchanges occurring between surface reservoirs (glaciers, lakes, rivers) and underground aquifers. Massive *permafrost melting* could have dramatic effects, especially on and around the northern hemisphere landmasses: ground destabilization and thermokarst developments (Chapter 3.3.3), deterioration of buildings and other surface constructions (roads, railway tracks, pipelines), methane release from palaeosols and coastal submarine sediments, and correlative increasing greenhouse effect (Demek, 1994).

Coastal changes Flat coastal regions strongly exposed to wind and swell hazards (Chapter 4) should particularly undergo the consequences of progressive warming and sea-level rise: sea water ingression on land, exacerbated erosion, flood- and storm-induced inundations, etc. (Paskoff, 2001). Global warming is also expected to cause important coastal biosystem disruptions: fauna and flora migrations, coral bleaching, mangrove deterioration, modification of fishing and leisure zones. From a general point of view, human activities, the density and implications of which are particularly strong in coastal domains, should be noticeably affected, especially in flat countries. Simulations performed about the Netherlands indicate that an increase by 2% of North Sea wind intensity triggered by global warming would be responsible for an increase by 50% of storm-induced damages to property and economic activities (Dorland et al., 1999).

11.2.5. Mitigation, Difficulties

Limitation of Greenhouse Gas Discharge

Mitigation of the man-induced part of global warming first implies reducing the release into the atmosphere of greenhouse gases, especially of CO_2. IPCC calculations presented in 2001 indicate that atmospheric CO_2 stabilization at the present level would necessitate both *reducing the gas emission below the 1990 values and keeping these conditions* for at least several decades. Time is therefore running out if concrete results are to be obtained in the century to come. Decisive advances in this field should be made during the present decade. *Ways to go about things are theoretically easy, but their practical implementation comes up against multiple technological, economic, demographic and political difficulties:*

- *Reduction in the burning of wood, coal, petroleum and natural gas* Such a measure would allow retaining most greenhouse gases in Earth's solid envelopes. It is nevertheless almost impossible to implement on a large scale both in industrialized countries, which are accustomed to massive energy consumption, and in poor countries, where the basic livelihood and development will necessitate using more and more energy. The refusal by the USA to sign the 1997 Kyoto agreement towards fighting global warming illustrates the complexity of CO_2 limitation problems and issues. Let us also recall that the possibility of solving the greenhouse gas question by burning fossil-fuel hydrogen instead of oxygen is presently totally unrealistic.

- *Augmentation of substitute energy production* Replacing fossil fuel energy by non-polluting power sources reduces greenhouse gas release in the atmosphere. Solar and wind energy, geothermal and hydrothermal power, and hydroelectricity are typically non-polluting substitutes to coal, oil and natural gas use. On the other hand nuclear power has proven to be efficient, non-polluting and relatively cheap; remember that due to its important nuclear power plant park, France is the least polluting country as far as greenhouse gases are concerned (Chapter 7). Unfortunately these various substitute power sources either are of dramatically insufficient efficiency for covering energy needs or are dreaded for their long-term potential consequences. Huge efforts are nevertheless deployed by several countries for increasing non-polluting power sources, namely through significant progress in aeolian energy productivity and investment cost.
- *Other measures* Some original but quantitatively accessory techniques may help to reduce the greenhouse gas discharge in the atmosphere:
  - *methane* and other biogas *recuperation* from organic waste dumps, and subsequent burning for domestic or urban heating (Chapter 9.4);
  - *chlorofluorocarbon replacement* by cooling and propelling gases devoid of radiation power;
  - *reduction of troposphere ozone* production through progressive modification of urban transportation.

Intensification of Greenhouse Gas Sequestration

Reforestation The more natural and efficient way for stocking carbon dioxide consists of *restoring the forest cover on a large scale and developing long-term deforestation–reforestation strategy.* Forested land is able to store much larger CO_2 amounts than cultivated land, which in turn has stronger carbon storage capacities than prairies and savannah soils. Increasing the world's forest blanket is unfortunately very difficult to implement, particularly in inter-tropical developing countries, where both the largest forests with most fragile soils and populations whose wood requirements are the most essential are located (Chapter 8).

Carbon stocking in soils Restoring the forest cover contributes to maintaining organic carbon in terrestrial soils. Carbon is the main organic component in soils, where it is protected from mineralization. Soils contain about 1,500 billion tons of carbon, which represent more than twice the carbon stored in world vegetation and constitute the largest ecosystem reserves. Soil carbon stocks may be augmented by appropriate cultivation practices, which in addition help fight soil erosion–land desertification phenomena and keep biological diversity:

- diminution of soil anthropogenic reworking and especially of ploughing rhythms and depth, in order to prevent organic matter oxidation at or near soil surface;
- protection against erosion and oxidation during fallow periods by installing surface layers of living or dead plants (i.e. mulch);
- planting in arid regions of long-root vegetables such as leguminous plants, which help carbon to be deeply stored in soil and therefore protected from oxidation.

Other measures Less natural and sometimes more risky techniques may be envisaged for storing carbon in earth sinks and therefore preventing its escape as CO_2 in the atmosphere (Williams et al., 1993):

- freshwater diversion from humid towards arid regions for aiding forest development;
- implementation of artificial peat bogs in depressed areas;
- stimulation of coral or algae growth by fertilizer addition to seawater;
- terrestrial plant burying within ground sediments for preventing oxidation, similarly to what happened in geological times for conditioning peat, lignite and coal formation;
- pumping of nutrient-rich deep marine water for enhancing bio-productivity and consequent carbon fixation.

Some less questionable measures, the reliability and efficiency of which remain somewhat imprecise, consist in stocking CO_2 within underground porous rocks, similarly to what is commonly carried out for natural gas (Chapter 5.2.1). Encouraging tests have been performed in North Sea and Canadian oil fields, where carbon dioxide is to be injected simultaneously with petroleum extraction.

Improvement in prediction Simulations of 21st-century Earth-warming characteristics still remain too vague to allow building robust hypotheses on the scale of a few human generations. This leads either to excessive optimism and even sloppiness, or to excessive dramatization destined to be amplified by both people and the media. It is essential to have *more reliable predictions*. Such an objective necessitates *intensifying scientific research efforts* on the basis of several converging actions:

- implementation of denser and better integrated networks of measurement stations in atmospheric, terrestrial and marine domains: Global Climate

Observation System (G*C*OS), Global *T*errestrial Observation System (G*T*OS), and Global *O*cean Observation System (G*O*OS);
- augmentation of calculation capacity by computer systems;
- gathering of complementary disciplinary expertise applied to global-scale research programmes;
- in-depth iterative comparisons of past and current climatic changes and related phenomena, in order to better anticipate potential environmental disruptions;
- strong moral and financial support by countries' federations for inciting researchers to further analyse, diagnose and help to prevent large-scale environmental bifurcations.

11.3. Acid Rain

11.3.1. Late 20th Century Deterioration of European Forests

Investigations performed during the 1980s on health conditions characterizing central and eastern European forests revealed that many trees, especially conifers, tended to wilt in various regions (Fig. 179, A). Most affected forests were located in Germany, Poland, the Czech Republic, Austria and Switzerland. Estimates revealed that 30–60% of trees were suffering damage such as needle or leave loss, malformation, reduced resistance to drought and insects, yellowing and drying out. Extensive pH measurements showed that *rainwater* was *abnormally acid*, particularly in central European regions (pH values of 4.2–4.3), as well as in some southern Scandinavian zones, where conifer forests were also affected (Fig. 179, B). In normal conditions the rainwater pH is close to 5.6. As anomalously high sulphate and nitrate concentrations had been measured in central European precipitation (Sand, 1990; see Holland & Petersen, 1995), rain acidity was tentatively attributed to the chemical transformation of industry-derived sulphur dioxide (SO_2) associated with nitrous oxides (NO_x). Abundant sulphur and nitrogen oxides in the atmosphere resulted from smoke discharge, especially by coal thermal power plants located in the British Midlands, Belgium, Western and, more so Eastern Germany and Poland. Prevailing westerly winds were responsible for widespread pollutant dispersal from thermal plant zones, including towards some countries where industrial activity was moderate (e.g. Scandinavia, Switzerland).

Coincidence between forest deterioration and rainwater acidity drew the attention of researchers, ecological associations and the media. Other cases were discovered that confirmed central European observations. In particular the diminution and sometimes the disappearance of some fish and other animal species in numerous Scandinavian and Canadian lakes, as well as plant

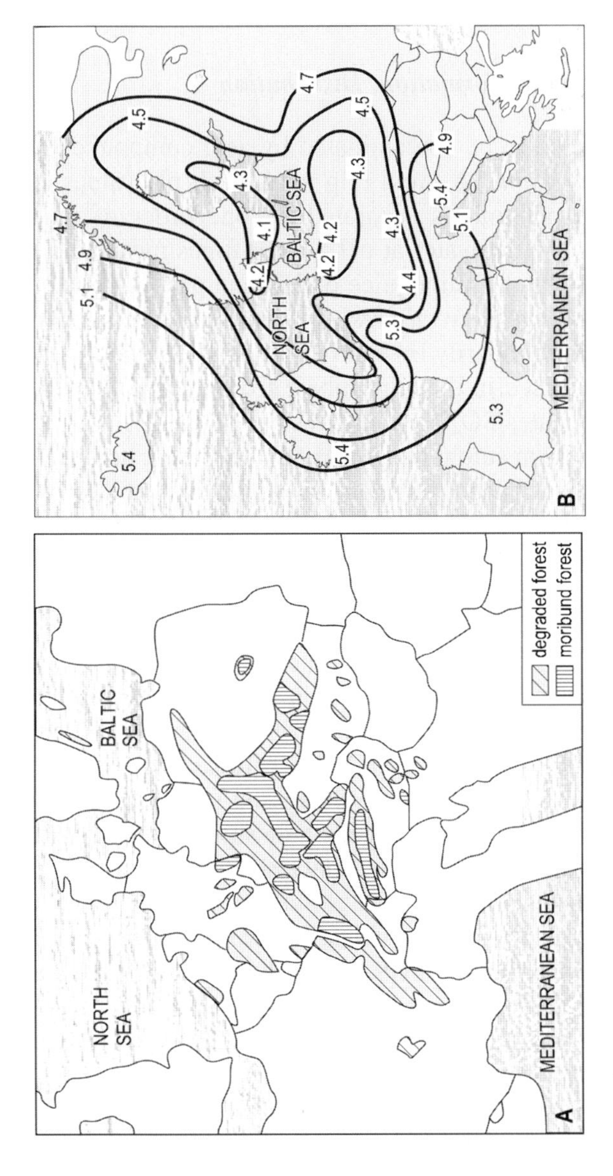

Figure 179: Deterioration states of Central European forests (A), and 1985 rain water pH values in the whole of Europe (B) (diverse sources; see Holland & Petersen, 1995.

deterioration in some northeastern American regions, proved that the acid rain phenomenon was widespread and could diversely affect natural biosystems. Various examples arose for water, soil, flora and fauna contamination in non-polluting countries situated on the leeward side of polluting countries.

11.3.2. Characterization, Formation, Distribution

Rainwater is considered to be contaminated by acid compounds emitted in the air when *pH is lower than 5.6*. Acid rain may reach pH values as low as 1.5; remember that each pH unit is separated by one order of magnitude from previous and next units (i.e. water at pH 5 is 10 times more acid than water at pH 6). Vinegar has a pH of 3, lime juice of 2.3, and gastric juice of 1.6.

Rainwater acidification indirectly results from fossil-fuel combustion, mainly by coal burning, and secondarily by that of hydrocarbons. Sulphur and nitrogen oxides are mainly released in the atmosphere from thermal power plants. They are oxidized into sulphates SO_4^{2-} and nitrates NO_3^-, which in turn combine with water vapour H_2O to form sulphuric acid H_2SO_4 and nitric acid HNO_3. Mixed in the air with cloud water and forming aerosols, fine acid droplets may be transported at variable, sometimes large distances, depending on wind intensity and atmosphere humidity. Sulphuric and nitric acids finally fall in the form of rain on ground soils, rocks and water.

An indirect proof of the major influence of fossil organic matter on acid rain formation arises from investigations performed on the geographic distribution of coal industry-derived residues discharged in the air as smoke, soot and dust. Sooty debris has been counted in about 30 European lakes, distributed from Svalbard to Spain, and from Ireland to Poland (Rose et al., 1999). The amount of coal micro-residues has dramatically increased in lake sediments during the last century, especially in central Europe, where sulphide emission from thermal coal plants has been serious (Fig. 180). These results also attest the cross-border dispersal of acid aerosols and rain.

On a worldwide scale, *particularly acid rainwater* is known to affect central Europe (pH 4.0–4.5), northeastern USA and adjacent regions of eastern Canada (pH 4.0–5.0), and to a lesser degree central-southern Russia and southwestern USA (pH 5.5). Most threatened regions are those directly and regularly subject to such rainfall (Fig. 181).

In addition to rainwater acidity the sensitivity of regions affected strongly depends on *chemical characteristics of soil and geological ground*. Rainwater acidity may easily and rapidly be buffered in regions where alkaline limestone and carbonated soils crop out largely, and where lake water exceeds pH values of 7. By contrast acid water effects tend to be exacerbated when rain falls on siliceous geological substrate such as granite, gneiss, schist or carbonate-

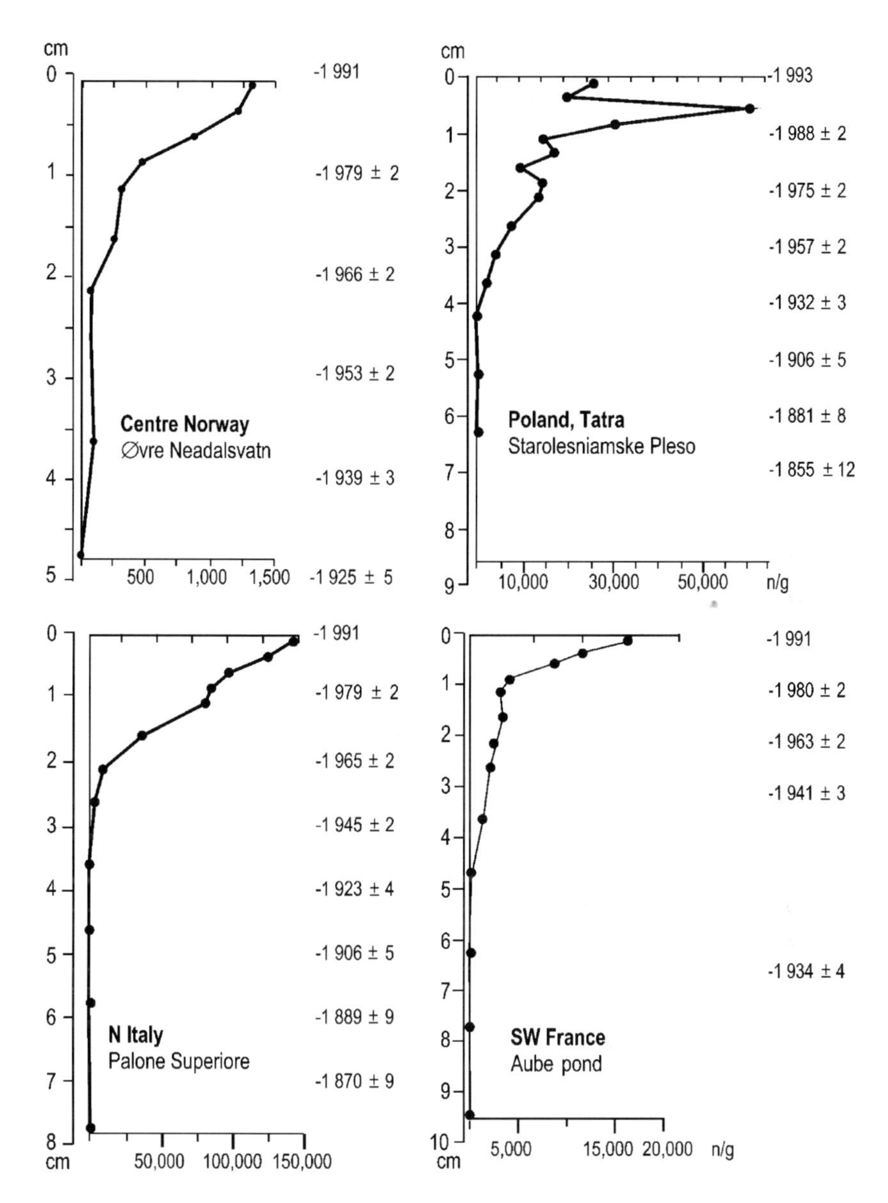

Figure 180: Concentration profiles of carbonaceous and coal micro-residues in sediments cored from four European lakes (after Rose et al., 1999). n/g, number of sooty particles per gram of dry sediment. Sediment deposition years are indicated on the right-hand side of each graph.

depleted soils. This is why Scandinavian lakes and soils, as well as German forests, where underground formations consist of granite and metamorphic

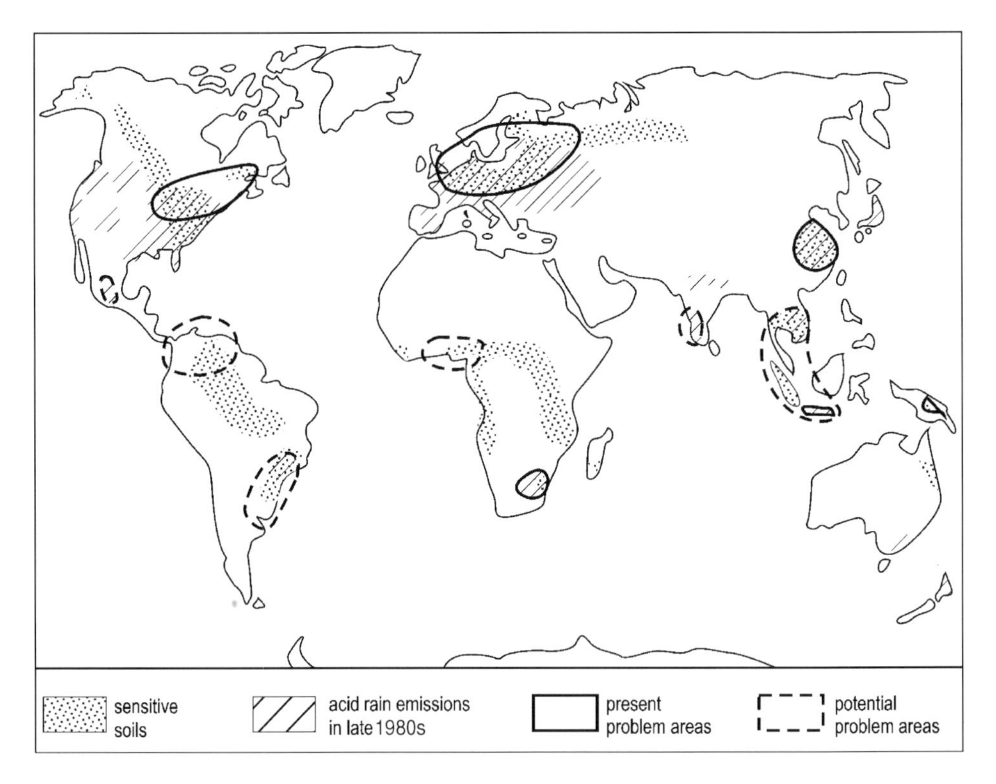

Figure 181: World distribution of soils sensitive to acid rain (dotted areas), of regions particularly threatened in late eighties (hatched areas), of currently most affected zones (full lines), and of potential problem areas (dashed lines) (after Rodhe et al.; see Pickering & Owen, 1997).

rocks, are strongly damaged despite the often moderately acid character of rainwater. In Canada and USA, the regions most sensitive to acid rain are the Appalachian and Rocky Mountains, where siliceous rocks and soils outcrop commonly.

11.3.3. Environmental Impact

Soils and Vegetation

Massive deforestation carried out in Europe and North America during the 19th and early 20th centuries caused some soil acidification through nitrous oxide release due to bacterial activity. Vast soil surfaces therefore became indirectly vulnerable to acid rain effects because of human action, even before important industrial discharge of SO_2 and NO_x took place. This caused the impact of acid rain to be amplified.

Exact mechanisms of acid rain action on soil and vegetation are still imperfectly understood. It seems that sulphurated and nitrous gases, as well as troposphere ozone, are responsible for some *photosynthesis disruption* and for *decreasing calcium and nutrient assimilation.* In addition acid substances seeping in the soil are susceptible to release of some toxic metals such as aluminium, which favour plant decay. Mid-mountain *conifer* species are especially sensitive to acid rain effects. Broad-leaved trees seem to be much less vulnerable, as shown by the densely growing vegetation in many urban parks (e.g. London, Paris, New York City). Generally, forests that have been damaged by acid rain display less aptitude for retaining soils against water and wind erosion.

In parallel to the buffering role played by carbonate rock and soil formations the water, soil and plant acidity may be caused or accentuated by local geological characteristics. *Natural acidification* may result mainly from *leaching of sulphur compounds* contained in subsurface geological terrains. Industrial activities are therefore not solely responsible for acid environmental disruptions. Statistical investigations conducted in Finland show that the industry-derived acid character of surface and subsurface water is restricted to Lapland in the northeastern-most part of the country, where many thermal power plants are established. Elsewhere in Finland, ground surface and water acidity mainly results from natural leaching of sulphides contained in glacial moraines, or of sulphates stored in post-glacial marine silt and clay (Lahermo et al., 1994).

Freshwater Fauna and Flora

Acid fallout is slightly harmful to air-living fauna but *strongly detrimental to aquatic organisms.* The pH of most rivers and lakes is usually between 6.5 and 8.5. When pH values drop below 5.0, fish and other animals tend to suffer diverse metabolic disorders; they mainly lose their skeleton rigidity because of depletion in calcium phosphate fixation. Lamellibranchs and gastropods can barely survive at pH 6, crayfish and mayfly at pH 5.5, trout and salamanders at pH 5, and perch, frogs and toads at pH 4.5. Aquatic plants that fix carbonates in their tissues also suffer from excess acid in freshwater. The lake aspect itself is usually not affected, acidified water being limpid and transparent.

Altogether the biological disruptions due to acid rain tend to be significant in lakes from granitic and metamorphic regions subject to noticeable sulphurated and nitrous oxide discharge. Estimates published in the late 20th century report that ecosystems undergo excessive acidification in several thousand lakes from Sweden, Norway and Scotland, that 16% of the lakes are affected in the Adirondacks (New York State, USA), and that excess acidity characterizes numerous lakes in Nova Scotia and Ontario (Canada). All these lakes are located under wind trajectories crossing industrial regions where large amounts of fossil fuels, especially coal, are currently burned.

Human Constructions and Human Health

Acid rain largely initiates "*stone diseases*" (Chapter 9.5.3). In particular sulphur dioxide (SO_2), water and carbonate minerals interact to cause gypsum precipitation ($CaSO_4$, $2H_2O$) in limestone and sandstone pores and fissures. These reactions are responsible for surface decay on countless buildings, statues and other constructions, mainly in urban and industrial polluted environments.

Human health itself is very slightly affected by acid water contact or ingestion. Indirect risks nevertheless arise by means of *respiratory diseases*. Deterioration of air quality in large conurbations is responsible for various sanitary risks, among which the consequences of acid rain may be crucial. The very dense fog that came down on London City from December 5–12 in 1952 indirectly caused the deaths by acute bronchitis of 4,000 people. Unusual "smog" developed that resulted from the presence above London of a stationary warm air mass. This mass was blocking cold above ground level, a 60-km-wide polluting cold cloud that was impregnated with sulphuric acid, mainly derived from industrial and domestic coal combustion.

11.3.4. Mitigation, Difficulties

Reduction of Noxious Gas Discharge

Alternative energy Replacing coal and hydrocarbon thermal energy by *increased production of solar, aeolian and nuclear power* helps to diminish sulphur and nitrogen oxides discharge in the atmosphere, as is the case for fuel combustion-derived greenhouse gases (Chapter 11.2.5). We have already seen that these substitute energies are presently unable to face the problems linked to world energy demand, safety, public opinion, etc.

Fossil-fuel selection Another means of limiting sulphur and nitrogen oxide discharge consists of selecting fossil fuels that contain few pollutants; this is the case of SO_2^- and NO_x-poor coal, light crude oil, and most natural gas types. Slightly air-polluting fuels can be chosen by performing appropriate laboratory analyses on lithological, mineralogical and geochemical characteristics of mine- and well-extracted materials. Such selection practices are unfortunately unworkable in a systematic way and on a worldwide scale, considering the exponentially growing need for exploiting new fossil fuels, many of which contain abundant sulphurated and nitrous components: impure coal, heavy crude oil, bituminous schist and tar sands.

Sulphide-rich coal has been and is still abundantly extracted and burned in Eastern Europe and Russia, where impure ores are widespread and are cheap to exploit. The question of fossil-fuel quality and especially of impurity in coal might become crucially important during the next decades in various countries from eastern and southeastern Asia, Africa, and South America, which are facing huge energy needs (Chapters 5, 8). Comparison of projected demographic curves and economic demand causes fear of future discharge of polluting gases in gigantic amounts. Some estimates report that 210 Mt of sulphur and 55 Mt of nitrogen could be annually released in the atmosphere around 2020, as a result of fossil fuel combustion (Mackenzie & Mackenzie, 1995). This would cause considerable increase in acid fallout damage to soils, water, flora, fauna and people.

Pollutant retention A third preventive method consists of preventing fuel impurities from being released into the atmosphere. Several measures may be taken for reaching this goal: setting up of SO_2 filters in thermal power plants, burning coal in the presence of lime or finely crushed limestone for neutralizing acid vapours, implementation of catalytic converters limiting nitrogen oxide emission, etc.

Such pollutant retention systems are generally efficient (Fig. 182) but expensive, which prevents using them in an extensive way. Sulphur oxide

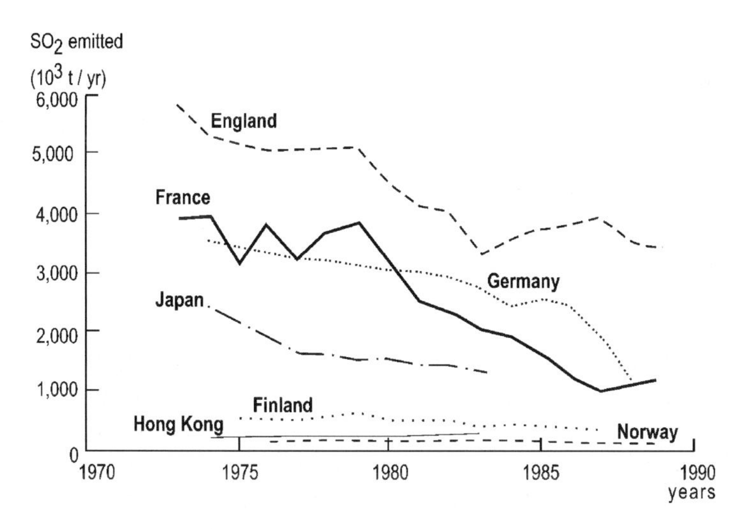

Figure 182: Evolution since 1970 of SO_2 atmospheric discharge in several industrialized countries (after Environmental Protection Agency; see Holland & Petersen, 1995). Values in thousand tons per year (10^3 t/yr).

emissions were reduced by 30% in the USA between 1970 and 1988, and by 39% in Japan between 1973 and 1984. In Europe a community protocol was established in 1980 to progressively reduce the SO_2 emissions by 30%. Catalytic silencers should reduce the NO_x release by 76%. An international conference held in 1984 at Helsinki under the aegis of the United Nations had led to a protocol for limiting SO_2 discharge on a worldwide scale. This protocol has been signed neither by Great Britain nor by the USA, where the majority of atmospheric pollutants are wind-dispersed outside their national frontiers; both these countries have nevertheless implemented active pollutant-discharge limitation programmes.

Statistical data gathered for some European and northeastern American ecosystems suggest that limitation of acid sulphur compounds' discharge in the atmosphere is of somewhat disputable efficiency. This seems due to the fact that SO_2 retention often determines also the retention of calcium compounds, the buffering effect of which is therefore reduced in rain droplets (Gimeno et al., 2001).

On the other hand, the positive effects of SO_2-limitation measures on water and soil acidity are often very slow because of the low mobility of sulphur compounds. Water and sediments of Big Moose Lake in northeastern USA became acidic as late as 70 years after the 1880 anthropogenic increase in SO_2 release in the atmosphere. The pH values, which used to be close to 5.7 under normal conditions, started dropping in 1950 only; pH values were 4.5 in 1980. Almost all the fish died between 1950 and 1965 (Berger, 1992). Similar, belated mechanisms occur in some polluted European lakes, which display very slow acidity reduction, although drastic limitation of SO_2 discharge has taken place for more than two decades. On a yearly scale it is frequently observed that water acidification downstream of mountainous regions develops with some delay, because of the release in springtime only of sulphuric acid stored in winter snow.

Treatment of Acidified Zones

The addition of carbonate for buffering acidified water has been experimented with in lakes from different countries (Canada, Sweden, USA); this method has proven to be disruptive, only slightly efficient and costly in the long term. A somewhat less-artificial process consists of *stimulating the primary productivity* by adding fertilizers to acid lake water. The development of phytoplankton tends to neutralize the water acidity through nutrients' stabilization and organic-matter consumption. Encouraging tests have been implemented in some English lakes, where no apparent ecosystem disruption has been registered.

Notice that buffering of acid rain-affected terrains may sometimes be naturally improved by alkaline fallout. The increasing frequency in northeastern Spain of atmospheric fallout of Saharan red dust tends to favour re-alkalination of plants and soils previously polluted by acid rain (Avila & Peñuelos, 1999).

11.4. Air Quality, Water Quality, Soil Reworking

11.4.1. Other Impacts of Atmospheric Discharge of Pollutants

Most anthropogenic emissions into the atmosphere of compounds other than greenhouse gases and sulphurous or nitrogenous oxides are of local or regional impact and therefore do not disrupt the global environment. Various processes are involved, which may combine to form significant risks:

- "heat islands" established above big conurbations are characterized by temperature, precipitation and nebulosity that are much higher than in adjacent zones. They result from intense urban and/or industrial energy release in the absence of wind and air circulation;
- "smog" consists of brownish-yellow, dust-loaded fog that is trapped above big cities or industrial complexes. It is determined by unusual temperature inversion, cold, dense air being trapped on the ground surface by warmer and stationary overlying troposphere air (Chapter 11.3.4);
- volatile organic compounds (VOC) consist of very fine and light particles and aerosols that result from incomplete fuel-burning and industrial processes involving photo-chemical reactions (e.g. solvent or paint manufacturing). Recurrent VOC release in the atmosphere may cause serious health risks. Flying ash dispersed by industrial smoke corresponds to similar processes and may be responsible for noticeable soil and water pollution (Chapter 9.3.3);
- low-altitude ozone (O_3) is rather abundantly produced in urban environments through functioning of the internal combustion engine and VOC, nitrogen oxide and air photo-chemical interactions. O_3 released in the troposphere participates in both increasing the greenhouse gas effect (Chapter 11.2.3) and deteriorating urban air quality.

Most of these mechanisms are responsible for *air quality deterioration*, the impact of which is restricted to local pollution of Earth's solid envelopes (e.g. industrial wasteland, soil and groundwater) and to building-stone and mortar surface degradation (Chapter 9.5). On the other hand, atmospheric deterioration may cause serious *health risks*: respiratory and cutaneous lesions, asbestosis, etc. (see also Chapter 5.3.4). Most important damages result from *industrial accidents* responsible for massive discharge in the atmosphere of toxic compounds and uncontrolled chemical reactions (e.g. AZF plant explosion on September 2001 at Toulouse, France). In December 1984 the accidental penetration of water in a methyl isocyanine reservoir at the Bhopal chemical plant (India) caused the formation of a 40-km^3-wide toxic cloud that was responsible for 2,800–4,000 fatalities, 20,000 injured people, and numerous pathological and genetic after-effects.

Heavy metal discharge in the atmosphere should be mentioned separately since it may cause widespread ground pollution. This is especially the case for lead and mercury, which may cause diverse diseases (see Table 44) and are discharged through human activity in large amounts relative to natural processes (Table 45). In addition both these elements may easily and widely be wind dispersed.

Anthropogenic lead (Pb) in the atmosphere has for a long time essentially resulted from its use as a fuel additive under tetraethyl form, which was emitted from exhaust systems as a combustion residue. As unleaded fuel is now mostly employed, the metal is being released in lesser amounts and principally proceeds from fusion, purification and industrial mining processes. The widespread dispersal and significant increase in lead concentration since the 19th century industrial revolution are attested by measurements performed in Greenland ice cores and the world's ocean water and sediments (Pye, 1987). The

Table 45: Natural and anthropogenic sources of atmosphere-released metals (after Lantzy & Mackenzie; see Merritts et al., 1997). Values in tons per year.

Element	**Natural emissions from geological processes A**	**Anthropogenic emissions from human activities B**	**Ratio A/B**
aluminium	48,900,000	7,200,000	0.15
iron	27,800,000	10,700,000	0.38
manganese	605,000	316,000	0.52
cobalt	7,000	4,400	0.63
chromium	58,000	94,000	1.6
nickel	28,000	98,000	3.5
copper	19,000	263,000	13.8
zinc	36,000	840,000	23.3
arsenic	2,800	78,000	27.9
silver	60	5,000	83.3
mercury	**40**	**11,000**	**275**
lead	**5,900**	**2,030,000**	**344**

large impact on ecosystems of lead atmospheric fallout is demonstrated by some specific data:

- lead concentration in penguin dejecta has augmented during the two last centuries, and especially since the mid-20th century. This points to noticeable metal accumulation through food chains, and could reflect some disruption in austral ecological systems (Liguang & Zhonging 2001);
- ^{206}Pb to ^{207}Pb ratio values have decreased since 1950 in lake sediments and forest soils of Sweden relative to values having characterized the three previous millennia, which reflect the increased effects of industrial global pollution (Brännwall et al., 2001).

Anthropogenic mercury (Hg), which is known for causing local and regional soil contamination, tends also to be dispersed on a global scale. This heavy metal may easily be removed from soils as long as it has not been chemically fixed. Its actual proportion in the atmosphere is 275 times higher than during the pre-industrial period; its residence time in the air ranges between 6 and 18 months.

Mercury is employed for various electrolytic processes, mining treatments and industrial developments. Its use was particularly widespread during 17th–19th century colonial times in Central and Southern America, due to gold and especially silver mining extraction and purification activities. The average discharge of mercury linked to silver extraction in Spanish America is estimated to have amounted to 612 t/year between 1580 and 1820 (Fig. 183). Sixty to

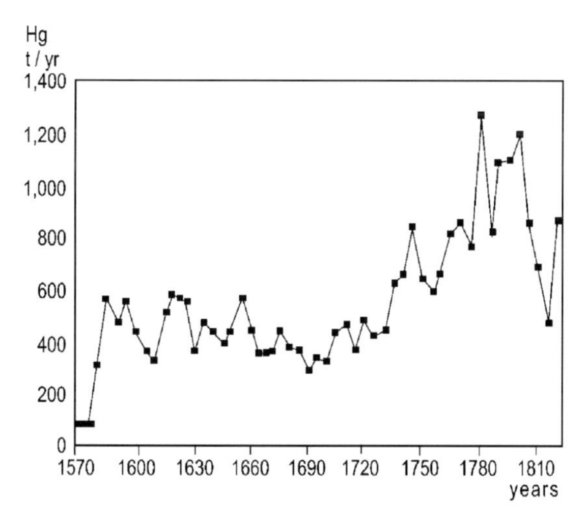

Figure 183: Estimate of mercury discharge levels due to silver ore mining and purification processes in colonial South America between 1580 and 1820 (diverse sources; see Nriagu, 1994.

sixty-five percent of the heavy-metal residues have been dispersed in the atmosphere; subsequent fallouts have polluted both soils and groundwater, from which remobilisation may presently still arise. The high mercury concentrations recorded worldwide in surface terrestrial environments are considered to partly result from ancient mining activities in the Americas (Nriagu, 1994).

Temporal variations of global heavy-metal distribution have been precisely measured through chemical analyses performed on ice layers that successively accumulated in very high latitude regions. A borehole drilled in the Central Greenland icecap has allowed quantification of the concentration of lead, cadmium, copper and zinc derived from atmospheric fallout between the late 18th and late 20th centuries (Candelone et al., 1995). Comparison with the metal distribution in ice deposited during pre-industrial times shows that the contents have started increasing strongly during the 19th century. Pb, Cd, Cu and Zn levels continued to increase in a more or less parallel way during the last century, and then started to decrease in the 1960s and 1970s. The heavy-metal concentrations at the end of the 20th century were again similar to those typical at the beginning of the century. These data lead to the following conclusions:

- the 19th century industrial revolution and its 20th century extension have actually been responsible for global heavy-metal contamination of planet surface formations;
- the heavy metals accumulated in the Greenland icecap between 1773 and 1992 are estimated to amount to 3,200 t of lead, 2,500 t of zinc, 200 t of copper, and 60 t of cadmium;
- mitigation measures implemented for a few decades for limiting the heavy metal discharge in the atmosphere have proved to be efficient on a global scale, and may help to significantly restore environmental quality.

11.4.2. Global Impact of Water Exploitation

Recurrent groundwater pumping leads to aquiferous rocks undergoing chronic subsidence movements that in the long term tend to propagate in overlying and adjacent terrains. Subsidence is usually of local extent (Chapter 3.2.2) but multiple and contiguous pumping may cause gravity movements to spread and affect widespread ground surfaces. Extensive groundwater extraction in North American eastern coastal plains is responsible for chronic subsidence movements from northern Florida to southern New England, over a distance of about 1,500 km. Subsidence rates average a few millimetres per year in the whole coastal domain, and may locally reach 2.5 cm/year (Davis, 1987; Sun et al., 1999). Recurrent pumping effects are locally amplified by the presence of swelling clay (smectite) and isostatic compensation movements. Resulting environmental impacts frequently comprise various types of surface damage to constructions, and the potential risk of marine ingression.

Development on a very large scale of continental water exploitation, river-course conversion for fluvial transportation, and artificial reservoir construction, strongly diminishes the hydrological and particle fluxes towards marine and natural lake basins (Chapter 6.4.1). The balance of land–ocean exchanges tends to become deeply disrupted due to the addition of diverse *anthropogenic developments that all favour continental trapping* of water, dissolved elements, and organic and inorganic terrestrial particles: water diversion for irrigation and consumption, evaporation of artificial water, massive particle sequestration behind constructed dams, etc. Some major rivers such as the Nile and Colorado have practically lost any capacity for feeding coastal regions with sediments and water. Countless river courses are no longer able to ensure their natural upstream-to-downstream transit role. The trapping rate of sedimentary particles along river basins through human actions is estimated to amount to 20% on a worldwide scale. The average water transit time in numerous hydrographical basins has been lengthened by about 1 year, and 61% of water used by Man (i.e. 3,750 km^3/year) evaporates before reaching the sea; this corresponds to a decrease by more than 5% of continental water supply to the ocean (IGPB, 2000). All these disruptions result in a general disequilibrium of land–sea natural hydrodynamic confrontation, for the benefit of marine forces (tides, swell, storms, hurricanes).

Flat coastal regions are exposed to the added effects of potential sea-level rise induced by global warming, continental particle shortage due to river trapping, and subsidence caused by excessive aquifer pumping. Such combined actions tend to favour inland penetration of seawater, exacerbating the effects of marine storms and intensifying coastal erosion.

Some regions subject to strong alternating rainfall may experience temporary neutralization of river discharge due to excess seawater ingression. Lake Taihu, situated in the Yangtze River delta in China, has been subject to a net retreat of freshwater during the last decades' dry seasons which was mechanically pushed inland by seawater. This *retrogression* progressively tends to extend both in space and time; it recently happened even during humid seasons due to the added effects of subsidence through aquifer pumping, upstream river-water diversion, and global sea level rise (Chen & Wang, 1999). Such phenomena, which cause strong disruption in deltaic ecosystems and sedimentary particle budget, are still of limited geographical extension but progressively tend to affect more widespread regions.

11.4.3. Large-scale Geomorphological Impact of Ground and Soil Use

Gigantic amounts of surface and subsurface materials have been displaced during the last millennia on Earth due to various human activities: uninterrupted soil exploitation; surface quarries and underground mines; river-course and lake conversion; construction of artificial communication ways, barriers and basins; domestic, urban and industrial settings and extensions; etc. Estimates performed globally indicate that the total amounts of rock and soil that have been displaced through human action during the last 5,000 years would allow building a mountain chain of 100 km length, 40 km width and 4,000 m height. Calculations show that this mountain of reworked material should get twice as long in only one century from now, if demographic expansion and inferred growing needs continue at their present pace (Hooke, 2000). World landscape upheaval has accelerated in an exponential way during the last few centuries, due to both strong population growth and enlarged human hold on nature (Fig. 184). Agricultural purpose and need for cultivated land were and still will be responsible for most of these soil and ground reworking actions. Global changes caused by such human activities affect primarily the geosciences domain. Much research work has still to be done for better quantifying, documenting, and understanding these phenomena.

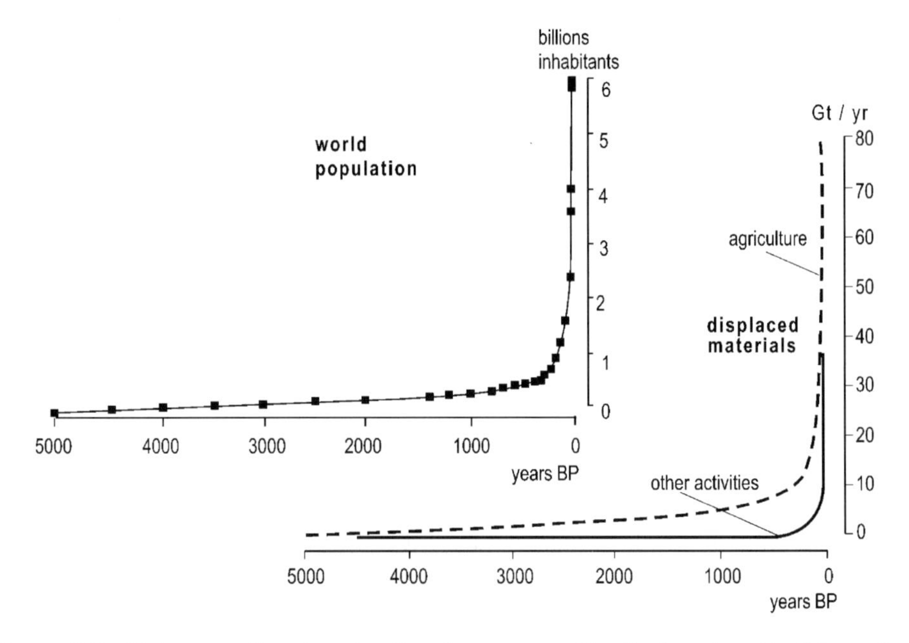

Figure 184: Estimates of rock and soil volumes annually displaced on Earth's surface over 5,000 years. Comparative distribution of world population (after Hooke, 2000). Solid material buildup given in billions tons per year (Gt/yr).

A Few Landmarks, Perspectives

- The factors potentially responsible for regional and global environmental disruptions have three possible origins: essentially natural major ocean–atmosphere exchanges; essentially anthropogenic acid rain or extensive exploitation of aquifer, river and water planes; and mixed greenhouse-gas emission, heavy-metal release, or geomorphological changes. Natural disruptions are still imperfectly understood and their mitigation is limited to prevention. Anthropogenic disruptions are well described and explained, and restoration measures are available; but mitigation is either very expensive or appear somewhat anachronistic. Global environmental deterioration due to mixed natural and human causes tends to be better understood and quantified, but its mitigation depends more on political decisions than on technical solutions. In all cases the disruptions and remedies are of transnational importance. Major disorders may affect fluid envelopes and have repercussions on surface solid envelopes: flooding and slope movements, droughts and desertification, soil erosion and contamination, and coastline weakening.
- Major disruptions of ocean–atmosphere global systems involve the Southern Oscillation, which controls the seasonal air pressure and marine current regime in the Equatorial Pacific. El Niño Southern Oscillation (ENSO) episodes consist of aperiodic, several 1–10-year-long events, marked by an eastwards migration of low pressure and warm water, and inducing a chain of climatic disturbances: strong rain, flooding, landslides and loss of marine productivity on the Pacific American side; drought, fire and desertification on the Indonesian side; and storms and cyclones in the intermediate Polynesian domain. Opposite disruptions sometimes occur, which are expressed by strong upwelling and dryness to the east and are called La Niña episodes. El Niño events may have meridian repercussions, namely in the Equatorial Atlantic, and also latitudinal repercussions responsible for unexpected flooding, warming and agricultural developments in temperate regions. Similar but less intense and dramatic disturbances result from the North Atlantic Oscillation (NAO), which results from pressure gradient changes between the Azores anticyclone and Iceland depression. ENSO and NAO are not synchronous but induce accentuated disturbances when their major phases occur simultaneously. Understanding and forecasting these major natural disruptions necessitate extending both marine and terrestrial observation and measurement networks.
- The Earth's surface radiation and energy balance result both from incident and reflected solar radiation and from infrared re-radiation by soils, atmosphere and cloud cover. The radiation power is responsible for a natural

greenhouse effect that is essential for allowing life on our planet. A global climatic warming, which is mainly proven by ice-cover retreat and mean sea-level rise, started developing about one century ago and obviously correlates with a greenhouse gas increase. Part of this increase indisputably proceeds from human activities, as testified by a parallel increase in temperature and industrial gas discharge (CO_2, CH_4, NO_x, O_3), by artificial greenhouse-gas release (CFCs), and by similar distribution of climatic and industrial particle discharge zones. The impact of global warming on Earth's solid formations potentially affects mainly the coastal zones subject to sea-level rise, as well as terrestrial regions where water balance is threatened by major climate disruptions. Few relationships arise between increasing greenhouse effects, El Niño episodes, and storm frequency and rhythm; global warming should even regionally induce climatic conditions favouring more efficient soil cultivation. Diminishing the human contribution to the greenhouse effect necessitates implementing measures of difficult practicability or depending on delicate international agreements: reduction of fossil-fuel consumption, development of alternative energies, massive reforestation and protection of organic soils, and strong limitation of pollutant discharge into the atmosphere. Additional research is needed for improving the prediction of large-scale environmental changes induced by global warming.

- Rainwater acidification is principally responsible for forest and freshwater fauna deterioration. Acid rain mainly results from atmospheric discharge by thermal power plants of sulphur and nitrogen oxides that are transformed into sulphuric and nitric acids. The acid rain impact is exacerbated in urban environments where SO_2 and NO_x are released in the air together with various other pollutants. Mitigation measures consist of diminishing the fossil-fuel consumption, developing alternative energies, and reducing pollutant gas emission. Geosciences are involved by acid rain through both building stone deterioration and the natural ability for rainwater to be buffered by carbonated soils and rocks. The question of acid rain fallout limitation comes within the competence of international agreements, the application of which is considerably complicated by the diversity of demographic, economical and political situations.
- Some other disruptions of surface terrestrial environments progressively become of general importance due to the multiple and repetitive character of human interventions. This is particularly the case of heavy metal and other contaminant atmospheric release, of coastal-zone erodibility through excessive aquifer pumping and sediment trapping along river courses, and of landscape reshaping by rock and soil extraction and exploitation. Noticeable efforts should be made for helping people to realize the environmental impact of such currently imperceptible, long-term changes.

Further Reading

Burroughs W. J., 2001. Climate change. A multidisciplinary approach, Cambridge University Press, 298 p.

Chapel A., Fieux M., Jacques G., Jacques J.-M., Laval K, Legrand M. & Le Treut H., 1996. Océans et atmosphères. Hachette, Paris, 160 p.

Drake F., 2000. Global warming: the science of climate change. Oxford University Press, 273 p.

Ellis D., 1989. Environment at risks. Case histories of impact assessment. Springer, 329 p.

Joussaume S., 1999. Climat d'hier à demain. CNRS éd., Paris, 143 p.

Mackenzie F. T. & Mackenzie J. A., 1995. Our changing planet: an introduction to earth system science and global environmental change. Prentice Hall, 387 p.

Moore P. D., Chaloner B. & Stott P., 1996. Global environmental change. Blackwell, 244 p.

Slaymaker O. & Spencer T., 1998. Physical geography and global environmental change. Longman, New York, 292 p.

Some Websites

http://www.agora21.org/refurl/changlob.htm: Called the "Web site of sustainable development", it contains most information concerning general environmental concerns as well as main institutions in charge of global environmental problems: Intergovernmental Panel on Climate Change, IPCC; national and international Agencies, documentation centres, etc.

http://www.agora21.org/mies/sommaire.html: site devoted to potential impacts of global warming, according to IPCC synthetic investigations

http://www.globalchange.org/: exhaustive information of world climate warming and global change, gathered by the Pacific Institute for Studies in Development, Environment, and Security. Connections are proposed with numerous other Websites, by the use of search engines

http://www.wmo.ch/indexflash.html: World Meteorological Organization site providing countless data on Earth's global environmental status, on world research programmes, on access means to other Websites, etc.

http://www.ogp.noaa.gov/enso/: detailed information on Southern Oscillation and El Niño–La Niña phenomena

http://www.epa.gov/globalwarming/ and http://www/epa/gov/airmarkets/: review of the mechanisms and evolutionary trends of global warming and of general airborne chemical pollution (acid rain, nitrous oxides, sulphur dioxide), presented under the aegis of the U.S. Environmental Protection Agency (EPA)

http://nsidc.org/NASA/SOTC: information on world ice- and snow-cover status and monitoring, from the perspective of global climate change; relations with sea-level change and permafrost evolution

Epilogue

1. The Earth Yesterday and Today

About 140 years ago the geologist Louis Figuier wrote approximately the following sentences in the preface of his textbook entitled *Earth before the Deluge — Picture of Nature* (*La Terre avant le Déluge — Tableau de la Nature*; Hachette 1864, 4th ed., 472 p.): "I am going to defend a strange idea. I am going to claim that the first book to put in children's hands should deal with natural history. Instead of catching young minds' admiring attention on La Fontaine's fables, Puss in Boots adventures, Peau d'âne history, or amorous adventures of Venus, we must turn it on innocent and simple natural sights: the structure of a tree, the perfection of mineral crystalline lines, the composition of the ground underfoot" *("Je vais soutenir une thèse étrange. Je vais prétendre que le premier livre à mettre entre les mains de l'enfance doit se rapporter à l'histoire naturelle; et qu'au lieu d'appeler l'attention admirative des jeunes esprits sur les fables de la Fontaine, les aventures du Chat botté, l'histoire de Peau d'âne, ou les amours de Vénus, il faut la diriger vers les spectacles naïfs et simples de la nature: la structure d'un arbre, la perfection des formes cristallines d'un minéral, l'arrangement intérieur des couches composant la terre que nous foulons sous nos pieds").* One may wonder how realistic would a scientist be considered at the dawn of the 21st century when introducing in a similar way a book on geological environments. All through this present volume, multiple disruptions have been described that affect the Earth and have been initiated either by natural processes or by human activities. Is it still easy today to contemplate on one's doorstep "innocent and simple natural sights"? Do countless regions and landscapes still deserve "admiring attention", even though they have been subject to human influence and have deteriorated increasingly during the last one and a half centuries?

Let us first remember that *man-induced disruptions of Earth exclusively involve the most external envelopes of our planet.* All deep Earth phenomena such as magnetism, gravity and electric fields, thermal exchanges and propagation, earthquakes or volcanic eruptions, are in no way affected. Earth's dynamic is also perfectly stable in its planetary environment. Fortunately Man is not able to disrupt the intra- and extra-terrestrial functioning of the planet. The often-painted picture of an Earth deeply disrupted by man activities sometimes comes close to being anthropomorphic.

It is nevertheless true that surface solid envelopes of our planet are the place where all exchanges with fluid envelopes (i.e. air, water) operate and therefore participate in an essential way to life processes. Disrupting surface solid envelopes deteriorates the biosphere that constitutes the essence of Earth.

Let us then recall that *the concept of "environmental disruptions" is somewhat relative and subjective, mainly when placed in a geological context*:

- *Earth's environments are in essence unstable and changeable* over the course of time. Throughout its history our planet has been subjected to progressive or acute, cyclic or aperiodic, local or global environmental modifications: sea-floor renewal inducing continent migration or distortion; climate warming, cooling or long-term stabilization; strong oxygen or carbon dioxide enrichment and depletion phases in the atmosphere; sea-level rise or drop at several tens or even hundreds of metres above or below the present level; flora and fauna migration over large latitudinal distances; etc. "Disrupting" physicochemical changes induced by solid and fluid Earth envelopes along geological times are largely responsible for both biological evolution and biodiversity (see Kershaw 1990). They are mainly the origin of the *Gaia* concept, which implies a unified and complementary evolution of Earth's solid, fluid and living components that are considered as forming an autonomous and self-regulated organism (see Lovelock, 2000).
- *People have induced or suffered some environmental changes during historical times that have been similar or greater than recent ones:* major earthquakes and volcanic eruptions, serious freshwater pollution responsible for disastrous epidemics, widespread urban contaminations, etc. Life on a large part of the planet has become more wholesome, longer and safer than in olden times. Such current conditions have led a growing number of people to reject the possibility of nature- or man-induced hazards. In some regions risks have become so rare that even the possibility of hazards is considered incongruous.
- *Sudden climate and other major natural disturbances* such as floods, storms, hurricanes or earthquakes *are often perceived in a distorted way* because of their impact on our lives and property, of their expansion by the media, and of our difficulty for integrating events long term.

Notwithstanding these reminders, the facts and events summarized in this book obviously show that *surface geoenvironments are currently subject to exceptionally numerous, intense and rapid disruptions*. The most unusual feature of these disruptions lies in their major *cause*. Many recent environmental disturbances are not of a physicochemical origin as were most geological crises, but of a biological origin and due to the *exponential growth and influence of a single species*, i.e. mankind. People have been exploiting and strongly modifying the various compartments of the surface lithosphere, as well as of the hydrosphere, atmosphere and biosphere for their needs and well-being. Human pressure has multiplied over a few decades due to the phenomenal increase in world population, and tends to be exacerbated in most surface environments: impoverished and eroded soils, actively excavated underground terrains, widespread re-shaping of landscapes, diverted and excessively extracted freshwater, more or less contaminated fluids and permeable rocks. *Similar situations have never been registered during the Earth's whole history.* Geologists, who are accustomed to deciphering complex geodynamical events and cycles, the nature and period of which are variable but always reproduced in a similar way, tend to lose some of their landmarks and face unusual situations. The expression "the past is not the key to the future" (Fyfe, 1998) sums up in an adequate way the uniqueness of the present situation relative to Earth's whole history.

2. Complexity of Earth–Human Interactions

For reasons of presentation, this volume has been cut in parts and chapters that present in separate sets the examination of either natural hazards and their influence on human activities, or anthropogenic influences disrupting the Earth's functioning. In fact *the diverse natural and human influences or influence clusters often interact with each other in a complex way.* They may constitute a *time–space continuum* that leads to various additional, subtractive, or even cancelled-out environmental effects. Let us recall a few examples:

- slope movements in southern Italy that are naturally triggered by river erosion or rainwater infiltration and become artificially accentuated through deforestation practices, slope re-profiling, and heavy-building construction (Alexander, 1992; Chapter 3);
- cumulative effects on the coastal freshwater balance south of Rome (Italy) either of years marked by natural drought and aquifer over-pumping, or of humid years during which excessive urbanization causes soil waterproofing and disastrous streaming (Chiocchini et al., 1999);
- Finnish freshwater and soil acidification induced either by natural drainage of sulphur-rich sedimentary rocks, by industry-derived acid rain fallout, or by a combination of these natural and anthropogenic effects (Lahermo et al., 1994; Chapter 11);

- freshwater, soil and coastal sediment contamination by heavy metals (Cd, Cu, Fe, Mn, Pb, Zn) in the Brisbane region, Australia, due to the addition of natural and human factors: concentration in water and soil because of drought and over-pumping, floods responsible for reworking, downstream dispersal and accumulation in channel deposits, etc. (Arakel, 1992);
- complex contamination of sediments of the Gulf of Saronikos (Greece) due to discharge from domestic, urban, industrial and harbour sources (Griggs et al., 1978; Chapter 10);
- dwindling of suspended and dissolved matter transfer along the Santa Cruz River system, Arizona, due to successive and cumulative anthropogenic actions: river canalisation, dam building for irrigation, drainage through channel diversion, water blockage by bridge piers, river confinement by dykes (Rhoads, 1991; Chapter 6).
- subsidence of Venice, Tokyo, Shanghai and other coastal urban regions subject to natural and human actions combining in a complementary or opposing way: sea-level rise, natural compacting, collapse through ground dissolution, overloading by building accumulation, subsidence through aquifer over-pumping, dwindling of river or marine sedimentary supply (Chapters 3, 6).

The respective parts of these natural and anthropogenic causes are difficult to assess quantitatively. Boomerang and domino effects occur frequently and are not easy to encompass. It seems crucial first to better characterize and understand the functioning of the diverse intervening processes and their interactions, and then only to quantify each process on a robust basis. This necessitates applying to surface environments an *interdisciplinary scientific approach*, in which geologists share their knowledge with physicists, chemists, biologists, economists, sociologists, etc. Integrated research programmes implemented by certain regions and countries, by the European Union and other international authorities, gradually help progress in this way. Such combined approaches nevertheless remain punctuated by serious difficulty due to persistent compartmentalization of disciplines, responsible for noticeable differences in vocabulary, working modes, and interpretation of reports.

3. Highly Fragile Geoenvironments

Some natural environments are slightly threatened by human actions. They mainly comprise the hard and compact rock outcrops devoid of economic interest, most very high latitude and altitude geological formations, many ice-covered terrains and preserved forest regions. On the other hand, some surface geo-environments are particularly vulnerable to anthropogenic modifications. They generally constitute *interface environments*:

- *Soils* Forming mostly a soft and thin blanket at the interface between the lithosphere, hydrosphere, atmosphere and biosphere, soils are universally exploited for land and livestock farming, as well as widely eroded and transformed through artificial practices. Insufficiently taken into account by geologists who often under-estimate their crucial role in terrestrial history and biosphere evolution, soils constitute the skin of our planet. They prove to be very vulnerable to environmental stresses and to a very slow renewal capacity.
- *River valleys* Traditional and privileged zones used by people for communication links, occupancy and constructions, valleys constitute recurrent places of confrontation of rivers and aquifers on one hand (e.g. floods, inundations) and of growing human influence on the other hand (e.g. agricultural, urban and industrial developments).
- *Estuaries, deltas and coastal zones* Land-to-sea transition zones and adjacent regions host about half of the world's population. They therefore tend to be dramatically exposed to human pressure, which combines with various natural, often aperiodic and intense natural influences that are both of continental and marine, and of aquatic and aeolian origins.
- *Aquifers* Forming freshwater underground reservoirs, aquiferous rocks are exposed on one hand to over-pumping and subsequent shortage and subsidence, and on the other hand to chemical contamination through vertical or lateral infiltration.
- *Climate transition zones* Regions where climatic changes occur over short latitudinal or altitudinal distances are potentially subject to strong and almost irreversible environmental bifurcations induced by soil exploitation or global changes. This is namely the case in deforestation, erosion and desertification in arid regions, and of permafrost melting, retreat and destabilization in Arctic regions.

4. A Few Rarely Admitted Facts

Similarly to all periods of human history, present times are marked by some erroneous believes that result from modes, irrationalism, information clumsiness, lack of long-term memory, corporate pressures, etc. Contrary to preceding ones, our period is characterized by an extraordinarily wide and rapid diffusion of sometimes-approximate information, by unreasonable amplification by the media, and by somewhat excessive manipulation of diverse social sectors. Let us recall some examples of often rarely-admitted facts that specifically concern geo-environment domains:

- Some *noxious or toxic chemical elements* are released naturally in a comparable or even greater proportion than by anthropogenic discharge. This is the case for lead, mercury, arsenic, fluorine, radon, sulphur dioxide and hydrocarbons.
- *Asbestos* indisputably presents a carcinogenic risk for miners and for workers handling this fibrous mineral without appropriate protection, but not for people living or working in well-kept, asbestos-containing buildings. Asbestos removal from normally maintained buildings is much more dangerous for human health and represents considerable waste of work, time and money.
- Radioactivity constitutes a basic natural phenomenon, which is responsible for providing at ground surface the heat diffused from Earth's internal envelopes, and which is indispensable for life and ecosystems functioning. *Nuclear power* presently constitutes one of the safer and less polluting energy sources. There is no renewable alternative energy that could currently take over from nuclear energy. Disposal of radioactive waste, the relative abundance of which is low, may be envisaged in some deep underground formations without any objective risks on a very long term geological scale.
- *Hydrocarbon pollution* due to accidents and shipwrecks represent only 5% of all hydrocarbon discharges at sea, 15% of which are naturally diffused from ocean bottom deposits. Oil spills have disastrous effects on a local and short-term scale; but released hydrocarbons are less dangerous for the natural environment than some chemicals discharged at sea, and are resorbed in the medium term through biological processes.
- *Inundations, groundwater flooding stages, storms, hurricanes, El Niño episodes*, etc, have increased neither in frequency nor in intensity during the last decades, as demonstrated by long-term series of data gathered in the past two centuries. It is quite natural that during flood, the major river courses extensively annexed for human activities become temporarily occupied by vast masses of water drained along the river basins. There is presently no argument proving a shift towards more catastrophic climatic disturbances induced by Earth's global warming.
- *Natural hazards* have not become more numerous and more dangerous than in previous times. Due to the spectacular growth of the human population and its excessive concentration in some sensitive regions of the planet, our species has become more vulnerable to some recurrent natural hazards, the consequences of which therefore become more disastrous. Western European countries are subject to natural hazards the violence and intensity of which are

very low, especially relative to developing countries situated in inter-tropical regions.

Fighting against distorted facts is uncomfortable and difficult to achieve because of frequent public exaggeration and media recurrence. Researchers often hesitate to commit themselves for guaranteeing their intellectual independence; but at the same time they tend to forget their social responsibility and potentially lose their credibility. In addition to the risk of more or less obscurantist shifts, the lack of clear scientific recommendation may lead to official decisions that are contrary to the general interest or at the origin of fruitless blockage. *Geologists should actively be involved* in informative discussions and debates regarding Earth-induced natural hazards, exploitation of soil and underground resources, and man-induced deterioration of Earth's surface formations.

5. A Few Patently Obvious Facts

Countless observations and measurements prove the *recent and continuous acceleration of deterioration of Earth's external envelopes due to growing human influence*. A few examples involving specifically the geosciences domain are:

- Huge exploitation from ground formations of *fossil fuels* progressively leads to dwindling of the available resources. Reserves of common hydrocarbons are guaranteed for a few decades only. Coal extraction may be planned for a few centuries, provided that world needs stop increasing in an exponential way and that other energy sources, including hydrocarbons and nuclear power, are not significantly reduced. Increasing the amounts of coal burned causes increasing discharge of pollutants, which in turn contributes to global warming, acid rain intensification, and atmospheric deterioration.
- If numerous *geomaterials and mineral ores* are still abundantly present in reachable underground formations and can be exploited for a few decades or even centuries, some precious or strategic metal reserves will become dramatically impoverished. Past and present mining and quarrying activities have resulted in huge ore removal. Estimates of the available reserves are often hardly reliable in the long term because of the periodic emergence of unexpected needs; in addition the local character of ore deposits may cause difficulties resulting from changing geopolitical context.
- *Fresh water* constitutes a non-exhaustible Earth resource but is both present in finite quantities and vulnerable to chemical pollution. Freshwater is on the way to becoming the most threatened natural resource. Water needs tend to increase proportionally to the population exponential growth. Excessive extraction of groundwater is responsible for local subsidence and sometimes

for widespread soil erosion, aridification and salination. Groundwater contamination frequently induces health problems and restrictive human uses. The increase in the number of artificial water planes in most river basins determines widespread evaporation processes and causes the general hydrological budget to loose its balance.

- Extensive developments along *river courses* and massive construction of *artificial reservoirs and lakes* have dramatically disrupted the world hydrographical systems. Less and less rivers are able to ensure their natural land-to-sea transfer role of rainwater, sedimentary particles, mineral elements and nutriments. This results in increasing disequilibria in the continent–ocean global balance.
- *Artificial erosion, weathering and waterproofing of soil surface formations* constitute major threats for the biosphere, and tend to be both amplified by accelerated deforestation–urbanization and to cause accelerated desertification. Forests can be reconstituted within a few decades, but soils constitute a non-renewable resource on the scale of several centuries.
- Countless *coastal zones* are over-populated, over-built, over-exploited and over-protected against natural hazards. Such an excessive human hold results in impending natural hydro-sedimentary phenomena, disrupting the land–sea exchanges, and favouring pollution or eutrophication.
- *Topographic re-shaping of land surfaces* by agricultural, mining, urban, industrial and tourist activities, as well as by the development of vast communication networks, has led to gigantic rock and soil displacements, which still tend to increase. Such continuous material reworking has progressively caused important landscape changes and environmental disruptions.
- *Industrial activities* determine innumerable *polluted wasteland and disposal areas.* Scoria, dust and gas release contribute to denaturing water and soils as well as deteriorating building stone, mainly through acidification phenomena.

6. Some Positive Reactions and Measures

The accelerated deterioration of the Earth's surface formation has given way since the last quarter of the 20th century to concrete realization by an increasing number of people. Evidence for such a change in our minds was first provided by the emergence and then the proliferation of associative and political groups, the position of which is sometimes excessive but nevertheless useful. We have progressively become aware of the fact that we form a whole with our environment, and that participating in environmental deterioration partly conditions the future of our own species.

Scientific research takes part in this general thinking and action process, as shown by the growing number of researchers concerned by recent "galloping" geology questions, and of papers devoted to *environmental geosciences*. For instance, three or four international journals dealing with environmental soil and earth sciences publish several thousands of pages each year. Papers in these journals tend to be of a gradually improving scientific level and are distributed through both traditional and electronic ways. Due to a mainly applied and local character, they are often marked by a moderate "impact factor"; but they are read and taken into account by numerous environmental activists and a growing number of geologists. On the sole question of radon, the journal *The Science of the Total Environment* (Elsevier) published in May 2001 a special issue forming a 387-page volume, including 37 scientific papers. In the same period, the journal *Environmental Geology* (Springer) reported the existence of 1,600 Internet sites dealing with environmental geosciences; by comparison the number of sites did not exceed about 100 at the end of the 1990s.

Various concrete advances in knowledge, often associated with reinforced legal measures, give recent evidence for the better consideration of hazards incurred both by man due to Earth's natural dynamic functioning, and by Earth's surface envelopes due to exponentially growing human activities:

- noticeable improvement of earthquake, volcanic eruption, landslide, and flooding forecasting (Chapters 1 to 4);
- active development of recycling processes and synthetic materials, which enables saving various natural underground resources (Chapter 5);
- continuous progress in rehabilitating open-air quarry regions and mining slag heaps or dumps, in renovating numerous mining zones after closure, in re-qualifying and reusing mine drainage water and materials stored in mining wasteland, in re-injecting processed water underground, and in reducing dust, soot and noxious gas emission (Chapters 5, 6, 10, 11);
- development of a nuclear "safety culture" that becomes worldwide, mainly thanks to the Soviet bloc opening up; active and concrete current investigations by several high-technology countries on the critical question of highly radioactive waste storage (Chapter 7);
- encouraging development of "clean" alternative energies. Exploitation of wind power represented worldwide 15,000 kW in 1981, 2,600,000 kW in 1992, and continues increasing in a significant way in Germany, Spain, Japan, USA, etc.; the European Union aims to cover in 2005 1% of its electricity needs (i.e. 68,000,000 kW) by aeolian energy. Solar power produced by photovoltaic cells jumped from 100 kW in 1971 to 58,000 kW in 1992, and still dramatically increases. Geothermal energy has evolved from 239,000 kW in 1950 to 9,400,000 kW in 1991. Hydroelectricity has grown from

44,600,000 kW in 1950 to 643,000,000 kW in 1991 (Mackenzie & Mackenzie 1995). The energy likely to be released by waste dumps (methane) and agricultural products (diverse biogas) is progressively taken into consideration and exploited;

- development in various countries of the efficient fight against soil erosion, of control of mineral fertilizer use and rehabilitation of organic fertilizers, of improvement of cultivation practices, and of "sustainable ecology" implementation (Chapter 8).

Development of prevention and environmental restoration measures is conducted in a particularly effective way by developed and high technology countries, whose population is almost stabilized and whose new needs are accessory rather than essential. In numerous developing countries, where population growth is very rapid, the perception of human-triggered or -amplified environmental risks still remains insufficient or of secondary importance; the ways of preventing natural hazards, reducing the exploitation of soil and underground resources, and limiting environmental deterioration, are still often poorly implemented.

7. Which Future?

At the moment of finishing writing a book on the evolution of man–environment interactions, it is difficult to dodge the question of the future. Escaping such a question seems particularly inappropriate in geoscience fields where researchers base their expertise on recent times for being able to better anticipate evolution in the near future.

Such projected exercises tend nevertheless to become considerably complicated because of the *growing interferences linking human activities and Earth natural functioning*. The geological evolution does not work alone any longer. Geologists may only contribute to the understanding of a system made more and more complex, in which problems are intimately mixed and come within the competence of earth and life sciences, but also of social and economic sciences. Some environmental amplifications, inversions or bifurcations may suddenly occur, the consequences of which are hardly predictable.

A typical example of such unexpected changes is provided by the *global climate evolution*. During the second half of the 20th century both geologist's and astronomer's investigations were converging to predict a world climatic cooling during the next millennia and tens of millennia; this cooling was the logical step following the current interglacial period that started around 10,000 years ago. During the two last decades, scientific evidence has been provided for an increasing greenhouse gas effect partly triggered by human activities, and responsible for an unexpected global warming; the consequences of this warming should last several hundred and perhaps 1,000 years (Chapter 11). Some people have supposed that this man-induced climate warming could at least partly compensate for the nature-induced climate cooling, even if the time scales involved differ greatly. What will happen exactly, and how will people be able to anticipate and fight an evolution they have contributed to triggering?

W. von Engelhardt, J. Goguel and four other precursors in environmental geology stated in 1976 that a period marked by a stabilization in the exploitation of Earth's resources should inevitably take place after the 20th century's frantic expansion (Introduction, 5). One must admit that no sign of such a change has appeared during the last 25 years. On the contrary, all indicators clearly point to a *spectacular acceleration of anthropogenic pressure on the environment*, which essentially results from an unprecedented population growth. Such an acceleration leads to reducing the expected duration and weakening the prediction of forthcoming events, and of course characterizes many other human activities: modes of transportation, means of communication and information, human penetration and invasion in all surface environments, personal and collective commitments, use of consumables (including books!), etc.

The addition of these various accelerating tendencies in current life renders most diagnoses null and void, and is not inclined towards optimistic forecasting. This worrying assessment is aggravated when noting that *terrestrial environment dependency* (Chapters 1 to 4), *exploitation* (Chapters 5 to 7) *and weakening* Chapters 8 to 11) *tend to be exacerbated in most populated, most disarmed and the poorest countries of the world.* Various water, soil and underground resources are no longer sufficient for meeting the basic needs of ever-larger populations. Most developing countries suffer strong backwardness in geo-environmental preservation, mitigation and restoration practices, which results in aggravating soil erosion, ground deterioration and widespread contamination. These countries frequently also depend on non-renewable, polluting or insufficiently controlled energy sources.

It is essential that a *pause in the expansion of human influence* be rapidly initiated. It would be exemplary that such a pause started in developed countries, where basic livelihood is not under threat. Perhaps some physicochemical or technological innovations will allow better valorising of the Earth's resources, as was the case so many times in historical and even prehistoric ages. But such a possibility remains difficult to envisage without reducing the exploitation of non-renewable and progressively exhausting resources, limiting extensive soil and water contamination, and diminishing the widespread artificial character of surface terrains. Human innovations may hardly be implemented on a global scale while global human pressure still increases dramatically. Basically Earth should not only be considered as a consumable item, for it serves as an essential support for all life types.

Geologists must progress in better understanding the functioning of natural and man-modified Earth environments. From that perspective they should continue to receive official support and enjoy social confidence. This implies that they put themselves actively in research partnerships with scientists and technicians from all other disciplinary fields concerned, integrate an increasing ethical dimension to their scientific approach, and accept communicating and involving themselves more widely in the environmental debate.

An important challenge faced by geologists consists of making people realize that *deterioration and rehabilitation of Earth's solid environments depend on time constants that are much longer than in most other investigation fields.* A forest, a biological community, or a disrupted biosystem may usually be restored and reconstituted within one or two human generations. This is never the case for soils, deep aquifers, fossil-fuel deposits or mineral ores. *Geosystems are not renewable on a human scale.* The concept of *very long term sustainable development* is therefore both characteristic and crucial in the domain of Earth Sciences.

At the same time, the very slow character of geological changes in many natural, hard rock environments leads to excluding the possibility for unexpected risks to happen. Natural hazards in some stable, old and compact rocky underground terrains are objectively nil, with a guaranteed stability of at least several million years. This should be thoroughly and carefully explained to both decision-makers and citizens, especially about the possibility for very long term waste storage in adequately chosen, deep geological formations. *In the same way as the concept of inevitable risks in some natural environments should be better accepted, the reality of non-existent objective risk in other fields should be more reasonably considered and admitted.*

Such messages towards human societies will be better transmitted and accepted when they are proved to be marked by professionalism and practical sense. In this connection, the detailed *knowledge and practice of the geological field* by environmental scientists should remain balanced with that of computer systems and software. Keeping close contact with nature certainly constitutes the best way for geologists to be impregnated by both strengths and weaknesses of terrestrial environments, and therefore to communicate a passion for planet Earth and respect for their fellow creatures.

References

Textbooks and general books are preceeded with an asterisk '*'.
References suggested at each chapter end for "Further Reading" are preceded with a dagger '†'.

Alexakhin, R. M. (1993). Countermeasures in agricultural production as an effective means of mitigating the radiological consequences of the Chernobyl accident. *The Science of the Total Environment*, **137**, 9–20.

Alexander, D. (1992). On the causes of landslides, human activities, perception, and natural processes. *Environmental Geology & Water Science*, **20**, 165–179.

†Alexander, D. (1993). *Natural disasters*. UCL Press, London, 632 p.

Al-Homoud, & Tahtamoni, W. W. (2000). SARETL: An expert system for probabilistic displacement-based dynamic 3-D slope stability analysis and remediation of earthquake triggered landslides. *Environmental Geology*, **39**, 848–874.

*Allègre, C. J. (1990). *Économiser la planète*. Fayard, Paris, 379 p.

*Allègre, C. J. (1993). *Écologie des villes, écologie des champs*. Fayard, Paris, 232 p.

*Allen, P. A. (1997). *Earth surface processes*. Blackwell, 404 p.

Alzieu, C. coord. (1999). *Dragages et environnement marin*. Public. Ifremer, Paris, 223 p.

André-Jehan, R., & Féraud, J. coord. (2001). Le radon. *Géochronique, Soc. géol. Fr & BRGM*, **78**, 8–25.

Andrews, J. E., Brimblecombe, P., Jickells, T. D., & Liss, P. S. (1995). *Introduction to environmental geochemistry*. Blackwell, 232 p.

Apambire, W. B., Boyle, D. R., & Michel, F. A. (1997). Geochemistry, genesis and health implications of fluoriferous groundwaters in the upper regions of Ghana. *Environmental Geology*, **33**, 13–26.

Apte, G. A., Price, P. N., Nero, A. V., & Revzan, K. L. (1999). Predicting New Hampshire indoor radon concentrations from geologic information and other covariates. *Environmental Geology*, **37**, 181–194.

Arakel, A. V., & Hongjun, T. (1992). Heavy metal geochemistry and dispersion pattern in coastal sediments, soil, and water of Kedron Brook floodplain area, Brisbane, Australia. *Environmental Geology & Water Science*, **20**, 219–231.

*Aswathanarayana, U. (1995). Geoenvironment, an introduction. Balkema, 270 p.

Avila, A., & Peñuelas, J. (1999). Increasing frequency of Saharan rains over northeastern Spain and its ecological consequences. *The Science of the Total Environment*, **228**, 153–158.

Banks, D., Younger, P. L., Arnesen, R. T., Iversen, E. R., & Banks, S. B. (1997). Mine-water chemistry: The good, the bad and the ugly. *Environmental Geology*, **32**, 157–174.

†Bardintzeff, J.-M. (1998). *Volcanologie*. Doin, Paris, 284 p.

Barnes, J. W. (1998). *Ores and minerals. Introducing economic geology*. Open University Press, 1988, 181 p.

*Barnier, M. (1992). *Atlas des risques majeurs*. Plon, Paris, 125 p.
Barton, C. D., & Karathanasis, A. D. (1999). Renovation of a failed constructed wetland treating acid mine drainage. *Environmental Geology*, **39**, 39–50.
*Bell, F. G. (1993). *Engineering geology*. Blackwell, 359 p.
*Bell, F. G. (1998). *Environmental geology*. Blackwell, 608 p.
*†Bell, F. G. (1999). *Geological hazards: Their assessment, avoidance and mitigation*. Spon, London, 648 p.
Benito, G., Machado, M. J., & Sancho, C. (1993). Sandstone weathering processes damaging prehistoric rock paintings at the Albarracin cultural park, NE Spain. *Environmental Geology*, **32**, 71–79.
*Bennett, M. R., & Doyle, P. (1997). *Environmental geology*. Wiley, 501 p.
*Berger, A., (1992). *Le climat de la Terre*. De Boeck Université, Bruxelles, 479 p.
Bernhard, C. C., Carbiener, R., Cloots, A. R., Groelicher, R. Schenck, Ch., & Zilliox, L. (1992). Nitrate pollution of groundwater in the Alsatian plain (France) — a multidisciplinary study of an agricultural area. *Environmental Geology & Water Science*, **20**, 125–137.
*Betz, F. Jr. (Ed.) (1975). *Environmental geology*. Dowden, Hutchinson & Ross, Stroudsburg, 390 p.
Bilham, R. (1996). Global fatalities from earthquakes in the past 2000 years, prognosis for the next 30. In: J. B. Rundle, D. L. Turcotte, & W. Klein (Eds), *Reduction & Predictability of Natural Disasters* (pp. 19–31). Addison-Wesley.
Birch, G., & Taylor, S. (1999). Source of heavy metals in sediments of the Port Jackson estuary, Australia. *The Science of the Total Environment*, **227**, 123–138.
†Blunden, J., & Reddish, A. (1991). *Energy, resources and environment*. Hodder & Stoughton, The Open University, 339 p.
*Bobrowsky, P. T. (Ed.) (2001). *Geoenvironmental mapping*. Balkema, 725 p.
Bodungen, B. von, & Turner, R. K. (Ed.) (2001). *Science and intergrated coastal management*. Wiley, Dahlem Series 85, 378 p.
†Bolt, B. A. (1999). *Earthquakes*. Freeman & Co, 4th ed., 366 p.
*Botkin, D., & Keller, E. (1995). *Environmental Science. Earth as a living planet*. Wiley, 627 p. + appendix.
Bourbonniere, R. A., & Meyers, P. A. (1996). Anthropogenic influences on hydrocarbon contents of sediments deposited in eastern Lake Ontario since 1800. *Environmental Geology*, **28**, 22–28.
†Bourgeois, J., Tanguy, P., Cogné, F., & Petit, J. (1996). *La sûreté nucléaire en France et dans le monde*. Polytechnica, Paris, 298 p.
†Bradshaw, A. D., Southwood, R. (Sir), & Warner, F. (Sir) (1992). *The treatment and handling of wastes*. Chapman & Hall, The Royal Society, London, 302 p.
†Brady, N. C., & Weill, R. R. (1999). *The nature and properties of soils*. Prentice Hall, 12th ed., 881 p.
*Brahic, A., Hoffert, M., Schaaf, A., & Tardy, M. (1999). *Sciences de la Terre et de l'Univers*. Vuibert, Paris, 634 p.
Brännwall, M. L., Kurkkio, H., Bindler, R., Emteryd, O., & Renberg, I. (2001). The role of pollution versus natural geological sources for lead enrichment in recent lake sediments and surface forest soils. *Environmental Geology*, **40**, 1057–1065.
Brookins, D. G. (1976). Shale as a repository for radioactive waste: The evidence from Oklo. *Environmental Geology*, **1**, 255–259.
Brookins, D. G. (1992). Background radiation in the Albuquerque, New Mexico, USA area. *Environmental Geology & Water Science*, **19**, 11–15.

*Brown, G. C., Hawkesworth, C. J., & Wilson, C. (Ed.) (1992). *Understanding the Earth.* Cambridge University Press, 551 p.

*Bryant, E. A. (1991). *Natural hazards.* Cambridge University Press, 294 p.

Bullock, S. E. T., & Bell, F. G. (1997). Some problems associated with past mining at a mine in the Witbank coalfield, South Africa. *Environmental Geology,* **33**, 61–71.

†Burroughs, W. J. (2001). *Climate change. A multidisciplinary approach.* Cambridge University Press, 298 p.

*Campy, M., & Macaire, J.-J. (1989). *Géologie des formations superficielles.* Masson, Paris, 433 p.

Candelone, J.-P., Hong, S., Pellone, C., & Boutron, C. F. (1995). Post-industrial revolution changes in large-scale atmospheric pollution of the northern hemisphere by heavy metals as documented in central Greenland snow and ice. *J. Geophys. Res. 100, D8,* **16**, 605–616.

Caristan, Y. (2000). Gérer notre planète. *Découverte,* **277**, 75–84.

†Casale, R., & Margotini, C. (Ed.) (1999). *Floods and landslides.* Springer, 373 p.

†Castany, G. (1982). *Principes et méthodes de l'hydrogéologie.* Dunod, Paris, 256 p.

Chambers, J. Ogilvy, R., Meldrum, P., & Nissen, J. (1999). 3D resistivity of buried oil- and tar-contaminated waste deposits. *Eur. J. Environ. Engin. Geophys.,* **4**, 3–15.

*Chamley, H. (1989). *Clay sedimentology.* Springer, 623 p.

Chamley, H. (2000). *Bases de sédimentologie.* Dunod, Paris, 178 p.

Chan, L. S., Yeung, C. H., Yim, W. W. S., & Or, O. L. (1998). Correlation between magnetic susceptibility and distribution of heavy metals in contaminated sea-floor sediments of Hong Kong harbour. *Environmental Geology,* **36**, 77–86.

†Chapel, A., Fieux, M., Jacques, G., Jacques, J.-M., Laval, K, Legrand, M., & Le Treut, H. (1996). *Océans et atmosphères.* Hachette, Paris, 160 p.

†Chapman, D. V. (1992). *Water quality assessment: A guide to the use of biota, sediments and water in environmental monitoring.* UNESCO Public., E. & F. N. Spon, London, 2nd ed., 626 p.

Chen, Z. Y., & Wang, Z. H. (1999). Yangtze delta, China: Taihu lake-level variation since the 1950s, response to sea-level rise and human impact. *Environmental Geology,* **37**, 333–339.

Chiocchini, U. Gisotti, G., Macioce, A., Manna, F., Bolasco, A., Lucarini, C., & Patrizi, G. M. (1999). Environmental geology problems in the Tyrrhenian coastal area of Santa Marinella, province of Rome, central Italy. *Environmental Geology,* **32**, 1–8.

Cigna, A. A. (1993). Environmental management of tourist caves. The examples of Grotta di Castellano and Grotto Grande del Vento, Italy. *Environmental Geology,* **21**, 173–180.

†Clark, R. B. (2001). *Marine pollution.* Oxford University Press, 5th ed., 237 p.

*†Cooke, R. U., & Doornkamp J. C. (1990). *Geomorphology in environmental management.* Oxford University Press, 2nd ed., 410 p.

†Costa, J. E., & Baker, V. R. (1981). *Surficial geology: Building with the Earth.* Wiley, 498 p.

Cozzarelli, I. M., Erganhouse, R. P., & Baedecker, M. J. (1990). Transformation of monoaromatic hydrocarbons to organic acids in anoxic groundwater environment. *Environmental Geology & Water Science,* **16**, 135–141.

*Craig, J. R., Vaughan, D. J., & Skinner, B. J. (1988). *Resources of the Earth.* Prentice Hall, 395 p.

Craw, D., Chappell, D., & Reay, A. (2000). Environmental mercury and arsenic sources in fossil hydrothermal systems, Northland, New Zealand. *Environmental Geology,* **39**, 875–887.

Cronin, S. J., Hedley, M. J., Neall, V. E., & Smith, R. G. (1998). Agronomic impact of tephra fallout from the 1995 and 1996 Ruapehu Volcano eruptions, New Zealand. *Environmental Geology,* **34**, 21–30.

Cuchi-Oterino, J. A., Rodriguez-Caro, J. B., & Garcia de la Noceda-Márquez, C. (2000). Overview of hydrogeothermics in Spain. *Environmental Geology*, **39**, 482–487.

Cuny, D., Van Haluwyn, C., & Pesch, R. (2001). Biomonitoring of trace elements in air and soil compartments along the major motorway in France. *Water, Air & Soil Pollution*, **125**, 273–289.

†Dagorne, A., & Dars, R. (1999). *Les risques naturels. La cyndinique*. PUF, coll. Que sais-je, Paris, 128 p.

Davies, B., & Archambeau, C. B. (1997). Geohydrological models and earthquake effects at Yucca Mountain, Nevada. *Environmental Geology*, **32**, 23–36.

Davis, G. H. (1987). Land subsidence and sea level rise on the Atlantic coastal plain of the United States. *Environmental Geology & Water Science*, **10**, 67–80.

Davis, R. A. Jr., Welty, A. T., Borrego, J., Morales, J. A., Pendon, J. G., & Ryan, J. G. (2000). Rio Tinto estuary (Spain), 5000 years of pollution. *Environmental Geology*, **39**, 1107–1116.

Davis, W. M., Cespedes, E. R., Lee, L. T., Powell, J. F., & Goodson, R. A. (1997). Rapid delineation of subsurface petroleum contamination using the site characterization and analysis penetrometer system. *Environmental Geology*, **29**, 229–237.

Demek, J. (1994). Global warming and permafrost in Eurasia: A catastrophic scenario. *Geomorphology*, **10**, 317–329.

†Detay, M. (1997). *La gestion active des aquifères*. Masson, Paris, 416 p.

Di Gregori, F., & Massoli-Novelli, R. (1992). Geological impact of some tailings dams in Sardinia, Italy. *Environmental Geology & Water Science*, **19**, 147–153.

†Domenico, P. A., & Schwartz, F. W. (1998). *Physical and chemical hydrogeology*. Wiley, 2nd ed., 506 p.

Dorland, C., Tol, R. S. J., & Palutikof, J. P. (1999). Vulnerability of the Netherlands and Northwest Europe to storm damage under climate change. *Climatic Change*, **43**, 513–535.

†Down, C. G., & Stocks, J. (1977). *Environmental aspects of mining*. Applied Science Publishers, London, 371 p.

†Drake, F. (2000). *Global warming: The science of climate change*. Oxford University Press, 273 p.

Drew, D. (1996). Agriculturally induced environmental changes in the Burren karst, Western Ireland. *Environmental Geology*, **28**, 137–144.

†Duchaufour, P. (1983). *Pédologie. 1 — Pédogenèse et classification*. Masson, Paris, 2ème éd., 491 p.

*Duplessy, J.-C. (1996). *Quand l'océan se fâche. Histoire naturelle du climat*. Odile Jacob, Paris, 277 p.

†Durieux, J. (2000). *Volcans*. CDrom, Syrinx, Paris.

†Eisenbud M., & Gesell T. F. (1997). *Environmental radioactivity: From natural, industrial and military sources*. Academic Press, 4th ed., 656 p.

†Ellis, D. (1989). *Environment at risks. Case histories of impact assessment*. Springer, 329 p.

Engelhardt, W. von, Goguel, J., King Hubbert, M., Prentice, J. E., Price, R. A., & Trümpy, R. (1976). Earth resources, time, and man. A geoscience perspective. *Environmental Geology*, **1**, 193–206.

†Evans, A. M. (1997). *An introduction to economic geology and its environmental impact*. Blackwell, 364 p.

†Eyles, N. (1997). *Environmental geology of urban areas*. Geol. Assoc. Canada, 590 p.

*Faurie, C., Ferra, C., Médori, P., & Devaux, J. (1998). *Écologie. Approche scientifique et pratique*. TecDoc, Paris, 4th ed., 339 p.

†Fisher, R. V., Heiken, G., & Hulen, J. B. (1997). *Volcanoes, crucibles of change*. Princeton University Press, 317 p.

†Flageollet, J.-C. (1989). *Les mouvements de terrain et leur prévention.* Masson, Paris, 224 p.

*Flawn, P. T. (1970). *Environmental geology: Conservation, land-use planning and resource management.* Harper & Row, New York, 313 p.

*Foley, D., McKenzie, G. D., & Utgard, R. O. (1999). *Investigations in environmental geology.* Prentice Hall, 2nd ed., 303 p.

†Foos, J., Rimbert, J.-N., & Bonfand, E. (1993–1995). *Manuel de radioactivité à l'usage des utilisateurs. Formascience, Orsay,* vol. 1, 197 p.; vol. 2, 310 p.; vol. 3 (& G. Lemaire), 350 p.

Fyfe, W. S. (1998). Toward 2000: The past is not the key to the future — challenges for the science of geochemistry. *Environmental Geology,* **33**, 92–95.

Garcia-Vallès, M., Vendrell-Saz, M., Molera, J., & Blasquez, F. (1998). Interaction of rock and atmosphere: Patimas on Mediterranean monuments. *Environmental Geology,* **36**, 137–149.

Gauri, K. L. (1990). Decay and preservation of stone in modern environments. *Environmental Geology & Water Science,* **15**, 45–54.

†Gérard, B. (Éd.) (1999). *Le littoral. Problèmes et pratiques de l'aménagement.* BRGM, Orléans: 351 p.

Gibson, P. J., Lyle, P., & George, D. M. (1996). Environmental applications of magnetometry profiling. *Environmental Geology,* **27**, 178–183.

Gimeno, L., Marin, E., del Teso, T., & Bourhim, S. (2001). How effective has been the reduction of SO2 emissions on the effect of acid rain on ecosystems? *The Science of the Total Environment,* **275**, 63–70.

Glade, Th. (1998). Establishing the frequency and magnitude of landslide-triggering rainstorm events in New Zealand. *Environmental Geology,* **35**, 161–174.

*Goguel, J. (1980). *Géologie de l'environnement.* Masson, Paris, 193 p.

Gottgens, J. F., Rood, B. E., Delfino, J. J., & Simmers, B. S. (1999). Uncertainty in paleoecological studies of mercury in sediment cores. *Water, Air & Soil Pollution,* **110**, 313–333.

Griggs, G. B., Grimanis, A. P., & Vassilaki Grimani, M. (1978). Bottom sediments in a polluted marine environment, upper Saronikos gulf, Greece. *Environmental Geology,* **2**, 97–106.

Grimalt, J. O., Ferrer, M., & Macpherson, E. (1999). The mine tailing accident in Aznalcollar. *The Science of the Total Environment,* **242**, 3–11.

Grousset, F. E., Jouanneau, J.-M., Castaing, P., Lavaux, G., & Latouche, C. (1999). A 70 year record of contamination from industrial activity along the Garonne river and its tributaries (SW France). *Estuarine, Coastal and Shelf Science,* **48**, 401–414.

Guillén, J., & Palanques, A. (1997). A historical perspective of the morphological evolution in the lower Ebro river. *Environmental Geology,* **30**, 174–180.

Gundlach, E. R., Ruby, C. H., Hayes, M. O., & Blount, A. E. (1978). The Urquiola oil spill, La Coruña, Spain, impact and reaction on beaches and rocky coasts. *Environmental Geology,* **2**, 131–143.

Gunn, J., & Bailey, D. (1993). Limestone quarrying and quarry reclamation in Britain. *Environmental Geology,* **21**, 167–172.

Haidouti, C. (1997). Inactivation of mercury in contaminated soils using natural zeolites. *The Science of the Total Environment,* **208**, 105–109.

Harland, B. J., Taylor, D., & Whither, A. (2000). The distribution of mercury and other trace metals in the sediments of the Mersey estuary over 25 years 1974–1998. *The Science of the Total Environment,* **253**, 45–63.

Harrell, J. A., Belsito, M. E., & Kumar, A. (1991). Radon hazards associated with outcrops of Ohio shale in Ohio. *Environmental Geology & Water Science,* **18**, 17–26.

Harris, P., Fichez, R., Fernandez, J. M., Gulterman, H., & Badie, C. (2001). Using geochronology to reconstruct the evolution of particulate phosphorus inputs during the past century in the Papeete lagoon (French Polynesia). *Oceanologica Acta*, **24**, 1–8.

Hayashi, H., & Ishii, M. (1989). Accidents due to oxygen deficiency and methane gas blow-off in Tokyo area, Japan. *Environmental Geology & Water Science*, **13**, 167–177.

†Hester, R. E., & Harrison, R. M. (1994). *Mining and its environmental impact*. Royal Society of Chemistry, Cambridge, 164 p.

†Hickin, E. J. (Ed.) (1995). *River geomorphology*. Wiley, 255 p.

*Holland, H. D., & Petersen, U. (1995). *Living dangerously. The Earth, its resources, and the environment*. Princeton University Press, 490 p.

Hooke, R. LeB. (2000). On the history of humans as geomorphic agents. *Geology*, **28**, 843–846.

Hopson, R. F. (1991). Potential impact on water resources from future volcanic eruptions at Long Valley, Mono county, California, USA. *Environmental Geology & Water Science*, **18**, 49–55.

Hubbert, M. K. (1974). *United States energy resources, a review as of 1972. Part I*. United States Senate, Washington, 267 p.

Hunt, C. O., Gilbertson, D. D., & Donahue, R. E. (1992). Palaeoenvironmental evidence for agricultural soil erosion from late Holocene deposits in the Montagnola Senese, Italy. In: M. Bell, & J. Bordman (Eds), *Present and Past Soil Erosion* (Oxbow Monograph 22, pp. 163–174).

IGBP (2000). *Global Change NewsLetter*, **44**, 28 p.

IGBP-WCRP (2000). *French NewsLetter*, **11**, 67 p.

Inbar, M., Ostera, H. A., Parica, C. A., Remesal, M. B., & Salani, F. M. (1995). Environmental assessment of 1991 Hudson volcano eruption ashfall effects on southern Patagonia region, Argentina. *Environmental Geology*, **25**, 119–125.

†Ingebritsen, S. E., & Sanford, W. E. (1998). *Groundwater in geologic processes*. Cambridge University Press, 365 p.

*Jackson, A. R. W., & Jackson, J. M. (1996). *Environmental science*. Longman, Harlow, 370 p.

Journal of Environmental Radioactivity (2000). Special Issue: The Tokai-mura Accident, **50**, 172 p.

†Joussaume, S. (1999). *Climat d'hier à demain*. CNRS ed., Paris, 143 p.

†Juteau, T., & Maury, R. (1997). *Géologie de la croûte océanique*. Masson, Paris, 367 p.

*Keller, E. A. (1999). *Introduction to environmental geology*. Prentice Hall, 383 p.

*Keller, E. A. (2000). *Environmental geology*. Prentice Hall, 562 p.

Keller, G., Schneiders, H., Schütz, M., Siehl, A., & Stamm, R. (1992). Indoor radon correlated with soil and subsoil radon potential — a case study. *Environmental Geology & Water Science*, **19**, 113–119.

*Kennett, J. P. (1982). *Marine geology*. Prentice-Hall, 813 p.

Kesel, R. H. (1989). The role of the Mississippi river in wetland loss in southeastern Louisiana, USA. *Environmental Geology & Water Science*, **13**, 183–193.

†Komar, P. D. (1998). *Beach processes and sedimentation*. Prentice Hall, 2nd ed., 544 p.

Kreitler, C. W., Akhter, S., & Donnelly, A. C. A. (1990). Hydrologic hydrochemical characterization of Texas Frio formation used for deep-well injection of chemical wastes. *Environmental Geology & Water Science*, **16**, 107–120.

Lahermo, P. W., Tarwainen, T., & Tuovinen, J.-P. (1994). Atmospheric sulfur deposition and streamwater quality in Finland. *Environmental Geology*, **24**, 90–98.

Lajczak, A. (1995). The impact of river regulation, 1850–1990, on the channel and floodplain of the Upper Vistula River, Southern Poland. In: E. J. Hickin (Ed.), *River Geomorphology* (pp. 209–233). Wiley.

†Lal, R., Kimble, J. Levine, E., & Stewart, B. E. (Ed.) (1995). *Soils and global change*. CRC, Lewis Publ., 440 p.

Lambrakis, N. J., Voudouris, K. S., Tiniakos, L. N., & Kallergis, G. A. (1997). Impacts of simultaneous action of drought and overpumping on Quaternary aquifers of Glafkos basin (Patras region, Western Greece). *Environmental Geology*, **29**, 209–215.

LaMoreaux, P. E. (1995). Worldwide environmental impacts from the eruption of Thera. *Environmental Geology*, **26**, 172–181.

LaMoreaux, P. E., & Newton, J. G. (1986). Catastrophic subsidence: An environmental hazard, Shelby County, Alabama. *Environmental Geology & Water Science*, **8**, 25–40.

Landa, E. R., & Gray, J. R. (1995). U.S. Geological Survey research on the environmental fate of uranium mining and milling wastes. *Environmental Geology*, **26**, 19–31.

†Lecomte, P. (1998). *Les sites pollués. Traitement des sols et des eaux souterraines*. TEC & DOC, Paris, 2th ed., 1998, 204 p.

†Ledoux, B. (1995). *Les catastrophes naturelles en France*. Payot, Paris, 455 p.

*†Lefèvre, C., & Schneider J.-L. (2003). *Risques naturels majeurs*. Gordon & Breach, 306 p.

Liguang, S., & Zhonqing, X. (2001). Changes in lead concentrations in Antarctic penguin droppings during the past 3,000 years. *Environmental Geology*, **40**, 1205–1208.

†van Loon, A. J. (2000). Reversed mining and reversed-reversed mining: The irrational context of geological disposal of nuclear waste. *Earth Sci. Rev.*, **50**, 269–276.

Lottermoser, B. G., Ashley, P. M., & Lawie, D. C. (1999). Environmental geochemistry of the Gulf Creek copper mine area, north-eastern New South Wales, Australia. *Environmental Geology*, **29**, 61–74.

Lottermoser, B. G., Schütz, U., Boenecke, J., Oberhänsli, R., Zolitschka, B., & Negendank, J. F. W. (1997). Natural and anthropogenic influences on the geochemistry of Quaternary lake sediments from Holzmaar, Germany. *Environmental Geology*, **31**, 236–247.

*Lovelock, J. (2000). *Gaia, a new look at life on Earth*. Oxford University Press, 148 p. (1rst ed. 1979).

*Lumsden, G. I. (Ed.) (1992). *Geology and the environment in Western Europe*. Oxford University Press, 325 p.

*Lundgren, L. W. (1999). *Environmental Geology*. Prentice Hall, 2nd ed., 511 p.

†MacKenzie, A. B. (2000). Environmental radioactivity: Experience from the 20th century — trends and issues for the 21st century. *The Science of the Total Environment*, **249**, 313–329.

*†Mackenzie, F. T., & Mackenzie, J. A. (1995). *Our changing planet: An introduction to earth system science and global environmental change*. Prentice Hall, 387 p.

†Madariaga, R., & Perrier, G. (1991). *Les tremblements de terre*. CNRS ed., Paris, 210 p.

Madrussani, G., Böhm, G., Vesnaver, A., & Schena, G. (1998–1999). Tomographic detection of cavities in mines for acid drainage control. *Eur. J. Environ. Engin. Geophys.*, **3**, 115–130.

†Mainguet, M. (1991). *Desertification. Natural background and human mismanagement*. Springer, 306 p.

†Marsily, G. de (1981). *Hydrogéologie quantitative*. Masson, Paris, 215 p.

Martin, P. (1997). *Ces risques que l'on dit naturels*. Edisud, Paris, 256 p.

Martinelli, G., Minissale, A., & Verruchi C. (1998). Geochemistry of heavily exploited aquifers in the Emilia-Romagna region (Po valley, northern Italy). *Environmental Geology*, **36**, 195–206.

Martinez, J. O., Pilkey, O. H. Jr., & Neal, W. J. (1990). Rapid formation of large coastal sand bodies after emplacement of Magdalena river jetties, Northern Colombia. *Environmental Geology & Water Science*, **16**, 187–194.

Massoud, Z., & Piboubès, R. (Éd.) (1994). *L'atlas du littoral de France*. J.-P. de Monza ed., Paris, 332 p.

†Maund, J. G., & Eddleston, M. (1998). Geohazards in Engineering Geology. *Geological Society of Engineering Geologists, spec. public.*, **15**, 448 p.

†McCall, G. J. H., Larning, D. J. C., & Scott, S. C. (Ed.) (1992). *Geohazards: Natural and man-made*. Chapman & Hall, London, 227 p.

†McGuire, W. J. (1998). Volcanic hazards and their mitigation. In: J. G. Maund, & M. Eddleston (Eds), *Geological Society of Engineering Geologists* (spec. public., vol. 15, pp. 79–95).

Menon, M. G., Gibbs, R. J., & Phillips, A. (1998). Accumulation of muds and metals in the Hudson River estuary turbidity maximum. *Environmental Geology*, **34**, 215–222.

*Merritts, D., de Wet, A., & Menking, K. (1997). *Environmental geology: An earth system science approach*. Freeman, 452 p.

Meybeck, M. (2001). *Transport et qualité des sédiments fluviaux: Variabilités temporelle et spatiale, enjeux de gestion*. Publication de la Société Hydrotechnique de France, 166th sess., 11–27.

Meyers, P. A., & Takeuchi, N. (1981). Environmental changes in Saginaw Bay, Lake Huron, recorded by geolipid contents of sediments deposited since 1800. *Environmental Geology*, **3**, 257–266.

*Middleton, N. (1999). *The global casino. An introduction to environmental issues*. Arnold, London, 2nd ed., 370 p.

*Miller, G. T. Jr. (2002). *Living in the environment*. Brooks/Cole, 12th ed., 757 p. + appendix.

Miller, J. R., Lechler, P. J., & Desilets, M. (1998). The role of geomorphic processes in the transport and fate of mercury in the Carson River basin, west-central Nevada. *Environmental Geology*, **33**, 249–262.

†Milnes, A. G. (1985). *Geology and radwaste*. Academic Press, 328 p.

*Montgomery, C. W. (2000). *Environmental geology*. McGraw Hill, 5th ed., 546 p.

†Morgan, R. P. C. (1986). *Soil erosion and conservation*. Longman Scientific & Technical, 298 p.

*†Moore, P. D., Chaloner, B., & Stott, P. (1996). *Global environmental change*. Blackwell, 244 p.

Mudd, G. M. (2000). Mound springs of the Great Artesian Basin in South Australia: A case study from Olympic Dam. *Environmental Geology*, **39**, 463–476.

Mulder, T., Savoye, B., Piper, D. J. W., & Syvitski, J. P. M. (1998). The Var system: Understanding Holocene sediment delivery processes and the geological record. In: M. Stoker, D. Evans, & A. Cramp (Eds), *Geological Processes on Continental Margins* (Geol. Soc. London, pp. 145–166).

Müller, J., Ruppert, H., Muramatsu, Y., & Schneider, J. (2000). Reservoir sediments — a witness of mining and industrial development (Malter Reservoir, eastern Erzgebirge, Germany. *Environmental Geology*, **39**, 1341–1351.

Münch, D. (1993). Concentration profiles of arsenic, cadmium, chromium, copper, lead, mercury, nickel, zinc, vanadium and polynuclear aromatic hydrocarbons (PAH) in forest soil beside an urban road. *The Science of the Total Environment*, **138**, 47–55.

*Murck, B. W., Skinner, B. J., & Porter, S. C. (1996). *Environmental Geology*. Wiley, 534 p.

*Murck, B. W., Skinner, B. J., & Porter, S. C. (1997). *Dangerous Earth: An introduction to geologic hazards*. Wiley, 299 p.

National Geographic (2000). *La colère des dieux: Séismes en Turquie, la faille anatolienne*, **2.7**, 32–75.

National Research Council (1993). *Solid-earth sciences and Society*. National Academy Press, Washington D. C., 346 p.

Noakes, J. E., Noakes, S. E., Dvoracek, D. K., Culp, R. A., & Bush, P. B. (1999). Rapid coastal survey of anthropogenic radionuclides, metals and organic compounds in surficial marine sediments. *The Science of the Total Environment*, **237/238**, 449–458.

Nord, A. G., & Holenyi, K. (1999). Sulphur deposition and damage on limestone and sandstone in Stockholm city buildings. *Water, Air & Soil Pollution*, **109**, 147–162.

†Nordstrom, K. F. (2000). *Beaches and dunes of developed coasts*. Cambridge University Press, 352 p.

Notcutt, G., & Davies, F. B. M. (1989). The environmental influence of a volcanic plume, a new technique of study, Mount Etna, Sicily. *Environmental Geology & Water Science*, **14**, 209–212.

Nriagu, J. O. (1994). Mercury pollution from the past mining of gold and silver in the Americas. *The Science of the Total Environment*, **149**, 167–181.

†Nuclear Energy (1995). *Who's afraid of atomic power? Understanding Global Issues*. European Schoolbooks Public., 18 p.

Oliveira, S. M. B. de, Melfi, A. J., Fostier, A. H., Forti, M. C., Favaro, D. I. T., & Boulet, R. (2001). Soils as an important sink for mercury in the Amazon. *Water, Air & Soil Pollution*, **26**, 321–337.

OSPAR (2001). *Qualitative Status Report, Region IV, Bay of Biscay and Iberian coast*. OsPar Commission, London, 134 p.

Otton, J. K., Zielinski, R. A., & Been, J. M. (1989). Uranium in Holocene valley-fill sediments, and uranium, radon and helium in waters, lake Tahoe-Carson range area, Nevada and California, USA. *Environmental Geology & Water Science*, **13**, 15–28.

*Owen, O. S., Chiras, D. D., & Reganold, J. P. (1998). *Natural resource conservation*. Prentice Hall, 7th ed., 594 p.

PAGES (Past Global Changes) (2000). Ecosystem processes and past human impacts. *News International Paleoscience Community, IGBP*, **8**, 1–35.

†Panizza, M. (1996). Environmental geomorphology. Elsevier, *Developments in Earth Surface Processes*, **4**, 268 p.

*Park, C. (1997). *The environment. Principles and applications*. Routledge, London, 598 p.

†Parker, A., & Rae, J. E. (Ed.) (1998). *Environmental interactions of clays*. Springer, 271 p.

†Paskoff, R. (1994). *Les littoraux. Impact des aménagements sur leur évolution*. Masson, Paris, 256 p.

Paskoff, R. (2001). *L'élévation du niveau de la mer et les espaces côtiers*. Institut Océanographique, Paris, 191 p.

*Pickering, K. T., & Owen L. A. (1997). *An introduction to global environmental issues*. Routledge, London, 2nd ed., 512 p.

†Pimentel, D. (1993). *World soil erosion and conservation*. Cambridge University Press, 349 p.

*Pipkin, B. W., & Trent, D. D. (2001). *Geology and the environment*. West Publishing Company, 3rd ed., 570 p.

†Price, M. (1996). *Introducing groundwater*. Chapman & Hall, 2nd ed., 278 p.

Pulido Bosch, A. Sanchez Martos, F., Martinez Vidal, J. L., & Navarrete, F. (1992). Groundwater problems in a semiarid area (low Andarax river, Almeira, Spain). *Environmental Geology & Water Science*, **20**, 195–204.

Purtscheller, F., Pirchl, T., Sieder, G., Stingl, V., Tessadri, T., Brunner, P., Ennemoser, O., & Schneider, P. (1995). Radon emanation from giant landslides of Koefels (Tyrol, Austria) and Langtang Himal (Nepal). *Environmental Geology*, **26**, 32–38.

†Pye, K. (1987). *Aeolian dust and dust deposits*. Academic Press, 334 p.

Pyle, D. M. (1997). The global impact of the Minoan eruption of Santorini, Greece. *Environmental Geology*, **30**, 59–61.

*Rahn, P. H. (1986). *Engineering geology: An environmental approach*. Prentice Hall, 646 p.

Rahn, P. H. (1992). A method to mitigate acid-mine drainage in the Shamokin area, Pennsylvania, USA. *Environmental Geology & Water Science*, **19**, 47–53.

Randazzo, G., Stanley, D. J., Di Geronimo, S. I., & Amore, C. (1998). Human-induced sedimentological changes in Manzala lagoon, Nile delta, Egypt. *Environmental Geology*, **36**, 235–258.

Rao, N. V. R., Rao, N., Rao, K. S. P., & Schuiling, R. D.1993. Fluorine distribution in waters of Nalgonda district, Andhra Pradesh, India. *Environmental Geology*, **21**, 84–89.

Rhoads, B. L. (1991). Impact of agricultural development on regional drainage in the lower Santa Cruz valley, Arizona, USA. *Environmental Geology & Water Science*, **18**, 119–135.

Richard, S., Arnoux, A., Cerdan, P., Reynouard, ??., & Horeau, V. (2000). Mercury levels of soils, sediments and fish in French Guiana, South America. *Water, Air & Soil Pollution, 124*, 221–244.

Rine, J. M., Berg, R. C., Shafer, J. M., Covington, E. R., Reed, J. K., Bennett, C. B., & Trudnak, J. E. (1998). Development and testing of a contamination potential mapping system for a portion of the General Separations area, Savannah River site, South Carolina. *Environmental Geology*, **35**, 263–277.

Robertson, J. B., & Ebberg, S. C. (1993). Technical considerations in extracting and regulating spring-water for public consumption. *Environmental Geology*, **22**, 52–59.

*Rogers, J. J. W., & Feiss, P. G. (1998). *People and the Earth*. Cambridge University Press, 338 p.

Rose, N. L., Harlock, S., & Appleby, P. G. (1999). The spatial and temporal distribution of spheroidal carbonaceous fly-ash particles (SCP) in the sediment records of European mountain lakes. *Water, Air & Soil Pollution*, **113**, 1–32.

Ross, M. (1995). The schoolroom asbestos abatement program: A public policy debacle. *Environmental Geology*, **26**, 182–188.

†Said-Jimenez, C. (Ed.) (1995). The deterioration of monuments. *The Science of the Total Environment*, **167**, Sp. Iss., 400 p.

†Salomons, W., Bayne, B. L., Duursma, E. K., & Förstner, U. (1988). *Pollution of the North Sea: An assessment*. Springer, 687 p.

Schmitt, J.-M., Thiry, M., & Van Oort, F. (Éd.) (2001). *Industrie minérale et environnement: Aspects géochimiques, hydrogéochimiques, biogéochimiques*. École des Mines de Paris, Mémoire des Sciences de la Terre 40, 167 p. + appendix.

†Sengupta, M. (1993). *Environmental impacts of mining: Monitoring, restoration, and control*. CRC Press, Lewis Publishers, 494 p.

Sidle, R. C., Kamil, I., Sharma, A., & Yamashita, S. (2000). Stream response to subsidence from underground coal mining in central Utah. *Environmental Geology*, **39**, 279–291.

Siegel, F. R., Kravitz, J. H., & Galasso, J. J. (2001). Arsenic and mercury contamination in 31 cores taken in 1965, St-Anna Trough, Kara Sea, Arctic Ocean. *Environmental Geology*, **40**, 528–542.

Simon, S. L., Graham, J. C., & Borchert, A. W. (1999). Concentrations and spatial distribution of plutonium in the terrestrial environment of the Marshall Islands. *The Science of the Total Environment*, **229**, 21–39.

†Singh, V. P. (Ed.) (1996). *Hydrology of disasters*. Kluwer, Dordrecht, 442 p.

†Slaymaker, O., & Spencer, T. (1998). *Physical geography and global environmental change*. Longman, New York, 292 p.

*Smith, K. (1992). *Environmental hazards*. Routledge, London, 324 p.

†Smith, K., & Ward, R. C. (1998). *Floods: Physical processes and human impacts*. Wiley, 408 p.

Smith, S. E. (1990). A revised estimate of the life span for lake Nasser. *Environmental Geology & Water Science*, **15**, 123–129.

Sowers, G. F. (1986). Correction and protection in limestone terrane. *Environmental Geology & Water Science*, **8**, 77–82.

Sprenke, K. F., Rember, W. C., Bender, S. F., Hoffmann, M. L., Rabbi, F., & Chamberlain, V. E. (2000). Toxic metal contamination in the lateral lakes of the Coeur d'Alene River valley, Idaho. *Environmental Geology*, **39**, 575–586.

Sun, H., Grandstaff, D., & Shagam, R. (1999). Land subsidence due to groundwater withdrawal, potential damage of subsidence and sea level rise in southern New Jersey, USA. *Environmental Geology*, **37**, 290–296.

Szczepanska, J., & Twardowska, I. (1999). Distribution and environmental impact of coal-mining wastes in Upper Silesia, Poland. *Environmental Geology*, **38**, 248–258.

*Tank, R. W. (Ed.) (1973). *Focus on environmental geology.* Oxford University Press, 474 p. (2nd ed. 1983, 549 p.).

†Testa, S. M. (1993). *Geological aspects of hazardous waste management.* CRC Press, Lewis Publishers, 537 p.

†Thomas, D. S. G., & Middleton, N. J. (1995). *Desertification: Exploding the myth.* Wiley, 194 p.

Thomas, R. L., Vernet, J.-P., & Frank, R. (1984). ΣDDT, PCBs and HCB in the sediments of lake Geneva and the upper Rhône river. *Environmental Geology*, **5**, 103–113.

†Tilling, R. I. (Ed.) (1989). *Volcanic hazards, short course in geology.* AGU, Washington, 1, 123 p.

Toy, T. J., Osterkamp, W. R., & Renard, K. G. (1993). Prediction by regression and intrarange data scatter in surface-process studies. *Environmental Geology*, **22**, 121–128.

Trentesaux, A., & Garlan, T. (Ed.) (2000). *Marine sandwaves dynamics.* Proc. Int. Workshop, Lille University, 240 p.

*Turk, J., & Thompson, G. R. (1995). *Environmental Geoscience.* Harcourt Brace Coll. Public., 423 p. + appendix.

*†Turner, B. L. II, Clark, W. C., Kates, R. W., Richards, J. F., Mathews, J. T., & Meyer, W. B. (1993). *The Earth as transformed by human action.* Cambridge University Press, 713 p.

Uri, N. D. (2001). A note on soil erosion and its environmental consequences in the United States. *Water, Air & Soil Pollution*, **129**, 181–197.

Uri, N. D., & Lewis, J. A. (1998). The dynamics of soil erosion in U.S. agriculture. *The Science of the Total Environment*, **218**, 45–58.

*Valdiya, K. S. (1987). *Environmental Geology, Indian context.* Tata McGraw-Hill Publishers, New Dehli, 583 p.

Veyret, Y. (1999). *Géoenvironnement.* Sedes — Géographie ed., Paris, 159 p.

†Viles, H., & Spencer, T. (1995). *Coastal problems. Geomorphology, ecology and society at the coast.* Edward Arnold, 350 p.

†Voight, B. (Ed.) (1978). Rockslides and avalanches. Elsevier, *Developments in Geotechnical Engineering*, **14A & 14B**, 833 & 850 p.

Voix du Nord (2001). Catastrophes maritimes au large des côtes de l'Ouest. Lille March 21, **1**.

Vörösmarty, C. J., Meybeck, M., Fekete, B., & Sharma, K. (1997). The potential impact of neo-castorization on sediment transport by the global network of rivers. Proc. *Rabat Symp. on Human Impact on Erosion & Sedimentation, IAHS Public*, **245**, 261–273.

*Waltham, A. C. (1994). *Foundations of engineering geology.* Blackie Academic & Professional, London, 88 p.

Weller, A., & Börner, F. D. (1996). Measurements of spectral induced polarization for environmental purposes. *Environmental Geology*, **27**, 329–334.

Williams, H. F. L. (1999). Sand-spit erosion following interuption of longshore sediment transport , Shamrock Island, Texas. *Environmental Geology*, **37**, 153–161.
Williams, M. (2001). Arsenic in mine waters, an international study. *Environmental Geology*, **40**, 267–278.
*Williams, M. A. J., Dunkerlley, D. L., De Deckker, P., Kershaw, A. P., & Stokes, T. J. (1993). *Quaternary environments*. Edward Arnold, 330 p.
†Williams, P. T. (2000). *Waste treatment and disposal*. Wiley, 417 p.
Williams, T. M. (1993). Particulate metal speciation in surficial sediments from Lock Dee, southwest Scotland, U.K. *Environmental Geology*, **21**, 62–69.
Wong, H. K. T., Gauthier, A., & Nriagu, J. O. (1999). Dispersion and toxicity of metals from abandoned gold mine tailings at Goldenwille, Nova Scotia, Canada. *The Science of the Total Environment*, **228**, 35–47.
Wuebbles, D. J., & Hayhoe, K. (2002). Atmospheric methane and global change. *Earth Sci. Rev.*, **57**, 177–210.
Zeng, E. Y., & Venkatesan, M. I. (1999). Dispersion of sediment DDTs in the coastal ocean off southern California. *The Science of the Total Environment*, **229**, 195–208.
Zhang, C., Selinus, O., & Kjellström, G. (1999). Discrimination between natural background and anthropogenic pollution in environmental geochemistry — exemplified in an area of south-eastern Sweden. *The Science of the Total Environment*, **244**, 129–140.
Zötl, J. G. (1995). Badgastein Spa, Austrian central Alps. *Environmental Geology*, **26**, 240–245.

Subject Index